Lecture Notes in Computer Science　　10859

Commenced Publication in 1973
Founding and Former Series Editors:
Gerhard Goos, Juris Hartmanis, and Jan van Leeuwen

More information about this series at http://www.springer.com/series/7409

Max Silberztein · Faten Atigui
Elena Kornyshova · Elisabeth Métais
Farid Meziane (Eds.)

Natural Language Processing and Information Systems

23rd International Conference on Applications
of Natural Language to Information Systems, NLDB 2018
Paris, France, June 13–15, 2018
Proceedings

 Springer

Editors
Max Silberztein
Université de Franche-Comté
Besançon
France

Elisabeth Métais
Conservatoire National des Arts et Métiers
Paris
France

Faten Atigui
Conservatoire National des Arts et Métiers
Paris
France

Farid Meziane
University of Salford
Manchester
UK

Elena Kornyshova
Conservatoire National des Arts et Métiers
Paris
France

ISSN 0302-9743   ISSN 1611-3349   (electronic)
Lecture Notes in Computer Science
ISBN 978-3-319-91946-1   ISBN 978-3-319-91947-8   (eBook)
https://doi.org/10.1007/978-3-319-91947-8

Library of Congress Control Number: 2018944303

LNCS Sublibrary: SL3 – Information Systems and Applications, incl. Internet/Web, and HCI

Printed on acid-free paper

This Springer imprint is published by the registered company Springer International Publishing AG
part of Springer Nature
The registered company address is: Gewerbestrasse 11, 6330 Cham, Switzerland

Preface

This volume of *Lecture Notes in Computer Science* (LNCS) contains the papers selected for presentation at the 23rd International Conference on Natural Language and Information Systems, held in Paris, France (CNAM), during June 13–15, 2018 (NLDB2018).

Since its foundation in 1995, the NLDB conference has brought together researchers, industry practitioners, and potential users interested in various application of natural language in the database and information systems field. The term "information systems" has to be considered in the broader sense of information and communication systems, including big data, linked data, and social networks. The field of natural language processing (NLP) has itself recently experienced several exciting developments. In research, these developments have been reflected in the emergence of neural language models (Deep Learning, Word2Vec) and a (renewed) interest in various linguistic phenomena, such as in discourse and argumentation theory (argumentation mining). Regarding applications, NLP systems have evolved to the point that they now offer real-life, tangible benefits to enterprises. Many of these NLP systems are now considered de facto offerings in business intelligence suites, such as algorithms for recommender systems and opinion mining/sentiment analysis as well as question-answering systems.

It is against this backdrop of recent innovations in NLP and its applications in information systems that the 23rd edition of the NLDB conference, NLDB 2018, took place. We have welcomed research and industrial contributions, describing novel, previously unpublished works on NLP and its applications across a plethora of topics. Indeed, this year NLDB had 99 submissions from which we selected 18 long papers (18% acceptance rate), 26 short papers, and nine posters, and each paper was reviewed by an average of three reviewers. The conference and Program Committee co-chairs had a final consultation meeting to look at all the reviews and make the final decisions.

This year, we had the opportunity to welcome two invited presentations from:

- Prof. Jean-Gabriel Ganascia (LIP6 Laboratory, Institut Universitaire de France)
- Dr. Serge Rosmorduc (Cedric Laboratory, Conservatoire National des Arts et Métiers, France)

We would like to thank all who submitted papers for reviews and for publication in the proceedings and all the reviewers for their time, effort, and for completing their assignments on time.

April 2018

Max Silberztein
Faten Atigui
Elena Kornyshova
Elisabeth Métais
Farid Meziane

Organization

General Chairs

Métais Elisabeth   Conservatoire des Arts et Métiers, France
Meziane Farid   University of Salford, UK
Silberztein Max   Université de Franche-Comté, France

Program Committee

Silberztein Max (Chair)   Université de Franche-Comté, France
Akoka Jacky   CNAM and TEM, France
Aras Hidir   FIZ Karlsruhe, Germany
Bajwa Sarwar   Imran, The Islamia University of Bahawalpur, Pakistan
Balakrishna   Mithun, Limba Corp, USA
Basile Pierpaolo   University of Bali, Italy
Béchet Nicolas   IRISA, France
Bordogna   Gloria, IREA, CNR, Italy
Bouma Gosse   Groningen University, The Netherlands
Bringay Sandra   LIRMM, France
Buitelaar Paul   Insight National University of Ireland, Galway
Caragea Cornelia   University of North Texas, USA
Christian Chiarcos   Universität Frankfurt am Main, Germany
Chiky Raja   ISEP, France
Cimiano Philipp   Universität Bielefeld, Germany
Comyn-Wattiau Isabelle   ESSEC, France
Frasincar Flavius   Erasmus University Rotterdam
Freitas André   University of Passau, Germany/Insight, Ireland
Ganguly Debasis   Dublin City University, Ireland
Gelbukh Alexander   CIC-IPN, Mexico
Guessoum Ahmed   USTHB, Algeria
Hacohen-Kerner Yaakov   Jerusalem College of Technology, Israel
Hahn Udo   University of Freiburg, Germany
Handschuh Siegfried   University of Passau, Germany
Herweg Michael   IBM, Germany
Horacek Helmut   Saarland University, Germany
Ienco Dino   IRSTEA, France
Ittoo Ashwin   HEC, University of Liege, Belgium
Johannesson Paul   Stockholm University, Sweden
Kapetanios Epaminondas   University of Westminster, UK
Kedad Zoubida   UVSQ, France
Kergosien Eric   GERiiCO, University of Lille, France
Kop Christian   University of Klagenfurt, Austria

Kordoni Valia	Humboldt University Berlin, Germany
Lafourcade Mathieu	LIRMM University College Ghent
Lefever Els	Ghent University, Belgium
Leidner Jochen	Thomson Reuters, USA
Lonsdale Deryle W.	Brigham Young University, USA
Lopez Cédric	Emvista, Montpellier, France
Mayr Heinrich C.	Alpen-Adria-Universität Klagenfurt, Austria
Mccrae John P.	National University of Ireland, Galway
Métais Elisabeth	CNAM, France
Meurs Marie-Jean	UQAM, Montreal, Canada
Meziane Farid	Salford University, UK
Mich Luisa	University of Trento, Italy
Montoyo Andres	Universidad de Alicante, Spain
Muñoz Rafael	Universidad de Alicante, Spain
Nguyen Le Minh	Japan Advanced Institute of Science and Technology, Japan
Panchenko Alexander	TU Darmstadt, Germany
Saint Dizier Patrick	IRIT-CNRS, France
Paulheim Heiko	University of Mannheim, Germany
Pei Yulong	Eindhoven University of Technology, The Netherlands
Picca Davide	Université de Lausanne, Switzerland
Prince Violaine	LIRMM, France
Qasemizadeh Behrang	Heinrich-Heine-Universität Düsseldorf, Germany
Rezgui Yacine	Cardiff University, UK
Riedl Martin	University of Hamburg, Germany
Roche Mathieu	Cirad, TETIS, France
Rosso Paolo	Technical University of Valencia, Spain
Saraee Mohamad	University of Salford, Manchester, UK
Sateli Bahar	Concordia University, Canada
Schneider Roman	IDS Mannheim, Germany
Schouten Kim	Erasmus University Rotterdam
Shaalan Khaled	The British University in Dubai, UAE
Storey Veda	Georgia State University, USA
Sugumaran Vijayan	Oakland University Rochester, USA
Teisseire Maguelonne	Irstea, TETIS, France
Thalheim Bernhard	Christian Albrechts University Kiel, Germany
Thelwall Mike	University of Wolverhampton, UK
Thirunarayan Krishnaprasad	Wright State University, USA
Tufis Dan	RACAI, Bucharest, Romania
Ureña Alfonso	Universidad de Jaén, Spain
Xu Feiyu	DFKI Saarbrücken, Germany
Zadrozny Wlodek	UNCC, USA
Zanzotto Fabio Massimo	University of Rome Tor Vergata, Italy

Organizing Committee Chairs

Atigui Faten   Conservatoire des Arts et Métiers, France
Kornyshova Elena   Conservatoire des Arts et Métiers, France

Webmaster

Lemoine Frédéric   Conservatoire des Arts et Métiers, France

Contents

Text Similarities and Plagiarism Detection

Text Classification

Information Mining

Recommendation Systems

Translation and Foreign Language Querying

Software Requirement and Checking

Opinion Mining and Sentiment Analysis in Social Media

A Proposal for Book Oriented Aspect Based Sentiment Analysis: Comparison over Domains

Tamara Álvarez-López[1,2](✉), Milagros Fernández-Gavilanes[1],
Enrique Costa-Montenegro[1], and Patrice Bellot[2]

[1] GTI Research Group, University of Vigo, Vigo, Spain
{talvarez,milagros.fernandez,kike}@gti.uvigo.es
[2] Aix Marseille Univ, Université de Toulon, CNRS, LIS, Marseille, France
patrice.bellot@univ-amu.fr

Abstract. Aspect-based sentiment analysis (ABSA) deals with extracting opinions at a fine-grained level from texts, providing a very useful information for companies which want to know what people think about them or their products. Most of the systems developed in this field are based on supervised machine learning techniques and need a high amount of annotated data, nevertheless not many resources can be found due to their high cost of preparation. In this paper we present an analysis of a recently published dataset, covering different subtasks, which are aspect extraction, category detection, and sentiment analysis. It contains book reviews published in Amazon, which is a new domain of application in ABSA literature. The annotation process and its characteristics are described, as well as a comparison with other datasets. This paper focuses on this comparison, addressing the different subtasks and analyzing their performance and properties.

Keywords: Aspect-based sentiment analysis · Book reviews
Datasets · Annotation · Evaluation

1 Introduction

People have access to a huge amount of information available online about a wide diversity of subjects and users can express their opinions about products or services by means of social networks or blogs. Analyzing the user-generated content on the Web became very interesting both for consumers and companies, helping them to better understand what people think about them and their products. For this reason, many studies arose in the fields of Opinion Mining and Sentiment Analysis [1], trying to extract the most relevant information in an automatic way. Most of these works aim at extracting the global sentiment from the whole text, by means of different approaches, which can be supervised [2], normally based on classifiers, or unsupervised, such as [3], where the authors aim at capturing and modeling linguistic knowledge by using rule-based techniques. In the last years many works can also be found applying deep learning techniques [4].

© Springer International Publishing AG, part of Springer Nature 2018
M. Silberztein et al. (Eds.): NLDB 2018, LNCS 10859, pp. 3–14, 2018.
https://doi.org/10.1007/978-3-319-91947-8_1

However, extracting the specific entities or aspects to which a sentiment is expressed provides a more specific insight of what people think about a particular product, and this is exactly the aim of the so-called ABSA. Several studies emerged in the ABSA field [5,6], as well as competitions like SemEval [7], where researchers are encouraged to submit and evaluate their systems over common datasets. Again we can find supervised approaches based on classifiers and conditional random fields (CRFS) [8] and unsupervised ones based on frequency studies or syntax dependencies [9].

For developing ABSA systems, annotated datasets are essential for training and testing. However, manually annotating the user-generated reviews is a tough task, time and resource consuming, and not many datasets are available. We can find the following ones. The first one and widely used in the literature is on electronic products [10], tagged with aspects and the sentiment associated. Then, [11] works with restaurant reviews, annotated with categories from a predefined list and their sentiments. Similarly annotated with predefined categories, we find [12], which is about movie reviews. Finally we have to mention the datasets created for SemEval [7], providing datasets for 8 languages and 7 domains, including restaurants, laptops, mobile phones, digital cameras, hotels and museums. Particularly for the book domain, only two works can be found on annotated corpora, one for Arabic language [13] and the other one for Portuguese [14].

Motivated by the few amount of resources found, we developed a new dataset for English language in the book domain, which was firstly presented in [15]. The aim of this paper is to provide a new glance of this dataset, its structure and annotation process, as well as to provide an evaluation benchmark for different subtasks in ABSA and to compare its performance to other datasets available, analyzing the particularities of each one.

The remaining of this article is structured as follows. In Sect. 2 the datasets under analysis are described. Section 3 presents the new dataset we created, including inter-annotator agreement levels and different statistics. In Sect. 4 the evaluation and comparison of all the datasets are shown. Finally, Sect. 5 provides some conclusions and future work.

2   ABSA Datasets

In this section we present the different datasets considered, all of them manually annotated for ABSA. Not much work on annotated datasets for this task can be found, so we use those which are publicly available, in electronics, restaurant and laptop domains, and were widely used in the literature.

2.1   Electronic Product Reviews

The *Customer Reviews Dataset* [10] contains customer reviews of five electronic products, bringing a total number of 314 reviews, collected from Amazon.com and C|net.com and manually annotated. The features for which an opinion is expressed are tagged, along with the sentiment, represented by a numerical value from +3 (most positive) to −3 (most negative). In this dataset[1] both explicit and implicit features (not explicitly appearing in the text) were annotated, however only the explicit ones were taken into account for this work. In Table 1 detailed information about this dataset is shown. For the next experiments, we will test each dataset separately, using the other four as training.

Table 1. Number of: reviews, sentences, aspects extracted, and aspects tagged as positive or negative, for each product in the electronics dataset

Data	Product	#Revs.	#Sent.	#Aspects	#Pos.Asp.	#Neg.Asp.
D1	Digital camera	45	597	256	205	51
D2	Digital camera	34	346	185	155	30
D3	Cell phone	41	546	309	230	79
D4	MP3 player	95	1716	734	441	293
D5	DVD player	99	740	349	158	191
	Total	314	3945	1833	1189	644

2.2 SemEval Datasets

In the following, the two datasets provided by SemEval workshop [7] for English language are examined, in the domains of restaurants and laptops. They both have a similar structure, containing several reviews annotated at sentence level. In every sentence the specific aspects, called Opinion Target Expressions (OTE), are identified; the aspect category to which the OTE belongs, chosen from a predefined list; and the polarity (*positive, negative* or *neutral*).

The list of categories designed for restaurants is formed by a combination of *entity#attribute* and it is composed of 12 different ones, such as *restaurant#prices, food#quality* or *ambience#general*. However, for the laptop dataset the categories are much more specific, combining 22 different entities with 14 attributes, obtaining a great number of possibilities. For the aim of this paper we shorten this list by regrouping the entities and attributes, so we can obtain a similar number of categories for all the datasets under evaluation. We only keep the entities *laptop, software, support* and *company*, while the entities *hardware, os, warranty* and *shipping* are removed due to their low frequency of appearance. Finally, the rest of the entities are grouped in *components*, as they all refer to different components of a laptop. About the attributes, we keep all of them, but only associated to the *laptop* entity. For the rest of entities, the attribute is always *general*. Like that we obtain a list of 13 categories.

Both datasets are divided into training and test and more detailed information is displayed in Table 2. The information shown about the laptop dataset belongs to the new annotations we created by summarizing the categories.

---

[1] Available online at https://www.cs.uic.edu/~liub/FBS/sentiment-analysis.html.

Table 2. Number of: reviews, sentences and categories, distinguishing between positive, negative and neutral ones, for restaurants and laptops

Domain		#Revs.	#Sent.	#Cat.	#Pos.Cat.	#Neg.Cat.	#Neu.Cat.
Restaurants	Train	350	2000	2506	1657	751	98
	Test	90	676	859	611	204	44
Laptops	Train	450	2500	2730	1547	1002	181
	Test	80	808	749	454	254	41

3   New Book Reviews Dataset

Performing ABSA on book reviews is very useful for different kind of users, including professional as well as non-expert readers, helping them when searching for a book which fits in certain requirements. It can be later applied to recommendation in digital libraries or electronic book shops.

In the following subsections we present the new dataset, the annotation process, its structure and some statistics. It is publicly available online[2] and it was previously introduced in [15], where some information about the annotation process and the dataset properties can be found. Throughout this work, we focus in providing additional information, as well as a baseline for its evaluation, comparing it to other datasets available for the same task.

3.1   Data Collection

For the construction of this dataset, 40 book records were selected randomly from the *Amazon/LibraryThing* corpus in English language provided by the Social Book Search Lab [16], a track which is part of the CLEF (Conference and Labs of the Evaluation Forum). Its main goal is to develop techniques to support readers in complex book search tasks, providing a dataset for the book recommendation task, consisting of 2.8 million book descriptions from Amazon, where each description contains metadata about the *booktitle*, *author* or *isbn*, as well as user generated content, like user ratings and reviews. However, it does not contain any annotations about the aspects or sentiments from the reviews.

Our dataset is composed by the textual content of the reviews associated to each of the 40 books selected, obtaining a total number of 2977 sentences from 300 reviews, which were annotated with aspects, categories and sentiment information.

3.2   Task Description

This new corpus is intended to cover different subtasks in the ABSA field:

---

[2] http://www.gti.uvigo.es/index.php/en/book-reviews-annotated-dataset-for-aspect-based-sentiment-analysis.

1. Aspect Extraction. The aim is to extract the specific features mentioned in the text, which are related to the authors, characters, writing quality, etc. In our dataset we differentiate between explicit and implicit aspects. Implicit aspects are those which are not explicitly written in the analyzed sentence, but can be figured out by the context or the previous sentences:

    e.g.: Sent 1 → When <u>Arthur</u> is suffering from asthma,... (target = Arthur)
    Sent 2 → Then, <u>he</u> starts seeing a large house. (implicitTarget = Arthur)

2. Category Detection. Each target detected in a sentence is classified at a more coarse-grained level, assigning a category from a predefined list. The categories defined for the book domain try to cover most of the features that readers mention in their reviews and are divided into two groups, the ones related to the book itself and those related to its content. In the first group: *general, author, title, audience* (type of readers which the book was written for), *quality* (about the writing style), *structure* (related to the chapters, index, etc.), *period* (when the book was written or published), *length* and *price*. Then the categories included in the second group are: *characters, plot, genre* (related to the literary genre) and *period* (when the plot takes place).

3. Sentiment Polarity. This last task consists of assigning a polarity to every aspect detected from three possible ones: *positive, negative* and *neutral*.

3.3   Annotation Process

In order to construct the annotated dataset and support all the tasks previously defined, different tags are attached at sentence level: *Out of Scope, Target, Occurrence, Implicit Target, Category* and *Polarity*. More information about them can be found in [15] and an example of an annotated sentence is shown in Fig. 1.

```
- <sentence id="000_0007175000_79:6">
  - <text>
      The characters are likeable, the plot is complicated yet compelling and the writing superb
    </text>
  - <Opinions>
      <Opinion category="CONTENT#CHARACTERS" occurrence="1" polarity="positive" target="characters"/>
      <Opinion category="CONTENT#PLOT" occurrence="1" polarity="positive" target="plot"/>
      <Opinion category="BOOK#QUALITY" occurrence="1" polarity="positive" target="writing"/>
    </Opinions>
  </sentence>
```

Fig. 1. Example of an annotated sentence from the corpus

The reviews selected for the dataset were annotated by three different annotators, researchers in the field of NLP and, in particular, in Sentiment Analysis from the University of Vigo. In order to ensure the consistency of the final corpus, some guidelines were given to the annotators, as well as training sessions to solve the doubts arisen. Finally, only those annotations in which at least two of the three annotators agreed were taken into account.

When integrating the results from the three annotators, one of the main difficulties found was to determine the boundaries of a particular target. For example, in the sentence *"I would recommend ages 8 and over"* one annotator can tag *ages 8 and over* as an aspect, whilst other can annotate just *ages 8*. For this situation we consider that both annotators agreed and the longest one is taken into account, as so they do the authors in [14] .

In order to calculate the inter-annotator agreement, the Dice coefficient (D_i) is used, instead of Cohen's Kappa coefficient, as the annotators may tag any word sequence from the sentences, which leads to a very large set of possible classes:

$$D_i = 2 \cdot \frac{|A_i \cap B_i|}{|A_i| + |B_i|} \tag{1}$$

where A_i, B_i are the aspects tagged by annotators A and B, respectively. In Table 3 the agreement between annotators A and B (A|B), A and C (A|C) and B and C (B|C) are shown, as well as the average of the three. The inter-annotator agreement study is also applied to the implicit aspects detected, the category identification, the polarity assigned to an aspect when there was already an agreement in annotating the aspect concerned and the pairs aspect-category, which means that both annotators extracted the same aspect and also assigned the same category to it.

Table 3. Detailed Dice coefficients for aspect, category, aspect+category, polarity, and implicit aspect annotations

Annotators	Aspect	Cat.	Asp.+Cat	Polarity	Implicit
A|B	0.76	0.72	0.65	0.77	0.36
A|C	0.76	0.74	0.68	0.77	0.37
B|C	0.74	0.70	0.63	0.71	0.37
Avg.	0.75	0.72	0.65	0.75	0.37

As we can see in Table 3, for the task of identifying the specific aspects, the 75% of agreement is reached, which is a good result due to the difficulty of the task. Then similarly, the annotators tag the same categories for a particular sentence with an agreement of 72%, regardless of the aspects extracted. When taking into account the tuple <aspect, category>, the Dice coefficient decreases to 65%, which is normal as in this case they have to agree simultaneously on two parameters. Once the annotators agree on a particular aspect, they also coincide in the polarity annotation with an agreement of 75%. Finally, the Dice measure obtained for the implicit aspect extraction task is 37%, which means a very poor agreement. This can be explained due to the complexity of detecting implicit aspects, as they do not appear written in the sentence, so it makes more difficult the annotation task. There is still much room for improvement in this last task and for the aim of this paper we will not take them into account for the experiments.

3.4   Dataset Characteristics

The complete dataset is composed of 300 reviews, belonging to 40 different books, with a total of 2977 sentences. For the following experiments the dataset was divided into training and test, selecting around the 25% of the reviews for testing and the rest for training, making sure that the reviews included in each one belong to different books, in order to avoid biased results. In Table 4 some additional information is shown.

Table 4. Number of: sentences, annotated aspects (explicit and implicit), and sentences tagged as *out of scope* for the book dataset

Dataset	#Sent.	#Explicit Asp.				#Implicit	#OutOfScope
		#P	#N	#NEU	Total		
Train	2219	726	296	1663	2685	265	469
Test	758	230	88	501	819	64	182

In Table 4 we can also see the polarity distribution (*positive* (P), *negative* (N) and *neutral* (NEU)) of explicit aspects, which is similar for training and test. Moreover, it can also be observed that the number of neutral aspects is quite higher that the rest of the polarities. This is due to the annotation schema, as in this dataset not only the opinionated aspects are annotated, but also those with no opinion associated (*neutral*). We found very common to mention characteristics of the book related to the plot or the characters in an informative way and annotating also this kind of aspects will be useful for future tasks, such as recommendation or summarization.

Finally, in Fig. 2 the distribution of the category annotations is shown for training and test, being similar across both of them. We find that the most common categories are *characters* and *plot*. However, categories like the *price* or the *length* of the book are not so usual in this kind of reviews.

4   Performance Evaluation: Comparative

In this section we present the evaluation results for the datasets previously described according to the three different tasks. A baseline system was designed for each task and applied to every dataset, analyzing and comparing the results.

For the evaluation of aspect term extraction and category detection we use precision, recall and F-measure, whilst for sentiment analysis we use the accuracy, following the same evaluation procedure as in the SemEval workshop [7].

4.1   Aspect Extraction Task

For this task the baseline system consists of CRFS, using the CRF++ tool [17]. We extract for each single word the following features: words, lemmas, bigrams,

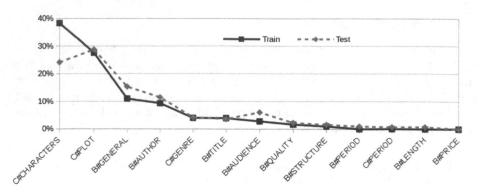

Fig. 2. Aspect category distribution in training and test sets for books

POS tags and entity recognition, obtained by means of Stanford CoreNLP tool [18]. We also extract the value of each two successive tokens in the the the range −2,2 (the previous and subsequent two words of actual word) for each feature. Moreover when the system identifies two or more consecutive words as aspects, we consider them as a single multiword aspect.

In Table 5 we can see the results obtained for the three datasets (the laptop dataset is not annotated with aspects, therefore it cannot be evaluated for this task), in terms of precision, recall and F-measure. It can be observed that applying the same baseline for all the datasets, the restaurant dataset obtains the highest F-measure. However, for electronics, and especially for books, we find quite low recall, what means that it is harder to extract all the aspects annotated, maybe due to the kind of texts which can be more complex in these two domains, in terms of the diversity of vocabulary.

Analyzing some properties of the datasets, we can see in Fig. 3 the relation between the number of sentences, the vocabulary size and the target extraction performance. The vocabulary size of a dataset is determined by the number of unique word unigrams that it contains. As we can see in Fig. 3, increasing the number of sentences does not always imply an increase in the vocabulary size. However, when the vocabulary size increases, the performance of the aspect extraction task becomes lower. With a higher amount of different words, there should also be more different aspects to detect.

If we inspect the list of aspects annotated in the test dataset for each domain, we find that for restaurants there are 312 different aspects, 278 different ones for electronics and 417 for books. Moreover, for the book dataset we find more terms which are considered as aspects just once in the whole dataset. For the electronics domain, even if there are less different aspects, the biggest difficulty is to differentiate when the same term should be considered as aspect or not. We find that the terms which are most frequently correctly detected (*true positive*) are usually the same as those which are most frequently not detected (*false negative*), as well as those that are extracted but should have not (*false positive*).

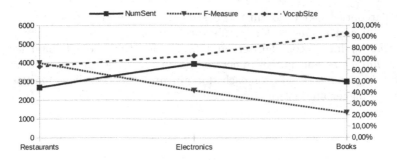

Fig. 3. Number of sentences, vocabulary size, and F-measure for the aspect extraction task for the different datasets

This situation arises a very interesting challenge for aspect extraction, as it is not easy to decide when a particular word should be considered as aspect or not, according to the specific context or sentence.

4.2 Category Detection Task

The baseline developed here is based on an ensemble of binary SVM classifiers with a linear kernel, one for each category. The library *libsvm*[3] was used, with the following binary features at sentence level: words, lemmas, POS tags and bigrams. Then, for each sentence, one, several or no categories can be found.

The datasets evaluated for this task are restaurants, laptops and books, as for electronics there are no category annotations, and the performance results, in terms of precision, recall and F-measure, can be seen in Table 5.

We can see again that the restaurant dataset obtains the highest results, while the worst performance is obtained for the book dataset. When the evaluation is performed for each category separately, some differences can be found for the three domains. For the restaurant dataset the F-measure results obtained for the different categories are more similar to each other than those obtained for the categories in the book domain. For laptop dataset, and especially for book dataset, we can find some categories with really low performance. While the worst performance in the restaurant domain is 23% for the *restaurant#miscellaneous* category, in laptops we find four different categories whose F-measure is lower than 25%, rising to eight categories in the book domain. However, we also have to highlight the low representation of certain categories in the book dataset, making it more difficult for the classifier to learn when to annotate them for a particular sentence.

In addition to this, we find the categorization in general more difficult for the book domain. In this dataset the category which achieves the highest F-measure is *characters*, as there are many sentences both in training and test annotated for it. However, the F-measure is still 50%, while in the restaurant domain the categories which are more accurately detected obtain results around 80%.

---

[3] https://www.csie.ntu.edu.tw/~cjlin/libsvm/.

The categories which obtain the best performance are *restaurant#general*, *food#quality* and *ambience* for restaurants; *laptop#general*, *laptop#price* and *laptop#quality* for laptop dataset; and finally for the book domain they are *general*, *characters* and *author*. These categories are also the most common in both training and test sets for each domain and the classifiers tend to work better with bigger amount of data.

Table 5. Precision, recall, and F-measure for aspect and category detection tasks

Dataset	Aspect			Category		
	Precision	Recall	F-measure	Precision	Recall	F-measure
Restaurants	0.69	0.64	0.66	0.72	0.64	0.68
Electronics	0.64	0.31	0.42	-	-	-
Laptops	-	-	-	0.66	0.40	0.50
Books	0.59	0.14	0.22	0.53	0.30	0.38

4.3 Aspect-Based Sentiment Analysis Task

For extracting the sentiment associated to each aspect, we consider a window of five words before and after the specific aspect. The window size was determined by performing several experiments varying this parameter, so the highest results were obtained when it was equal to five. Then, we add the polarities of all the words included, considered as a numerical value and extracted from a sentiment lexicon, SOCAL [19]. This dictionary is composed of a list of words with a polarity associated, expressed on a scale between -5 (most negative) and $+5$ (most positive). Then, if the addition of the polarities divided by the number of words with sentiment associated is higher than 1, the target is considered positive; if it is lower than -1 it is tagged as negative; and neutral otherwise.

This task is applied to electronics, restaurant and book datasets at aspect level. For laptop dataset, as there are no annotated aspects, we extract the sentiment of the whole sentence and assign it to the annotated categories. In Table 6 we can see the results obtained. It can be observed that the weighted average accuracy is quite similar for all the datasets. The highest results are obtained for positive aspects. One of the reasons is the high amount of positive aspects in relation to negative or neutral in most of the datasets. In the restaurant domain, the 71% of the aspects are positive, whilst only 5% of them are tagged as neutral. Similar percentages can be obtained from the laptop domain. In electronics, 65% of the aspects are positive and the other 35% are negative. Finally, for the book dataset it can be seen that the accuracy is similar for the three classes. In this case the 61% of the aspects are neutral, 28% are positive and only 10% are negative. However, if we take into account the F-measure instead of accuracy, we observe that the highest results are obtained for neutral aspects (70%), followed by positive (58%) and negative (24%).

Table 6. Accuracy for aspect polarity extraction task

Dataset	Positive	Negative	Neutral	Weight.Avg.
Restaurants	0.72	0.52	0.32	0.66
Electronics	0.68	0.51	-	0.62
Laptops	0.6	0.5	0.35	0.55
Books	0.57	0.56	0.57	0.57

5 Conclusions

The aim of this paper was to provide a common evaluation benchmark for the ABSA task and its different subtasks: aspect extraction, category detection and sentiment analysis at the aspect level. Different datasets available in the literature were studied and compared, belonging to electronics, restaurant and laptop domains. Moreover, different characteristics and the annotation process were described for the new dataset in the domain of books, not yet explored in the state of the art. Then, different baselines were proposed for the evaluation and comparison. The aim here was not to provide a new approach for ABSA, but to analyze the distinctive features of reviews from different domains and how they affect to the ABSA performance, as well as to perform the evaluation of the new dataset developed.

As future work we plan to continue the research in the book domain and develop new baselines which fit better for this kind of reviews, which seem to present bigger challenges for ABSA. Moreover, we would like to work in the integration of aspect extraction from book reviews for improving book recommendation systems, introducing them as new inputs, so that the system could apply a reranking of the recommendation list according to this new information.

Acknowledgments. This work is supported by Mineco grant TEC2016-C2-2-R, Xunta de Galicia grant GRC and the French ANR program "Investissements d'Avenir" EquipEx DILOH (ANR-11-EQPX-0013).

References

1. Liu, B.: Sentiment analysis and opinion mining. Synth. Lect. Hum. Lang. Technol. **5**, 1–167 (2012)
2. Pak, A., Paroubek, P.: Twitter as a corpus for sentiment analysis and opinion mining. In: 7th International Conference on Language Resources and Evaluation, pp. 1320–1326. LREC (2010)
3. Fernández-Gavilanes, M., Álvarez-López, T., Juncal-Martínez, J., Costa-Montenegro, E., González-Castaño, F.J.: Unsupervised method for sentiment analysis in online texts. Expert Syst. Appl. **58**, 57–75 (2016)
4. Dos Santos, C.N., Gatti, M.: Deep convolutional neural networks for sentiment analysis of short texts. In: 25th International Conference on Computational Linguistics, pp. 69–78. COLING (2014)

5. Schouten, K., Frasincar, F.: Survey on aspect-level sentiment analysis. IEEE Trans. Knowl. Data Eng. **28**, 813–830 (2016)
6. Akhtar, M.S., Gupta, D., Ekbal, A., Bhattacharyya, P.: feature selection and ensemble construction: a two-step method for aspect based sentiment analysis. Knowl.-Based Syst. **125**, 116–135 (2017)
7. Pontiki, M., Galanis, D., Papageorgiou, H., Androutsopoulos, I., Manandhar, S., AL-Smadi, M., Al-Ayyoub, M., Zhao, Y., Qin, B., De Clercq, O., Hoste, V., Apidianaki, M., Tannier, X., Loukachevitch, N., Kotelnikov, E., Bel, N., Jiménez-Zafra, S.M., Eryigit, G.: SemEval-2016 task 5: aspect based sentiment analysis. In: Workshop on Semantic Evaluation (SemEval-2016), pp. 19–30. ACL (2016)
8. Jakob, N., Gurevych, I.: Extracting opinion targets in a single and cross domain setting with conditional random fields. In: Conference on Empirical Methods in Natural Language Processing, pp. 1035–1045 (2010)
9. Poria, S., Cambria, E., Ku, L., Gui, C., Gelbukh, A.: A rule-based approach to aspect extraction from product reviews. In: 2nd Workshop on Natural Language Processing for Social Media, pp. 28–37 (2014)
10. Hu, M., Liu, B.: Mining and summarizing customer reviews. In: 10th ACM SIGKDD International Conference on Knowledge Discovery and Data Mining, pp. 168–177. ACM (2004)
11. Ganu, G., Elhadad, N., Marian, A.: Beyond the stars: improving rating predictions using review text content. In: 12th International Workshop on the Web and Databases, pp. 1–6 (2009)
12. Thet, T.T., Na, J.C., Khoo, C.S.G.: Aspect-based sentiment analysis of movie reviews on discussion boards. J. Inf. Sci. **36**, 823–848 (2010)
13. Al-Smadi, M., Qawasmeh, O., Talafha, B., Quwaider, M.: Human annotated arabic dataset of book reviews for aspect based sentiment analysis. In: 3rd International Conference on Future Internet of Things and Cloud, pp. 726–730 (2015)
14. Freitas, C., Motta, E., Milidiú, R., César, J.: Sparkling Vampire... lol! annotating opinions in a book review corpus. New Language Technologies and Linguistic Research: A Two-Way Road, pp. 128–146. Cambridge Scholars Publishing, Newcastle upon Tyne (2014)
15. Álvarez-López, T., Fernández-Gavilanes, M., Costa-Montenegro, E., Juncal-Martínez, J., García-Méndez, S., Bellot, P.: A book reviews dataset for aspect based sentiment analysis. In: 8th Language & Technology Conference, pp. 49–53 (2017)
16. Koolen, M., et al.: Overview of the CLEF 2016 social book search lab. In: Fuhr, N., Quaresma, P., Gonçalves, T., Larsen, B., Balog, K., Macdonald, C., Cappellato, L., Ferro, N. (eds.) CLEF 2016. LNCS, vol. 9822, pp. 351–370. Springer, Cham (2016). https://doi.org/10.1007/978-3-319-44564-9_29
17. Kudo, T.: CRF++: Yet another CRF toolkit (2005). http://crfpp.sourceforge.net/
18. Manning, C.D., Surdeanu, M., Bauer, J., Finkel, J., Bethard, S.J., McClosky, D.: The stanford CoreNLP natural language processing toolkit. In: Association for Computational Linguistics (ACL) System Demonstrations, pp. 55–60 (2014)
19. Taboada, M., Brooke, J., Tofiloski, M., Voll, K., Stede, M.: Lexicon-based methods for sentiment analysis. Comput. Linguist. **37**, 267–307 (2011)

Stance Evolution and Twitter Interactions in an Italian Political Debate

Mirko Lai[1,2(✉)], Viviana Patti[1], Giancarlo Ruffo[1], and Paolo Rosso[2]

[1] Dipartimento di Informatica, Università degli Studi di Torino, Turin, Italy
{lai,patti,ruffo}@di.unito.it
[2] PRHLT Research Center, Universitat Politècnica de València, Valencia, Spain
prosso@dsic.upv.es

Abstract. The number of communications and messages generated by users on social media platforms has progressively increased in the last years. Therefore, the issue of developing automated systems for a deep analysis of users' generated contents and interactions is becoming increasingly relevant. In particular, when we focus on the domain of online political debates, interest for the automatic classification of users' stance towards a given entity, like a controversial topic or a politician, within a polarized debate is significantly growing. In this paper we propose a new model for stance detection in Twitter, where authors' messages are not considered in isolation, but in a diachronic perspective for shedding light on users' opinion shift dynamics along the temporal axis. Moreover, different types of social network community, based on retweet, quote, and reply relations were analyzed, in order to extract network-based features to be included in our stance detection model. The model has been trained and evaluated on a corpus of Italian tweets where users were discussing on a highly polarized debate in Italy, i.e. the 2016 referendum on the reform of the Italian Constitution. The development of a new annotated corpus for stance is described. Analysis and classification experiments show that network-based features help in detecting stance and confirm the importance of modeling stance in a diachronic perspective.

Keywords: Stance · Political debates · Homophily · Twitter

1 Introduction

Nowadays, social media are gaining a very significant role in public debates. Political leaders use social media to directly communicate with the citizens, and citizens often take part in the political discussion, by supporting or criticizing their opinions or proposals. Therefore, social media provide a powerful experimental tool to deduce public opinion's mood and dynamics, to monitor political

The work of the last author was partially funded by the Spanish MINECO under the research project SomEMBED (TIN2015-71147-C2-1-P).

sentiment, and in particular to detect users' stance towards specific issues, like political elections or reforms, and their evolution during the debate and the related events [3]. Online debates are featured by specific characteristics. As observed by Adamic and Glance, web users tend to belong to social communities segregated along partisan lines [1]. Albeit the scientific debate is still open, some recent studies suggest that the so called "echo chambers" and "filter bubbles" effects tend to reinforce people's pre-existing beliefs, and they also filter and censure divergent ones [13].

In this study we examine the political debate in Twitter about the Italian constitutional referendum held on December 4, 2016 in Italy. To carry on our analysis, we first collected a dataset of about 1M of Italian tweets posted by more than 100 K users between November 24 and December 7, 2016, about the Italian constitutional referendum. Then, we extended the collection by retrieving retweets, quotes, and replies, aiming at a representation of political communication through different types of social networks. Furthermore, we divided our dataset in four temporal phases delimited by significant events occurred around the consultation period, for analyzing the dynamism of both users' stance and social relations. We manually annotated the evolution of the users' stance towards the referendum of 248 users, creating a corpus for stance detection (SD), i.e. the task of automatically determining whether the author of a text is in favour, against, or neutral towards a given target [9]. On this corpus, we were able to analyze the relations that occur among users not only considering the social network structure, but also the users' stance. Based on this analysis we propose a new model for SD in Twitter featured by two main characteristics: (i) network-based features have been included in the model, which result from the analysis of different types of social network communities, based on retweet, quote, and reply relations; (ii) authors' messages are not considered in isolation, but in a diachronic perspective. The major contributions of this work are:

1. *A new resource.* We developed a manually annotated corpus for SD about an Italian political debate, CONREF-STANCE-ITA henceforth. Such kind of resource is currently missing for Italian, in spite of the growing interest in the SD witnessed by the recent shared tasks proposed for English [9], Spanish and Catalan [12].
2. *Stance detection.* We propose a new SD model including a set of features based on social network knowledge. Experiments show that analyzing users' relations helps in detecting stance.
3. *Stance diachronic evolution.* Our analysis on the debate provides some evidence that users reveal their stance in different ways depending on the stage of the debate; in particular, our impression is that users tend to be less explicit in expressing their stance as the outcome of the vote approaches.
4. *Network analysis.* Users tend to communicate with similar users, and a strong signal of homophily by stance among supporters and critics of the reform has emerged. Moreover, users having different opinions on the referendum often communicate using *replies*: a significant number of replies posted among ideologically opposed users occurs in the corpus.

The rest of the paper is organized as follows. In Sect. 2 we briefly discuss the related work. In Sect. 3 we describe the development of the corpus, and its characteristics in terms of social network. In Sect. 4 we describe our SD model and the classification experiments. Section 5 concludes the paper.

2 Related Work

Political Sentiment and Stance Detection. Techniques for sentiment analysis and opinion mining are often exploited to monitor people's mood extracting information from users' generated contents in social media [11]. However, especially when the analysis concerns the political domain [3], a recent trend is to focus on finer-grained tasks, such as SD, where the main aim is detecting users' stance towards a particular target entity. The first shared task on SD in Twitter held at *SemEval 2016*, Task 6 [9], where is described as follows: "Given a tweet text and a target entity (person, organization, movement, policy, etc.), automatic natural language systems must determine whether the tweeter is *in favor* of the target, *against* the given target, or whether neither inference is likely". Standard text classification features such as n-grams and word embedding vectors were exploited by the majority of the participants of the task. The best result was obtained by a deep learning approach based on a recurrent neural network [14].

Machine learning algorithms and deep learning approaches were also exploited in a second shared task held at *IberEval 2017* on gender and SD in tweets on Catalan Independence, with a focus on Spanish and Catalan [12]. With regard to SD, participating teams exploited different kinds of features such as bag of words, bag of parts-of-speech, n-grams, word length, number of words, number of hashtags, number of words starting with capital letters, and so on. The best result was obtained by a support vector machine (SVM) classifier exploiting three groups of features: *Stylistic* (bag of: n-grams, char-grams, part-of-speech labels, and lemmas), *Structural* (hashtags, mentions, uppercase characters, punctuation marks, and the length of the tweet), and *Contextual* (the language of each tweet and information coming from the URL in each tweet) [5].

Political Debates and Diachronic Perspective. Recently, Lai et al. [6] explored stance towards BREXIT at user level by aggregating tweets posted by the same user on 24-h time windows. This shows how stance may change after relevant events, a finding supported by the work of Messina et al. analysing the same debate [8]. A way to represent a dynamic system is aggregating empirical data over time considering different size of time-windows. Albeit the aggregation time window size is often dictated by the availability of data gathered and this issue has often been neglected in the literature, the importance of the choice of time-windows needs to be considered [4].

Political Debate and Social Media. The huge amount of users generated data allows researchers to observe social phenomena with computational tools in an unprecedented way. Despite social media ease the access to a range of several conflicting views, some works suggest that the existence of the so called "echo chambers" (i.e., when users are exposed only to information from like-minded ones) and "filter bubbles" (i.e., when content is selected by algorithms according to the user's previous behaviors) can have both positive and negative effects in online and offline forms of political participation [1,13]. Lazarsfeld and Merto theorized that homophily is involved [7] after the observation that people tend to bond in communities with others who think in similar ways, regardless of any differences in their status characteristics (i.e. gender, age, social status). Recent works shed some light on the relation between social media network structure and sentiment information extracted from posted contents. For example, Lai et. al. [6] reported some preliminary results showing that a strong relation exists between user's stance and friend-based social media community the user belongs, studying the English debate on BREXIT.

3   The ConRef-STANCE-ita Corpus

3.1   Data Collection and Diachronic Perspective

Twitter is a microblogging platform where users post short messages called *tweets*. Users can share with their *followers* (users who follow them) the tweets written by other users; this type of shared tweets is known as *retweets*. Furthermore, users can add their own comments before retweeting making a tweet a *quote*. Moreover, it is possible to answer to another person's tweet, generating a so called *reply*. Replying to other replies makes possible the development of longer *conversation threads*, including direct and nested replies.

Researches on Twitter are made easy by the Twitter's REST and Streaming APIs, a set of clearly defined Web services that allow the communication between the Twitter platform and developers. All APIs return a message in JSON, a cross-platform data-interchange format. Also for these reasons, we chose Twitter as platform to gather our experimental data.

Collection. We collected tweets on topic of the Referendum held in Italy on December 4, 2016, about a reform of the Italian Constitution. On Sunday 4 December 2016, Italians were asked whether they approve a constitutional law that amends the Constitution to reform the composition and powers of the Parliament, the division of powers between the State, the regions, and other administrative entities. This referendum was source of high polarization in Italy and the outcome caused a sort of political earthquake[1]. The data collection consists of four steps:

1. About 900 K tweets were collected between Nov. 24th and Dec. 7th through the Twitter's Stream API, using as keywords the following hastags: #referendumcostituzionale, #iovotosi, #iovotono[2].

---

[1] The majority of the voters rejected the reform causing the resignation of Matteo Renzi, the Prime Minister that assumed full responsibility for the referendum defeat.
[2] #constitutionalreferendum, #Ivoteyes, #Ivoteno.

2. The source tweet from each retweet was recovered by exploring the tweet embedded within the JSON field *retweeted_status*. Then, we used the *statuses/retweets/:id* Twitter REST API in order to collect all retweets of the each retweeted tweet present in the dataset.
3. We recovered the quoted tweet of each quote exploring the embedded tweet within the JSON field *quoted_status*.
4. We retrieved *conversation threads* recursively resorting to the Twitter REST API *statuses/show/:id*, by using, as parameter, the *id* specified in the field *in_reply_to_status_id* of each replied tweet.

Through these steps, we have thus enlarged the available number of tweets (more than 2 M) w.r.t. those gathered by the Twitter Stream API alone (about 900 K). Therefore, we extended the number of possible relations between users (retweets, quotes, and replies) involved in the debate through steps 2, 3 and 4 for deeper analyzing social media networks.

Diachronic Perspective. Using the same methodology described in [6], we divided the collected tweets in four discrete temporal phases, each one delimited by significant daily spikes of tweets. The spikes correspond to events occurred leading up to the referendum, as it is shown in Fig. 1. We thus consider the following four 72-h temporal phases:

– "The Economist" (EC): The newspaper *The Economist* sided with the "yes" campaign of the referendum (tweets retrieved between 2016-11-24 00:00 and 2016-11-26 23:59).
– "Demonstration" (DE): A demonstration supporting the "no" campaign of the referendum had been held in Rome exactly one week before the referendum (tweets retrieved between 2016-11-27 00:00 and 2016-11-29 23:59).
– "TV debates" (TD): The Italian Prime Minister, Matteo Renzi, who supported the "yes" campaign of the referendum, participated to two influential debates on TV (tweets retrieved between 2016-11-30 00:00 and 2016-12-02 23:59).
– "Referendum outcome" (RO): The phase includes the formalization of the referendum outcome, and the resignation of the Italian Prime Minister (tweets between 2016-12-04 00:00 and 2016-12-06 23:59).

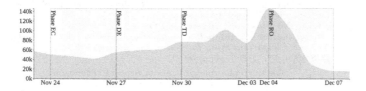

Fig. 1. Daily frequency of tweets and the discrete division in temporal phases.

3.2　Annotation for Stance

We applied to our data the same annotation schema previously exploited at the shared tasks proposed at SemEval 2016 [9] and IberEval 2017 [12] for annotating stance in English, Spanish and Catalan tweets. Here three labels were considered: FAVOR, AGAINST, NONE. The annotation guidelines provided to the annotators follow.

From reading the following tweets, which of the options below is most likely to be true about the tweeter's stance or outlook towards the reform subjected to the Italian Constitutional referendum?

- **FAVOR**: We can infer from the tweet that the tweeter supports the reform.
- **AGAINST**: We can infer from the tweet that the tweeter is against the reform.
- **NONE**: We can infer from the tweet that the tweeter has a neutral stance towards reform or there is no clue in the tweet to reveal the stance of the tweeter towards the reform.

Stance at User Level. We followed the same approach described in [6], where the stance is at user level rather than at tweet level. This means that we deduced the stance from multiple texts written by the same user rather than considering the stance of a single text. We define a *triplet* as a set of three tweets written by the same user in a single temporal phase. The triplet includes: one tweet, one retweet and one reply. This means that each user, for which we annotated the stance, may be a connected node in a network of relations of both retweet or reply. The users who wrote at least one tweet, one retweet, and one reply (a triplet) in each temporal phase are 248. The annotated corpus consists of 992 triplets (248 users by 4 temporal phases). For example, a single user wrote the tweet, the retweet, and the reply highlighted by the black bullet. The reply message in the triplet includes also the related tweet (marked with a white bullet) written by another user.

TWEET	• Travaglio: "Il 2 dicembre grande serata nostra Costituzione in diretta streaming" #ioDicoNo URL via @fattoquotidiano *(Travaglio: "The 2th December a great night for our Constitution in streaming live" #ISayNo URL through @fattoquotidiano)*
RETWEET	•RT @ComitatoDelNO: Brava @GiorgiaMeloni che ricorda a @matteorenzi di (provare a) dire la verità almeno 1 volta su 10! *(RT @NOCommittee: well done @GiorgiaMeloni who reminds to @matteorenzi to (try to) say the truth at least 1 time over 10!)*
REPLY	•@angelinascanu @AntonellaGramig @Rainbowit66 per la poltrona. La cosa più cara a voi del #bastaunSi #IoDicoNo #IoVotoNO #vergognaPD *(@angelinascanu @AntonellaGramig @Rainbowit66 for their seats. The most important thing for you of the #justaYES #ISayNo #IVoteNO #shamePD)*
↪ TO	oGià dovrebbe spiegare...ma la risposta si conosce. Il 4 dicembre #bastaunSi #IoVotoSI URL *(He already should explain... but the answer is known. The 4 December #justaYES #IVoteYES URL)*

Manual Annotation. Two native Italian speakers, domain experts, provided two independent annotations on all the 992 triplets. For what concerns the triplets for which an agreement between the first two annotators was not achieved, we recurred to CrowdFlower[3], a crowd-sourcing platform. We exploited 100 tweets as test questions in order to evaluate the CrowdFlower annotators. We required that annotators were native Italian speakers living in Italy. The annotators have been evaluated over the test questions and only if their precision was above 80% they were included in the task. A further annotator was required unless at least 60% of the previous annotators agreed on the stance of a given triplet. We required a maximum of 3 additional annotators in addition to the 2 domain experts, regarding ambiguous triples. Overall, each triplet was annotated by at least 2 annotators to a maximum of 5.

Agreement. We calculated the inter-annotation agreement (IAA) as the number of annotators who agree over the majority label divided by the total number of annotators for each single triplet. This type of inter-annotator agreement was proposed by Mohammad et al. [10] to overcome the problem of calculating agreement over a set of documents annotated by a different number of annotators. The IAA calculated over all 992 triplets is 74.6%. Finally, we discharged triplets annotated by 5 annotators having less than 3 annotators in agreement on the same label. We named the Twitter with the stance about the Constitutional reform as CONREF-STANCE-ITA, and it consists of 963 triplets.

Label Distribution. Table 1 shows the label distribution over temporal phases in the CONREF-STANCE-ITA[4]. The percentage of triplets labeled as AGAINST is higher than the rest of labels. This is in tune with the final oucome of the referendum (59.12% vote "no")[5]. The frequency of the label NONE over the different temporal phases is another interesting point. As we can see, the distribution of this label constantly increases from phase EC to phase RO.

Table 1. Label distribution

LABEL	EC	DE	TD	RO	OVERALL
AGAINST	72.7%	72.7%	71.5%	62.8%	69.9%
FAVOR	19.8%	18.3%	16.9%	14.0%	17.2%
NONE	**6.2%**	**9.1%**	**11.6%**	**22.3%**	12.3%
Disagreement	1.2%	0%	0%	0.8%	0.5%

---

[3] http://www.crowdflower.com.
[4] ConRef-STANCE-ita and code available at: https://github.com/mirkolai/Stance-Evolution-and-Twitter-Interactions.
[5] https://en.wikipedia.org/wiki/Italian_constitutional_referendum,_2016.

We also explored if users' stance changes over time. We find that 66.8% of the users were labeled with the same stance in all three intervals (55.0% AGAINST, 10.9% FAVOR, 0.8% NONE). For what concerns users that change stance across different time intervals, about 12% of them varies annotated stance in the last phase (10% AGAINST → NONE; 2.5% FAVOR → NONE). Similar observations were made in [6], while investigating English tweets on the UK European Union membership referendum debate (BREXIT).

3.3 Social Media Networks Communities

Networks Science has applications in many disciplines due to networks (or graphs) that are able to represent complex relations among involved actors. Those relations are usually called *edges* and the actors are *nodes*. A network is *weighted* when each edge is characterized with a numerical label that reflects the strength of the connection between two nodes. Therefore, the network is *unweighted* when there is no difference between edges, i.e., all weights are equals to one.

In this work, we represent the relations among Twitter users involved in the Constitutional Referendum debate in the form of graphs. We extracted social media network communities from each graph using the Louvain Modularity algorithm [2]. Then, we examined the structure of four types of communication networks focusing on the dynamism of interactions and the percentage of *uncross-stance* relations (edges between two users with the same stance) for each type of communication. Table 2 shows the dimensions of each graph in each temporal phase.

Table 2. Graphs' dimension for each temporal phases.

	RETWEET		QUOTE		REPLY	
	nodes	*edges*	*nodes*	*edges*	*nodes*	*edges*
Overall	94,445	405,843	24,976	69,240	20,936	41,292
EC	25,793	83,134	6,907	13,574	6,236	8,651
DE	28,015	98,717	7,577	15,665	6,663	9,714
TD	33,860	127,593	9,599	22,479	8,801	14,046
RO	63,805	158,243	14,919	21,977	8,497	10,832

Retweet. First, we consider the retweet-based networks. We gathered the retweet list of 649,306 tweets. We created a directed graph for each temporal phase. In particular, an edge between two users exists if one user retweeted a text of the other user during a defined temporal phase. The Louvain Modularity algorithm find about 800 communities for each temporal phase (except for the phase RO where about 1100 communities exist). About 90% of users belong to less than 20 communities.

Quote. We also considered the quote-based networks. We created a directed graph for each temporal phase. An edge between two users exists if one user quotes the other within a defined temporal phase. The four quote-based networks contain about 500 distinct communities (except for phase RO where about 800 communities exist). 1% of the communities contains about 50% of users.

Reply. Finally, we considered the reply-based networks. We recursively gathered the replied tweets of 81,321 replies. The recovered replies are 103,559 at the end of the procedure. Then, we created a directed graph for each temporal phase. In particular, an edge between two users exists if one user replies the other during a defined temporal phase. The communities extracted from the reply-based network are about 700 for each temporal phase (except for phase RO where about 1500 communities exist). There are many communities that contain very few users, indeed only the 2% of the communities contains more than 10 users.

3.4   Relations and Stance

Here, we analyze the relations that occur among users not only considering the network structure, but also the users' stance. Table 3 shows the percentage of "uncross-stance" relations (edges between two users with the same stance) considering only users annotated with the labels AGAINST or FAVOR. We considered both unweighted and weighted graphs, where the strength of the connection is the number of interactions (retweet, quote, or reply) between two users within the same temporal phase. Following, we evaluated the percentage of "uncross-stance" relations for each of the four network types.

Table 3. The percentage of uncross-stance relations among users.

	RETWEET		QUOTE		REPLY	
	unweighted	*weighted*	*unweighted*	*weighted*	*unweighted*	*weighted*
Overall	98.6%	99.1%	**94.8%**	**97.6%**	**81.9%**	**77.3%**
EC	98.1%	98.9%	94.0%	96.9%	82.0%	71.9%
DE	99.7%	99.8%	96.1%	97.9%	83.2%	81.0%
TD	98.6%	99.4%	93.9%	97.7%	81.2%	78.9%
RO	**97.5%**	**97.6%**	96.3%	97.9%	80.9%	77.1%

Retweet. First, we analyzed the reply-based network. The considered 3,099 relations are respectively distributed on the four temporal phases as follows: 749, 885, 989, and 476. The column RETWEET in Table 3 shows the percentage of uncross-stance retweets in the retweet-based network. The users usually retweet only tweets belonging to users having the same stance (98.6% and 99.1% overall respectively for unweighted and weighted graphs). There are no significant differences between unweighted and weighted graphs. Notably, the percentage of uncross-stance relations slightly decreases in the phase RO.

Quote. Then, we considered networks based on quote relations. We performed the analysis over 717 relations (respectively 183, 179, 247, and 108 for each temporal phase). The column QUOTE in Table 3 shows the percentage of uncross-stance quotes over the temporal phases. There are no significant differences between temporal phases, but the percentage of uncross-stance relations varies between unweighted and weighted graphs (from 94.8% to 97.6% overall).

Reply. Finally, we analyzed the reply-based network. 662 relations are distributed over the four temporal phases as follows: 172, 173, 207, and 110. The column REPLY in Table 3 shows the percentage of uncross-stance in both unweighted and weighted for each temporal phase. There are no significant differences between temporal phases, but the percentage of uncross-stance replies significantly varies between unweighted and weighted graphs (in particularly from 81.9% to 77.3% overall). Moreover, here we find a signal that uncross-stance relations is not the whole story.

4   Experiments

We propose a new SD model relying on a set of new features, which exploits SVM as machine learning algorithm in a supervised framework. As evaluation metrics, we use two macro-average of the F_{micro} metrics i.e. F_{avg} and $F_{avg_{AF}}$. The first one computes the average among f-AGAINST, f-FAVOR, and F-NONE F_{micro} metrics. The second one, proposed in both SemEval 2016 Task 6 and IberEval 2017 SD tasks [9,12], computes the average between f-AGAINST and f-FAVOR F_{micro} metrics. We compare our results with two baselines such as: unigrams, bigrams and trigrams Bag of Words using SVM (BoW) and Majority Class ($MClass$). We compute the two metrics performing a five-cross validation on the CONREF-STANCE-ITA corpus employing each combination of the following features:

- **Bag of Hashtags** (BoH) and **Bag of Mentions** (BoM): hashtags/mentions as terms for building a vector with binary representation. These features use the texts contained in the tweet, the retweet, and the reply belonging to the triplet.
- **Bag of Hashtags+** ($BoH+$) and **Bag of Mention+** ($BoM+$): tokens extracted from the hashtags/mentions as terms for building a vector with binary representation. We segmented hashtags in tokens using the *greedy algorithm* attempting to find the longest word from a list of about 10M words extracted from Wikipedia's Italian pages. We consider as token the lemma of the verb *to vote* when an inflection of this verb is found. For what concerns mentions, tokens are the result of the *name* splitting, using space as separator. Names are extracted from the *User Object* field *name* of the mentioned user. The feature uses the texts contained in the tweet, the retweet, and the reply belonging to the triplet.
- **Bag of Hashtags+ Replies** ($BoH+R$) and **Bag of Mentions+ Replies** ($BoM+R$): These features are similar to $BoH+$ and $BoM+$, but they use information from the conversation thread, by exploiting the text of the replied

tweet belonging to the triplet. A different prefix has been used in order to differentiate these tokens from the ones belonging to *BoH+* and *BoM+* feature.

The combination of *BoH+*, *BoM+*, and *BoH+R* (afterwards *TCon*) achieved the highest results, F_{avg} 0.76 and $F_{avg_{AF}}$ 0.85. Notably, removing *BoH+R* from *TCon*, $F_{avg_{AF}}$ declines to 0.83 and F_{avg} declines to 0.69. The model probably is benefiting from the opposition of stance between reply and replied tweets.

Network-Based Features. In order to study the impact of knowledge of the social network for each network's type, we introduced three new features that consider the community which the user belongs to: **Retweet Communities** (*CRet*), **Quote Communities** (*CQuo*), and **Reply Communities** (*CRep*) respectively. In particular, considering the temporal phase $tp \in \{EC, DE, TD, RO\}$, N binary variables exist, one for each of the N detected communities in the retweet-based, quote-based, or reply-based networks. The variable set to one corresponds to the community to which the users who wrote the triplet belongs in the given temporal phase tp. Fig. 2 shows the combination of the three network-based features with *TCon*. As we can see, the combination of *TCon*, *CRet*, and *CQuo* achieved the highest value for both F_{avg} and $F_{avg_{AF}}$ (0.79 and 0.90, respectively) by improving the results obtained using only the *TCon* features (0.76 and 0.85, respectively). Nevertheless, adding the *CRep* feature does not improve neither F_{avg} and $F_{avg_{AF}}$.

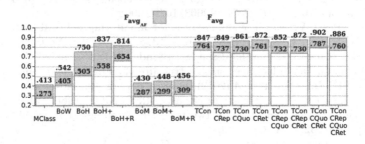

Fig. 2. F measured achieved adding network-based features to *TCon*. $F_{avg_{AF}}$: average between f-AGAINST and f-FAVOR F_{micro} metrics. F_{avg}: average among f-AGAINST, f-FAVOR, and F-NONE F_{micro} metrics.

5   Discussion and Conclusion

In this work we created a manually annotated Italian corpus for addressing SD from a diachronic perspective, which allows us to shed some light on users' opinion shift dynamics. We observed that in this debate, users tend to be less explicit on their stance as the outcome of the vote approaches. Analyzing the relations among users, we also observed that the retweet-based networks achieved the highest percentage of uncross-stance relations (percentage very close to 100%). This is a signal that Twitter's users retweet almost exclusively tweets they agreed on. Very high percentage of uncross-stance were achieved also by the quote-based networks. The variation between unweighted and weighted graphs could mean that users mainly quote users they agree on. Therefore, it is more likely to be in agreement when the number of quotes connecting two users increases. Interestingly, the opposite is happening on reply-based networks, where we can observe a higher percentage of communications between users with different stances. These observations led us to propose a new model for SD, which includes three new network-based features. The performed experiments show that adding *CRet* and *CQuo* features to content-based features considerably improve the accuracy of SD. We are guessing that when homophily is observed, the user's awareness of being a member of a community can ease user' stance prediction. This does not happen in *CRep*: although the users mainly reply to other users with a similar opinion, we observe about 20% of cross-stance edges. This is a particularly interesting case where inverse homophily (or also heterophily) could be observed. It will be matter of future investigations.

References

1. Adamic, L.A., Glance, N.: The political blogosphere and the 2004 u.s. election: divided they blog. In: Proceedings of the 3rd International Workshop on Link Discovery, LinkKDD 2005, pp. 36–43. ACM, New York (2005)
2. Blondel, V.D., Guillaume, J.L., Lambiotte, R., Lefebvre, E.: Fast unfolding of communities in large networks. J. Stat. Mech. Theory Exp. **10**, 10008 (2008). https://doi.org/10.1088/1742-5468/2008/10/P10008
3. Bosco, C., Patti, V.: Social media analysis for monitoring political sentiment. In: Alhajj, R., Rokne, J. (eds.) ESNAM, pp. 1–13. Springer, New York (2017)
4. Krings, G., Karsai, M., Bernhardsson, S., Blondel, V.D., Saramäki, J.: Effects of time window size and placement on the structure of an aggregated communication network. EPJ Data Sci. **1**(1), 4 (2012). https://doi.org/10.1140/epjds4
5. Lai, M., Cignarella, A.T., Hernández Farías, D.I.: ITACOS at ibereval2017: detecting stance in Catalan and Spanish tweets. In: Proceedings of IberEval 2017, vol. 1881, pp. 185–192. CEUR-WS (2017)
6. Lai, M., Tambuscio, M., Patti, V., Ruffo, G., Rosso, P.: Extracting graph topological information and users' opinion. In: Jones, G.J.F., Lawless, S., Gonzalo, J., Kelly, L., Goeuriot, L., Mandl, T., Cappellato, L., Ferro, N. (eds.) CLEF 2017. LNCS, vol. 10456, pp. 112–118. Springer, Cham (2017). https://doi.org/10.1007/978-3-319-65813-1_10
7. Lazarsfeld, P.F., Merton, R.K.: Friendship as a social process: a substantive and methodological analysis. In: Berger, M., Abel, T., Page, C. (eds.) Freedom and Control in Modern Society, pp. 18–66. Van Nostrand, NY (1954)
8. Messina, E., Fersini, E., Zammit-Lucia, J.: All atwitter about Brexit: Lessons for the election campaigns. https://radix.org.uk/work/atwitter-brexit-lessons-election-campaigns (2017). Accessed 28 Jan 2018
9. Mohammad, S., Kiritchenko, S., Sobhani, P., Zhu, X., Cherry, C.: Semeval-2016 task 6: detecting stance in tweets. In: Proceedings of SemEval-2016, pp. 31–41. ACL, San Diego (2016)

10. Mohammad, S.M., Sobhani, P., Kiritchenko, S.: Stance and sentiment in tweets. ACM TOIT **17**(3), 26:1–26:23 (2017). https://doi.org/10.1145/3003433
11. Pang, B., Lee, L.: Opinion mining and sentiment analysis. Found. Trends Inf. Retr. **2**(1–2), 1–135 (2008). https://doi.org/10.1561/1500000011
12. Taulé, M., Martí, M.A., Rangel, F.M., Rosso, P., Bosco, C., Patti, V., et al.: Overview of the task on stance and gender detection in tweets on Catalan independence at IberEval 2017. In: Proceedings of IberEval 2017, vol. 1881, pp. 157–177. CEUR-WS (2017)
13. Theocharis, Y., Lowe, W.: Does Facebook increase political participation? evidence from a field experiment. Inf. Commun. Soc. **19**(10), 1465–1486 (2016)
14. Zarrella, G., Marsh, A.: Mitre at semeval-2016 task 6: transfer learning for stance detection. In: Proceedings of SemEval-2016, pp. 458–463. ACL, San Diego (2016)

Identifying and Classifying Influencers in Twitter only with Textual Information

Victoria Nebot[1(✉)], Francisco Rangel[2,3], Rafael Berlanga[1], and Paolo Rosso[2]

[1] Departamento de Lenguajes y Sistemas Informáticos,
Universitat Jaume I, Campus de Riu Sec, 12071 Castellón, Spain
romerom@uji.es
[2] PRHLT Research Center, Universitat Politècnica de València,
Camino de vera s/n, 46022 Valencia, Spain
[3] Autoritas, Av. Puerto 267-5, 46011 Valencia, Spain

Abstract. Online Reputation Management systems aim at identifying and classifying Twitter influencers due to their importance for brands. Current methods mainly rely on metrics provided by Twitter such as followers, retweets, etc. In this work we follow the research initiated at RepLab 2014, but relying only on the textual content of tweets. Moreover, we have proposed a workflow to identify influencers and classify them into an interest group from a reputation point of view, besides the classification proposed at RepLab. We have evaluated two families of classifiers, which do not require feature engineering, namely: deep learning classifiers and traditional classifiers with embeddings. Additionally, we also use two baselines: a simple language model classifier and the "majority class" classifier. Experiments show that most of our methods outperform the reported results in RepLab 2014, especially the proposed Low Dimensionality Statistical Embedding.

Keywords: Online reputation management
Influencers classification · Author categorization
Textual information · Twitter

1 Introduction

The rise of Social Media has led to the emergence and spread out of new ways of communicating, without borders or censorship. Nowadays, people can connect with other people anywhere and let them know their opinion about any matter. Furthermore, it is happening massively. The collective imaginary that is formed by the opinions expressed about a brand (organisational or personal) is what makes up its (online) reputation. That is why Online Reputation Management (ORM) has proliferated in parallel with the final goal of taking control of the online conversation to mold it at will. The first step is to find out where people are expressing their opinions about the brand, and what these opinions are. In recent years, research has been very prolific in sentiment analysis, irony

M. Silberztein et al. (Eds.): NLDB 2018, LNCS 10859, pp. 28–39, 2018.
https://doi.org/10.1007/978-3-319-91947-8_3

detection, stance detection, and so forth. However, one of the main interests for brands is to know who is behind these opinions. Concretely, to know who are the influencers[1] of their sector and what kind of influence they may generate (what kind of authority/auctoritas[2] they have over what they say).

1.1 Related Work

Companies are interested in identifying influencers and classifying them. This problem has been addressed in Twitter in the framework of the RepLab[3] shared task in 2014 [2], as a competitive evaluation campaign for ORM.

At the RepLab the three approximations which obtained the best results were UTDBRG, LyS and LIA. In UTDBRG [1], the underlying hypothesis was that influential authors tweet actively about hot topics. A set of topics was extracted for each domain of tweets and a time-sensitive voting algorithm was used to rank authors in each domain based on the topics. They used quantitative, stylistic and behavioral features extracted from tweet contents. In LyS [14], features were extracted considering PoS tagging and dependency parsing. They used specific (such as URLs, verified account tag, user image) and quantitative (number of followers) profile meta-data. Finally, the authors of LIA [4] modeled each user based on the textual content, together with meta-data associated to her tweets.

Most of the current approaches rely on topology-based features. For example, the authors in [3] used a logistic regression method where all tweets from each user belonging to a given class were merged to create one large document. They employed 29 features such as user activity, stylistic aspects, tweets characteristics, profile fields, occurrence-based term weighting, and local topology. That is, they considered also features such as local topology (size of the friends set, size of the followers set, etc.) that are traditionally used by researchers from social network analysis. In this vein, the Collective Influence algorithm [10], which has been recently optimised [11], uses the "percolation" concept to isolate and identify the most influential nodes in a complex network. Recent investigations [7] also combine textual with topological features to find which tweets are worth spreading. Some findings suggest that influencers use more hashtags and mentions, as well as they follow more people on average.

However, our approximation aims at addressing the problem of identifying and classifying influencers in Twitter only on the basis of the textual information available in the tweets, without taking into consideration meta-data nor topological information from Social Network Analysis. We aim at investigating whether only textual features allow to find and classify such influencers in Twitter.

---

[1] Influencer is a user (person or brand) who may influence a high number of other people.
[2] In ancient Rome, auctoritas was the general level of prestige a person had. Due to that, authority in this context means prestige.
[3] http://www.clef-initiative.eu/track/replab.

1.2 Our Approach

In RepLab, several different categories are proposed to classify influencers. However, some of these categories are very low populated and the overall distribution is very imbalanced. Furthermore, industry usually considers two types of
influence: *(i)* the capacity of influencing others with their own opinions (influ-
ence/authority); and *(ii)* the ability to spread a message to the users with whom
the user interacts (betweenness). Due to that, we have focused on an alternative
way to detect and classify a subgroup of authority. Concretely, journalists who
may spread a message, and professionals who have the authority. The proposed
approach is based on the schema shown in Fig. 1. In a first step, influencers are
identified. Then, the identified influencers are classified in two ways. On the one
hand, they are classified according to the RepLab taxonomy. On the other hand,
influencers are firstly classified as belonging or not to the authority group and
then, they are disaggregated into professionals or journalists.

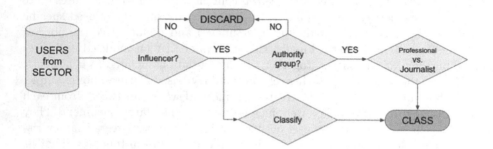

Fig. 1. Workflow to identify and classify influencers.

The remainder of this paper is organized as follows. Section 2 describes our
methodology. Section 3 discusses the experimental results, whereas the conclusions are drawn in Sect. 4.

2 Methodology

In this section, the RepLab 2014 shared task corpus is described. Then, we
present the methods we propose to address the problem. Finally, the evaluation
measures are discussed.

2.1 RepLab'14 Shared Task Corpus

The data collection contains over 7,000 Twitter profiles (all with at least 1,000
followers) that represent the automotive, banking and miscellaneous domains.
We focus on the first two domains. Each profile contains the last 600 tweets
published by the author at crawling time. Reputation experts performed manual annotations for two subtasks: Author Categorization and Author Ranking.

First, they categorized profiles as company, professional, celebrity, employee, stockholder, journalist, investor, sportsman, public institution, and ngo. Those profiles that could not be classified into one of these categories, were labeled as undecidable. In addition, reputation experts manually identified the opinion makers and annotated them as Influencer. Because the Author Categorization task is evaluated only over the profiles annotated as influencers in the gold standard, the exact number of profiles in training/test is shown in Table 1[4].

Table 1. RepLab'14 corpus statistics.

Class	Automotive			Banking			Total
	Training	Test	Total	Training	Test	Total	
Undecidable	454	-	454	556	-	556	1,010
Professional	312	358	670	279	286	565	1,235
Journalist	202	171	373	258	231	489	862
Company	94	119	213	51	33	84	297
Sportsmen	49	36	85	8	4	12	97
Celebrity	27	24	51	33	7	40	91
NGO	21	5	26	78	83	161	187
Public Inst.	12	3	15	27	30	57	72
Employee	3	8	11	1	6	7	18
Total	1,174	724	1,898	1,291	680	1,971	3,869

2.2 Methods

In this section we present all the approaches that we have evaluated and compared, along with their running parameters and configuration. We combine both traditional machine learning and deep learning methods with different word embedding techniques. We have classified the approaches based mainly on the type of embeddings they use, as we would like to explore the impact of the different types of embeddings on the classification tasks described above. We have also included a language model approach that was meant to serve as a baseline but has surprisingly given good results.

Glove. Glove [12] is a word embedding model used for learning vector representations of words. Such representations are able to encode semantic relations between words, like synonyms, antonyms, or analogies. Therefore, they have been widely used for NLP tasks. It is a count-based model based on matrix factorization. We have evaluated the following approach using this type of word vectors:

---

[4] We did not include Stockholder and Investor classes because they did not have data for Automotive and in Banking they did not have either training or test data.

– LSTM with Glove Twitter embeddings (LSTM+Glove)

Long short-term memory (LSTM) [5] is a deep learning technique that addresses the problem of learning long range dependencies and thus, is a state-of-the-art semantic composition model for a variety of text classification tasks.

In our experiment, we use a combination of the Glove pre-trained word vectors on Twitter[5] and the learned vectors for the RepLab dataset as input sequences for LSTM. In particular, each tweet is represented as the sum[6] of its word embeddings of dimension 50. In order to generate longer sequences per user, we generate sequences of five consecutive tweets. The configuration of the network is as follows: LSTM has 128 hidden units, we add a dropout layer with rate 0.5 and a softmax layer. We set a batchsize of 128 and train during 50 epochs.

Doc2vec. Doc2vec [8] is an extension of Word2vec [9]. Word2vec is another word embedding model but it differs from Glove in that it is a predictive model and uses a neural network architecture. Doc2vec learns to correlate labels and words, rather than words with other words. The purpose is to create a numeric representation of a document, regardless of its length. We have evaluated the following approaches using Doc2vec:

- **Logistic Regression with doc2vec (LR+d2v)**
- **Support Vector Machines with doc2vec (SVM+d2v)**
- **Multi Layer Perceptron with doc2vec (MLP+d2v)**
- **Convolutional Neural Network with doc2vec (CNN+d2v)**

We have evaluated both traditional machine learning methods (i.e., LR and SVM) and deep learning methods (i.e., MLP and CNN).

In the experiments, we consider each tweet as a document and obtain its vector representation of dimension 50. All the tweets of a user are aggregated by summing[7] the tweet vectors. This way, we obtain one feature vector per user that encapsulates the semantics of all her tweets. The feature vectors are passed to different classifiers for training. LR is configured with regularization 1e-5 and SVM uses a linear kernel. MLP is composed by 5 layers, a dense layer with 64 neurons, a dropout layer with rate 0.5, another dense layer and dropout layer with the same configuration and the output softmax layer. The batchsize is set to 128 and we train it for 20 epochs. For the CNN we basically use the configuration shown in [6].

Low Dimensionality Statistical Embedding (LDSE). LDSE[8] [13] represents documents on the basis of the probability distribution of the occurrence

---

[5] https://nlp.stanford.edu/projects/glove/.
[6] Sum has empirically given better results than average or concatenation.
[7] Sum has empirically given better results than average.
[8] Previously in other tasks as Low Dimensionality Representation (LDR).

of their words in the different classes. The key concept of LDSE is a weight, representing the probability of a term to belong to one of the different categories: influencer vs. non-influencer, interest-group vs. others, professional vs. journalist. The distribution of weights for a given document should be closer to the weights of its corresponding category. Formally, we represent the documents following the next three steps:

Step 1. We calculate the *tf-idf* weights for the terms in the training set D and build the matrix Δ, where each row represents a document d, each column a vocabulary term t, and each cell represent the *tf-idf* weight w_{ij} for each term in each document. Finally, $\delta(d_i)$ represents the assigned class c to the document i.

$$\Delta = \begin{bmatrix} w_{11} & w_{12} & ... & w_{1m} & \delta(d_1) \\ w_{21} & w_{22} & ... & w_{2m} & \delta(d_2) \\ ... & ... & ... & ... \\ w_{n1} & w_{n2} & ... & w_{nm} & \delta(d_n) \end{bmatrix}, \tag{1}$$

Step 2. Following Eq. 2, we obtain the term weights $W(t, c)$ as the ratio between the weights of the documents belonging to a concrete class c and the total distribution of weights for that term.

$$W(t, c) = \frac{\sum_{d \in D/c = \delta(d)} w_{dt}}{\sum_{d \in D} w_{dt}}, \forall d \in D, c \in C \tag{2}$$

Step 3. Following Eq. 3, these term weights are used to obtain the representation of the documents.

$$d = \{F(c_1), F(c_2), ..., F(c_n)\} \sim \forall c \in C, \tag{3}$$

Each $F(c_i)$ contains the set of features showed in Eq. 4, with the following meaning: *(i)* average value of the document term weights; *(ii)* standard deviation of the document term weights; *(iii)* minimum value of the weights in the document; *(iv)* maximum value of the weights in the document; *(v)* overall weight of a document as the sum of weights divided by the total number of terms of the document; and *(vi)* proportion between the number of vocabulary terms of the document and the total number of terms of the document.

$$F(c_i) = \{avg, std, min, max, prob, prop\} \tag{4}$$

As can be seen, this representation reduces the dimensionality to only six features per class by statistically embedding the distribution of weights of the document terms.

Finally, these weights are learned with a machine learning algorithm. Concretely[9]:

---

[9] We have tested several machine learning algorithms and finally we report the ones with the best results.

- **Naive Bayes & BayesNet** for influencers identification in the Automotive and the Banking domains, respectively.
- **Naive Bayes** for authority group detection.
- **SVM** for journalist vs. professional discrimination.
- **SVM** for RepLab task on classification.

Language Model (LM). Language models represent documents as a generative process of emitting words from each document $\mathbf{d}$, denoted $P(w_i|\mathbf{d})$. From the training set we can estimate the probabilities that each class c_j generates words w_i by simply applying Maximum Likelihood and Laplace Smoothing. The resulting model, denoted with $P(w_i|c_j)$, and the document model can be then factorized to get an estimate of the probability of each class to be generated by the document as follows:

$$P(c_j|\mathbf{d}) = \sum_{w_i} P(c_j|w_i) \cdot P(w_i|\mathbf{d})$$

Note that $\mathbf{d}$ can be either a tweet or a collection of tweets associated to a user. In this paper, $\mathbf{d}$ refers to the collection of user tweets. The resulting distribution $P(c_j|\mathbf{d})$ provides us the ranking of classes associated to each user to perform the classification.

2.3   Evaluation Measures

In RepLab, because the number of opinion makers is expected to be low, the influencers task is modeled as a search problem rather than as a classification problem. Thus, the evaluation measure selected is MAP (Mean Average Precision). For the categorization task, the measure selected to be able to compare us against RepLab results is Accuracy, which for multiclass settings, is equal to micro-averaged precision. However, we also compute the macro averaged version to see the effectiveness on the smaller classes.

3   Experiments and Discussion

This section presents the results of the different tasks defined in Fig. 1. Firstly, the identification of influencers. Then, the detection of authority users and their classification in journalists or professionals. Finally, the classification of the influencers in the RepLab taxonomy.

3.1   Identification of Influencers

The first task is to identify whether a user is an influencer or not, which corresponds to the Author Ranking task at RepLab. Therefore, the results shown in Table 2 can be compared to the best ones obtained in the competition. As can be seen, the LDSE approach obtains the best results, followed by LR+d2v.

Both of them outperform the best results obtained in the RepLab competition by more than 20% on average, demonstrating their competitiveness for the task. It has to be remarked that this was obtained on the basis of textual information of tweets only, without considering meta-data and Social Network Analysis information such as it was done by the best teams at RepLab and posteriorly by Cossu et al. [3].

Table 2. Identification of influencers (MAP score). RepLab'14 best results were obtained by UTDBRG in automotive and LyS in banking.

	Automotive	Banking	Average
LSTM+Glove	0.663	0.654	0.659
MLP+d2v	0.674	0.718	0.696
CNN+d2v	0.785	0.718	0.752
LR+d2v	0.861	**0.816**	0.839
SVM+d2v	0.833	0.784	0.809
LDSE	**0.874**	0.810	**0.842**
LM	0.865	0.526	0.696
RepLab'14	0.720	0.520	0.620
Cossu'15	0.803	0.668	0.735

In this task, except in the case of LM in Banking, traditional approaches obtain a better performance than deep learning approaches, with average differences between 5.7% and 18.3%.

3.2 Identification of the Authority Group

In order to separate the authority group from the rest of users, we have grouped journalist and professional into one group, and the remaining classes in the second group. As shown in Table 3, the resulting dataset is imbalanced towards the authority group.

Table 3. Authority vs. others corpus statistics.

Class	Automotive			Banking			Total
	Training	Test	Total	Training	Test	Total	
Authority group	514	529	1,043	537	517	1,054	2,067
Others	206	195	401	198	163	361	762
Total	720	724	1,444	700	878	1,415	2,829

Results are shown in Table 4. The best average results have been obtained by LDSE, which also demonstrates its robustness against the imbalance between classes with values about 77% in both micro and macro average. Deep learning methods such as MLP+d2v and CNN+d2v, and to a lesser extent LSTM+GLove, albeit they obtain a similar micro precision, show a trend to bias to the majority class, as shown by the low macro precision.

Table 4. Identification of the authority group.

	Automotive		Banking		Average	
	P-micro	P-macro	P-micro	P-macro	P-micro	P-macro
LSTM+Glove	0.724	0.472	0.760	0.589	0.742	0.531
MLP+d2v	0.731	0.365	0.760	0.380	0.746	0.373
CNN+d2v	0.731	0.365	0.760	0.380	0.746	0.373
LR+d2v	0.733	0.641	0.785	0.702	0.759	0.672
SVM+d2v	**0.751**	0.677	0.774	0.681	0.763	0.679
LDSE	0.745	**0.767**	**0.801**	**0.784**	**0.773**	**0.776**
LM	0.606	0.432	0.716	0.520	0.661	0.476
Majority class	0.731	0.365	0.760	0.38	0.746	0.373

3.3 Discriminating Between Journalists and Professionals

Once the authority influencers are identified, they should be separated into the corresponding class: journalist or professional. The corresponding dataset is also imbalanced towards the professional class, as shown in Table 5.

Table 5. Journalists vs. Professionals corpus statistics.

Class	Automotive			Banking			
	Training	Test	Total	Training	Test	Total	Total
Professional	312	358	670	279	286	565	1,235
Journalist	202	171	373	258	231	489	862
Total	514	529	1,043	537	517	1,054	2,097

The obtained results are shown in Table 6. Again, the LDSE obtains the highest results for both micro and macro precisions. In this task, traditional approaches also obtain higher results than deep learning approaches, besides less bias to the majority class.

3.4 Author Categorization

In this section we discuss the results of the classification of the influencers in the RepLab taxonomy. Table 7 shows our results compared with the best accuracies

Table 6. Discrimination between Journalist vs. Professional

	Automotive		Banking		Average	
	P-micro	P-macro	P-micro	P-macro	P-micro	P-macro
LSTM+Glove	0.673	0.574	0.625	0.619	0.649	0.597
MLP+d2v	0.677	0.338	0.553	0.277	0.615	0.308
CNN+d2v	0.677	0.338	0.584	0.617	0.630	0.478
LR+d2v	**0.745**	0.720	0.551	0.710	0.648	0.715
SVM+d2v	0.728	0.686	0.749	0.748	0.738	0.717
LDSE	0.742	**0.746**	**0.756**	**0.755**	**0.749**	**0.751**
LM	0.687	0.614	0,671	0.679	0.679	0.646
Majority class	0.677	0.338	0.553	0.277	0.615	0.308

Table 7. Results of the categorization task of the RepLab'14 shared task. RepLab'14 best results were obtained by LIA in both Automotive and Banking domains.

	Automotive		Banking		Average	
	P-micro	P-macro	P-micro	P-macro	P-micro	P-macro
LSTM+Glove	0.488	0.153	0.463	0.196	0.476	0.174
MLP+d2v	0.495	0.062	0.340	0.043	0.417	0.052
CNN+d2v	0.495	0.062	0.335	0.058	0.415	0.060
LR+d2v	0.524	0.228	0.594	0.330	0.559	0.279
SVM+d2v	0.526	**0.243**	0.579	0.279	0.553	0.261
LDSE	0.562	0.195	0.568	0.282	0.565	0.238
LM	**0.699**	0.205	**0.760**	**0.394**	**0.730**	**0.300**
Majority class	0.475	0.237	0.410	0.205	0.443	0.221
RepLab'14	0.450	-	0.500	-	0.475	-

reported in RepLab and the majority class prediction. This experiment includes all the classes defined in the dataset[10].

Results show that deep learning approaches (LSTM, MLP, CNN) have comparable performance to the best results of RepLab'14 in terms of micro precision. However, macro precision reveals poor performance regarding small classes. In particular, LSTM classifies great part of the test dataset to the majority class, basically following the training distribution. MLP+d2v and CNN+d2v also obtain very poor performance because they classify all the test data to the dominant class of the training set.

---

[10] As the Undecidable class only appears in training set, we have removed it to avoid noise in the training phase.

Traditional machine learning approaches using embeddings outperform the best results of the RepLab'14 competition, whose results are very similar to the majority class baseline. It is worth noting the scores given by LM, which was initially included as a baseline, but clearly outperforms the rest of the approaches in this task. LM has a similar performance in all the tasks and its performance is not downgraded in this multiclass classification task, unlike the other approaches. This is due to the fact that LM builds the model for each class independently.

Once more, we should highlight that our approach, which only uses textual information from tweets, obtains more than 25% better results than LIA where meta-data information associated to tweets was also used.

4   Conclusions

Companies are interested in identifying influencers and classifying them. This task was addressed in the framework of RepLab. In this paper, we were interested in approaching the problem both on the basis of deep learning techniques and with traditional machine learning. We compared the obtained results with the best systems in RepLab and the more recent approach of Cossu et al. On the contrary of the above systems, which used also meta-data and information from Social Network Analysis, we used only textual information and outperformed their results by more than 20% in influencer identification and more than 25% in influencer classification.

It is noteworthy that deep learning approaches obtained worse results than traditional techniques in all the previous tasks, where their results were biased to the majority class. Probably the low number of training samples (around 500 without undecidable class) explains part of these results. Low Dimensionality Statistical Embedding (LDSE) obtained the best results in almost all the tasks and it shows its robustness against corpus imbalance.

Due to the interest of the industry on identifying different kinds of influence, we have proposed an alternative methodology: first, we identify the influencer; then, we determine if the influencer belongs to the authority group; finally, we classify the influencer as journalist or professional. The results obtained show that the best approach (LDSE) is competitive: more than 84% identifying influencers, and 77% and 75% classifying them respectively in the authoritative group, and as the type of influencer. Nevertheless, future work should focus on combining it with generative models like LM as it shows best results in the categorization task, which is the only multi-classification task.

Acknowledgements. The work of the last author was funded by the SomEMBED TIN2015-71147-C2-1-P MINECO research project. Authors from Universitat Jaume I have been funded by the MINECO R&D project TIN2017-88805-R.

References

1. AleAhmad, A., Karisani, P., Rahgozar, M., Oroumchian, F.: University of Tehran at RepLab 2014. In: Proceedings of the Fifth International Conference of the CLEF Initiative (2014)
2. Amigó, E., Carrillo-de-Albornoz, J., Chugur, I., Corujo, A., Gonzalo, J., Meij, E., de Rijke, M., Spina, D.: Overview of RepLab 2014: author profiling and reputation dimensions for online reputation management. In: Kanoulas, E., Lupu, M., Clough, P., Sanderson, M., Hall, M., Hanbury, A., Toms, E. (eds.) CLEF 2014. LNCS, vol. 8685, pp. 307–322. Springer, Cham (2014). https://doi.org/10.1007/978-3-319-11382-1_24
3. Cossu, J.-V., Dugué, N., Labatut, V.: Detecting real-world influence through Twitter. In: 2015 Second European Network Intelligence Conference (ENIC), pp. 83–90. IEEE (2015)
4. Cossu, J.-V., Janod, K., Ferreira, E., Gaillard, J., El-Bèze, M.: Lia@ RepLab 2014: 10 methods for 3 tasks. In: Proceedings of the Fifth International Conference of the CLEF Initiative (2014)
5. Hochreiter, S., Schmidhuber, J.: Long short-term memory. Neural Comput. 9(8), 1735–1780 (1997)
6. Kim, Y.: Convolutional neural networks for sentence classification. In: Moschitti, A., Pang, B., Daelemans, W. (eds.) Proceedings of the 2014 Conference on Empirical Methods in Natural Language Processing, EMNLP 2014, 25–29 October 2014, Doha, Qatar, A meeting of SIGDAT, A Special Interest Group of the ACL, pp. 1746–1751. ACL (2014)
7. Lahuerta-Otero, E., Cordero-Gutiérrez, R.: Looking for the perfect Tweet. The use of data mining techniques to find influencers on Twitter. Comput. Hum. Behav. 64, 575–583 (2016)
8. Le, Q., Mikolov, T.: Distributed representations of sentences and documents. In: Proceedings of the 31st International Conference on International Conference on Machine Learning - Volume 32, ICML 2014, pp. II-1188-II-1196 (2014). JMLR.org
9. Mikolov, T., Chen, K., Corrado, G., Dean, J.: Efficient estimation of word representations in vector space. CoRR abs/1301.3781 (2013)
10. Morone, F., Makse, H.A.: Influence maximization in complex networks through optimal percolation. Nature 524(7563), 65 (2015)
11. Morone, F., Min, B., Bo, L., Mari, R., Makse, H.A.: Collective influence algorithm to find influencers via optimal percolation in massively large social media. Sci. Rep. 6, 30062 (2016)
12. Pennington, J., Socher, R., Manning, C.D.: Glove: global vectors for word representation. In: EMNLP, pp. 1532–1543. ACL (2014)
13. Rangel, F., Franco-Salvador, M., Rosso, P.: A low dimensionality representation for language variety identification. In: Gelbukh, A. (ed.) CICLing 2016. LNCS, vol. 9624. Springer, Cham (2018). https://doi.org/10.1007/978-3-319-75487-1_13. arXiv:1705.10754
14. Vilares, D., Hermo, M., Alonso, M.A., Gómez-Rodríguez, C., Vilares, J.: LyS at CLEF RepLab 2014: creating the state of the art in author influence ranking and reputation classification on Twitter. In: Proceedings of the Fifth International Conference of the CLEF Initiative (2014)

Twitter Sentiment Analysis Experiments Using Word Embeddings on Datasets of Various Scales

Yusuf Arslan[1(✉)], Dilek Küçük[2], and Aysenur Birturk[1]

[1] Middle East Technical University, Ankara, Turkey
{yusuf.arslan,birturk}@ceng.metu.edu.tr
[2] TÜBİTAK Energy Institute, Ankara, Turkey
dilek.kucuk@tubitak.gov.tr

Abstract. Sentiment analysis is a popular research topic in social media analysis and natural language processing. In this paper, we present the details and evaluation results of our Twitter sentiment analysis experiments which are based on word embeddings vectors such as word2vec and doc2vec, using an ANN classifier. In these experiments, we utilized two publicly available sentiment analysis datasets and four smaller datasets derived from these datasets, in addition to a publicly available trained vector model over 400 million tweets. The evaluation results are accompanied with discussions and future research directions based on the current study. One of the main conclusions drawn from the experiments is that filtering out the emoticons in the tweets could be a facilitating factor for sentiment analysis on tweets.

Keywords: Sentiment analysis · Twitter · Word embeddings
word2vec · doc2vec

1 Introduction

Sentiment analysis is a well-defined and well-studied topic in social media analysis and natural language processing (NLP). As presented in the related survey papers [11,14], considerable research effort has been devoted to solve this problem and, more recently, deep learning methods have also been employed to improve the sentiment analysis performance [21].

On the other hand, word embedding vectors like word2vec [12] have been extensively utilized for NLP problems. These problems include sentiment analysis [16,18–20] on microblog posts, named entity recognition [4], and text classification [10].

Bag-of-words approach has been utilized for a long time in various domains in NLP. In bag-of-words method, each word is equidistant to each other. However, word2vec method utilizes the distances between words. In doc2vec method, document information is also utilized [12].

In this paper, we present the settings and evaluation results of our sentiment analysis experiments on tweets using features based on word embeddings. Polarity classes (positive and negative) at different granularities are considered during these experiments. Widely-employed Keras API [5] is used during the experiments where word2vec and doc2vec [9] features are employed by an ANN-based classifier. An ANN-based classifier is chosen since it is good at adaptive learning, parallelism, pattern learning, sequence recognition, fault tolerance, and generalization [17]. The experiments are carried out using three automatically-annotated versions of two datasets and a model previously trained on a considerably large third tweet dataset. The results are accompanied with discussions and future research plans.

Our main contribution in the paper is the use of publicly available tools and datasets during the experiments and, particularly, the use of the word2vec model of a publicly-available and large tweet set (of 400 million tweets in English) [7] for sentiment analysis purposes.

The rest of the paper is organized as follows: In Sect. 2, first we present the details of the six datasets and the trained model used during the sentiment analysis experiments and then describe the details of the actual experiments on these datasets. Section 3 includes discussions of the results of these experiments together with future research prospects. Finally, Sect. 4 concludes the study with a summary of the main points of the presented study.

2 Sentiment Analysis Experiments Using Word Embeddings

In this section, we first present the settings (in terms of test/train datasets and trained model) of our sentiment analysis experiments and next, we present the evaluation results of the actual experiments.

2.1 Sentiment Datasets and Model Used in the Experiments

Sentiment analysis experiments are carried out using the following seven resources: two publicly-available datasets, four subsets derived from these datasets, and the vector model of another larger tweet set. The details of these resources are provided below:

1. The first dataset used for both training and testing purposes is the set of 1,6 million tweets in English which were automatically annotated as positive or negative, based on the existence of positive and negative emoticons [6]. This dataset is commonly referred to as Stanford Twitter Sentiment (STS) dataset. In our experiments, STS dataset without emoticons (which will be referred to as STS-w/oe) has been used as training and testing dataset after splitting, as will be presented in the upcoming subsection.
2. A subset of up to 50,000 tweets is extracted from the STS dataset, by automatically processing them with sentiment dictionaries (as given in [2]) to

obtain a fairly balanced set of highly positive and highly negative tweets. This second dataset is created in order to alleviate the effects of inaccurate annotations during the original automatic annotation of the STS dataset as reported in the corresponding study [6]. Hence, although we have not manually annotated the dataset to calculate the actual accuracies, we expect this subset to have less errors in the sentiment annotations. Because, as stated above, only those tweets labelled with the polarity classes of highly positive and highly negative are included in this subset, and it is likely that these annotations have higher accuracy compared to that of the annotations with the polarity classes of positive and negative. This dataset also does not include emoticons and will henceforth be referred to as STS-50K-w/oe dataset. It has also been used for training and testing purposes as will be clarified.

3. The third dataset used for both training and testing purposes is the set of 1,578,612 tweets from [13]. The dataset is publicly available and as it contains emoticons, we will refer to it as TN-w/e. The dataset has been used as training and testing datasets after splitting, as will be presented in the upcoming subsection.

4. Similar to the second dataset above, a subset of up to 50,000 tweets is extracted from the TN dataset, by automatically processing them to obtain a fairly balanced set of highly positive and highly negative tweets. This dataset will be referred to as TN-50K dataset. It has also been used for training and testing purposes as will be clarified.

5. A random subset of up to 50,000 tweets is also extracted from the STS-50K-w/oe dataset. This dataset will be referred to as STS-50K-random-w/oe, and has been used for training and testing purposes during the experiments.

6. Similarly, a random subset of up to 50,000 tweets is also extracted from the TN-w/e dataset. This dataset will be referred to as TN-50K-random-w/e, and has been used for training and testing purposes.

7. We have used the word2vec word embeddings model for 400 million English tweets, as presented in [7] where the model has been used to improve named entity recognition on tweets. To the best of our knowledge, this model which was trained on a considerably large tweet corpus, has not previously been used for sentiment analysis purposes. Prior to the model creation, the tweets have been preprocessed to use replacement tokens for mentions, URLs, and numbers [7]. This preprocessing scheme has also been employed on the previous three dataset versions to increase the comparability of the results. This model will be referred to as Model-400M in the rest of this paper and will be used as the trained model only, since the actual tweets are not publicly available.

Statistics regarding the six dataset versions, in terms of the tweets annotated as positive or negative and total number of tweets, are summarized in Table 1.

2.2 Sentiment Analysis Experiments

In the experiments, word embeddings are generated by use of Gensim library [15]. Keras neural networks API [5] with Tensorflow backend [1] is used. However, for

Table 1. Statistical information on the dataset versions.

Dataset	Positive tweets	Negative tweets	Total
STS-w/oe	800,000	800,000	1,600,000
TN-w/e	790,185	788,443	1,578,612
STS-50K-w/oe	10,449	18,144	28,593
TN-50K-w/e	17,859	28,235	46,094
STS-50K-random-w/oe	10,449	18,144	28,593
TN-50K-random-w/e	17,859	28,235	46,094

these initial experiments reported in this paper, an ANN classifier implementation in Keras is used instead of the deep learning algorithms. Similar experiments with deep learning algorithms such as those based on Long Short-Term Memory (LSTM) [8] will be utilized within the course of future work.

Our sentiment analysis framework consists of one input layer, one hidden layer and one output layer.

In the hidden layer, ReLu (Rectified Linear Units) method is employed. This method functions mimics the neurons in human body. If its input is less than 0 then it outputs 0, and if the input is greater than 0 then it outputs raw value. The function of it can be seen in following formula:

$$f(x) = max(x, 0) \tag{1}$$

In the output layer we have implemented the sigmoid function for binary classification, which can be expressed with the following formula:

$$S(x) = 1/(1 + (e^{-x})) \tag{2}$$

The sigmoid function is especially used to predict the probability as 0 or 1 based on the model and this feature makes it suitable to be used in binary classification. It can be implemented both in hidden and output layers and it is implemented in the output layer of our neural nets.

In our model, the neural nets are trained for 20 epochs which means that each training sample are examined 20 times by the model to detect the underlying patterns.

In our Twitter sentiment analysis experiments, either a dataset version listed in Table 1 is utilized for training and testing purposes or the trained model (Model-400M) [7] is used for training and a dataset listed in Table 1 is used during the testing phase. The settings in each of the experiments differ in the following aspects:

- The dataset version/trained model used during training and the dataset version used during testing,
- Either word2vec or doc2vec vectors are used as features

Table 2. The evaluation results of sentiment analysis experiments.

Train dataset/model	Test dataset	Accuracy using	
		Word2Vec	Doc2Vec
TN-50K-w/e	TN-50K-w/e	81.4%	**82.1%**
STS-50K-w/oe	STS-50K-w/oe	83.4%	**83.6%**
TN-50K-random-w/e	TN-50K-random-w/e	**66.9%**	64.7%
STS-50K-random-w/oe	STS-50K-random-w/oe	**67.6%**	67.1%
TN-w/e	TN-w/e	**79.1%**	79.0%
STS-w/oe	STS-w/oe	**79.6%**	79.3%
TN-w/e	TN-50K-w/e	83.6%	**85.3%**
TN-w/e	TN-50K-random-w/e	76.3%	76.3%
TN-w/e	STS-50K-w/oe	**87.5%**	86.1%
TN-w/e	STS-50K-random-w/oe	76.9%	**77.0%**
STS-w/oe	TN-50K-w/e	84.7%	**85.2%**
STS-w/oe	TN-50K-random-w/e	76.7%	**77.4%**
STS-w/oe	STS-50K-w/oe	88.5%	**88.8%**
STS-w/oe	STS-50K-random-w/oe	**77.5%**	77.2%
Model-400M	TN-w/e	78.4%	-
Model-400M	STS-w/oe	78.8%	-
Model-400M	TN-50K-w/e	83.9%	-
Model-400M	TN-50K-random-w/e	75.4%	-
Model-400M	STS-50K-w/oe	86.3%	-
Model-400M	STS-50K-random-w/oe	74.0%	-

In all of the experiments excluding the latter ones using Model-400M, the 9/10 of the training set has been randomly chosen as the actual training data and 1/10 of the test dataset has been randomly chosen as the actual test dataset. In all experimental settings, three trials are conducted and the arithmetic averages of the accuracies of these three trials are reported as the accuracy.

The test settings and sentiment analysis accuracies obtained as the results of these experiments are presented in the Table 2. In the first column, the name of the dataset/model used during training is given and in the second column the test dataset is given. The latter two columns present the actual accuracies using word2vec and doc2vec vectors as features, respectively.

3 Discussion of the Results

The following conclusions can be drawn from the evaluation results of the Twitter sentiment analysis experiments presented in the previous section:

- The sentiment analysis results are considerably higher for both versions of STS-50K which contains automatically extracted highly positive and highly negative sentiment annotations. Although this set is not a gold standard, since it is further processed to increase its accuracy compared to the other datasets, it is a comparatively cleaner dataset which leads to higher accuracies, reaching up to ∼88% and to ∼86% using Model-400M during training.
- Considering the existence of emoticons in the tweets of the datasets in the training and test sets, it is hard to draw a sound conclusion regarding this issue. In some cases, the existence of emoticons in the training and test sets leads to slight decreases but such differences may not be statistically significant. However, for particular cases like the experiments on TN-50K-w/e and STS-50K-w/oe (in the first two rows of Table 2), the sentiment analysis results on the dataset without the emoticons (STS-50K-w/oe) are considerably higher with both word2vec and doc2vec features.
- Using the vectors formed within Model-400M leads to moderate results comparable to the results of the other settings. It is a promising outcome since this model includes only word2vec vectors without any sentiment-specific feature on a considerably large tweet dataset.

Future work based on the current study includes the following:

- Conducting experiments using deep learning algorithms with word2vec and doc2vec features on the STS, TN, and derived datasets used in the current paper, as well as on other publicly available sentiment datasets. In these experiments, widely-employed resources for sentiment analysis such as SentiWordNet [3] can be used.
- The experiments presented in the current paper can be replicated on tweets in other languages as well, in order to observe the corresponding performance rates. While doing this, the peculiarities of these languages should be taken into account and the necessary preprocessing steps should be employed to handle these peculiarities.
- Performing statistical tests to present the statistical significance of the attained evaluation results as well as the results of those of the prospective experiments.
- Further research on the processing of emoticons for sentiment analysis can be conducted with an intention to improve the overall system performance.

4   Conclusion

Sentiment analysis is a significant NLP problem and word embedding models such as word2vec and doc2vec are recent and important resources for the algorithms tackling with this problem as well as other NLP problems. In this paper, we present the settings and evaluation results of our sentiment analysis experiments on two publicly available and commonly-used datasets and two datasets at different scales derived from these datasets. The experiments are based on word2vec and doc2vec features and also utilize a previously created publicly available model over 400 million tweets. Our results indicate that filtering out emoticons seems to be an improving factor on clean annotated datasets, and word embeddings based vectors over large tweet sets lead to favorable sentiment analysis accuracies. Future work includes using deep learning algorithms and performing statistical tests to reveal the statistical significance of the reported results.

References

1. Abadi, M., Agarwal, A., Barham, P., Brevdo, E., Chen, Z., Citro, C., Corrado, G.S., Davis, A., Dean, J., Devin, M., Ghemawat, S., Goodfellow, I., Harp, A., Irving, G., Isard, M., Jia, Y., Jozefowicz, R., Kaiser, L., Kudlur, M., Levenberg, J., Mané, D., Monga, R., Moore, S., Murray, D., Olah, C., Schuster, M., Shlens, J., Steiner, B., Sutskever, I., Talwar, K., Tucker, P., Vanhoucke, V., Vasudevan, V., Viégas, F., Vinyals, O., Warden, P., Wattenberg, M., Wicke, M., Yu, Y., Zheng, X.: TensorFlow: large-scale machine learning on heterogeneous systems (2015). https://www.tensorflow.org/. Software available from tensorflow.org
2. Arslan, Y., Birturk, A., Djumabaev, B., Küçük, D.: Real-time lexicon-based sentiment analysis experiments on Twitter with a mild (more information, less data) approach. In: 2017 IEEE International Conference on Big Data, BigData 2017, pp. 1892–1897 (2017)
3. Baccianella, S., Esuli, A., Sebastiani, F.: SentiWordNet 3.0: an enhanced lexical resource for sentiment analysis and opinion mining. In: LREC, vol. 10, pp. 2200–2204 (2010)
4. Chiu, J.P., Nichols, E.: Named entity recognition with bidirectional LSTM-CNNs. arXiv preprint arXiv:1511.08308 (2015)
5. Chollet, F.: Keras (2015). https://github.com/fchollet/keras
6. Go, A., Bhayani, R., Huang, L.: Twitter sentiment classification using distant supervision. CS224N Project Report, Stanford 1(12) (2009)
7. Godin, F., Vandersmissen, B., De Neve, W., Van de Walle, R.: Multimedia Lab @ ACL WNUT NER Shared Task: named entity recognition for Twitter microposts using distributed word representations. In: Workshop on Noisy User-generated Text (WNUT), pp. 146–153 (2015)
8. Greff, K., Srivastava, R.K., Koutník, J., Steunebrink, B.R., Schmidhuber, J.: LSTM: a search space Odyssey. IEEE Trans. Neural Netw. Learn. Syst. 28(10), 2222–2232 (2017)
9. Le, Q., Mikolov, T.: Distributed representations of sentences and documents. In: International Conference on Machine Learning, pp. 1188–1196 (2014)
10. Lilleberg, J., Zhu, Y., Zhang, Y.: Support vector machines and word2vec for text classification with semantic features. In: IEEE 14th International Conference on Cognitive Informatics and Cognitive Computing (ICCI* CC), pp. 136–140 (2015)
11. Liu, B., Zhang, L.: A survey of opinion mining and sentiment analysis. In: Aggarwal, C., Zhai, C. (eds.) Mining Text Data, pp. 415–463. Springer, Boston (2012). https://doi.org/10.1007/978-1-4614-3223-4_13
12. Mikolov, T., Sutskever, I., Chen, K., Corrado, G.S., Dean, J.: Distributed representations of words and phrases and their compositionality. In: Advances in Neural Information Processing Systems, pp. 3111–3119 (2013)
13. Naji, L.: Twitter sentiment analysis training corpus (dataset) (2017). http://thinknook.com/twitter-sentiment-analysis-training-corpusdataset-2012-09-22/. Accessed 06 Feb 2016

14. Pang, B., Lee, L., et al.: Opinion mining and sentiment analysis. Found. Trends®
    Inf. Retr. **2**(1–2), 1–135 (2008)
15. Řehůřek, R., Sojka, P.: Software framework for topic modelling with large cor-
    pora. In: Proceedings of the LREC 2010 Workshop on New Challenges for NLP
    Frameworks, pp. 45–50. ELRA, Valletta, Malta (May 2010). http://is.muni.cz/
    publication/884893/en
16. dos Santos, C., Gatti, M.: Deep convolutional neural networks for sentiment anal-
    ysis of short texts. In: 25th International Conference on Computational Linguistics
    (COLING), pp. 69–78 (2014)
17. Sharma, A., Dey, S.: A document-level sentiment analysis approach using artificial
    neural network and sentiment lexicons. ACM SIGAPP Appl. Comput. Rev. **12**(4),
    67–75 (2012)
18. Tang, D., Wei, F., Yang, N., Zhou, M., Liu, T., Qin, B.: Learning sentiment-specific
    word embedding for Twitter sentiment classification. In: 52nd Annual Meeting of
    the Association for Computational Linguistics (Volume 1: Long Papers), vol. 1,
    pp. 1555–1565 (2014)
19. Xue, B., Fu, C., Shaobin, Z.: A study on sentiment computing and classification of
    Sina Weibo with word2vec. In: IEEE International Congress on Big Data (BigData
    Congress), pp. 358–363 (2014)
20. Zhang, D., Xu, H., Su, Z., Xu, Y.: Chinese comments sentiment classification based
    on word2vec and SVMperf. Expert Syst. Appl. **42**(4), 1857–1863 (2015)
21. Zhang, L., Wang, S., Liu, B.: Deep learning for sentiment analysis: a survey. arXiv
    preprint arXiv:1801.07883 (2018)

A Deep Learning Approach for Sentiment Analysis Applied to Hotel's Reviews

Joana Gabriela Ribeiro de Souza, Alcione de Paiva Oliveira$^{(\boxtimes)}$,
Guidson Coelho de Andrade, and Alexandra Moreira

Universidade Federal de Viçosa, Viçosa, MG 36570-900, Brazil
joana.souza@ufv.br, alcione@gmail.com

Abstract. Sentiment Analysis is an active area of research and has presented promising results. There are several approaches for modeling that are capable of performing classifications with good accuracy. However, there is no approach that performs well in all contexts, and the nature of the corpus used can exert a great influence. This paper describes a research that presents a convolutional neural network approach to the Sentiment Analysis Applied to Hotel's Reviews, and performs a comparison with models previously executed on the same corpus.

Keywords: Sentiment analysis · Convolutional neural network
Hotel's reviews

1 Introduction

Sentiment Analysis has become an important decision-making tool and several machine learning techniques have been applied with the objective of obtaining analyzes with a good degree of prediction. A model that has commonly been used as a baseline for Sentiment Analysis is the bag of words model, commonly applied to a generative technique, such as Naive Bayes [2]. Even though a good technique, word bag models and Naive Bayes, does not produce the best results in all situations, and discriminative models, usually it has better results in relation to accuracy [6]. One of the discriminative techniques that has been most currently applied in the of Sentiment Analysis is the use of Artificial Neural Nets. The great recent growth of the neural networks approach as a machine learning tool was mainly due to the development of hardware and learning algorithms that enabled the implementation of networks with multiple layers called deep learning [4]. Regardless of the technique used, it is fundamental that there are datasets that can be used to feed these machine learning algorithms. One of the possible sources for generating data set for Sentiment Analysis is the extraction of user reviews of products and services such as travel and hotel booking sites. People conduct reviews on a variety of Web sites, such as on social networks, forums, and specialized websites. There are some example of websites such as TripAdvisor

© Springer International Publishing AG, part of Springer Nature 2018
M. Silberztein et al. (Eds.): NLDB 2018, LNCS 10859, pp. 48–56, 2018.
https://doi.org/10.1007/978-3-319-91947-8_5

(http://www.tripadvisor.com), and Booking (http://www.booking.com), among others that receive ratings related to tourism services. This type of site usually has a reserved area so the traveler can evaluate the location and service they have enjoyed using a star selector where the evaluator usually selects between one and five stars and can make a more detailed comment. When observing a series of evaluations made by people of hotels in Rio de Janeiro (the corpus used in this work), it is possible to notice that several people evaluate the hotel with high scores, but their commentary presents several problems found in the establishment. This type of information can not be acquired if only the assessment made by the number of stars is considered. Thus, analyzing the content of the evaluation becomes an important activity for the task of identifying the opinions expressed by the guests.

This paper describes a research that presents a convolutional neural network approach to the Sentiment Analysis Applied to Hotel's reviews written in Brazilian Portuguese for the city of Rio de Janeiro, and performs a comparison with models previously executed on the same corpus. Reviews were taken from tripadvisor site. This paper is organized as follows: the next section presents researches previously developed that are related to this work; Sect. 3 describes the corpus used; Sect. 4 describes the model created to perform the analysis; Sect. 5 presents the results obtained; and Sect. 6 presents the final conclusions.

2 Related Works

Context is an important aspect when it comes to review analysis. Context helps to get more information and disambiguate the meaning of the terms used by the reviewer. Taking that into account Aciar [1] presented a method that uses rules for the identification of preferences and context of hotel reviews taken from TripAdvisor. The author selected 100 reviews to conduct her experiment. After baseline tests and 50 other reviews that were also manually classified by specialists, F-measure values of 88% and 91% were obtained for rules that identify context and preferences respectively. Our work does not seek to identify the context, but the focus is to determine the polarity of the reviews, although this information can serve as basis for future work that improves the determination of polarity.

Naive Bayes classifier is a method used by several researchers to perform Sentiment Analysis as in [9]. Our work performs an analysis of the same dataset used by [9], which allows us to carry out a performance comparison between the two researches. Sentiment Analysis was carried out using as dataset the comments, in Brazilian Portuguese, made on the TripAdvisor site about hotels in Rio de Janeiro city. The tests used were lemmatization, polarity inversion (after the word "no" the subsequent words receives a reverse polarity) and Laplace smoothing. The best result obtained 74.48% of accuracy by removing the stop-words and performing the inversion of polarity, lemmatization and Laplace smoothing. However, the accuracy for the neutral and negative values was 38.24% and 56.11% respectively. Classifying only the positive and negative classes using the same number of evaluations for both classes, the precision increased considerably from 90.83% to 97.9% and from 56.11% to 78.69% in the positive and negative classes respectively. In our work we did not remove accents, punctuation, did not do lemmatization and did not convert the words into lowercase, in order to take advantage of aspects that may influence the refinement of the task of sentiment analysis. Even though we did not perform this pre-processing, we obtained relevant results.

[5] developed a neural language model called ConvLstm. The architecture of its model is based on convolutional neural network (CNN). To obtain better results, the model has an LSTM recurrence layer as an alternative to the grouping layer in order to efficiently capture the long term dependency. The model is more compact in comparison with networks that only adopt convolutional layers. The model was validated with two sentiment databases: the Stanford Sentiment Treebank and the Stanford Large Movie Review Dataset. In relation to this work the similarity is in the choice of using databases with short texts, but the dataset is not directed to the tourism domain.

3   The Corpus

Details on the creation of the corpus were presented in [9]. It consists of a set of 69,075 reviews about hotels located at Rio de Janeiro city, written in Brazilian Portuguese. The reviews were taken from the TripAdvisor site, and the most recent review date is from October 31, 2016.

Before applying the sentiment analysis techniques over the data, we first performed a pre-processing in the corpus to correct some issues. Empty lines, repeated punctuation characters, parentheses, and meaningless characters were removed. Also, because the tool that extracted comments from the site truncated them into 60 words, it was necessary to remove unfinished sentences. After these normalization the corpus resulted in 3,298,395 words and 51,408 types. The next stage of pre-processing was the removal of the stop-words except for the word "no", since it's relevant when one want to extract an opinion from a particular topic. A common technique is to invert the polarity of the words in a sentence that occurs after the word no, although this is not always the appropriate method [7]. After the withdrawal of the stop-words, the corpus resulted in 2,158,008 words and 42,284 types.

4   The Deep Learning Approach

The deep learning approach (DL) to natural language processing (NLP) has become quite popular in the last decade [4]. According to [4], this was due to the ease of model development, advances in hardware for parallel computing, and the creation of learning algorithms for neural networks with a large number of layers. Particularly in the case of NLP the popularity is also due to the adherence of the technique to a large number of problems such as automatic translation, document summarization, syntactic and semantic annotation of lexical items, speech recognition, text generation, and sentiment analysis.

However, in the case of NLP text analysis, in order to adopt the DL approach, it is necessary to represent the words in a numerical form so that it can serve as input to the neural network. One way of numerically capturing contextual information and word relationships is through the use of dense vectors that encode these relationships through their positioning in an n dimensional space. There are some techniques for coding words in the form of dense vectors, with the best known being Word2vec [10] and GloVe (global vectors) [11]. As it is considered the state of the art in the representation of words in the form of vectors, GloVe was the methodology used to represent the words in our work. Preliminary tests showed that 100-dimensional GloVe vectors were sufficient for our purposes. They were generated over the entire corpus, after the preprocessing phase, but before the removal of the stop-words. We used the entire corpus in order to capture as much statistical information as possible about the context of the words aiming to create a vector that was best representative.

Another preparation step was the division of the corpus into 2 parts, one with the score given by the travelers and the other with the comments. For simplification instead of using the five possible classes (horrible, bad, reasonable, very good and excellent), we mapped this five classes into just three classes (negative, neutral and positive) according to the polarities presented in [8]. In our mapping we have grouped the "horrible" and "bad" classes as the negative class, the "reasonable" class became the neutral class and the "very good" and "excellent" classes became the positive class.

For the development of the neural network model for a performance of the sentiment analysis, the Keras framework was used. Keras [3] is an open source neural network library written in Python and it was chosen due to its high level API, and because it was developed with a focus on enabling fast experimentation. Keras is capable of running on top of TensorFlow, CNTK, or Theano, using them as backend. According to its documentation, Keras assists in the development of deep learning models, providing high-level building blocks, so it is not necessary to perform low-level operations such as product tensor and convolutions. Nevertheless, it needs a specialized and optimized tensor manipulation library to do so, serving as the "backend mechanism". In this work we used TensorFlow as backend. It is an open-source symbolic tensor manipulation framework developed by Google. We picked TensorFlow as it is actively maintained, robust, and flexible enough to be able to develop models target for both CPU and GPU.

In the case of sentences or sequence of words, the preferred neural network architecture is the recurrent neural networks and their variations, such as LSTM (long short-term memory) and GRU (Gated recurrent units). However, in the case of sentences with a known maximum size, convolutional networks can also be applied, due to the fact that they are able retain locality information. For these reasons, the convolutional networks were the type of networks chosen for the development of our model. After several tests the structure of the convolutional network that produced the best results presented the layers shown is described below.

The first layer of the network is the layer that receives the embedded sentences. It should be sized so that it can receive the sentence with the maximum size multiplied by the size of the vector of each word. Thus, in our case, the embedding layer has to have the dimension $(MaxLen, VectorSize)$, where $MaxLen$ is the review maximum size (i.e. 80), and $VectorSize$ is the size of the GloVe vector of each word, which in this case was established 100. Tests with 200-size vectors were performed, but the results were not better and lead to a longer training time. The tests showed a structure with ten hidden layers, being four convolutional of one dimension, four pooling layers and two dense layers was able to produce good results without demanding high processing time. We also added three DropOut layer aiming to avoid overfitting. For the pooling layers, we used as the basis of calculation the MaxPooling. At the output of each convolution layer we applied a Rectified Linear Unit (ReLU) activation function. The two dense layers emitted their results through a hyperbolic tangent activation function. The output layer had a single neuron using the sigmoid activation function, so the result would be a number between 0 and 1. For simplification purposes, we normalized the outputs to distribute them in each class according to the given interval: between 0 and 0.25 was assigned to the negative class which identification was 0. An output greater than 0.25 and up to 0.75 was attributed to the neutral class, identified as 1. It was attributed to the positive class, identified by 2, a result greater than 0.75.

We compiled the model with the Adam optimizer, and binary cross-entropy were used as the loss function. For training, we divided the corpus into two parts, 80% for training and 20% for testing. We performed several experiments in order to define the model parameters.

5   Results

The tests were carried out employing five different sub-corpus of the original corpus aiming to balance the numbers of reviews in each class: *Corpus 1* (68078 reviews divided into 4863 negative, 12117 neutral and 52098 positive); *Corpus 2* (29097 reviews divided into 4863 negative and 12117 neutral and positive); *Corpus 3* (14589 reviews divided equally into each class); *Corpus 4* (9726 reviews with the same amount of negatives and positives); and *Corpus 5* (56960 reviews divided into 4863 negative and 52098 positive). All sub-corpus, except *Corpus 1*, had their reviews shuffled randomly to keep data from being concentrated, and the training and test split could capture examples from all classes involved. We did several tests to determine the number of training times for the model, and we identified that 50 epochs yield good results, and increasing the number of epochs did not result in greater accuracy, only a longer processing time.

Additionally, we tested word vectors (generated as GloVe vectors) of size 50, 100 and 200. The results presented here were obtained with the vector of size 100 as it achieved better results than vectors of size 50, and less processing time than those of size 200. On the other hand, when performing tests with vectors of size 200 we obtained results similar to those of size 100, but with a longer processing time. The results presented here correspond to the values obtained using a convolutional neural network architecture with the layout and data shape previously described.

The results obtained with our approach exceed those presented by [9]. Table 1 shows the precision and recall, and the confusion matrix generated using Corpus 1. Observing the confusion matrix, one can see that the greatest challenge is to identify the neutral class. The classes imbalance is one of the factors that influence this difficulty. Additionally, the neutral class ends up having a fuzzy boundary, which causes some difficulties. Another aspect that we can highlight is the normalization of the previous labels. The previous "bad" and "very good" classes ended up being classified as negative and positive respectively, but the neutral class has a broader range (between 0.25 and 0.75). The recall is relatively low for the negative and neutral classes, but the accuracy of our model was better than the work of [9]. The result shows an increase of 16.71% and 12.92% accuracy of the negative and neutral classes, respectively, and a result very similar to that found for the positive class accuracy (around 90%).

Table 1. Results for Corpus 1

	Precision and recall		Confusion matrix		
	Precision	Recall	Negative	Neutral	Positive
Negative	72.82%	56.78%	526	1380	79
Neutral	51.16%	56.19%	163	1013	913
Positive	90.47%	89.93%	33	1013	9384

Table 2 shows the precision and recall found using Corpus 2. In this corpus the number of positive reviews has been reduced to the same number of neutrals. We can note that the performance for neutral reviews has improved considerably, and remained close to that found in the negative and positive classes compared to Corpus 1. With this balancing we obtained the best accuracy to identify the neutral class, above the results found in related works, pointing out that our approach was able to better classify the neutral class correctly, as can be observed in the confusion matrix presented in Table 2.

Using Corpus 3, where all classes were balanced, we obtained the results presented in Table 3, indicating a higher hit rate for the positive and negative classes in relation to the other two corpus. However, in relation to the neutral class, there was an improvement over the results obtained in Corpus 1, but we lost precision in relation to Corpus 2. It is possible to observe that the results obtained in our work surpass the results obtained by [9] in all classes, with improvement of 1.54%, 29.19% and 25.99% for the positive, neutral and negative classes, respectively. With this we were able to demonstrate that our network

Table 2. Results for Corpus 2

	Precision and recall		Confusion matrix		
	Precision	Recall	Negative	Neutral	Positive
Negative	73.09%	61.85%	627	361	27
Neutral	73.36%	82.05%	215	1976	217
Positive	89.27%	84.19%	20	358	2016

is able to correctly classify most of the reviews, even the neutral class, which according to the literature is harder to identify and is often dismissed as having difficult identification.

Table 3. Results for Corpus 3

	Precision and recall		Confusion matrix		
	Precision	Recall	Negative	Neutral	Positive
Negative	82.1%	73.44%	710	246	10
Neutral	67.43%	79.78%	145	758	61
Positive	92.37%	86.63%	10	121	852

We also performed experiments using only negative and positive reviews, as [9] did. Using Corpus 4, which contains the same number of reviews of each class, the obtained accuracy of the model was 95.74%. In relation to [9] the precision for the positive class was 1.86% smaller but in relation to the negative class we had an expressive improvement of 16.78%. Finally, we used Corpus 5 which contain a high imbalance between the negative and positive classes. It is noticed that even with the imbalance, we obtained a slightly higher precision for the negative class (from 78.69% to 88.94% in relation to the classification using the same number of reviews of each class performed by [9]) and a precision of 98.31% for the positive class.

6   Conclusions

A challenging aspect of working with Sentiment Analysis is related to the fact that opinions are subjective. Evaluations made by users may present inconsistencies. For instance, in the corpus there are evaluations with positive scores while in the corresponding description the user mostly points out the negative aspects or suggestions to the place visited. This phenomenon can cause the classifier to label the review as negative or neutral according to the input text, but its class is positive due to the score given by the user. This type of occurrence is discussed in [12], where they performed a study that analyzed the polarity of the evaluations performed by people compared to the results obtained by three different methods of sentiment analysis.

The results obtained with the tests performed in different corpus setup showed that the convolutional neural network achieved considerably better results than the study used in the comparison. We also took advantage of the maximum information available in the corpus. Unlike [9] we did not perform lemmatization, withdraw the punctuation marks, normalize the text in lower-case, classify the polarity of the previously words or manual classify the text. Also, using CNN eliminate the need for creating feature rules that can be difficult to develop and validate as in [1]. We also covered three reviews classes.

CNN is often applied to corpus that has as many as billions of tokens, but even for a corpus that can be considered relatively small, the results were satisfactory. Future works may use a larger corpus and perform the tests again to observe the accuracy of the approach for massive data sets. Using CNN for sentence classification is an interesting approach when one is working with small sentences, but as the size of sentences grows, recurrent network should be a more adequate approach.

Acknowledgments. This research is supported in part by the funding agencies FAPEMIG, CNPq, and CAPES.

References

1. Aciar, S.: Mining context information from consumers reviews. In: Proceedings of Workshop on Context-Aware Recommender System, vol. 201. ACM (2010)
2. Agarwal, B., Mittal, N.: Prominent Feature Extraction for Sentiment Analysis. Springer, Cham (2016). https://doi.org/10.1007/978-3-319-25343-5
3. Chollet, F.: Keras. GitHub (2015). https://github.com/fchollet/keras
4. Goldberg, Y.: Neural network methods for natural language processing. Synth. Lect. Hum. Lang. Technol. **10**(1), 1–309 (2017)
5. Hassan, A., Mahmood, A.: Deep learning approach for sentiment analysis of short texts. In: 3rd International Conference on Control, Automation and Robotics (ICCAR), pp. 705–710 (2017)
6. Jurafsky, D., Martin, J.H.: Speech and Language Processing: An Introduction to Natural Language Processing, Computational Linguistics, and Speech Recognition, 2nd edn. Prentice Hall, Upper Saddle River (2008)
7. Kasper, W., Vela, M.: Sentiment analysis for hotel reviews. In: Computational Linguistics-Applications Conference, vol. 231527, pp. 45–52 (2011)
8. Liu, B.: Sentiment analysis and opinion mining. Synth. Lect. Hum. Lang. Technol. **5**(1), 1–167 (2012)
9. Martins, G.S., de Paiva Oliveira, A., Moreira, A.: Sentiment analysis applied to hotels evaluation. In: Gervasi, O., et al. (eds.) ICCSA 2017. LNCS, vol. 10409, pp. 710–716. Springer, Cham (2017). https://doi.org/10.1007/978-3-319-62407-5_52
10. Mikolov, T., Chen, K., Corrado, G., Dean, J.: Efficient estimation of word representations in vector space. arXiv:1301.3781 [cs] (2013)

11. Pennington, J., Socher, R., Manning, C.D.: GloVe: global vectors for word representation. In: Empirical Methods in Natural Language Processing (EMNLP), pp. 1532–1543 (2014)
12. Valdivia, A., Luzón, M.V., Herrera, F.: Sentiment analysis on tripadvisor: are there inconsistencies in user reviews? In: Martínez de Pisón, F.J., Urraca, R., Quintián, H., Corchado, E. (eds.) HAIS 2017. LNCS (LNAI), vol. 10334, pp. 15–25. Springer, Cham (2017). https://doi.org/10.1007/978-3-319-59650-1_2

Automatic Identification
and Classification of Misogynistic
Language on Twitter

Maria Anzovino[1], Elisabetta Fersini[1(✉)], and Paolo Rosso[2]

[1] University of Milano-Bicocca, Milan, Italy
`fersini@disco.unimib.it`
[2] PRHLT Research Center, Universitat Politècnica de València, Valencia, Spain

Abstract. Hate speech may take different forms in online social media. Most of the investigations in the literature are focused on detecting abusive language in discussions about ethnicity, religion, gender identity and sexual orientation. In this paper, we address the problem of automatic detection and categorization of misogynous language in online social media. The main contribution of this paper is two-fold: (1) a corpus of misogynous tweets, labelled from different perspective and (2) an exploratory investigations on NLP features and ML models for detecting and classifying misogynistic language.

Keywords: Automatic misogyny identification · Hate speech
Social media

1 Introduction

Twitter is part of ordinary life of a great amount of people[1]. Users feel free to express themselves as if no limits were imposed, although behavioural norms are declared by the social networking sites. In these settings, different targets of hate speech can be distinguished and recently women emerged as victims of abusive language both from men and women. A first study [9] focused on the women's experience of sexual harassment in online social network, reports the women perception on their freedom of expression through the *#mencallmethings* hashtag. More recently, as the allegations against the Hollywood producers were made public, a similar phenomenon became viral through the *#metoo* hashtag. Another investigation was presented by Fulper et al. [4]. The authors studied the usefulness of monitoring contents published in social media in foreseeing sexual crimes. In particular, they confirmed that a correlation might lay between the yearly per capita rate of rape and the misogynistic language used in Twitter. Although the problem of hate speech against women is growing rapidly,

---

[1] https://www.statista.com/statistics/282087/number-of-monthly-active-twitter-users/.

ⓒ Springer International Publishing AG, part of Springer Nature 2018
M. Silberztein et al. (Eds.): NLDB 2018, LNCS 10859, pp. 57–64, 2018.
https://doi.org/10.1007/978-3-319-91947-8_6

most of the computational approaches in the state of the art are detecting abusive language about ethnicity, religion, gender identity, sexual orientation and cyberpedophilia. In this paper, we investigate the problem of misogyny detection and categorization on social media text. The rest of the paper is structured as follows. In Sect. 2, we describe related work about hate speech and misogyny in social media. In Sect. 3, we propose a taxonomy for modelling the misogyny phenomenon in online social environments, together with a Twitter dataset manually labelled. Finally, after discussing experiments and the obtained results in Sect. 4, we draw some conclusions and address future work.

2   Misogynistic Language in Social Media

2.1   Hate Speech in Social Media

Recently, several studies have been carried out in the attempt of automatically detecting hate speech. The work presented in [15] makes a survey of the main methodologies developed in this area. Although linguistic features may differ within the various approaches, the classification models implemented so far are supervised. Furthermore, a great limit is that a benchmark data set still does not exist. Hence, the authors of [11] built and made public a corpus labelled accordingly to three subcategories (hate speech, derogatory, profanity), where hate speech is considered as a kind of abusive language. In [17] the authors described a corpus that was labelled with tags about both racial and sexist offenses. A recent study about the distinction of hate speech and offensive language has been presented in [8].

2.2   Misogyny in Social Media

Misogyny is a specific case of hate speech whose targets are women. Poland, in her book about cybermisogyny [13], remarked among the others the problem of online sexual harassment. She deepened the matter about Gamergate occurred in 2014 primarily bursted on *4chan* and then spread across different social networking sites. Gamergate was an organized movement which seriously threatened lives of women belonging to the video games industry. The harassment took place firstly online, then it degenerated offline. This episode confirms that the cybermisogyny does exist, thus it is necessary to put effort in trying to prevent similar phenomena. To the best of our knowledge only a preliminary exploratory analysis of misogynous language in online social media has been presented in [7]. The authors collected and manually labelleld a set of tweets as positive, negative and neutral. However, nothing has been done from a computational point of view to recognize misogynous text and to distinguish among the variety of types of misogyny.

Table 1. Examples of text for each misogyny category

Misogyny category	Text
Discredit	I've yet to come across a nice girl. They all end up being bitches in the end #WomenSuck
Stereotype	I don't know why women wear watches, there's a perfectly good clock on the stove. #WomenSuck
Objectification	You're ugly. Caking on makeup can't fix ugly. It just makes it worse!
Sexual Harassment	Women are equal and deserve respect. Just kidding, they should suck my dick
Threats of Violence	Domestic abuse is never okay.... Unless your wife is a bitch #WomenSuck
Dominance	We better not ever have a woman president @WomenSuckk
Derailing	@yesallwomen wearing a tiny skirt is "asking for it". Your teasing a (hard working, taxes paying) dog with a bone. That's cruel. #YesAllMen

Table 2. Unbalanced composition of misogyny categories within the dataset

Misogyny category	# Tweets
Discredit	1256
Sexual Harassment and Threats of Violence	472
Stereotype and Objectification	307
Dominance	122
Derailing	70

3   Linguistic Misogyny: Annotation Schema and Dataset

3.1   Taxonomy for Misogynistic Behaviour

Misogyny may take different forms in online social media. Starting from [13] we designed a taxonomy to distinguish between misogynous messages and, among the misogynous ones, we characterized the different types of manifestations. In particular, we modeled the followed misogynistic phenomena:

1. *Discredit:* slurring over women with no other larger intention.
2. *Stereotype and Objectification:* to make women subordinated or description of women's physical appeal and/or comparisons to narrow standards.
3. *Sexual Harassment and Threats of Violence:* to physically assert power over women, or to intimidate and silence women through threats.
4. *Dominance:* to preserve male control, protect male interests and to exclude women from conversation.
5. *Derailing:* to justify abuse, reject male responsibility, and attempt to disrupt the conversation in order to refocus it.

A representative text for each category is reported in Table 1.

3.2 Dataset

In order to collect and label a set of text, and subsequently address the misogyny detection and categorization problems, we start from the set of keywords in [7] and we enriched these keywords with new ones as well as hashtags to download representative tweets in streaming. We added words useful to represent the different misogyny categories, when they were related to a woman, as well as when implying potential actions against women. We also monitored tweets in which potential harassed users might have been mentioned. These users have been chosen because of their either public effort in feminist movements or pitiful past episodes of harassment online, such as Gamergate. Finally, we found Twitter profiles who declared to be misogynistic, i.e. those user mentioning hate against women in their screen name or the biography. The streaming download started on 20th of July 2017 and was stopped on 30th of November 2017. Next, among all the collected tweets we selected a subset querying the database with the co-presence of each keyword with either a phrase or a word not used to download tweets but still reasonable in picturing misogyny online. The labeling phase involved two steps: firstly, a gold standard was composed and labeled by two annotators, whose cases of disagreement were solved by a third experienced contributor; secondly, the remaining tweets were labeled through a majority voting approach by external contributors on the CrowdFlower platform. The gold standard has been used for the quality control of the judgements throughout the second step. As far as it concerns the gold standard, we estimated the level of agreement among the annotators before the resolution of the cases of disagreement. The kappa coefficient [3] is the most used statistic for measuring the degree of reliability between annotators. The need for consistency among annotators immediately arises due to the variability among human perceptions. This interagreement measure can be summarized as:

$$k = \frac{observed\,agreement - chance\,agreement}{1 - chance\,agreement} \tag{1}$$

However, considering only this statistic is not appropriate when the prevalence of a given response is very high or very low in a specific class. In this case, the value of kappa may indicate a low level of reliability even with a high observed proportion of agreement. In order to address these imbalances caused by differences in prevalence and bias, the authors of [2] introduced a different version of the kappa coefficient called prevalence adjusted bias-adjusted kappa (PABAK). The estimation of PABAK depends solely on the observed proportion of agreement between annotators:

$$PABAK = 2 \cdot observed\,agreement - 1 \tag{2}$$

A more reliable measure for estimating the agreement among annotators is PABAK-OS [12], which controls for chance agreement. PABAK-OS aims to avoid the peculiar, unintuitive results sometimes obtained from Cohen's Kappa, especially related to skewed annotations (prevalence of a given label). Given the great unbalance among the misogyny categories aforementioned within the dataset employed, we applied the PABAK-OS metric: with regards to the misogyny detection we obtained 0.4874 PABAK-OS, whereas for the misogyny categorization 0.3732 PABAK-OS. Thus, on one side the level of agreement on the presence of misogynistic language is *moderate* as it is within the range [0.4;0.6], and on the other side the level of agreement about the misogynistic behavior encountered is *fair* as it is within the range [0.2;0.4]. These measures are reasonable, because the classification of the misogynistic behavior is more complex than the identification of the misogynistic language. The final dataset is composed of 4454 tweets[2], balanced between misogynous vs no-misogynous. Considering the proposed taxonomy, the misogynous text have been distinguished as reported in Table 2.

4   Methodology

4.1   Feature Space

Misogyny detection might be considered as a special case of abusive language. Therefore, we chose representative features taking into account the guidelines suggested in [11]:

1. **N-grams:** we considered both *character* and *token n-grams*, in particular, from 3 to 5 characters (blank spaces included), and tokens as unigrams, bigrams and trigrams. We chose to include these features since they usually perform well in text classification.
2. **Linguistic:** misogyny category classification is a kind of stylistic classification, which might be improved by the use of quantitative features [1,5]. Hence, for the purpose of the current study we employed the following stylistic features:
   (a) *Length of the tweet in number of characters*, tweets labelled as sexual harassment are usually shorter.
   (b) *Presence of URL*, since a link to an external source might be an hint for a derailing type tweet.
   (c) *Number of adjectives*, as stereotype and objectification tweets include more describing words.
   (d) *Number of mentions of users*, since it might be useful in distinguishing between individual and generic target.
3. **Syntactic:** we considered Bag-Of-POS. Hence, unigrams, bigrams and trigrams of Part of Speech tags.
4. **Embedding:** the purpose of this type of features is to represent texts through a vector space model in which each word associated to a similar context lays close to each other [10]. In particular, we employed the *gensim* library for Python [14] and used the pre-trained model on a Twitter dataset[3] that was made public.

---

[2] The dataset has been made available for the IberEval-2018 (https://amiibereval2018.wordpress.com/) and the EvalIta-2018 (https://amievalita2018.wordpress.com/) challenges.

[3] https://www.fredericgodin.com/software/.

Table 3. Accuracy performance for misogynistic language identification

Features combination	RF	NB	MPNN	SVM
Char n-grams	0.7930	0.7508	0.7616	0.7586
Token n-grams	0.7856	0.7432	0.7582	**0.7995**
Embedding	0.6893	0.6834	0.7041	0.7456
Bag-of-POS	0.6064	0.6031	0.6017	0.5997
Linguistic	0.5831	0.6098	0.5963	0.5348
Char n-grams, Linguistic	0.7890	0.7526	0.7443	0.7627
Token n-grams, Linguistic	0.7739	0.7164	0.7593	0.7966
Embedding, Linguistic	0.6830	0.5878	0.7014	0.6556
Bag-of-POS, Linguistic	0.6069	0.6286	0.5997	0.5799
All Features	0.7427	0.7730	0.7613	0.7739

Table 4. Macro F-measure performance for misogyny category classification

Features combination	RF	NB	MPNN	SVM
Char n-grams	0.3429	0.31953	0.3591	0.3548
Token n-grams	0.3172	0.38177	0.3263	**0.3825**
Embedding	0.2905	0.14510	0.1734	0.3635
Bag-of-POS	0.2629	0.24625	0.2810	0.2628
Linguistic	0.2292	0.14424	0.1544	0.2125
Char n-grams, Linguistic	0.3458	0.27725	0.3637	0.3651
Token n-grams, Linguistic	0.2794	0.14991	0.3097	0.3841
Embedding, Linguistic	0.2945	0.22272	0.1661	0.2636
Bag-of-POS, Linguistic	0.2423	0.23096	0.2585	0.2221
All Features	0.2826	0.20266	0.3697	0.3550

4.2   Experimental Investigation

Measures of Evaluation. Since the nature of the dataset, we chose different evaluation metrics in regard to the classification task. Thus, for the Misogynistic Language identification we considered the accuracy as the dataset is balanced in that respect. On the contrary, the representation of each misogyny category is greatly unbalanced, and therefore the macro F-measure was used.

Supervised Classification Models. Linear Support Vector Machine (SVM), Random Forest (RF), Naïve Bayes (NB) and Multi-layer Perceptron Neural Network (MPNN) were used[4], being these models the most effective in text categorization [16].

---

[4] We employed the machine learning package scikit-learn: http://scikit-learn.org/stable/supervised_learning.html.

Experiments. We firstly carried out 10-fold cross validation experiments with each type of features, where character n-grams and token n-grams are evaluated individually, next the linguistic features in conjunction with each of the remaining ones, and finally the all features were considered[5].

Results. In Table 3 we report the results that have been obtained with the classifiers employed in the misogyny identification. Further, in Table 4 the results that have been obtained in the misogynistic behavior classification are shown. As far as it concerns the misogynistic language identification, the performances reached by the classifiers chosen are close to each other. On the contrary, about the misogyny classification they may differ substantially. Token n-grams allow to achieve competitive results, especially for the task of misogynistic language identification in Twitter where indeed using all the features decreases marginally the obtained accuracy[6]. Results about misogyny classification show the difficulty of recognizing the different phenomena of misogyny. No matter that, token n-grams obtained competitive results if compared when also linguistic features were employed. Although the investigated features show promising results, additional sets related to skip character [6] could be considered in the future.

5   Conclusions and Future Work

The work presented is the first attempt in detecting misogynistic language in social media. Our aim was twofold: identifying whether a tweet is misogynous or not, and also classifying it on the basis of misogynistic language that was employed. Moreover, a taxonomy was introduced in order to study the different types of misogynistic language behaviour. Last but not least, a dataset containing tweets of the different categories was built and it will be made available to the research community to further investigate this sensitive problem of automatic identification of misogynistic language. Although the misogyny identification and classification is still in its infancy, the preliminary results we obtained show a promising research direction. As future work, the increasing level of aggressiveness will be tackles through a multi-label classification approach. Final challenging task would be how to deal with irony in the context of misogyny identification.

Acknowledgements. The work of the third author was partially funded by the Spanish MINECO under the research project SomEMBED (TIN2015-71147-C2-1-P).

---

[5] When training the considered classifiers, we didn't apply any feature filtering or parameter tuning.

[6] Results obtained with All Features are statistically significant (Student t-test with p-value equal to 0.05).

References

1. Argamon, S., Whitelaw, C., Chase, P., Hota, S.R., Garg, N., Levitan, S.: Stylistic text classification using functional lexical features: research articles. J. Am. Soc. Inf. Sci. Technol. **58**(6), 802–822 (2007)
2. Byrt, T., Bishop, J., Carlin, J.B.: Bias, prevalence and kappa. J. Clin. Epidemiol. **46**(5), 423–429 (1993)
3. Cohen, J.: A coefficient of agreement for nominal scales. Educ. Psychol. Measur. **20**(1), 37–46 (1960)
4. Fulper, R., Ciampaglia, G.L., Ferrara, E., Ahn, Y., Flammini, A., Menczer, F., Lewis, B., Rowe, K.: Misogynistic language on Twitter and sexual violence. In: Proceedings of the ACM Web Science Workshop on Computational Approaches to Social Modeling (ChASM) (2014)
5. HaCohen-Kerner, Y., Beck, H., Yehudai, E., Rosenstein, M., Mughaz, D.: Cuisine: classification using stylistic feature sets and/or name-based feature sets. J. Assoc. Inf. Sci. Technol. **61**(8), 1644–1657 (2010)
6. HaCohen-kerner, Y., Ido, Z., Ya'akobov, R.: Stance classification of tweets using skip char Ngrams. In: Altun, Y., Das, K., Mielikäinen, T., Malerba, D., Stefanowski, J., Read, J., Žitnik, M., Ceci, M., Džeroski, S. (eds.) ECML PKDD 2017. LNCS (LNAI), vol. 10536, pp. 266–278. Springer, Cham (2017). https://doi.org/10.1007/978-3-319-71273-4_22
7. Hewitt, S., Tiropanis, T., Bokhove, C.: The problem of identifying misogynist language on Twitter (and other online social spaces). In: Proceedings of the 8th ACM Conference on Web Science, pp. 333–335. ACM, May 2016
8. Davidson, T., Warmsley, D., Macy, M., Weber, I.: Automated hate speech detection and the problem of offensive language. In: Proceedings of the 12th International AAAI Conference on Web and Social Media (2017)
9. Megarry, J.: Online incivility or sexual harassment? Conceptualising women's experiences in the digital age. In: Women's Studies International Forum, vol. 47, pp. 46–55. Pergamon (2014)
10. Le, Q., Mikolov, T.: Distributed representations of sentences and documents. In: International Conference on Machine Learning, pp. 1188–1196, January 2014
11. Nobata, C., Tetreault, J., Thomas, A., Mehdad, Y., Chang, Y.: Abusive language detection in online user content. In: Proceedings of the 25th International Conference on World Wide Web, pp. 145–153. International World Wide Web Conferences Steering Committee (2016)
12. Parker, R.I., Vannest, K.J., Davis, J.L.: Effect size in single-case research: a review of nine nonoverlap techniques. Behav. Modif. **35**(4), 303–322 (2011)
13. Poland, B.: Haters: Harassment, Abuse, and Violence Online. University of Nebraska Press, Lincoln (2016)
14. Rehurek, R., Sojka, P.: Software framework for topic modelling with large corpora. In: Proceedings of the LREC 2010 Workshop on New Challenges for NLP Frameworks (2010)
15. Schmidt, A., Wiegand, M.: A survey on hate speech detection using natural language processing. In: Proceedings of the Fifth International Workshop on Natural Language Processing for Social Media. Association for Computational Linguistics, Valencia, Spain, pp. 1–10 (2017)
16. Sebastiani, F.: Machine learning in automated text categorization. ACM Comput. Surv. (CSUR) **34**(1), 1–47 (2002)
17. Waseem, Z., Hovy, D.: Hateful symbols or hateful people? predictive features for hate speech detection on Twitter. In: SRW@ HLT-NAACL, pp. 88–93 (2016)

Assessing the Effectiveness of Affective Lexicons for Depression Classification

Noor Fazilla Abd Yusof$^{(\boxtimes)}$, Chenghua Lin, and Frank Guerin

Computing Science, University of Aberdeen, Aberdeen, UK
{noorfazilla.yusof,chenghua.lin,f.guerin}@abdn.ac.uk

Abstract. Affective lexicons have been commonly used as lexical features for depression classification, but their effectiveness is relatively unexplored in the literature. In this paper, we investigate the effectiveness of three popular affective lexicons in the task of depression classification. We also develop two lexical feature engineering strategies for incorporating those lexicons into a supervised classifier. The effectiveness of different lexicons and feature engineering strategies are evaluated on a depression dataset collected from LiveJournal.

Keywords: Depression analysis · Affective lexicon · Language analysis

1 Introduction

Depression is one of the most common mental disorders that can affect people of all ages. It is the leading cause of disability and requires significant health care cost to treat effectively [1]. Compared to the traditional clinical consultation, mental health studies based on social media present several advantages [2]. For instance, social media sites provide great venues for people to share their experiences, vent emotion and stress, which are useful for medical practitioners to understand patients' experiences outside the controlled clinical environment. In addition, information captured during clinical consultation generally reflects only the situation of the patient at the time of care. In contrast, data collected from social media is dynamic, thereby providing opportunities for observing and recognising critical changes in patients' behaviour [2].

There is a large body of work using social media for depression related studies, e.g., predicting postpartum depression in new mothers [3], identifying depression-indicative posts [4], and analysing the causes of depression [5]. These works normally leverage various features for the depression classification task such as lexical features (e.g., affective lexicons and linguistic styles), behavioural markers (e.g. time-posting), and social-networking features (e.g. engagement, number of followers). Among these features, lexical features are the most commonly used one. However, their effectiveness in supporting the task of depression classification is relatively unexplored.

In this study, we investigate the effectiveness of three popular affective lexicons in the task of depression classification, namely, ANEW [6], MPQA [7],

© Springer International Publishing AG, part of Springer Nature 2018
M. Silberztein et al. (Eds.): NLDB 2018, LNCS 10859, pp. 65–69, 2018.
https://doi.org/10.1007/978-3-319-91947-8_7

and SentiWordNet [8]. These lexicons are incorporated into a Multinomial Naive Bayes (MNB) classifier via two different lexical feature engineering strategies, i.e., (1) include a data instance (i.e., a post) for training only if it contains as least one word from the lexicon, and (2) constrain the training set feature space to the feature space of the lexicon. We test the effectiveness of different lexicons and feature engineering strategies based on a depression dataset collected from LiveJournal. Experimental results show that SentiWordNet outperforms the other lexicons in general, and gives the best result. SentiWordNet achieved 84.4% accuracy using feature engineering Strategy 1, which is significantly better than the baseline (MNB without incorporating any sentiment lexicons).

2   Methodology

In this section, we describe our approach for detecting depressive text by combining a supervised learning model, namely Multinomial Naive Bayes (MNB) [9], with different lexical feature engineering techniques. Our hypothesis is that by leveraging external affective lexicon resources with appropriate feature engineering techniques, the performance of a MNB supervised classifier in predicting depressive text can be significantly enhanced. Specifically, we have explored two lexical feature engineering strategies as detailed below.

- **Strategy 1**: Include a data instance (e.g., a post) for training only if it contains at least one word from the lexicon. This strategy filters out posts that are potentially not relevant to depression.
- **Strategy 2**: Constrain the training set feature space to the features of the affective lexicon. This strategy significantly reduces the feature space of the training set by dropping the words that do not appear in the affective lexicon.

Please note that we only apply the feature engineering strategies to the training set in order to perform a fair comparison to the baseline, i.e., the original MNB without incorporating any affective lexicons. The test set for the baseline model and our models are identical. Our feature engineering strategies are simple, but are adequate to achieve the main goal of this paper: i.e. to assess the effectiveness of various affective lexicons for depression classification. The feature engineering strategies provide a natural means for incorporating lexicon resources into a MNB supervised classifier.

3   Experimental Setup

Dataset. We conduct our experiment based on a real-world dataset collected from LiveJournal[1], which consists of 29,686 depressive posts and 87,830 normal (non-depressive) posts. The depressive posts are collected from the depression communities (i.e., identified by searching communities whose name contains the

---

[1] http://www.livejournal.com.

keyword *depress* and its inflections, e.g.,*depressed, depression*), whereas the normal posts are randomly selected from communities not related to depression. The total number of posts in the dataset is 117,516.

Affective Lexicons. We have investigated three affective lexicons, i.e., ANEW, MPQA[2] and SentiWordNet[3]. ANEW is a set of 2,476 words that are rated by three emotional dimension (valence, arousal, and dominance). SentiWordNet is a lexical resource that contains 7,247 words which are assigned three sentiment scores (positivity, negativity, and objectivity). MPQA contains 7,319 words for opinions and other private states, e.g., beliefs, emotions, speculations, etc.

Setting. We ran our experiments using the Multinomial Naive Bayes (MNB) algorithm in Weka as it is computationally efficient and often provides competitive performance for text classification problems [9]. Also note that we randomly split the data into 80-20 fractions for training and testing. We repeat the process 5 times and report the averaged results for all models and the baseline.

4   Results

We analyse the statistics of affective lexicons coverage as shown in Table 1. The top row in Table 1 shows the number of words from an affective lexicon that have appeared in the corpus; and bottom row shows the percentage of words in the corpus that are covered by an affective lexicon. We can see that the statistics for both ANEW and SentiWordNet are quite similar, i.e., around 2,500 words of the lexicon have appeared in the dataset, covering 4% of the words in the dataset. The MPQA lexicon, in contrast, gives the highest coverage of about 9% for the entire dataset.

Table 1. Statistics of affective lexicon.

	ANEW	MPQA	SentiWordNet
# of matched affective words	2,430	5,781	2,699
Lexicon coverage	4%	9%	4%

Table 2 shows the classification performance of affective lexicons over two lexical feature engineering strategies. It can be observed that SentiWordNet outperforms ANEW and MPQA under both feature engineering strategies. In particular, SentiWordNet achieves the best overall accuracy of 84.4% with strategy 1, which is about 2.6% higher than the baseline (i.e., 81.8%). This improvement is significant according to a paired T-test with $p < 0.005$. This result is slightly surprising as one would expect the MPQA lexicon to perform best as it has the highest coverage on the dataset. In addition we did a further experiment using Support Vector Machines (SVM) with RBF kernel; SentiWordNet performed

---

[2] http://mpqa.cs.pitt.edu.
[3] http://sentiwordnet.isti.cnr.it/.

Table 2. Classification performance of MNB with different lexical feature engineering strategies.

Models	Strategy 1				Strategy 2			
	Pre.	Rec.	F1	Acc.	Pre.	Rec.	F1	Acc.
Baseline	63.6	79.8	70.0	81.8	63.6	79.8	70.0	81.8
ANEW	64.1	78.7	69.8	81.9	64.9	57.9	60.3	80.3
MPQA	64.2	78.6	69.9	82.0	60.1	30.5	40.3	77.2
SentiWordNet	68.4	81.3	73.5	84.4	68.8	49.3	57.0	81.0
Combined	69.2	82.1	75.1	**85.3**	70.2	53.2	60.5	**82.8**

best with 84.7% accuracy with Strategy 1. By combining all the three affective lexicons together (i.e, see 'Combined' in Table 2), we see a further performance boost 85.3%. When comparing the two different feature engineering strategies, we observe that Strategy 2 gives worse performance than Strategy 1. In fact, none of the models incorporating affective lexicons (apart from the combined model) can outperform the baseline. This might be due to the fact that constraining the training set feature space to the features of the affective lexicon has excluded many useful features, leading to a significant drop in model performance. To conclude, SentiWordNet gives the best performance for depression classification based on our feature engineering Strategy 1.

5   Conclusion

In this paper, we investigate the effectiveness of three popular affective lexicons (i.e., ANEW, MPQA, SentiWordNet) in the task of depression classification. We also develop two lexical feature engineering strategies for incorporating those lexicons into a supervised classifier. Our experimental results show that affective lexicons are useful for depression classification, and SentiWordNet is more effective than the other two affective lexicons under both lexical feature engineering strategies. In the future, we would like to conduct experiments on more datasets and test new feature engineering strategies, e.g., to apply topic models.

Acknowledgments. This work is supported by the award made by the UK Engineering and Physical Sciences Research Council (Grant number: EP/P005810/1).

References

1. Smith, K., Renshaw, P., Bilello, J.: The diagnosis of depression: current and emerging methods. J. Compr. Psychiatry **54**, 1–6 (2013)
2. Inkster, B., Stillwell, D., Kosinski, M., Jones, P.: A decade into Facebook where is psychiatry in the digital age. J. Lancet Psychiatry **01** (2016)

3. De Choudhury, M., Gamon, M., Counts, S., Horvitz, E.: Characterizing and predicting postpartum depression from shared Facebook data. In: CSCW, pp. 626–638 (2014)
4. Resnik, P., Armstrong, W., Claudino, L., Nguyen, T., Nguyen, V., Boyd-graber, J.: Beyond LDA: exploring supervised topic modeling for depression-related language in Twitter. In: CLPsych, pp. 99–107 (2015)
5. Yusof, N.F.A., Lin, C., Guerin, F.: Analysing the causes of depressed mood from depression vulnerable individuals. In: DDDSM Workshop at IJCNLP, pp. 9–17 (2017)
6. Bradley, M., Lang, P.: Affective Norms for English Words (ANEW): instruction manual and affective ratings. Technical report C-2, University of Florida (2010)
7. Wilson, T., Wiebe, J., Hoffmann, P.: Recognizing contextual polarity in phrase-level sentiment analysis. In: EMNLP-HLT, pp. 347–354 (2005)
8. Baccianella, S., Esuli, A., Sebastiani, F.: SentiWordNet 3.0: an enhanced lexical resource for sentiment analysis and opinion mining. In: LREC 10, pp. 2200–2204 (2010)
9. Kibriya, A.M., Frank, E., Pfahringer, B., Holmes, G.: Multinomial naive bayes for text categorization revisited. In: Webb, G.I., Yu, X. (eds.) AI 2004. LNCS (LNAI), vol. 3339, pp. 488–499. Springer, Heidelberg (2004). https://doi.org/10.1007/978-3-540-30549-1_43

Semantics-Based Models and Applications

Toward Human-Like Robot Learning

Sergei Nirenburg[1(✉)], Marjorie McShane[1], Stephen Beale[1],
Peter Wood[1], Brian Scassellati[2], Olivier Magnin[2],
and Alessandro Roncone[2]

[1] Language-Endowed Intelligent Agents Lab,
Rensselaer Polytechnic Institute, Troy, NY, USA
nirens@rpi.edu
[2] Social Robotics Lab, Yale University, New Haven, USA

Abstract. We present an implemented robotic system that learns elements of its semantic and episodic memory through language interaction with its human users. This human-like learning can happen because the robot can extract, represent and reason over the meaning of the user's natural language utterances. The application domain is collaborative assembly of flatpack furniture. This work facilitates a bi-directional grounding of implicit robotic skills in explicit ontological and episodic knowledge and of ontological symbols in the real-world actions by the robot. In so doing, this work provides an example of successful integration of robotic and cognitive architectures.

Keywords: Learning based on language understanding
Integration of robotic and cognitive architectures
Memory management in artificial intelligent agents

1 Setting the Stage

The ability of today's well-known artificial intelligent agents (Siri, Alexa, IBM's Watson, etc.) to react to vocal and written language communication relies ultimately on a variety of sophisticated methods of manipulating semantically uninterpreted textual strings. This methodology offers a solid breadth of coverage at a societally acceptable level of quality for select application areas. Still, as an approach to modeling human-level capabilities, this methodology has well-known limitations that are due to the dearth of knowledge resources and processing engines needed for extracting and representing the various kinds of meaning conveyed through natural language (NL) texts.

Two characteristics set out system apart from the few recent systems that actually address aspects of extracting and representing linguistic meaning [5, 13]. First, it simultaneously addresses the challenges of (a) learning-oriented language-based human-robotic interaction, (b) symbol grounding, (c) linguistic meaning extraction, and (d) the enhancement and management of the episodic, semantic and procedural memory of a robot/agent. Second, the language processing component of the system and its associated knowledge resources address a broader set of meaning-related language phenomena, described at a finer grain-size of analysis.

© Springer International Publishing AG, part of Springer Nature 2018
M. Silberztein et al. (Eds.): NLDB 2018, LNCS 10859, pp. 73–82, 2018.
https://doi.org/10.1007/978-3-319-91947-8_8

To implement language-based learning in a social robotics environment, we must address the co-dependence among three capabilities: language understanding, learning, and task-oriented physical, mental (reasoning) and verbal action. Language understanding and action require knowledge, while learning, achieved through language understanding and reasoning, automatically adds to that knowledge.

2   Learning How to Build a Chair

The system we describe is a social robot collaborating with a human user. The experimental domain is furniture assembly (e.g., [4]), widely accepted as useful for demonstrating human-robot collaboration on a joint activity. Roncone et al. [13] report on a Baxter robot supplied with high-level specifications, represented in the HTN formalism [3], of basic actions implementing chair-building tasks. Using a keyboard command or pressing a button, the user could trigger the execution of basic actions by triggering the operation of low-level task planners that the robot could directly execute. The robot could not reason about its actions, which were stored in its procedural memory as uninterpreted skills. The system described here integrates the robotic architecture of [13] with the OntoAgent cognitive architecture [8]. The integrated system allows the robot to (a) learn the semantics of initially uninterpreted basic actions; (b) learn the semantics of operations performed by the robot's human collaborator when they are described in natural language; (c) learn, name and reason about meaningful groupings and sequences of actions and organize them hierarchically; and (d) integrate the results of learning with knowledge stored in its semantic and episodic memory and establish their connections with the robot's procedural memory.

The core prerequisite for human-like learning is the ability to automatically extract, represent and use the meaning of natural language texts – utterances, dialog turns, etc. This task is notoriously difficult: to approach human-level capabilities, intelligent agents must account for both propositional and discourse meaning; interpret both literal and non-literal (e.g., metaphorical) meaning; resolve references; interpret implicatures; and, particularly in informal genres, deal with stops and starts, spurious repetitions, production errors, noisy communication channels and liberal (if unacknowledged) use of the least effort principle (e.g., [12]) by speakers and hearers. The language understanding module of our system, OntoSem [9], uses an ontology (semantic memory) of some 9,000 concepts with on average of 16 properties each; an episodic memory of remembered instances of ontological concepts; a semantic lexicon for English covering about 25,000 lexical senses; and a frame-oriented formalism suitable for representing the semantics of robotic actions, natural language utterances and results of the robot's processing of other perceptual modalities (e.g., interoception, [8]).

The Process. At the beginning of the learning process, the robot can (a) visually *recognize* parts of the future chair (e.g., the seat and the tool (screwdriver) to be used, (b) *generate* meaning representations (MRs) of user utterances and physical actions and (c) *perform* basic programmed actions: GET(object) from storage area to workspace, HOLD(object) and RELEASE(object). The user teaches the robot three types of things (implemented in three learning modules of Fig. 1): (a) *concept grounding*: the connection between basic actions and MRs of utterances that describe them; (b) *legal sequences* of basic actions forming complex actions; and (c) *augmenting the robot's memory* with descriptions of newly learned complex actions and objects.

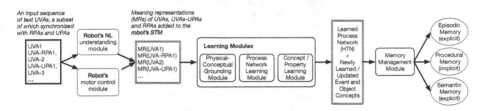

Fig. 1. The core learning process. Input is a sequence of user verbal actions (UVAs) which explain user physical actions (UPAs) and issue commands to the robot, thus verbalizing robot's physical actions (RPAs), which facilitates grounding the former in the latter. UVAs are interpreted into uniform meaning representation and provide input to grounding, process network and concept/property learning modules (LMs). The memory management module (MMM) incorporates the results of learning into the episodic and semantic memories of the robot and mutually grounds RPAs in the robot's procedural memory and corresponding concepts in its semantic memory.

Suppose the user issues the call for the robot to execute GET(screwdriver) *and* the utterance *Now you will fetch a screwdriver*. The physical-conceptual grounding learning module (LM) will link this procedure call with the representation of change-location-1 in the MR that Ontosem produces for the example utterance (Fig. 2), thus linking the robotic and the cognitive architectures. This is done by adding the property PM-LINK with the filler GET(screwdriver) to the ontological concept instance CHANGE-LOCATION-1. The immediate purpose of this linking of the robotic and the cognitive architectures is to make the robot capable in its subsequent functioning to (a) trigger basic actions autonomously on the basis of language input alone and (b) learn complex event sequences by just being told, without having to actually perform the actions comprising the complex event.

The robot can learn legal sequences of basic actions by understanding the user's utterances in their context. The robot organizes action sequences hierarchically and makes sure that any non-terminal nodes in the resulting process network represent meaningful complex actions. If the robot does not have specifications for these complex actions in its stored knowledge, it *learns new concepts for them*, on the basis of the MRs obtained by processing the relevant user utterances. We treat joint tasks as complex tasks and require the system to decompose them into subtasks carried out by each of the team members. Basic individual tasks include Robot Physical Actions (RPAs), User Physical Actions (UPAs) and User Verbal Actions, UVA. The RPAs and UPAs appear as terminal nodes in the process network being learned. The robot's activity that includes all the kinds of learning it does as well as updating its memory structures comes under the rubric of Robot Mental Action, RMA. Due to space constraints we cannot illustrate a complete process of assembling a chair (even the shortest version of the process numbers over 150 steps). So, we present a small subset of this

SPEECH-ACT-1
type	command
scope	CHANGE-LOCATION-1
producer	*speaker*
consumer	ROBOT-0
time	time-0 ; time of speech

CHANGE-LOCATION-1
agent	ROBOT-0
theme	SCREWDRIVER-1
effect	BESIDE
	(AGENT.LOCATION
	THEME.LOCATION)
time	> time-0
token	*fetch*
from-sense	move-v2

HUMAN-1
agent-of	CHANGE-LOCATION-1
token	*you*
from-sense	you-n1

SCREWDRIVER-1
theme-of	CHANGE-LOCATION-1
token	*screwdriver*
from-sense	screwdriver-n1

Fig. 2. Meaning representation for the utterance *Now you will fetch a screwdriver* (simplified).

process – assembling the third of the four legs of the chair – accompanied by associated robotic learning, as illustrated in Fig. 3. All UVAs are first analyzed and their meanings are represented as MRs. UVA1 signals the beginning of the subsequence and, together with UVA7, marks the boundaries of the complex action. All the RPAs and the UPA occurring within this span, in the order of their occurrence, will form the set of the terminal nodes in the subset of the overall process network, becoming children of the non-terminal designating the complex action of building the right back leg. Once this (sub)hierarchy is constructed, the non-terminal node at its root must be named. As the robot assembles the back leg for the first time, it learns the composition of this complex action (RMA1) and labels the parent node of this small subhierarchy with the name of the concept ASSEMBLE-RIGHT-BACK-LEG. It also learns the new object-type concept RIGHT-BACK-LEG, whose existence is the EFFECT of the above action (RMA2) and updates the concept CHAIR by adding RIGHT-BACK-LEG as a filler of that concept's HAS-OBJECT-AS-PART property (RMA3). The newly learned concepts are illustrated in Fig. 4. The results of the operation of the process network LM are recorded in the HAS-EVENT-AS-PART property of a result of the operation of the concept LM. At this stage in the process, the fillers of some of the properties in the concepts are tentative and are expected to be modified/tightened at the memory management stage.

Memory Management. Knowledge learned by the robot during each session with a human trainer (such as the sequence in Fig. 3) must be remembered so they can be used in subsequent functioning. Mutual grounding of basic actions and corresponding ontological events is recorded both in the robot's procedural memory (by augmenting the procedures implementing the robot's basic motor actions with links to their corresponding concepts in semantic memory) and in its semantic memory (by adding pm-links, see above). Newly learned process sequences and objects (such as

UVA1	We will now build the right back leg
UVA2	Get another foot bracket
RPA1	GET(bracket-foot)
RPA2	RELEASE(bracket-foot)
UVA3	Get the right back bracket
RPA3	GET(bracket-back-right)
RPA4	RELEASE(bracket-back-right)
UVA4	Get and hold another dowel
RPA5	GET(dowel)
RPA6	HOLD(dowel)
UVA5	I am mounting the third set of brackets on a dowel
UPA1	The user affixes the foot and the right back brackets to the dowel
UVA6	Finished. Release the dowel
RPA7	RELEASE(dowel)
UVA7	OK, done assembling right back leg
RMA1	Learns action subsequence ASSEMBLE-RIGHT-BACK-LEG
RMA2	learns the object RIGHT-BACK-LEG with BRACKET-FOOT, BRACKET-BACK-RIGHT and DOWEL as fillers of HAS-OBJECT-AS-PART slot of RIGHT-BACK-LEG
RMA3	Adds RIGHT-BACK-LEG as a filler of HAS-OBJECT-AS-PART of CHAIR

Fig. 3. Assembling the right back leg of the chair.

```
ASSEMBLE-RIGHT-BACK-LEG
    IS-A                        PHYSICAL-EVENT
    AGENT                       HUMAN, ROBOT
    THEME                       RIGHT-BACK-LEG
    INSTRUMENT                  SCREWDRIVER
    HAS-EVENT-AS-PART           GET(ROBOT, BRACKET-FOOT)
                                RELEASE(USER, BRACKET-FOOT)
                                GET(ROBOT, BRACKET-BACK-RIGHT)
                                RELEASE(USER, BRACKET-BACK-RIGHT)
                                GET(ROBOT, DOWEL)
                                HOLD(ROBOT, DOWEL)
                                MOUNT(USER, {BRACKET-FOOT, BRACKET-BACK-RIGHT}, DOWEL)
    PART-OF-EVENT               ASSEMBLE-CHAIR
    EFFECT                      RIGHT-BACK-LEG    //default effects are events; if filler of effect is
                                                  //an object, this means the effect is its existence
RIGHT-BACK-LEG
    IS-A                        CHAIR-PART
    HAS-OBJECT-AS-PART          BRACKET-FOOT, BRACKET-BACK-RIGHT, DOWEL
    PART-OF-OBJECT              CHAIR
```

Fig. 4. Concepts learned as a result of processing the sequence in Fig. 3.

assemble-right-back-leg and right-back-leg of Fig. 4) must be incorporated in the robot's long-term semantic and episodic memories. Due to space constraints, in what follows we give an informal description of the process. A more comprehensive description describing the relevant algorithms in full detail will be published separately.

For each newly learned concept, the memory management module (MMM) first determines whether this concept should be (a) added to the robot's semantic memory or

(b) merged with an existing concept. To make this choice, the MMM uses an extension of the concept matching algorithm of [2, 10]. This algorithm is based on unification, with the added facility for naming concepts and determining their best position in the hierarchy of the ontological world model in the robot's long-term semantic memory.

As an illustration, suppose, the concept matching algorithm has determined that the newly learned concept ASSEMBLE-RIGHT-BACK-LEG must be added to the robot's semantic memory. In this case, the algorithm also suggests the most appropriate position for the concept in the ontological hierarchy. This is determined by comparing (a) the inventory and (b) sets of fillers of the properties defined for the new concepts and for the potential parents of the new concepts in the ontological hierarchy. In the example illustrated in Fig. 4, the LM used the safest, though the least informative, filler (PHYSICAL-EVENT) for the IS-A property of ASSEMBLE-RIGHT-BACK-LEG. To determine the appropriate parent, the algorithm traverses the ontological hierarchy from PHYSICAL-EVENT down until it finds the closest match that does not violate recorded constraints. Figure 5 shows the immediate children of PHYSICAL-EVENT (triangles next to concepts indicate that they are non-terminals) in our current ontology. Figure 6 illustrates the fact that our ontology supports multiple inheritance – notably, the concept ASSEMBLE is a descendant of both CREATE-ARTIFACT and MANUFACTURING-ACTIVITY, while CREATE-ARTIFACT, by virtue of being a child of the concept PRODUCE, is a descendant of both PHYSICAL-EVENT and SOCIAL-EVENT. This mechanism facilitates an economical representation of different aspects of the meaning of ontological concepts and allows the representation of both differences and similarities among concepts.

Fig. 5.

Returning to the example of incorporating ASSEMBLE-RIGHT-BACK-LEG into the robot's semantic memory, the MMM attempts to make the newly learned concept a child of the semantically closest existing concept, to be able to inherit the values of as many of its

```
⊿ all
  ⊿ event
    ⊿ physical-event
      ⊿ produce
        ⊿ create-artifact
          assemble
    ⊿ social-event
      ⊿ work-activity
        ⊿ manufacturing-activity
          assemble
        ⊿ produce
          ⊿ create-artifact
            assemble
```

Fig. 6.

properties as possible without the extra work of determining them one by one). This attempt fails because the filler of the AGENT property in ASSEMBLE, ASSEMBLY-LINE-WORKER, is more constrained than the corresponding filler in the newly learned ASSEMBLE-RIGHT-BACK-LEG, which is HUMAN or ROBOT. The algorithm backtracks and succeeds in making ASSEMBLE-RIGHT-BACK-LEG a child of CREATE-ARTIFACT (and, thus, a sibling of ASSEMBLE). Suppose now that a concept RIGHT-BACK-LEG already exists in the ontology.

If this concept is as illustrated in Fig. 7, then, after comparing this concept with the newly learned concept (see Fig. 4), the MMM will, instead of adding a new (possibly renamed) concept to the robot's semantic memory, just add an *optional* filler BRACKET-BACK-RIGHT to the HAS-OBJECT-AS-PART property of the standing concept of Fig. 7, thus merging the standing and the newly learned knowledge. If, however, the standing concept is as illustrated in Fig. 8, then, because of the mismatch of the fillers of part-of properties between the newly learned and the standing concept, the MMM will yield two new concepts (Figs. 9 and 10). Note the need of modifying the names of the concepts. An important case of merging several versions of a concept in one representation is the system's ability to represent the content of an action's HAS-EVENT-AS-PART property as an HTN, augmented with the means of ex pressing temporal ordering, optionality and valid alternative action sequences. Semantic memory stores the robot's knowledge of concept types. So, for example, it will contain a description of what the robot knows about chairs and chair legs. This knowledge will be used to feed the reasoning rules the robot will use while processing language, learning and making decisions. To make the robot more human-like, we also support reasoning by analogy. For this purpose, the MMM records sequences of RPAs, UPAs and UVAs that the robot represents and carries out during specific sessions of interacting with specific users in the robot's long-term episodic memory. The contents of the episodic memory will also support the robot's ability to "mind-read" its various users [2, 6] and, as a result, to be able to anticipate their needs at various points during joint task execution as well as interpret their UVAs with higher confidence.

```
RIGHT-BACK-LEG
    IS-A                  CHAIR-PART
    HAS-OBJECT-AS-PART BRACKET-FOOT,
                         DOWEL
    PART-OF-OBJECT       CHAIR
```

Fig. 7. A possible RIGHT-BACK LEG.

```
RIGHT-BACK-LEG
    IS-A                  SOFA-PART
    HAS-OBJECT-AS-PART    BRACKET-FOOT,
                          DOWEL
    PART-OF-OBJECT        SOFA
```

Fig. 8. Another RIGHT-BACK LEG.

```
RIGHT-BACK-LEG-CHAIR
    IS-A                  CHAIR-PART
    HAS-OBJECT-AS-PART    BRACKET-FOOT,
                          DOWEL,
                          BRACKET-BACK-RIGHT
    PART-OF-OBJECT        CHAIR
```

Fig. 9. The right-back leg of a chair.

```
RIGHT-BACK-LEG-SOFA
    IS-A                  SOFA-PART
    HAS-OBJECT-AS-PART    BRACKET-FOOT,
                          DOWEL
    PART-OF-OBJECT        SOFA
```

Fig. 10. The right-back leg of a sofa.

Discussion. The system presented concentrates on robotic learning through language understanding. This learning results in extensions to and modifications of the three kinds of robotic memory – the explicit semantic and episodic memory and the implicit (skill-oriented) procedural memory. The expected practical impact of the ability to learn and reason will include the robot's ability to (a) perform complex actions without the user having to spell out a complete sequence of basic and complex actions; (b) reason about task allocation between itself and the human user; and (c) test and verify its knowledge through dialog with the user, avoiding the need for large numbers of training examples required by learning by demonstration only. The inability of the state-of-the-art deep learning-based systems to provide human-level explanations is a well-known constraint on the utility of such systems. The cognitive robots we develop will still be capable of sophisticated reasoning by analogy but will be also capable of explaining their decisions and actions. Finally, our approach to learning does not depend on the availability of "big data" training materials. Instead, we model the way people learn since early childhood and throughout their lives – by being taught using natural language.

Language Understanding. In the area of cognitive systems, language understanding is not to be equated with current mainstream natural language processing (NLP), a thriving R&D area that methodologically is almost entirely "knowledge-lean" [7]. Some of the more application-oriented projects in cognitive systems support their language processing needs with such knowledge-lean methods, thus agreeing to a lower level of quality in exchange for broader coverage and faster system development. Longer-term, more theoretically motivated projects seek to develop explanatory models of human-level language processing that require knowledge (see [8] for a discussion]). The knowledge in such models supports not only language understanding but also reasoning and decision-making [11]. Indeed, deep language analysis requires knowledge that is not overtly mentioned in the text or dialog. To be able to interpret language input, a language understanding agent must, thus, be able to reason about the world, the speech situation and other agents present in it. It must also routinely make decisions about the interpretation of components of the input, about what is implied by the input and what is omitted from it, and about whether and, if so, to what depth to analyze the input.

Memory Modeling. Another theoretical contribution of our work is overt modeling of the robot's memory components. These components include an implicit memory of skills and explicit memories of concepts (objects, events and their properties) and of instances of sequences of events (episodes, represented in our system as HTNs). The link established between the implicit and explicit layers of memory allows the robot to reason about its own actions. Scheutz et al. [15] discuss methodological options for integrating robotic and cognitive architectures and propose three "generic high-level interfaces" between them – the perceptual interface, the goal interface and the action interface. In our work, the basic interaction between the implicit robotic operation and explicit cognitive operation is supported by interactions among the three components of the memory system of the robot.

Next Steps. The first enhancement of the current learning system will consist in demonstrating how, after RPAs are mutually grounded in ontological concepts, the robot will be able to carry out commands or learn new action sequences by acting on UVAs, without any need for direct triggering through software function calls or hardware operations. Next, we intend to add text generation capabilities, both to allow the robot a more active role in the learning process (by asking questions) and to enrich interaction during joint task performance with a human user. Another novel direction of work will involve adapting to particular users – modeling robots' individuality and related phenomenological ("first-person" view) aspects of its internal organization and memory, developing and making use of mindreading capabilities [1, 6] that will in turn facilitate experimentation in collaboration among agents with different "theories of minds of others."

Acknowledgements. This work was supported in part by Grant N00014-17-1-221 from the U. S. Office of Naval Research. Any opinions or findings expressed in this material are those of the authors and do not necessarily reflect the views of the Office of Naval Research.

References

1. Bello, P.: Cognitive foundations for a computational theory of mindreading. Adv. Cogn. Syst. **1**, 1–6 (2011)
2. English, J., Nirenburg, S.: Ontology learning from text using automatic ontological-semantic text annotation and the web as the corpus. In: Proceedings of the AAAI 07 Spring Symposium on Machine Reading (2007)
3. Erol, K., Hendler, J., Nau, D.S.: HTN planning: complexity and expressivity. Proceedings of AAAI 94 (1994)

4. Knepper, R.A., Layton, T., Romanishin, J., Rus, D.: IkeaBot: an autonomous multi-robot coordinated furniture assembly system. In: IEEE International Conference on Robotics and Automation (2013)
5. Lindes, P., Laird, J.: Toward integrating cognitive linguistics and cognitive language processing. In: Proceedings of ICCM 2016 (2016)
6. McShane, M.: Parameterizing mental model ascription across intelligent agents. Interact. Stud. 15(3), 404–425 (2014)
7. McShane, M.: Natural language understanding (NLU, not NLP) in cognitive systems. In: AI Magazine Special Issue on Cognitive Systems (2017)
8. McShane, M., Nirenburg, S.: A knowledge representation language for natural language processing, simulation and reasoning. Int. J. Semant. Comput. 6(1), 3–23 (2012)
9. McShane, M., Nirenburg, S., Beale, S.: Language understanding with ontological semantics. Adv. Cogn. Syst. 4, 35–55 (2016)
10. Nirenburg, S., Oates, T., English, J.: Learning by reading by learning to read. In: Proceedings of the International Conference on Semantic Computing (2007)
11. Nirenburg, S., McShane, M.: The interplay of language processing, reasoning and decision-making in cognitive computing. In: Biemann, C., Handschuh, S., Freitas, A., Meziane, F., Métais, E. (eds.) NLDB 2015. LNCS, vol. 9103, pp. 167–179. Springer, Cham (2015). https://doi.org/10.1007/978-3-319-19581-0_15
12. Piantadosi, S.T., Tily, H., Gibson, E.: The communicative function of ambiguity in language. Cognition 122, 280–291 (2012)
13. Roncone, A., Mangin, O., Scassellati, B.: Transparent role assignment and task allocation in human robot collaboration. In: Proceedings of International Conference on Robotics and Automation, Singapore (2017)
14. Scheutz, M., Krause, E., Oosterveld, B., Frasca, T., Platt, R.: Spoken instruction-based one-shot object and action learning in a cognitive robotic architecture. Proceedings of AAMAS 17 (2017)
15. Scheutz, M., Harris, J., Schmermerhorn, P.: Systematic integration of cognitive and robotic architectures. Adv. Cogn. Syst. 2, 277–296 (2013)

Identifying Argumentative Paragraphs: Towards Automatic Assessment of Argumentation in Theses

Jesús Miguel García-Gorrostieta[✉] and Aurelio López-López

Computational Sciences Department, National Institute of Astrophysics,
Optics and Electronics, Tonantzintla, Puebla, Mexico
{jesusmiguelgarcia,allopez}@inaoep.mx

Abstract. The academic revisions by instructors is a critical step of the writing process, revealing deficiencies to the students, such as lack of argumentation. Argumentation is needed to communicate clearly ideas and to convince the reader of the stated claims. This paper presents three models to identify argumentative paragraphs in different sections (Problem Statement, Justification, and Conclusion) of theses and determine their level of argumentation. The task is achieved using machine learning techniques with lexical features. The models came from an annotated collection of student writings, which served for training. We performed experiments to evaluate argumentative paragraph identification in the sections, reaching encouraging results compared to previously proposed approaches. Several feature configurations and learning algorithms were tested to reach the models. We applied the models in a web-based argument assessment system to provide access to students, and instructors.

Keywords: Computer-assisted argument analysis
Academic writing · Argumentation studies · Annotated theses corpus

1 Introduction

Argumentation is a crucial skill in academic writing. This ability consists of formulating an argumentation structure that provides numerous arguments to support each claim stated in academic theses. An argument is a set of statements (i.e. premises) that individually or collectively provide support to a claim (conclusion). There are several computational tools which employ diagrams to analyze the argumentation [9,11]. Such tools have been conceived and developed to help in the visual mapping and analysis of arguments, assisting students to achieve a deeper understanding of their construction. For review of essays, systems are available [2,16] to provide support in several linguistic dimensions. However, we have not found systems aimed to analyze automatically argumentation in larger academic works such as theses. Theses are often written in college as one of the final requirements, being thus quite important for students. So,

© Springer International Publishing AG, part of Springer Nature 2018
M. Silberztein et al. (Eds.): NLDB 2018, LNCS 10859, pp. 83–90, 2018.
https://doi.org/10.1007/978-3-319-91947-8_9

assessment models for theses are needed to support students in mastering this skill.

In this paper, we present three models for argument assessment. We propose models for the sections of theses with most abundant argumentation: Problem Statement, Justification, and Conclusion. The models identify argumentative paragraphs using machine learning techniques with representations of several lexical features and determine the level of argumentation in text. To apply machine learning, we created a corpus of thesis sections with annotated argumentative paragraphs. We annotated 300 sections and performed experiments on one hundred examples of each section, to automatically identify paragraphs with arguments, showing better efficacy than other previously reported approaches. The models are implemented in an Internet-based system to support students in the analysis and assessment of their argumentation during their thesis writing.

The paper is structured as follows. In Sect. 2, we show related work of argument identification found so far. We describe our architecture for the argument assessment system in Sect. 3. In Sect. 4, we present an exploratory analysis of the annotated corpus. Section 5 reports the result of the three models for argument identification and assessment using the corpus. Finally, we conclude our work and point out possible future research directions in Sect. 6.

2   Related Work

Several researchers have addressed the task of argument assessment. We assume as a preliminary step for arguments assessment the argument identification task. In this section, we present previous research in argument mining and assessment. In particular, we focus on the tasks of corpus creation and argument detection. We require these tasks beforehand, to develop later on the argument assessment models. Argument mining involves automatic argument extraction from unstructured text. First, we address the corpus creation to validate the efficacy of the proposed methods. As observed in the literature, most researchers in the field of argument analysis create their annotated corpus, adhering to certain argumentative scheme. We found few annotated corpus available for our purpose.

Corpus creation is done in different types of text, as well as in various domains. In [12], an experimental collection was built with ten legal documents from the European Court of Human Rights corpus, with annotated premises and conclusions. In this corpus, the level of agreement between the two annotators is a Kappa of 0.80. In [7], the authors created a corpus with 24 scientific articles in education for the sections of introduction and discussion. Four participants annotated argument components as premises or conclusions, and four relationships (support, attack, sequence and detail) between the components, with an average in the level of agreement of Fleiss Kappa of 0.41. Thus, we noticed that obtaining acceptable levels of annotation agreement in texts is a complex task, that depends on an appropriate annotation guide and regularly monitoring annotators during the corpus construction. For our research, the closest kind of document are scientific articles since theses share a similar complex structure and scientific terminology. However, undergrad theses have a longer extension per section, and student writings present often several argumentative errors. Besides, scientific articles are written by more experienced researchers.

Once a corpus is built, it is necessary to detect the presence of arguments in paragraphs, sentences, or clauses. In [14], an automatic identification of argumentative sentences in Araucaria corpus was performed. They represented sentences with features like combinations of pairs of words, verbs and text statistics before applying a Naive Bayes classifier, achieving a 73.75% of accuracy. Working with legal texts of the ECHR corpus [13], an 80% of accuracy was reached. In this domain, texts have a specific structure which allows lawyers to identify easily arguments. Other approach was reported [4], extracting a set of discourse markers and features based on mood and tense of verbs. They achieved an F1-measure of 0.764 with a decision tree classifier.

3   System Architecture

This section describes the architecture of the argument assessment system (Fig. 1). The system contains a web interface to capture the student text and report back evaluation results (already online[1] for demonstration purposes). The system identifies the proportion of argumentative paragraphs, assigns a level of this proportion, and provides a textual feedback to the student.

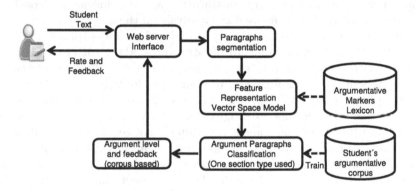

Fig. 1. System architecture

We based our approach on specific processes of argument mining methodology [15]. First, a paragraph segmentation is required. Secondly, as shown in Fig. 1, we generate a particular feature representation (vector space model) for each section type to represent each paragraph. Then, we supply this representation to the argumentative paragraph classifier, where we employ machine learning to identify argumentative paragraphs. Once we identify all the argumentative

---

[1] https://ccc.inaoep.mx/~jesusmiguelgarcia/detection_arg_par/.

paragraphs, this information is used for argument level detection. This module computes the proportion of argumentative paragraphs, to assess the level of the text, according to our corpus statistics. Finally, a textual feedback is supplied to the student, according to the level achieved in the text. Our goal is to provide an assessment along with recommendations, to support students with a clear identification of paragraphs with arguments, so help them to improve the argumentation in their writings.

In the input interface used to assess the sections of the academic thesis, the user can submit his plain text for analysis. Then, he can select a section type, i.e. Problem Statement, Justification, or Conclusion. And then, he can request the evaluation by clicking a button to assess the text. As a result, the identified argumentative paragraphs are indicated in blue and non-argumentative in red, along a proportion of argumentative paragraphs found. A textual feedback is also provided, and is defined depending on the level achieved by the student.

4   Corpus Creation and Analysis

Corpus analysis is done to understand the argumentative characteristics in writings of undergraduate and graduate level. For this analysis, we used the collection Coltypi [5] consisting of 468 theses and research proposals in computer and information technologies, in Spanish. This corpus has undergraduate (TSU[2] and Bachelor) and graduate level (M. Sc. and Ph.D.) texts. According to [10], the sections of the problem statement, justification and conclusions are considered highly argumentative, so we focused the analysis on them.

To conduct the annotation study, we formulated an annotation guide, where we described the different argumentative structures, along with examples. Two annotators with experience reviewing theses were instructed to read the title and objectives of the thesis first, and then move to identify and mark all argumentative paragraphs. We did the annotation of 100 sections of each type.

The analysis of inter-annotator agreement (IAA) was done considering the paragraphs with observed agreement (i.e. different of zero) in the identification of argumentative components (e.g. premises or conclusions). A total of 890 paragraphs were used to analyze the IAA with Cohen Kappa [3]. So, we obtained values of 0.867, 0.935, and 0.817 for Problem Statement, Justification and Conclusion, respectively. These IAAs were "Almost perfect", according to [8].

As shown in Table 1, most sections have more than half of the paragraphs with arguments. We selected only the paragraphs where the two annotators agreed. This reduces the number of paragraphs to 837, that once analyzed, led to 565 argumentative paragraph (i.e. 68%). From this, we observed that a large proportion of paragraphs in theses have arguments. One characteristic of this corpus is that Conclusion includes more paragraphs per section, when compared to Problem Statement or Justification. Moreover, we observed a higher number of paragraphs with arguments in Conclusion.

---

[2] Advanced College-level Technician degree, study program offered in some countries.

Table 1. Class distribution per section

	Problem statement	Justification	Conclusion	Total
Paragraphs with arguments	164	151	250	565 (68%)
Paragraphs without arguments	93	81	98	272 (32%)

We employed the class distribution of argumentative paragraphs of the three sections to calculate the intervals for argumentative levels: low, medium and high. These intervals were based on the proportion of argumentative paragraphs found in one hundred sections per section type. To calculate the boundary between low and medium level, we take half standard deviation minus the mean. For the boundary between medium and high level, we add to the mean, half of the standard deviation. So, the intervals were: Problem Statement (Low: 0–36%, Medium: 37–77%, High: 78–100%), Justification (Low: 0–42%, Medium: 43–82%, High: 83–100%), and Conclusion (Low: 0–48%, Medium: 49–83%, High: 84–100%).

5   Argumentative Paragraph Identification

The identification of paragraphs with arguments was approached as a binary classification for each paragraph, i.e. identify if it has arguments or not. The corpus used for this experiment is presented in Table 1. We built a model for each section type. The classification was performed using Weka [6]. In particular, we applied Support Vector Machine (SVM), Random Forest (RF), Naive Bayes (NB) and Simple Logistic Regression (SLR) classifiers since they have been previously used in argument mining. We employed a stratified 10-fold cross-validation.

Feature Representations. The first representations taken as baseline to compare our proposed models were those approaches of Florou [4] and Moens [14]. The Florou representation consists of discourse markers and mood and tense of verbs. The Moens representation is built of combinations of all possible pairs of words, main verbs, and text statistics. We also compare our models with lexical and semantic features as detailed below.

The lexical representation taken also as baseline consists of bigrams for consecutive pairs of terms in paragraph, including punctuation marks (; : , .). The semantic representation built for comparison consisted of word embeddings, specifically those of Polyglot [1], trained with Spanish Wikipedia, containing words with vectors which represent the word meaning. The number of word embeddings for Spanish is 100,004, with vector size of 64. We computed the average of word embedding vectors contained in each paragraph.

We performed a set of experiments with several types of features to identify the best representation for each section type. The sections had different characteristics and purpose. Justification has more arguments in shorter paragraphs since the purpose is to indicate the importance of the research. In contrast, the Conclusion, with a higher number of paragraphs, describes a summary of the research and point out future work. The section with fewer arguments was Problem Statement; we observed in undergrad theses problem statements with shorter paragraphs to indicate only the problem to be solved using few words. Since the three sections showed different characteristics, we explored a set of feature types such as n-grams from one to three, argumentative markers categories (i.e. justification, explanation, deduction, refutation, or conditional), ordered word pairs, word embeddings, lemmas, part-of-speech (POS) tags, and grammatical category (initial of POS tag). We experimented with the combinations of the feature types using frequency and *tf-idf* (term frequency - inverse document frequency).

The best representation for Conclusion includes the following features: (1) unigrams, i.e. all single terms in the paragraph, (2) bigrams, (3) ordered word pairs combinations per sentence and (4) categories of argumentative markers, i.e. the number of markers in each category found in the paragraph using our markers lexicon. We computed a *tf-idf* weight for this representation. The representation for Problem Statement was less elaborated than that of Conclusion and consists of: (1) the frequency of unigrams and (2) the frequency of bigrams, in both cases only those with a frequency higher than five. The Justification is represented with: (1) the frequency of unigrams, (2) frequency of bigrams, and (3) categories of argumentative markers.

Results. For the different representations extracted, we trained the classifiers with the data set for that purpose and applied them to the test data set. The classifiers trained for each representation were: RF, SLR, SVM and NB. Table 2 shows the averages of F-measure, recall, precision and accuracy of each representation for the ten folds. We present the best combination of classifier and feature set obtained by feature selection with information gain for each representation. As we can notice, our representations in the three section types achieved the best accuracy and F-measure for identification of paragraphs with an argument. For Problem Statements in our representation, we applied a RF classifier using only attributes with information gain (i.e. IG > 0). The baseline consisting of bigrams and RF showed the best recall, just above the recall of our proposed representation. In our representation for Justifications, we employed SLR classifier combined with the first 100 features with information gain. Finally, we utilized a SVM classifier with a feature selection of 75 attributes with information gain for Conclusions model. We observed in our three proposed representations, an accuracy above 81.7%. We noticed word embeddings had a better efficacy in Conclusions, above the bigram baseline. With these results, we provide support to recur to the proposed combination of representation and classifier to perform the identification of argumentative paragraphs in the corresponding section.

In our three models, an important part was the lexical features as unigrams and bigrams, which consists of argumentative and domain words. Argumentative words were considered in argument categories in two of our representations. We observed that argumentative terms exhibit information gain, indicating that they

Table 2. Classification of argumentative paragraphs results. FS stands for feature selection criteria

Section type	Representation	Classifier	FS	F-measure	Recall	Precision	Accuracy
Problem statement	Bigrams	RF	IG > 0	0.876	**0.909**	0.849	0.833
	Word embeddings	NB	None	0.799	0.779	0.827	0.754
	Florou	RF	IG > 0	0.805	0.824	0.798	0.747
	Moens	RF	IG > 0	0.84	0.873	0.816	0.791
	Our representation	RF	IG > 0	**0.884**	0.897	**0.874**	**0.848**
Justification	Bigrams	RF	100	0.829	0.847	0.817	0.775
	Word embeddings	RF	None	0.817	**0.921**	0.736	0.733
	Florou	SLR	IG > 0	0.834	0.867	0.806	0.776
	Moens	RF	None	0.834	0.88	0.799	0.771
	Our representation	SLR	100	**0.883**	0.88	**0.893**	**0.849**
Conclusion	Bigrams	NB	50	0.822	0.774	**0.882**	0.758
	Word embeddings	RF	None	0.864	0.97	0.779	0.776
	Florou	RF	None	0.867	0.944	0.802	0.788
	Moens	RF	None	0.864	**0.985**	0.769	0.771
	Our representation	SVM	75	**0.875**	0.874	0.879	**0.817**

contribute for classification. The argument marker "ya-que" (since in English) achieved the best information gain among the three section which suggests students used often this marker to introduce a premise to justify a conclusion. We also observe the argument category of deduction in the Justifications and Conclusions. The words in this category were associated with the argumentative component conclusion, for example, "por lo tanto" (therefore) expressed to indicate a conclusion. Also the category justification came out in Conclusions, revealed by the use of markers such as "si" (if) or "porque" (because).

6 Conclusion

The system presented in this paper aims to support students to improve their argumentative writing by identifying which paragraphs do not seem to have arguments. After using the system and following suggestions, this can also benefit academic instructors by reviewing improved writings, with better content.

As observed in the experimental collection, there were enough arguments in theses to exploit. From corpus analysis, we realized that more than half of the paragraphs include arguments, so is important to make further progress in building systems that support the argumentative assessment of this type of texts.

According to the results, the best accuracy and F-measure observed in our experiments to identify paragraphs with arguments was achieved by our three proposed feature representations. We found patterns with information gain used to indicate a premise and patterns utilised to express conclusions. The patterns reflect the uses of argumentation in the academic texts.

In the future, we plan to work on the identification of argumentative components (premises, conclusions) to indicate precisely to students their deficiencies.

Acknowledgement. The first author was partially supported by CONACYT, México, under scholarship 357381.

References

1. Al-Rfou, R., Perozzi, B., Skiena, S.: Polyglot: distributed word representations for multilingual NLP. In: Proceedings of the 17th Conference on Computational Natural Language Learning, pp. 183–192. ACL, Sofia, August 2013
2. Burstein, J., Chodorow, M., Leacock, C.: Criterion^{SM} online essay evaluation: an application for automated evaluation of student essays. In: Proceedings of the 15th Conference on Innovative Applications of Artificial Intelligence, pp. 3–10 (2003)
3. Cohen, J.: A coefficient of agreement for nominal scales. Educ. Psychosoc. Meas. **20**(1), 37–46 (1960)
4. Florou, E., Konstantopoulos, S., Koukourikos, A., Karampiperis, P.: Argument extraction for supporting public policy formulation. In: Proceedings of the 7th Workshop on Language Technology for Cultural Heritage, Social Sciences, and Humanities, pp. 49–54 (2013)
5. González-López, S., López-López, A.: Colección de tesis y propuesta de investigación en tics: un recurso para su análisis y estudio. In: XIII Congreso Nacional de Investigación Educativa, pp. 1–15 (2015)
6. Hall, M., Frank, E., Holmes, G., Pfahringer, B., Reutemann, P., Witten, I.H.: The weka data mining software: an update. ACM SIGKDD Explor. Newsl. **11**(1), 10–18 (2009)
7. Kirschner, C., Eckle-Kohler, J., Gurevych, I.: Linking the thoughts: analysis of argumentation structures in scientific publications. In: Proceedings of the 2nd Workshop on Argumentation Mining, pp. 1–11. ACL, June 2015
8. Landis, J.R., Koch, G.G.: The measurement of observer agreement for categorical data. Biometrics **33**(1), 159–174 (1977)
9. Loll, F., Pinkwart, N.: LASAD: flexible representations for computer-based collaborative argumentation. Int. J. Hum. Comput. Stud. **71**(1), 91–109 (2013)
10. López Ferrero, C., García Negroni, M.: La argumentación en los géneros académicos. In: Actas del Congreso Internacional La Argumentación, pp. 1121–1129. Universidad de Buenos Aires, Buenos Aires (2003)
11. Lynch, C., Ashley, K.D.: Modeling student arguments in research reports. In: Proceedings of the 4th Conference on Advances in Applied Human Modeling and Simulation, pp. 191–201. CRC Press, Inc. (2012)
12. Mochales, R., Moens, M.F.: Study on the structure of argumentation in case law. In: Proceedings of the 2008 Conference on Legal Knowledge and Information Systems, pp. 11–20. IOS Press (2008)
13. Mochales, R., Moens, M.F.: Argumentation mining. Artif. Intell. Law **19**(1), 1–22 (2011)
14. Moens, M.F., Boiy, E., Palau, R.M., Reed, C.: Automatic detection of arguments in legal texts. In: Proceedings of the 11th International Conference on Artificial Intelligence and Law, pp. 225–230. ACM (2007)
15. Peldszus, A., Stede, M.: From argument diagrams to argumentation mining in texts: a survey. Int. J. Cogn. Inform. Nat. Intell. (IJCINI) **7**(1), 1–31 (2013)
16. Roscoe, R.D., Allen, L.K., Weston, J.L., Crossley, S.A., McNamara, D.S.: The writing Pal intelligent tutoring system: usability testing and development. Comput. Compos. **34**, 39–59 (2014)

Semantic Mapping of Security Events to Known Attack Patterns

Xiao Ma[1], Elnaz Davoodi[1], Leila Kosseim[1(⊠)], and Nicandro Scarabeo[2]

[1] Department of Computer Science and Software Engineering, Concordia University,
1515 Ste-Catherine Street West, Montréal, Québec H3G 2W1, Canada
maxiao0215@gmail.com, ed.davoodi@gmail.com, leila.kosseim@concordia.ca
[2] Hitachi Systems Security Inc., 955 Boulevard Michéle-Bohec Suite 244, Blainville,
Québec J7C 5J6, Canada
nicandro.scarabeo@hitachi-systems-security.com

Abstract. In order to provide cyber environment security, analysts need to analyze a large number of security events on a daily basis and take proper actions to alert their clients of potential threats. The increasing cyber traffic drives a need for a system to assist security analysts to relate security events to known attack patterns. This paper describes the enhancement of an existing Intrusion Detection System (IDS) with the automatic mapping of snort alert messages to known attack patterns. The approach relies on pre-clustering snort messages before computing their similarity to known attack patterns in Common Attack Pattern Enumeration and Classification (CAPEC). The system has been deployed in our partner company and when evaluated against the recommendations of two security analysts, achieved an f-measure of 64.57%.

Keywords: Semantic similarity · Clustering · Cyber security

1 Introduction

With the increasing dependence on computer infrastructure, *cyber security* has become a major concern for organizations as well as individuals. *Cyber security* refers to the collection of tools, approaches and technologies which are used to prevent unauthorized behavior in cyber environments [1]. In order to detect and prevent harmful behaviour, sensors are typically installed in computer networks. Each sensor is equipped with several security systems, such as Intrusion Detection Systems (IDS) or Asset Detection Tools [2]. These systems perform network traffic analysis in real-time, detect suspicious activities and generates security events. Snort [3] is a widely used IDS system installed in many sensors [4]. By capturing and decoding suspicious TCP/IP packets, snort generates messages regarding network traffic data to facilitate the task of security analysts to recognize suspicious behaviours and act accordingly [3]. Figure 1 shows seven examples snort messages.

Recognizing suspicious behaviours from snort messages is difficult as the messages are typically short. Thus, today the task is still mostly performed by human

© Springer International Publishing AG, part of Springer Nature 2018
M. Silberztein et al. (Eds.): NLDB 2018, LNCS 10859, pp. 91–98, 2018.
https://doi.org/10.1007/978-3-319-91947-8_10

1. *APP-DETECT Apple OSX Remote Mouse usage*
2. *FILE-IDENTIFY RealPlayer skin file attachment detected*
3. *SQL generic sql exec injection attempt - GET parameter*
4. *FILE-OTHER XML exponential entity expansion attack attempt*
5. *MALWARE-OTHER Win.Exploit.Hacktool suspicious file download*
6. *BROWSER-PLUGINS MSN Setup BBS 4.71.0.10 ActiveX object access*
7. *BROWSER-IE Microsoft Internet Explorer asynchronous code execution attempt*

Fig. 1. Examples of 7 snort messages

security experts. However, because of the increasing volume of network traffic, the workload of security analysts has become much heavier and the possibility of not detecting a security risk has become critical. In order to allow security analysts to better assess risks, the automatic mapping of security events to attack patterns is desired. The goal of this paper is to enhance the performance of an existing Intrusion Detection System with the automatic mapping of snort alert messages to known attack patterns.

2   Previous Work

An Intrusion Detection System (IDS) is an essential part of a typical Security Information and Event Management System (SIEM). An IDS is responsible for detecting unauthorized behaviours by matching security events to known network and host-based attack patterns stored in knowledge bases [5]. If an attack pattern does not exist in the knowledge base, the IDS cannot detect the related malicious behaviour [6]. Therefore, it is necessary to keep this knowledge base up to date as new attack patterns are discovered. Much research in this area has thus focused on the automatic curation of knowledge bases of attack patterns.

As many novel attack patterns or cyber security related concepts are discussed online in forums and chat rooms (e.g. [7]), several natural language processing techniques have been proposed to extract security concepts from these unstructured texts (e.g. [8–12]). In particular, a system to extract cyber security concepts from semi-structured and unstructured texts, such as online documents and the National Vulnerability Database (NVD) [13], was introduced in [7]. It consists of an SVM classifier to identify cyber security related texts, a Wikitology [14] and a named entity tagger [15] to extract security concepts [10,12,16].

In addition to work on mining known attack patterns from the Web, the research community has also curated and made publicly available several resources of vulnerability, cyber attack patterns and security threats. These include the U.S. National Vulnerability Database (NVD) [13], Common Weakness Enumeration (CWE) [17] and Common Attack Pattern Enumeration and Classification (CAPEC) [18]. The CWE inventories software weakness types; while the NVD is built upon the CWE to facilitate data access. On the other hand, CAPEC not only provides common vulnerabilities, but also contains attack steps that hackers mostly use to explore software weaknesses as well as the suggested mitigation.

To our knowledge, not many novel techniques have been proposed to automatically match security events messages to known attack patterns based directly on their natural language descriptions. [4] proposed a system based on a KNN classifier to address this issue. However no formal evaluation of the quality of the output is provided. Our work builds upon this work by formally evaluating its performance with real cyber security data and enhancing it to improve its recall.

3   Original System

The approach of [4] was used as our baseline system. Its purpose is to map security events to known attack patterns in CAPEC [18] based solely on their natural language descriptions.

3.1   Description of the Original System

The system of [4] uses snort alert messages as input and tries to identify the n most relevant CAPEC fields to propose to cyber security analysts. CAPEC (Common Attack Pattern Enumeration and Classification [18]) is a publicly available knowledge base containing 508 known attack patterns. As shown in Fig. 2, an attack pattern in CAPEC is composed of various fields that are described in natural language. These include a *Summary* of the attack pattern, *Attack Steps Survey, Experiments, Attack Prerequisites etc.* On average, each CAPEC pattern contains 10 fields for a total of 5,096 fields.

Since the snort messages are relatively short (only 8 tokens on average), not much information can be extracted from them to be used in building automated model for detecting the attacks. To overcome this limitation, they are first expanded by replacing common domain keywords with a richer description. To do this, security experts analyzed 32,246 snort alert messages, and identified 68 important terms. For each term, they then wrote a description of about 5 words. Figure 3 shows the expansion of 3 such terms. By replacing domain terms in the original snort messages with their longer descriptions, the length of each snort message increased from an average of 8 words to an average of 15 words. To map these expanded security events to specific fields in CAPEC, both snort messages and CAPEC fields are pre-processed (tokenized, stop words removed and stemmed), and unigrams, bigrams and trigrams are used as features. Frequency variance (FV) is then used to filter out features that appear either too often or too rarely. Using TF-IDF and the cosine similarity, the distance between snort messages and each CAPEC field is then computed. Finally, the 3 most similar CAPEC fields that have a distance smaller than some threshold t, are selected.

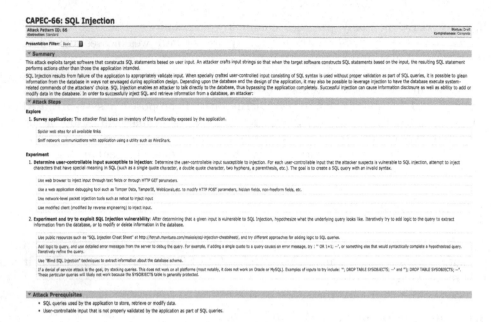

Fig. 2. Example of a CAPEC attack pattern

3.2 Evaluation of the Original System

In [4], the system was only evaluated in terms of coverage, i.e. the number of snort messages that were matched to at least one CAPEC field. However, the quality of the recommended CAPEC fields were not evaluated. To address this issue, we created a gold-standard dataset by asking 2 cyber security experts to evaluate the mapping of 3,165 snort messages mapped to at most 5 CAPEC fields. This gave rise to 16,826 mappings. Each mapping was annotated by the 2 experts with one of 3 levels of quality:

- *a correct mapping*: the analysts could use the CAPEC field directly as a solution,
- *an acceptable mapping*: the analysts could use the CAPEC field in order to generate a solution, or
- *a wrong mapping*: the recommended CAPEC field was not useful.

Term	Expanded Description
file-identify	*file extension file magic or header found in the traffic*
server-webapp	*web based applications on servers*
exploit-kit	*exploit kit activity*

Fig. 3. Examples of domain term expansion

Table 1. Statistics of the gold-standard dataset

Tag	Number of mappings
Correct	9,222
Acceptable	5,496
Wrong	2,108
Total	**16,826**

Table 1 shows statistics of the gold-standard. As the table shows, 9,222 mappings were labelled as correct; 5,496 mappings were tagged as acceptable and 2,108 mapping were judged wrong.

The output generated by the original system was then evaluated against the gold-standard dataset. We used the same 3,165 snort messages and evaluated the overlapping answers; mappings provided by the system that were not included in the gold-standard data were therefore not evaluated.

Following the recommendation of our security analysts, recall was deemed more important than precision. Indeed, in this domain, it is preferable to alert clients too often with false alarms than to miss potential cyber threats. To account for this, we used lenient definitions of precision (P^L) and recall (R^L) where *acceptable* mappings are considered *correct*. In addition, two values of β were used to compute the f-measure: $\beta = 1$, and $\beta = 0.5$, where recall is more important than precision. This gave rise to two f-measures: F_1^L and $F_{0.5}^L$.

When evaluated against the gold-standard dataset, the original system[1] achieved a lenient recall R^L of 35.22%, a lenient precision of P^L of 97.96%, an $F_{0.5}^L$ of 72.23% and an F_1^L of 51.82%. Although precision was high, recall was particularly low.

4   Enhancing the Original System

After analyzing the result of the original system, we noticed that many snort messages share similar content, hence it would seem natural that they be mapped to similar CAPEC fields. To ensure this, we experimented with clustering the snort messages prior to mapping them. Each snort message within a cluster is then mapped to the same CAPEC field. Specifically, snort messages are first expanded (see Sect. 3.1), then clustered into n clusters using k-means clustering. All messages in the same cluster are then concatenated into a single long message, and the resulting longer message is then mapped using the same approach as in the original system.

We experimented with various numbers of clusters (n) which as a side-effect also varied the length of the resulting message to map. The trade-off is that a larger number of clusters (n) should lead to a greater number of possible CAPEC fields being mapped to each snort message, but should also lead to a shorter message and sparser representation.

---

[1] with the parameters $FV = 0.98$ and $t = 0$ (see Sect. 3.1).

Table 2. Results of the system with different cluster numbers

System	n	P^L	R^L	$F_{0.5}^L$	F_1^L	Snort length
Original	n/a	97.96%	35.22%	72.23%	51.82%	15
Clustering	5000	85.66%	47.93%	74.01%	61.47%	94
Clustering	4000	85.83%	47.70%	73.99%	61.32%	117
Clustering	**3000**	95.86%	48.46%	80.18%	64.38%	157
Clustering	**2000**	95.40%	48.18%	79.77%	64.03%	235
Clustering	**1000**	95.71%	48.18%	79.94%	64.09%	470
Clustering	**500**	95.39%	48.80%	80.10%	64.57%	941
Clustering	100	93.80%	36.69%	71.53%	52.75%	4,707
Clustering	50	94.13%	36.23%	71.34%	52.33%	9,415
Clustering	20	94.17%	36.11%	71.25%	52.20%	23,539

4.1 Results and Analysis

Table 2 shows the results of the system with and without clustering as part of the pre-processing for various values of n. As shown in Table 2, the best configurations in terms of f-measure are when using clustering with values of n between 500 and 3000. With these values, the results are not statistically different, with $F_{0.5}^L \approx 80\%$ and $F_1^L \approx 64\%$. Recall itself has reached $\approx 48\%$ from a low 35.22% in the original system. Recall from Sect. 3 that the average length of CAPEC fields is 214 words. Hence using $n = 2000$ allows us to bring the average size of snort messages (235 words) at par with the size of CAPEC fields.

Table 2 also shows the trade-off between the use of a smaller number of clusters (smaller n) which leads to a smaller number of possible output CAPEC fields and the use of a larger n which leads to a sparser snort representation. Hence leading to lower f-measures with $n \geq 2000$ and $n \leq 100$.

5   Conclusion and Future Work

In this paper, we described an enhancement to the approach proposed by [4] to maps security events to related attack fields in CAPEC. We have shown that by expanding domain terms and clustering snort messages prior to mapping them, the recall of the approach can be increased significantly without much loss in precision.

As future work, we plan to investigate the use of automatic snort expansion, by using existing knowledge bases such as the Common Weakness Enumeration [17] rather than relying on hand-written term expansion. In addition, it would be interesting to look beyond the mapping of individual snort messages, but try to identify and match entire patterns/groups of snort messages as an indication of possible cyber attacks.

Acknowledgement. The authors would like to thank the anonymous reviewers for their feedback on the paper. This work was financially supported by an Engage Grant from the Natural Sciences and Engineering Research Council of Canada (NSERC).

References

1. Schatz, D., Bashroush, R., Wall, J.: Towards a more representative definition of cyber security. J. Digital Forensics Secur. Law **12**(2), 8 (2017)
2. Ashoor, A.S., Gore, S.: Importance of intrusion detection system (IDS). Int. J. Sci. Eng. Res. **2**(1), 1–4 (2011)
3. Roesch, M.: Snort: lightweight intrusion detection for networks. In: Proceedings of the 13th Conference on System Administration, LISA 1999, Seattle, Washington, USA, pp. 229–238, November 1999
4. Nicandro, S., Fung, B.C.M., Khokhar, R.H.: Mining known attack patterns from security-related events. Peer J. Comput. Sci. **1**, e25 (2015)
5. Buczak, A.L., Guven, E.: A survey of data mining and machine learning methods for cyber security intrusion detection. IEEE Commun. Surv. Tutor. **18**(2), 1153–1176 (2016)
6. More, S., Matthews, M., Joshi, A., Finin, T.: A knowledge-based approach to intrusion detection modeling. In: Proceedings of the IEEE Symposium on Security and Privacy Workshop (SPW), San Francisco, California, USA, pp. 75–81. IEEE, May 2012
7. Mulwad, V., Li, W., Joshi, A., Finin, T., Viswanathan, K.: Extracting information about security vulnerabilities from web text. In: Proceedings of the IEEE/WIC/ACM International Conference on Web Intelligence and Intelligent Agent Technology (WI-IAT), Lyon, France, vol. 3, pp. 257–260. IEEE, August 2011
8. Atallah, M.J., McDonough, C.J., Raskin, V., Nirenburg, S.: Natural language processing for information assurance and security: an overview and implementations. In: Proceedings of the 2001 Workshop on New Security Paradigms, Ballycotton, County Cork, Ireland, pp. 51–65, September 2001
9. Raskin, V., Hempelmann, C.F., Triezenberg, K.E., Nirenburg, S.: Ontology in information security: a useful theoretical foundation and methodological tool. In: Proceedings of the 2001 Workshop on New Security Paradigms, Cloudcroft, New Mexico, pp. 53–59. ACM (2001)
10. Undercoffer, J., Joshi, A., Finin, T., Pinkston, J.: Using DAML+ OIL to classify intrusive behaviours. Knowl. Eng. Rev. **18**(3), 221–241 (2003)
11. Undercoffer, J., Joshi, A., Pinkston, J.: Modeling computer attacks: an ontology for intrusion detection. In: Vigna, G., Kruegel, C., Jonsson, E. (eds.) RAID 2003. LNCS, vol. 2820, pp. 113–135. Springer, Heidelberg (2003). https://doi.org/10.1007/978-3-540-45248-5_7
12. Undercoffer, J., Pinkston, J., Joshi, A., Finin, T.: Proceedings of the IJCAI Workshop on Ontologies and Distributed Systems, Acapulco, Mexico, pp. 47–58, August 2004
13. National Cyber Security Division. National Vulnerability Database (NVD) (2017). https://nvd.nist.gov
14. Finin, T., Syed, Z.: Creating and exploiting a web of semantic data. In: Filipe, J., Fred, A., Sharp, B. (eds.) Agents and Artificial Intelligence, pp. 3–21. Springer, Berlin Heidelberg (2011)

15. Nadeau, D., Sekine, S.: A survey of named entity recognition and classification. Lingvisticae Investigationes **30**(1), 3–26 (2007)
16. UMBC Ebiquity. Index of /ontologies/cybersecurity/ids. (2014). http://ebiquity.umbc.edu/ontologies/cybersecurity/ids/
17. MITRE. Common Weakness Enumeration (CWE) (2017). https://cwe.mitre.org/index.html
18. MITRE. Common Attack Pattern Enumeration and Classification (CAPEC) (2017). https://capec.mitre.org/

Neural Networks Based Approaches

Accommodating Phonetic Word Variations Through Generated Confusion Pairs for Hinglish Handwritten Text Recognition

Soumyajit Mitra[✉], Vikrant Singh, Pragya Paramita Sahu, Viswanath Veera, and Shankar M. Venkatesan

Samsung R&D Institute India - Bangalore, Bangalore, India
{s.mitra,vkr.singh,pragya.sahu,viswanath.v,s.venkatesan}@samsung.com

Abstract. On-line handwriting recognition has seen major strides in the past years, especially with the advent of deep learning techniques. Recent work has seen the usage of deep networks for sequential classification of unconstrained handwriting recognition task. However, the recognition of "Hinglish" language faces various unseen problems. Hinglish is a portmanteau of Hindi and English, involving frequent code-switching between the two languages. Millions of Indians use Hinglish as a primary mode of communication, especially across social media. However, being a colloquial language, Hinglish does not have a fixed rule set for spelling and grammar. Auto-correction is an unsuitable solution as there is no correct form of the word, and all the multiple phonetic variations are valid. Unlike the advantage that keyboards provide, recognizing handwritten text also has to overcome the issue of mis-recognizing similar looking alphabets. We propose a comprehensive solution to overcome this problem of recognizing words with phonetic spelling variations. To our knowledge, no work has been done till date to recognize Hinglish handwritten text. Our proposed solution shows a character recognition accuracy of 94% and word recognition accuracy of 72%, thus correctly recognizing the multiple phonetic variations of any given word.

Keywords: Handwriting recognition
Colloquial language recognition · Phonetic spelling variations
Transliteration

1 Introduction

With increasing number of phones and tablets that support a pen based input, real-time recognition of handwritten text has become an important area of study. Deep-learning techniques have achieved major improvements in sequential learning tasks. Using a Bidirectional Long Short Term Memory Networks (Bi-LSTM) with Connectionist Temporal Classification (CTC) cost functions have shown

© Springer International Publishing AG, part of Springer Nature 2018
M. Silberztein et al. (Eds.): NLDB 2018, LNCS 10859, pp. 101–112, 2018.
https://doi.org/10.1007/978-3-319-91947-8_11

extremely high recognition accuracy for writer independent unconstrained handwriting in English. Extensive work has also been done for recognition of various regional languages, using both deep networks as well as Hidden Markov Model (HMM) based models [3,10,13].

Hinglish, however, presents various previously unseen challenges. With rising literacy rate and connectivity in India, increasing number of people are switching to Hinglish as a mode of communication both in social media and in daily life. The ease of typing Indic languages using an English keyboard has also increased the use of Hinglish words in conversations. This language involves a hybrid mixing of the two languages during conversations, individual sentences and even words. For example, a sentence stating "She was frying the spices when the phone rang" can be written as "She was bhunno-ing the masala-s jab phone ki ghantee bajee", where "bhunno", "masala", "jab", "ghantee" and "bajee" are Hindi words written in English characters. This language is so commonly used that David Crystal, a British linguist at the University of Wales, projected in 2004 that at about 350 million, the world's Hinglish speakers may soon outnumber native English speakers [11].

Being a colloquially developed language, Hinglish does not have any specific grammar or spelling rules. With the absence of a direct phoneme to character mapping between the two languages, different users use varied English characters to denote the same sound. For example, both "ph" and "f" have the same sound, due to which the same Hindi word can be written as both "phayda" as well as "fayda". These semantic differences are highly prevalent when writing in Hinglish. Simple auto-correction techniques cannot be used here because in the absence of a rule set, both "phayda" and "fayda" are valid words, and one should not be replaced or auto-corrected to the other.

Additionally, handwriting recognition faces a major problem with the recognition of out-of-vocabulary (OOV) words. Recognition rate of OOV words is comparatively lower than in-vocab words. Decoding modules that recognize strokes into words find the most probable word available in the dictionary to present as the output. In cases where a similar word (i.e. with a difference of just 1 or 2 characters, commonly seen in the word variations for Hinglish) is present in the dictionary, the presented output would not be what the user wrote, but what the dictionary contains. Thus, recognizing Hinglish text using previously known techniques often results in displaying words that are contained in the dictionary, and not necessarily what the user wrote.

In our paper we present a solution to this problem. We discuss a modified method of decoding to ensure that common possible phonetic variations of a word can be accounted for. We have also compared our results with a deep model without the modified decoding, to demonstrate the high difference in accuracies. For all comparisons, we are only taking into account a Bi-LSTM and CTC based network, as various works have proven this deep network to be more efficient at recognizing English characters than a statistical network. Our results thus show the improvement that our modified decoding brings into an already deep network. The rest of the paper is divided as follows: Sect. 2 provides related work on handwriting recognition across various languages. Section 3 explains the background on bidirectional LSTM, CTC and decoding. Our proposed method is described in Sect. 4 and experimental results and simulations on these are discussed in Sect. 5. Finally, we conclude the paper in Sect. 6.

2   Related Work

Handwriting recognition has been widely studied over the past few decades [1,2,8,10,12]. Mainly there are two methodologies for handwriting recognition: on-line and off-line While off-line recognition deals with images of handwritten text, on-line recognition is concerned with mapping a temporal sequence of pen-tip co-ordinates to a sequence of characters. Our method uses this idea of on-line handwriting recognition.

One such approach of on-line handwriting recognition is the writer independent model based HMMs [8]. Another approach *cluster generative statistical dynamic time warping* (CSDTW) uses a hybrid of HMMs with Dynamic Time Warping (DTW). It treats writing variations by combining clustering and generative statistical sequence modeling [1]. A third method uses support vector machine with Gaussian DTW kernel [2]. All these methods, however, only work for isolated character recognition and are useful generally for smaller vocabulary sized dataset.

Recurrent neural networks (RNNs) is a promising alternative which does not suffer from this limitation (performance under large vocabulary dataset). However, traditional networks require pre-segmented data which limits their application. In order to train RNNs for labeling unsegmented sequence data directly, a CTC network [4] is used. In this paper, we use a bidirectional LSTM with CTC output layer [7] for our task of handwriting recognition.

A language model is used to weight the probabilities of a particular sequence of words. Generally we want the output to be a sequence of words from the dictionary, in accordance to a predefined grammar. This limits the direct applicability of this for Hinglish, where people tend to use many inconsistent spellings for a particular word. So, even for a word present in the dictionary, accuracy becomes poor if the user writes it in a different but valid spelling. To tackle this problem, we use an idea derived from the work proposed in [9], where the authors learned the most common spellings deviations found in English transliteration of Hindi words as compared to a standard transliteration scheme.

3   Background

3.1   Bi-LSTM with CTC

LSTMs are a special type of Recurrent Neural Networks designed to overcome the *vanishing gradient problem* faced by general networks. A LSTM memory block consists of sets of recurrently connected cells which are controlled by three multiplicative gates: input, forget and output gate. Moreover, Bidirectional LSTMs are capable of accessing past as well as future contexts, which is quite helpful in case of handwriting recognition. To overcome the requirement of using pre-segmented data, a CTC output layer is used [4]. It directly outputs a probability distribution over label sequences without the need for segmented data.

A CTC output layer contains one more unit than there are labels, the additional unit corresponding to observing a 'blank' or 'no label'. Let us define this set as L. For a T length input sequence $\mathbf{x}$, the network output is an element (known as path) of the set L^T. The conditional probability of a path π is given by

$$p(\pi|\mathbf{x}) = \prod_{t=1}^{T} y_{\pi_t}^t \tag{1}$$

where y_k^t is the activation of the kth output unit at time t. A operator $\mathcal{B}$ maps all the paths onto label sequences by removing repeated labels and blanks. The conditional probability of some labeling $\mathbf{l}$ is given as

$$p(\mathbf{l}|\mathbf{x}) = \sum_{\pi \in \mathcal{B}^{-1}(\mathbf{l})} p(\pi|\mathbf{x}) \tag{2}$$

The most likely label $\mathbf{l}^*$ for an input sequence $\mathbf{x}$ is given as

$$\mathbf{l}^* = \operatorname*{argmax}_{\mathbf{l}} \quad p(\mathbf{l}|\mathbf{x}) \tag{3}$$

3.2   Decoding: External Grammar Integration

In order to ensure that the output from the LSTM is a sequence of dictionary words, the probabilities in (3) can be made conditioned on some probabilistic grammar G, as well as the input sequence $\mathbf{x}$:

$$\mathbf{l}^* = \operatorname*{argmax}_{\mathbf{l}} \quad p(\mathbf{l}|\mathbf{x}, G) \tag{4}$$

Using a bi-gram model, the CTC Token Passing Algorithm as described by Graves et al. in [5] allows us to find such $\mathbf{l}^*$.

The probability of a labeling is non zero only if it forms a sequence of words belonging to the dictionary (that G corresponds to). This causes a problem in case of Hinglish language due to inconsistent spelling of a particular word. Thus a traditional decoding is incapable of handling such spelling variations commonly found in "Hinglish".

4   Proposed Approach

We propose a handwriting recognition system incorporated with modified decoding to enhance recognition rate and deal with the problem of inconsistent spelling usage for Hinglish language. For the purpose of handwriting recognition, we experimented with the same theory mentioned above by using a Bidirectional LSTM model with CTC cost function [6]. We developed on the ideas proposed by Graves in their Token Passing Algorithm [5] to cope with the multiple phonetic spelling variations. The following subsections describes our solution modules. We also present the complete flow diagram of our solution in Fig. 1.

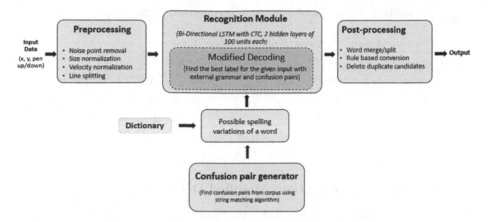

Fig. 1. Flow chart of the proposed methodology to recognize Hinglish handwritten text

4.1 Input

The system receives a set of points as inputs i.e. x-coordinate, y-coordinate and pen-up/pen-down information. This transforms each input data point into a 3-dimensional vector.

4.2 Preprocessing

Handwriting styles can vary significantly from person to person. In order to make the system robust to variation in height, width, slant, skew and any other geometric variations, the input signal is normalized with respect to its mean and standard deviation. Additionally, to handle different writing speeds, the data points are re-sampled and interpolated to obtain points at regular intervals. Further, any noise points that were inadvertently entered by the user are detected and removed. For multi-line inputs, an approximate "dividing" line is calculated by linear regression. Finally, the base and corpus line of the text are calculated by applying a linear regression through the minimas and maximas respectively, of all strokes.

4.3 Handwriting Recognition: LSTM Training

Our Bi-LSTM network contains two hidden layers, each of 100 units. For training with unsegmented data, we use a CTC output layer comprising of 195 units (including the 'blank' labeling). The network was trained on labeled data of 0.15 million words from 100 writers (1500 words from each writer) with a `learning rate` of 0.0001 and `momentum` of 0.9. All hyper parameters like learning rate were chosen after empirical experimentation. The CTC forward backward algorithm was used to update the network weights.

Table 1. Examples of commonly encountered confusion pairs

Confusion pair	Probability of occurrence	Example
(a,)	0.306	ladaka, ladka
(w,v)	0.0998	hawa, hava
(q,k)	0.021	aashiq, aashik
(en, ein)	0.019	zarooraten, zarooratein
(on,o)	0.016	ladkiyon, ladkiyo

4.4   Handwriting Recognition: Modified Decoding

The problem of recognizing a Hinglish word written in a spelling different from the standard transliterated form available in the dictionary can be solved if the common phonetic variations of that word can be identified.

We refer to such variations in a word as confusion pairs (a, b), where a and b are group of phonetically similar characters, one often being replaced by the other. For example, *khusbu* and *khusboo* are two different commonly used spellings of the same Hindi word. This generates the confusion pair (oo, u).

Probabilistic Generation of Confusion Pairs. Confusion pairs were initially generated from a corpus of $\approx 20,000$ sentences, by comparing each variation of the word with its standard transliterated form. A string matching algorithm was used for this comparison. Using dynamic programming, a character wise comparison from the start to end, was done between the words. This resulted in a substring pair that were frequently replaced or 'confused' with each other. Replacing one element of the pair with the other results in the phonetic variation of the same word [9].

Generated confusion pairs were then ranked according to the probability of its occurrence derived from the corpus. Some of the commonly used confusion pairs, along with their probability of occurrence, are given in Table 1. These confusion pairs are used to generate all phonetic word variations of any given Hinglish word.

The proposed technique is not limited to use in Hinglish. It can be extended to any scenario wherein multiple phonetic word variations have to be considered, and output should not be limited to the predefined dictionary.

Decoding. For each word w, we define $S(w)$ as the set containing all the spelling variations of the word. This set was generated using the concept of confusion pairs described above. It was also observed that there were many such pairs that only occur as suffix. As a result, the search space was limited to only the cases wherein an element of the pair occurred as a suffix (at the end of the word). This property, reduced the processing time as well. For example, the pair (on,o) occurs only as a suffix and hence while generating word variations, *on* was

---

Algorithm 1. Modified Decoding

```
 1: Initialisation:
 2: 𝒟′ = 𝒟
 3: for  words w ∈ 𝒟 do
 4:     for w̃ ∈ S(w) do
 5:         𝒟′∪ = w̃
 6:         tok(w̃, 1, 1) = (ln(y_b^1), (w̃))
 7:         tok(w̃, 2, 1) = (ln(y_{w̃_1}^1), (w̃))
 8:         if |w̃| = 1 then
 9:             tok(w̃, −1, 1) = tok(w̃, 2, 1)
10:         else
11:             tok(w̃, −1, 1) = (−∞, ())
12:         end if
13:         tok(w̃, s, 1) = (−∞, ()) for all s ≠ −1
14:     end for
15: end for
16: Pseudo code:
17: for t = 2 to T do
18:     sort tokens tok(w, −1, t − 1) by ascending score
19:     for  words w ∈ 𝒟 do
20:         S(w) ← set of spelling variations of w
21:         for w̃ ∈ S(w) do
22:             tok(w̃, 0, t).score = ln(p(w̃|w*)) + tok(w*, −1, t − 1).score
23:             tok(w̃, 0, t).history = w̃ + tok(w*, −1, t − 1).history
24:             for segment s = 1 to |w̃′| do
25:                 P = {tok(w̃, s, t − 1), tok(w̃, s − 1, t − 1)}
26:                 if (w̃′_s ≠ blank)&(s > 2)&(w̃′_{s−2} ≠ w̃′_s) then
27:                     add tok(w̃, s − 2, t − 1) to P
28:                 end if
29:                 tok(w̃, s, t) = max score token in P
30:                 tok(w̃, s, t).score+ = ln(y_{w̃′_s}^t)
31:             end for
32:             tok(w̃, −1, t) = max_score {tok(w̃, |w̃′|, t), tok(w̃, |w̃′| − 1, t)}
33:         end for
34:     end for
35: end for
36: Find output token tok*(w, −1, T) with highest score at time T
37: Output tok*(w, −1, T).history
```

---

replaced by o only when it occurred as a suffix of a word. For N such variations generated corresponding to a particular word, each variation was assigned a word probability of p/N, where p is the word probability of the standard transliterated form present in our dictionary $\mathcal{D}$. It is to be noted that the corpus used to create $\mathcal{D}$ for recognition and for finding confusion pairs were different. The dictionary $\mathcal{D}$ for recognition was created by transliterating an existing Hindi dictionary.

We present our method of decoding in Algorithm 1. We build upon the ideas presented by Graves et al. in [5]. The token passing algorithm is constrained by the usage of a fixed dictionary set, primarily predicting only those outputs present in the dictionary. We modify the decoding to take into account the most highly probable phonetic variations of the words as well. This ensures that words are not incorrectly 'auto-corrected' to its dictionary form. These words are created using the generated confusion pairs, by replacing one substring present in the confusion pair list with the other.

Thus, for all such generated variations in $S(w)$ we run the CTC Token Passing Algorithm [5]. In Algorithm 1, for every word w, a modified word w' is defined with blanks added at the beginning and end and between each consecutive pair of labels. A token $tok = (score, history)$ is a pair consisting of real valued $score$ and a $history$ of previously visited words. Each token corresponds to a particular path through the network outputs and the score of that particular token is defined as the log probability of that path. The basic idea of the token passing algorithm is to pass along the highest scoring tokens at every word state and to maximize over these to find the next state highest scoring tokens. For any spelling variant ($\tilde{w} \in S(w)$) corresponding to a word $w \in \mathcal{D}$, a token is added, thus increasing the search space along a new path corresponding to $\tilde{w}$. This allows a non-zero probability of $\tilde{w}$ to be part of the output sequence.

Using the above described process, our BiLSTM-CTC system recognizes a list of possible candidates (l) that maximizes the value of $p(\mathbf{l}|\mathbf{x}, G^*)$, $\mathbf{x}$ being the input sequence (described in Sect. 3.2). On account of our modified decoding technique, G^* takes into account the words phonetically varied from the ones in the dictionary. This candidate list is then sent further for post-processing.

The variations in handwriting recognition time and accuracy with the number of confusion pairs used are given in the next section.

4.5   Post-processing

Before the candidate list is output to the user, it passes through a module of post-processing to remove any other inadvertently caused errors. This module checks for any erroneous word merging or splitting issues by checking the distance between the bounding boxes of consecutive words. The module also checks for the mis-recognition of symbols like ',', ':', ';' etc. using stroke size information. Updated candidates are then output to the user.

5   Experiments and Results

5.1   Data Sets

Our model was initially trained to recognize English characters on a dataset of approximately 0.15 million English words from 100 different writers. The model converged after around ≈35 epochs. For testing, we collected handwritten Hinglish test data from 100 different users, each providing 100 samples. Each sample contained one sentence or phrase written in Hinglish, of the form represented in Fig. 2. To ensure that different writing styles were recorded, we ensured that the users were from different parts of India, representing a different way of pronouncing and writing the same words. As, there is no publicly available dataset of Hinglish handwritten text, we were limited to using our collected dataset for all testing purposes.

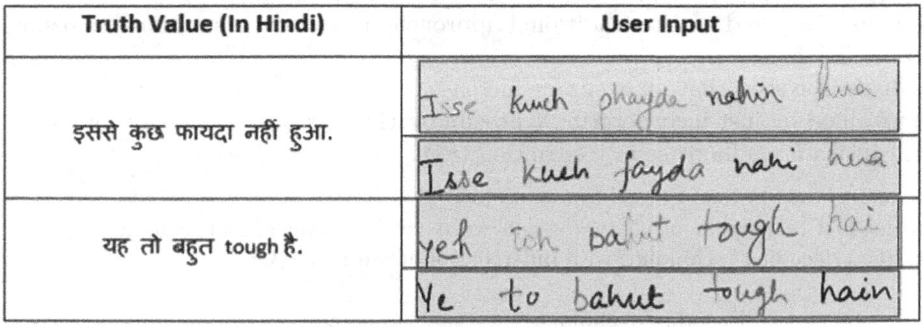

Truth Value (In Hindi)	User Input
इससे कुछ फायदा नहीं हुआ.	*Isse kuch phayda nahin hua*
	Isse kueh fayda nahi hua
यह तो बहुत tough है.	*yeh toh bahut tough hai*
	Ye to bahut tough hain

Fig. 2. Samples of the test data

5.2   Evaluation Metrics

According to our knowledge, no previous work exists for Hinglish handwritten text recognition. Past research has proven that a BiLSTM-CTC model provides the best recognition accuracy. We intend to use the same model and incorporate our modified decoding techniques to show a significant increase in word recognition accuracy. Thus, we compare our solution with modified decoding to a system with the existing decoding module (CTC Token Passing Algorithm as described by Graves in [5]), both being deep models.

Word accuracy is simply defined as the percentage of correctly recognized words in the test data set. However, for the calculation of character accuracy, the following formula is used:

$$Char\_Accuracy = \frac{N - S - I - D}{N} * 100\% \qquad (5)$$

where N represents the total number of characters in data, S denotes the number of substitutions, I represents the number of insertions and D represents the number of deletions required to convert the predicted output to the known truth value.

5.3   Experimental Results

In this paper, we propose a robust, all-inclusive Hinglish handwriting recognition system with modified decoding module. We also demonstrate how our solution overcomes problems associated with the recognition of colloquially developed languages with undefined rule-sets.

Table 2 represents the variation in character and word level accuracies between the proposed and traditional CTC Token Passing decoding [5] methodologies. We also provide the results obtained while varying the number of confusion pairs used to generate word variations.

Figure 3 denotes the increase in time with increasing number of word variations as compared to the traditional approach (tested on a standard Samsung Note-5 device).

It can be seen that an accuracy increase of ≈8.5% for word recognition, can be attained by just incorporating 5 possible variations from the confusion pairs, with just a 5 ms increase in recognition time.

Table 2. Comparison of character and word level accuracies between existing and modified decoding techniques with different confusion pairs (CP)

Decoding technique		Accuracy	
		Character	Word
Existing (CTC token passing)		92.8%	61.98%
Modified	Top 5	93.70%	70.36%
	Top 10	93.84%	71.02%
	Top 15	94.03%	72.27%

Errors that occur during Hinglish recognition is due to different spellings of the same word. This spelling difference is usually the variation of 1 or 2 characters, with all multiple word variations being a valid possibility. The BiLSTM - CTC model is highly efficient at correctly recognizing characters in general, which is reflected in the already high character recognition accuracy of the existing model (BiLSTM-CTC with CTC Token Passing Algorithm). However, the mis-recognition of a few characters causes misclassification of the complete word, leading to a low word recognition rate. Proposed modified decoding technique incorporating confusion pairs, is able to solve this problem of word misclassification. Our method shows an improved word recognition accuracy of nearly 10% as compared to existing text recognition module, while the increase in character recognition accuracy is around 1.5% (Table 2). The considerably smaller increment in character recognition accuracy as compared to the word recognition accuracy is intuitive. The effect of a misclassifying a word, on character accuracy is nominal as only a small percentage of characters is responsible for the mis-classification.

We also present the recognition time across increasing number of word variations as well as for the original decoding in Fig. 3 (tested on a Samsung Note-5 device). As can be noted, this time increase is minimal and will not affect user experience. The increment in word recognition accuracy is much more significant than a slight increase in time.

Another major advantage of using the proposed solution is the possibility of teaching systems to learn local language variations, even when huge data-sets are unavailable. Our current model learned to recognize English characters by being trained on a general English data-set, while it learned the language intricacies from a Hindi dictionary and corpus. Our decoding module combined these two variate systems to correctly recognize Hindi words written in English script.

While we present all our experiments and results for the specific use case of Hinglish, the proposed technique can be used over any deep model architecture for any language, to significantly increase word recognition accuracies, especially in the case of writing one language using the script of another. The usage of our solution, thus, should not seem to be limited to just recognizing Hinglish handwritten text using the shown BiLSTM-CTC network. The ideas proposed here can be extended to any similar situation.

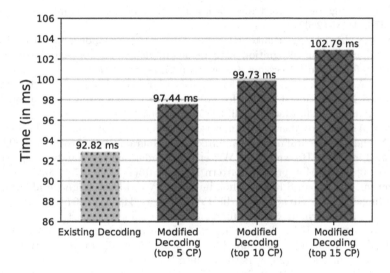

Fig. 3. Variation in recognition time with increasing number of included confusion pairs (CP)

6   Conclusion

In this work, we provide a comprehensive solution for recognizing locally developed hybrid languages. Our proposed solution is more accurate than using English trained models with negligible increase in recognition time. We propose our modified decoding technique over popularly used BiLSTM-CTC deep models for a significant increase in word recognition accuracy, without a major increase in time. However, its usage is not limited to the model used here, and the proposed ideas can be extended for use over any deep model for text recognition.

Currently, we are working on personalizing the solution based on past user writing habits, such that the spelling that he/she uses is always the top suggestion. We also plan to work on real-time translation of Hinglish to English as well as Hindi.

References

1. Bahlmann, C., Burkhardt, H.: The writer independent online handwriting recognition system frog on hand and cluster generative statistical dynamic time warping. IEEE Trans. Pattern Anal. Mach. Intell. **26**(3), 299–310 (2004)
2. Bahlmann, C., Haasdonk, B., Burkhardt, H.: Online handwriting recognition with support vector machines-a kernel approach. In: Proceedings of the Eighth International Workshop on Frontiers in Handwriting Recognition, 2002, pp. 49–54. IEEE (2002)
3. Bharath, A., Madhvanath, S.: Hidden Markov Models for online handwritten Tamil word recognition. In: ICDAR, pp. 506–510. IEEE (2007)
4. Graves, A., Fernández, S., Gomez, F., Schmidhuber, J.: Connectionist temporal classification: labelling unsegmented sequence data with recurrent neural networks. In: Proceedings of the 23rd International Conference on Machine Learning, pp. 369–376. ACM (2006)
5. Graves, A., Liwicki, M., Bunke, H., Schmidhuber, J., Fernández, S.: Unconstrained on-line handwriting recognition with recurrent neural networks. In: Advances in Neural Information Processing Systems, pp. 577–584 (2008)
6. Graves, A., Liwicki, M., Fernández, S., Bertolami, R., Bunke, H., Schmidhuber, J.: A novel connectionist system for unconstrained handwriting recognition. IEEE Trans. Pattern Anal. Mach. Intell. **31**(5), 855–868 (2009)
7. Graves, A., Schmidhuber, J.: Framewise phoneme classification with bidirectional LSTM and other neural network architectures. Neural Netw. **18**(5), 602–610 (2005)
8. Hu, J., Lim, S.G., Brown, M.K.: Writer independent on-line handwriting recognition using an HMM approach. Pattern Recogn. **33**(1), 133–147 (2000)
9. Jhamtani, H., Bhogi, S.K., Raychoudhury, V.: Word-level language identification in bi-lingual code-switched texts. In: PACLIC, pp. 348–357 (2014)
10. Plamondon, R., Srihari, S.N.: Online and off-line handwriting recognition: a comprehensive survey. IEEE Trans. Pattern Anal. Mach. Intell. **22**(1), 63–84 (2000)
11. Baldauf, S.: A Hindi-English jumble, spoken by 350 million (2004). https://www. csmonitor.com/2004/1123/p01s03-wosc.html
12. Tappert, C.C., Suen, C.Y., Wakahara, T.: The state of the art in online handwriting recognition. IEEE Trans. Pattern Anal. Mach. Intell. **12**(8), 787–808 (1990)
13. Zhou, X.D., Yu, J.L., Liu, C.L., Nagasaki, T., Marukawa, K.: Online handwritten Japanese character string recognition incorporating geometric context. In: Ninth International Conference on Document Analysis and Recognition, ICDAR 2007, vol. 1, pp. 48–52. IEEE (2007)

Arabic Question Classification Using Support Vector Machines and Convolutional Neural Networks

Asma Aouichat[⊠], Mohamed Seghir Hadj Ameur, and Ahmed Geussoum

NLP and Machine Learning Research Group (TALAA), Laboratory for Research in Artificial Intelligence (LRIA), University of Science and Technology Houari Boumediene (USTHB), Algiers, Algeria
{aaouichat,aguessoum}@usthb.dz, mohamedhadjameur@gmail.com

Abstract. A Question Classification is an important task in Question Answering Systems and Information Retrieval among other NLP systems. Given a question, the aim of Question Classification is to find the correct type of answer for it. The focus of this paper is on Arabic question classification. We present a novel approach that combines a Support Vector Machine (SVM) and a Convolutional Neural Network (CNN). This method works in two stages: in the first stage, we identify the coarse/main question class using an SVM model; in the second stage, for each coarse question class returned by the SVM model, a CNN model is used to predict the subclass (finer class) of the main class. The performed tests have shown that our approach to Arabic Questions Classification yields very promising results.

Keywords: Arabic · Question classification
Support Vector Machines · Word embeddings
Convolutional Neural Networks

1 Introduction

The aim of a Question Answering System (QAS) to process an input question posed in natural language so as to extract a short, correct answer which then gets returned to the user. Most of QASs consist of three main modules which are pipelined. First, the Question Processing module starts by analyzing the question so as to identify its class - and this is usually done by means of text classification methods - and a set of most important features. Once this is done, the extracted information gets into the Document Processing module the aim of which is to search (on the web or in a given corpus) for a set of relevant documents which probably contain the right answer to the input question. Finally, the third module of a QAS is the Answer Extraction module which extracts the correct answer to the user's question by making use of the information and documents returned by the previous two modules.

© Springer International Publishing AG, part of Springer Nature 2018
M. Silberztein et al. (Eds.): NLDB 2018, LNCS 10859, pp. 113–125, 2018.
https://doi.org/10.1007/978-3-319-91947-8_12

The quality of a QAS is directly affected by the accuracy of the question classification, a step in the Question Processing module. Indeed, if the question is classified under an incorrect category, this will affect the answer extraction step and it will most likely produce a wrong type of answer. For several languages such as English, this module has received much attention [10,17,20]. This is not the case for the Arabic language in which little research work has been done on this problem. It is worth mentioning also the severe lack of question-answering resources for the Arabic language. As such the approaches to question classification that have been proposed for Arabic have mostly been tested on very small corpora, usually a few hundred instances. Hence, even though such systems may produce good results on the small datasets on which they were developed, there is no guarantee that the results will be as good on larger datasets. To this end, and in order to test our approach on a substantially larger dataset, we have enriched the TALAA-AFAQ corpus developed in [3] by manually translating a subset of the UIUC question classification dataset. The questions in these corpora are classified based on the taxonomy proposed in Li and Roth [13] in which they defined a two-layered taxonomy. This hierarchy contains 6 coarse classes (ABBREVIATION, ENTITY, DESCRIPTION, HUMAN, LOCATION and NUMERIC VALUE) and 50 finer classes, where each coarse class can be further refined into a well-defined set of finer classes. For instance, the Location coarse class can have Country, City, Mountain, etc., as finer classes.

Encouraged by the impressive results that have been achieved on the task of sentence classification using Convolutional Neural Networks (CNNs) [6], we present in this paper, a novel approach to the classification of Arabic questions, which combines Support Vector Machines (SVMs) and Convolutional Neural Networks. The module works in two stages: in the first stage an SVM model is used to determine the coarse (i.e. main) class of the question. Then, given that coarse class, a CNN model is used to identify its finer class. Our intuition is that a pre-trained word embedding that can preserve the relative meaning of words will be better suited to handle the task of finer classification in which each finer class involves way fewer instances.

The rest of the paper is organized as follows: Sect. 2 presents the research work that has been done on the Arabic question classification task. Section 3 introduces the structure of Arabic question. Section 4 details the approach we propose while Sect. 5 presents the test data, the results, and an analysis thereof. Finally, in the conclusion section, we summarize our contribution and point out some directions for possible future improvements.

2   Related Work

Few research works having tackled Arabic question classification, we first present the most relevant ones and then introduce some important contributions to English question classification.

Al Chalabi et al. [2] proposed a method for the classification of Arabic questions using regular expressions and context-free grammars. The method was tested on a set of 200 Arabic questions and the authors reported a recall of 93% and a precision[1] of 100%. Ali et al. [8] proposed a question classification system for "Who", "Where" and "What" questions. Their contribution is based on the use of SVMs with TF-IDF Weighting. They tested their proposal with n-grams of varying lengths on 200 questions about Hadith (sayings of the prophet Muhammad) and reported that in the case of 2-grams, the best classification was obtained with an F1-score of 87.25%. Abdenasser et al. [1] used an SVM-based classifier to classify questions on the holy Quran and reported an accuracy of 77.2%. Their data set consisted of 230 Quranic domain questions. We note that all of these contributions were tested on very small datasets whose size did not exceed 300 questions.

A good number of systems have been proposed to address the question classification problem for the English language. Kim et al. [11] proposed a Convolutional Neural Network (CNN) that uses a static pretrained word vectors for the task of text classification. They showed that very little hyperparameter tuning can lead to a great deal of improvement. The results of their experimentations performed on several benchmarks showed some very promising results. Silva et al. [20] presented a method that starts by matching a question against some hand-crafted rules and then use the matched rules features in the learning-based classifier. Huang et al. [10] proposed the use of a head word feature and presented two approaches to augment the semantic features of such head words using WordNet. In addition, they adapted Lesk's Word Sense Disambiguation (WSD) algorithm to optimize the hypernym depth feature. Their linear SVM and Maximum Entropy models reached accuracies of 89.2% and 89.0% respectively. Nyberg et al. [16] proposed a set of handcrafted rules to boost the performance of a syntactic parser; their system, Javelin, achieved an accuracy of 92% on the TREC questions test set. Given the results that have been reached on Arabic Question Classification so far and the fact that they have all been trained/tested on very small data sets, our work presents a combined SVM-CNN model which is trained on a much larger dataset, which yields a much more reliable model for Arabic question classification.

3 The Structure of Arabic Questions

The Arabic alphabet consists of 28 letters each of which can be combined with one of four diacritics that can be omitted, the correct reading of the words having then to be done according to the context. Arabic is known to be a morphology-rich language, which results in a large number of possible word forms. However, unlike most languages where the syntax of a sentence changes to form a question, in Arabic, a question is simply a sentence that starts with a question word known as أداة استفهام (Interrogative Particle) and finishes with "?".

---

[1] A closer scrutiny of the patterns that were used in [2] has shown that they do not cover all the possible variations of uses of Interrogative Patterns in different contexts and settings.

Table 1. Samples of structures of Arabic questions

حصل آحمد عبد السلام على جائزة نوبل في الفيزياء عام ١٩٧٩	
Muhammad Abdus Salam received the Nobel prize in physics **in the year 1979**	
Type of information	**Question**
Asking for a Name	من حصل على جائزة نوبل للسلام عام ١٩٧٩؟
	Who received the Nobel prize in physics in the year 1979
Asking for a Date	متى حصل آحمد عبد السلام على جائزة نوبل في الفيزياء ؟
	When Muhammad Abdus Salam received the Nobel prize in physics[a]

[a]Though this sentence is not correct in English, it reflects the word ordering of its Arabic equivalent.

Table 1 presents the construction of two Arabic questions: the first one "Who won the Nobel Peace Prize in 1979?" asks for a Name and the second one "When did Mohamed Abdus Salam win the Nobel Prize?" asks for a Date. It clearly appears from the sentence that the Named Entity is removed from the declarative sentence and the Interrogative Particle is introduced at the start of the sentence in each case.

We distinguish different categories of questions: (1) factoid questions, (2) list questions, (3) Definition questions, (4) and the category Others which groups the remaining question types such as Yes/No, arguments, etc. To define these categories, several interrogative particles (IPs) can be used such as: من (man, who), كم (kem, how many/much), أين (ayna, where), متى (mata, when), ما (maa, what is), أي (ayy, which), كيف (kaifa, how), هل (hal, question particle for yes/no questions).

Some types of questions are directly recognized by identifying their Interrogative Particles. For instance, من (man, who) asks for the Name of a person(s); أين (ayna, where) asks for a Location; متى (mata, when) asks for a Time/Date; كم (kem, how many/much) asks for a Numeric value(s); كيف (kaifa, how) asks for a description; هل (hal) poses a Yes/No question. Using POS-tag sequences, such types of questions can be identified by some rules that describe each expected answer type. However, the main challenge is with those Interrogative Particles which can express different types of questions depending on the context. Consider, for instance, ما (maa, what is). This Interrogative Particle can be used to ask for a Name as in ما اسم أهم شركة ؟ (What is the name of the most important company?) . It can also be used to ask for a Date as in ما تاريخ اليوم ؟ (what is the date of today ?). It can additionally be used to ask for a Description ما طريقة صنع الكعكة ؟ (What is the way to make a cake?). The same applies to different types of questions and their Interrogative Particles.

4   Design of the SVM-CNN Arabic Question Classification System

Figure 1 presents the architecture of the SVM-CNN Arabic Question Classification System. The system works in two phases: first, an SVM-based model is used to predict the coarse class of the question. Then, a CNN model is used to predict its finer class. The CNN uses word embedding vectors that we have trained on a large Arabic corpus. Each of the two stages is detailed in subsequent sections.

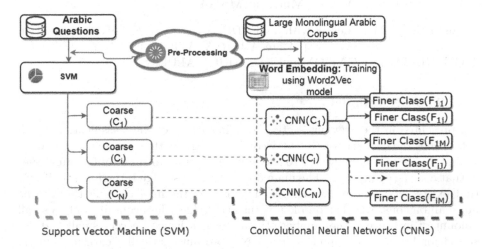

Fig. 1. General architecture of the SVM-CNN Arabic question classification system

4.1   Preprocessing and Normalization

In the preprocessing step, we apply word tokenization using the Polyglot tokenizer[2] then we remove all the Arabic diacritical marks. We also normalize all the letter variants to their most basic from. For instance, the letters أ and إ will be normalized to ا, which will help reduce the number of word forms in the dataset. We also define a regular expression to annotate each date or year entities with a "Date" label and the same thing for numbers which we tag as "NBR". The aim of this stage is to ignore the difference between various occurrences of the same type of entity in the classification system. For instance, if the question contains the date "01/02/2018" or "03/03/2014", the classification system will consider both as just Date.

4.2   Arabic Word Embedding

Our word embeddings are trained using the word2vec model [14] by means of the Gensim library[3]. The word2vec model represents each individual word as

---

[2] https://github.com/aboSamoor/polyglot.
[3] https://radimrehurek.com/gensim/.

belonging to a fixed-size vector of words that are relatively similar to others in terms of their occurrences in sentences. The training was carried out using a subset of $1,513,097$ sentences from the Arabic United Nations Parallel Corpus [7]. We empirically reached the following Gensim parameters: the size of the embedding vectors was set to 100 , the minimal word frequency was set to 10 and the number of iterations to 5. This resulted in a vocabulary of $850,797$ unique words.

4.3   The Support Vector Machine Model

We used the SVM model [9] with a TF-IDF weighting to classify each question into one of 6 coarse classes that are ABBREVIATION, ENTITY, DESCRIPTION, HUMAN, LOCATION and NUMERIC VALUE.

4.4   The CNN Model

We presented in the previous section, the first phase of the classification which determines the coarse class. In the second stage, given that coarse class we want to find its finer class. So we try to classify the question into one of the subclasses of that coarse (main) class. The main difficulty of this stage is the huge similarity that exists between the finer classes. This is why the finer classification can be quite ambiguous. Another problem is that each finer class contains only a small amount of instances. The number of features will hence be very small and not suited for bag-of-words models. The CNN can automatically learn meaningful features to identify the finer class for each question. Figure 2 represents the general architecture of the CNN.

The following is performed:

Padding. Each question will be padded to the same length k. When the length of the question is larger than k, we keep only the first k words of the question. For the questions that have less than k words a special token called "pad" is used to fill the remaining tokens up to a length of k.

Word Embedding. Each word will be represented as an embedding vector of size $\mathbb{R}^d$, thus any given question q_i is represented as a fixed-size matrix of dimension $\mathbb{R}^{kd}$.

Convolution. The first layer in our network applies a convolution filter (F) to the input data, where F is a matrix of dimension R^{fd} and f is the length of the filter. This filter will be applied to each region of q, producing a feature-map C of dimension R^l where $l = k - f + 1$. Each feature $r_i \in C_i$ is obtained by applying F to a window $w_i :_{i+f-1}$ of size f using the following equation [12]:

$$r_i = \Phi(F \Delta w_i :_{i+f-1} + b) \tag{1}$$

where Φ is a nonlinear activation function and b$\in$R is a bias. To generate a feature map C $= \{r_1, r_2, ..., r_l\}$, we apply this filter to each window of size f in q.

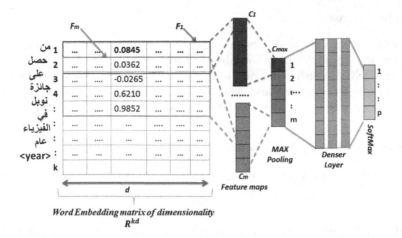

Fig. 2. General architecture of the Arabic question classification CNN

Pooling. Pooling is then used to perform a special down-sampling on the feature-map. One common method of pooling is known as max-pooling [19] which extracts only the most important feature C_{max} from C:

$$C_{max} = max\{C\} = max\{r_1, r_2, ..., r_l\}. \tag{2}$$

The usage of max-pooling selects only the most important m features.

Fully connected layer. To receive the output of the max-pooling layer, multiple fully-connected layers are stacked together (Fig. 2). Finally, the final layer uses a Softmax activation function to output a probability distribution over the finer classes for the main class that was found in the first (SVM) phase.

5   Test Results

This section provides an in-depth discussion of the tests we have carried out. As explained above, the classification is carried out in two consecutive stages (two levels): first, we build an SVM classifier for the main class of the question. Then, given that main class, we perform another CNN-based classification to identify the finer class of the question for that specific main class. Thus, our tests will also follow these two stages: first, we investigate the use of multiple classification models to identify the main class, and then we examine the case of the finer class.

We start this section by presenting some statistics about the data we have used and the preprocessing that we have incorporated. Then, we provide all the hyper parameters that have been used. Finally, we address the two aforementioned key tests.

5.1   The Dataset

Our dataset is an extended version of the dataset proposed by Aouichat and Guessoum [3]. The enrichment was obtained through the translation of instances from the UIUC English question classification dataset[4]. The statistics of the number of questions, tokens and types in the main and finer classes before and after extending the dataset are provided in Table 2.

5.2   Models Hyperparameters

In this section, we provide the detailed hyper parameters of all the models we have used.

Bag-of-words Models. We have considered three bag-of-words models for Support Vector Machines (SVM) [22], Maximum Entropy (MaxEnt) [21] and Random Forests (RF) [4]. These models have been represented using vector-space formulation with the TF-IDF weighting [18].

We have used the Python Scikit-learn machine learning library to implement all our models[5]. For all the models, the unigram, bigram and trigram features have been investigated. The feature counts for all the bag-of-words models are provided in Table 3.

CNN Models. For our CNN models we have used an embedding dimension of 100 units, and a vocabulary size of 10, 000 words. We have set the maximum sequence length to 20 tokens and have incorporated three filtering kernels for which the window sizes are 3, 5 and 7 respectively, each with 64 filters. Moreover, we have used four dense layers with 100 units each, and all these layers have been used with a Rectified Linear Unit (ReLU) [15] activation function. Dropout layers with a dropout-rate of $p = 0.4$ have been incorporated between each two consecutive dense layers. A mini-batch of size 128 has been used. The training was done by means of Stochastic Gradient Descent (SGD) with the Adam optimization function [12]. The CNN models have been trained using the Nvidia GTX 1070 GPU with 8 GB of DDR5 memory. The Keras deep learning library [5][6] has been used to implement all the models using the Tensorflow backend.

5.3   Results and Analyses

The test results for all the classification systems when predicting the main classes are presented in Table 4.

As shown in Table 4, the best results were obtained when using the SVM models followed by CNN and the worst results were obtained when using the

---

[4] The dataset is available at http://cogcomp.org/Data/QA/QC/.
[5] http://scikit-learn.org/.
[6] https://keras.io/.

Table 2. Statistics about the extended version of the Arabic question classification dataset proposed by Aouichat and Guessoum [3]

Class	Original dataset			Extended dataset		
	Questions	Tokens	Types	Questions	Tokens	Types
ABBREVIATION						
Expression	-	-	-	51	360	141
Abbreviation	-	-	-	16	131	57
HUMAN						
GROUPE	11	103	75	184	1966	812
Individual	899	7171	2317	1128	9979	2980
LOCATION						
Mount	2	14	12	21	176	65
State	137	843	371	202	1467	566
Country	121	691	310	275	2182	813
NUMERIC						
Money	18	163	114	80	784	338
Date	81	621	327	278	2188	862
Speed	2	24	21	8	59	34
Count	276	2925	1111	723	6039	2000
ENTITY						
Instrument	-	-	-	9	77	45
Word	-	-	-	26	268	127
Currency	-	-	-	4	29	16
DESCRIPTION						
definition	-	-	-	216	1282	466
Reason	-	-	-	183	1649	766
description	-	-	-	234	2100	839
Manner	-	-	-	202	1600	746
Total	**1547**			**3840**		

Table 3. Statistics about the features used for all the bag-of-words models for the identification (classification) of the main class

N-grams	Features count
Unigrams	5970
Bigrams	17102
Trigrams	28975

Table 4. The test results for all the classification systems.

Models	Precision	Recall	F-measure	Accuracy
SVM-uni	0.9272	0.9071	0.9170	0.9170
SVM-bi	0.9375	0.9227	0.9300	0.9227
SVM-tri	0.9292	0.9071	0.9180	0.9071
Rand-forest-uni	0.9182	0.9090	0.9021	0.9001
Rand-forest-bi	0.9391	0.9290	0.9226	0.9192
Rand-forest-tri	0.9123	0.9013	0.8913	0.8906
MaxEnt-uni	0.8712	0.8585	0.8648	0.8585
MaxEnt-bi	0.8640	0.8758	0.8698	0.8758
MaxEnt-tri	0.8635	0.8741	0.8687	0.8741
CNN	0.9350	0.9218	0.9249	0.9283

Maximum Entropy classifier. An increase of about 1% to 2% has been observed when the bigram features were used and these gave slightly better results than with tri-grams. The performance of the CNN model was better than those of the MaxEnt and Rand-forest models with a margin that varies between 2 to 6%.

After predicting the main class, we have used another CNN model to predict its finer class. The test results when predicting the finer class for each main class are provided in Table 5.

Table 5. The test results when predicting the finer class for each main class

CNN Models	Precision	Recall	F-measure	Accuracy
CNN-Abb	0.9166	0.8888	0.8917	0.8888
CNN-Desc	0.9507	0.8571	0.8943	0.8571
CNN-Enty	0.9375	0.9166	0.9198	0.9166
CNN-Hum	0.9901	0.976	0.9828	0.9761
CNN-Loc	0.9083	0.5939	0.6903	0.5939
CNN-Num	0.9844	0.9837	0.9839	0.9837
Full system	0.9055	0.8932	0.8865	0.8932

As shown in Table 5, the classification precisions for the finer class are all above 85% except for the Location class. The accuracy of the full architecture (the main and finer classes) is 0.82%.

5.4 Errors and Discussion

In this section, we will provide a quick glance at the misclassification errors made by our CNN system for the finer class. The confusion matrix is provided in

Table 6 for finer classes of the main class Location produced by the classification system (CNN-Loc).

Table 6. Confusion matrix for the prediction made by the CNN-Loc model

	AB	EXP
AB	6	0
EXP	2	10

The diagonal of the matrix in Table 6 contains the correctly predicted instances, and the values outside the diagonal represent the numbers of misclassified instances. We can see that the system makes only a few classification mistakes. For instance, for the finer EXP two instances were wrongly classified as AB.

6   Conclusion and Future Work

In this paper, we have focused on the question classification module in an Arabic Question Answering System. We have presented a novel approaches that combines a Support Vector Machine with a Convolutional neural network for Arabic Question classification using Arabic word embeddings. The aim of our approach being to extract the finer class of an input question, we start by identifying the coarse class using an SVM model and, at a second stage, and for each coarse class returned by the SVM model, a CNN model is applied to extract the finer class.

In order to train and test our proposal on a substantially large dataset, we have enriched the TALAA-AFAQ corpus developed in [3] by manually translating a subset of the UIUC English question classification dataset. The results obtained in this work were very promising yet the room for improvement remains open. As a future work, we plan to further enrich the TALAA-AFAQ and to create a common test set to be used by NLP researchers working on this task, and we also plan to integrate our work into a full Arabic Question Answering System.

References

1. Abdelnasser, H., Ragab, M., Mohamed, R., Mohamed, A., Farouk, B., El-Makky, N., Torki, M.: Al-Bayan: an Arabic question answering system for the Holy Quran. In: Proceedings of the EMNLP 2014 Workshop on Arabic Natural Language Processing (ANLP), pp. 57–64 (2014)
2. Al Chalabi, H.M., Ray, S.K., Shaalan, K.: Question classification for Arabic question answering systems. In: International Conference on Information and Communication Technology Research (ICTRC), Abu Dhabi, United Arab Emirates, pp. 310–313. IEEE (2015)

3. Aouichat, A., Guessoum, A.: Building TALAA-AFAQ, a corpus of Arabic FActoid question-answers for a question answering system. In: Frasincar, F., Ittoo, A., Nguyen, L.M., Métais, E. (eds.) NLDB 2017. LNCS, vol. 10260, pp. 380–386. Springer, Cham (2017). https://doi.org/10.1007/978-3-319-59569-6_46
4. Breiman, L.: Random forests. Mach. Learn. 45(1), 5–32 (2001)
5. Chollet, F., et al.: Keras: https://github.com/fchollet/keras (2015)
6. Dahou, A., Xiong, S., Zhou, J., Haddoud, M.H., Duan, P.: Word embeddings and convolutional neural network for arabic sentiment classification. In: The 26th International Conference on Computational Linguistics: Proceedings of COLING 2016, Technical Papers, pp. 2418–2427 (2016)
7. Eisele, A., Chen, Y.: MultiUN: a multilingual corpus from united nation documents. In: Tapias, D., Rosner, M., Piperidis, S., Odjik, J., Mariani, J., Maegaard, B., Choukri, K., Chair, N.C.C. (eds.) Proceedings of the Seventh Conference on International Language Resources and Evaluation, pp. 2868–2872. European Language Resources Association (ELRA), May 2010
8. Hasan, A.M., Zakaria, L.Q.: Question classification using support vector machine and pattern matching. J. Theor. Appl. Inf. Technol. 87(2), 259–265 (2016)
9. Hearst, M.A., Dumais, S.T., Osuna, E., Platt, J., Scholkopf, B.: Support vector machines. IEEE Intell. Syst. Appl. 13(4), 18–28 (1998)
10. Huang, Z., Thint, M., Qin, Z.: Question classification using head words and their hypernyms. In: Proceedings of the Conference on Empirical Methods in Natural Language Processing, 25–27 October (Sat-Mon) in Waikiki, Honolulu, Hawaii, pp. 927–936. Association for Computational Linguistics (2008)
11. Kim, Y.: Convolutional neural networks for sentence classification. arXiv preprint arXiv:1408.5882 (2014)
12. Kingma, D., Ba, J.: Adam: A method for stochastic optimization. arXiv preprint arXiv:1412.6980 (2014)
13. Li, X., Roth, D.: Learning question classifiers. In: Proceedings of the 19th International Conference on Computational Linguistics-Volume 1, Taipei, Taiwan, 24 August–01 September, pp. 1–7. Association for Computational Linguistics (2002)
14. Mikolov, T., Sutskever, I., Chen, K., Corrado, G., Dean, J.: Distributed representations of words and phrases and their compositionality. In: Proceedings of the 26th International Conference on Neural Information Processing Systems - Volume 2, NIPS 2013, pp. 3111–3119. Curran Associates Inc., USA (2013). http://dl.acm.org/citation.cfm?id=2999792.2999959
15. Nair, V., Hinton, G.E.: Rectified linear units improve restricted boltzmann machines. In: Proceedings of the 27th International Conference on Machine Learning (ICML-2010), pp. 807–814 (2010)
16. Nyberg, E., Mitamura, T., Carbonnell, J., Callan, J., Collins-Thompson, K., Czuba, K., Duggan, M., Hiyakumoto, L., Hu, N., Huang, Y.: The JAVELIN question-answering system at TREC 2002. NIST SPEC. PUBL. SP 251, 128–137 (2003)
17. Ray, S.K., Singh, S., Joshi, B.P.: A semantic approach for question classification using WordNet and Wikipedia. Pattern Recognit. Lett. 31(13), 1935–1943 (2010)
18. Salton, G., Wong, A., Yang, C.S.: A vector space model for automatic indexing. Commun. ACM 18(11), 613–620 (1975). https://doi.org/10.1145/361219.361220
19. Scherer, D., Müller, A., Behnke, S.: Evaluation of pooling operations in convolutional architectures for object recognition. In: Diamantaras, K., Duch, W., Iliadis, L.S. (eds.) ICANN 2010. LNCS, vol. 6354, pp. 92–101. Springer, Heidelberg (2010). https://doi.org/10.1007/978-3-642-15825-4_10

20. Silva, J., Coheur, L., Mendes, A.C., Wichert, A.: From symbolic to sub-symbolic information in question classification. Artif. Intell. Rev. **35**(2), 137–154 (2011)
21. Skilling, J.: Maximum Entropy and Bayesian Methods. Springer Science & Business Media, Netherlands (1988)
22. Vapnik, V.: The Nature of Statistical Learning Theory. Springer science & business media, New York (2013)

A Supervised Learning Approach for ICU Mortality Prediction Based on Unstructured Electrocardiogram Text Reports

Gokul S. Krishnan[(✉)] and S. Sowmya Kamath

Department of Information Technology, National Institute of Technology Karnataka,
Surathkal 575025, India
gsk1692@gmail.com, sowmyakamath@nitk.edu.in

Abstract. Extracting patient data documented in text-based clinical records into a structured form is a predominantly manual process, both time and cost-intensive. Moreover, structured patient records often fail to effectively capture the nuances of patient-specific observations noted in doctors' unstructured clinical notes and diagnostic reports. Automated techniques that utilize such unstructured text reports for modeling useful clinical information for supporting predictive analytics applications can thus be highly beneficial. In this paper, we propose a neural network based method for predicting mortality risk of ICU patients using unstructured Electrocardiogram (ECG) text reports. Word2Vec word embedding models were adopted for vectorizing and modeling textual features extracted from the patients' reports. An unsupervised data cleansing technique for identification and removal of anomalous data/special cases was designed for optimizing the patient data representation. Further, a neural network model based on Extreme Learning Machine architecture was proposed for mortality prediction. ECG text reports available in the MIMIC-III dataset were used for experimental validation. The proposed model when benchmarked against four standard ICU severity scoring methods, outperformed all by 10–13%, in terms of prediction accuracy.

Keywords: Unstructured text analysis
Healthcare analytics
Clinical Decision Support Systems · Word2Vec · NLP
Machine Learning

1 Introduction

Identifying individuals who are at risk of death while admitted to hospital Intensive Care Units (ICUs) is a crucial challenge facing critical care professionals. Extensive and continuous clinical monitoring of high-risk patients is often required, which, given the limited availability of critical care personnel

M. Silberztein et al. (Eds.): NLDB 2018, LNCS 10859, pp. 126–134, 2018.
https://doi.org/10.1007/978-3-319-91947-8_13

and equipment, is very expensive. Several mortality risk scoring systems are in use currently in ICUs that rely on certain patient-specific diagnostic and physiological factors identified by medical experts, extracted from structured health records (EHRs), to calculate mortality risk. However, studies have reported that their performance in actual prediction is quite low when compared to more recent non-parametric models based on data mining and Machine Learning (ML) [4,18]. As structured EHRs are put together manually with extensive human effort, a lot of context information contained in clinician's notes might be lost [17]. Another significant issue is the limited adoption rate of structured EHRs in developing countries, thus necessitating the use of alternate methods to obtain patient-specific information [22]. Most existing Clinical Decision Support System (CDSS) applications [4,10,18] depend on the availability of clinical data in the form of structured EHRs. However, in developing countries, clinical experts and caregivers still rely on unstructured clinical text notes for decision making. Unstructured clinical notes contain abundant information on patients' health conditions, physiological values, diagnoses and treatments, are yet to be explored for predictive analytics applications like mortality risk prediction and disease prediction. Such unstructured clinical text represent a significant volume of clinical data, which has remained largely unexploited for building predictive analysis models. Big data analytics and ML can help in developing better CDSSs with significant man-hour and medical resource savings [2].

Traditional methods for computing ICU mortality comprise of the standard severity scoring systems currently in use in hospitals. APACHE (Acute Physiology And Chronic Health Evaluation) [11] and SAPS (Simplified Acute Physiological Score) [6], along with their different variants; SOFA (Sequential Organ Failure Assessment) [21] and OASIS (Oxford Acute Severity of Illness Score) [8] are popular severity scoring models used for computing an ICU patient's mortality score using their clinical data. Other approaches [4,5,10,16,18] focus on application of ML techniques like decision trees, neural networks and logistic regression to structured EHR data for predicting mortality scores. Free text clinical notes written by medical personnel possess a significant volume of patient-specific knowledge, that is expressed in natural language. Several researchers proposed methods [1,14,19,23] for making use of such data for various purposes like data management, patient record classification and event prediction using ML, Hidden Markov Models, genetic algorithm and several other natural language processing (NLP) and data mining techniques.

In this paper, an ICU Mortality prediction model that utilizes patients' unstructured Electrocardiogram (ECG) text reports is proposed. We adopt the Word2Vec word embedding models for vectorizing and modeling the syntactic and semantic textual features extracted from these reports. An unsupervised data cleansing technique designed for identifying and removing anomalous data and special cases is used for optimizing the data representation. Further, a neural network architecture called Extreme Learning Machine (ELM) is trained on the ECG data for mortality risk prediction. The proposed model when benchmarked against existing traditional ICU severity scoring methods, SAPS-II,

SOFA, OASIS and APS-III, achieved an improved accuracy of 10–13%. The remainder of this paper is organized as follows: In Sect. 2, we describe the process of designing the proposed prediction model. Section 3 presents results of the experimental validation, followed by conclusion and references.

2 Proposed ICU Mortality Prediction Model

The methodology adopted for the design of the proposed mortality prediction model is composed of several processes, which are depicted in Fig. 1.

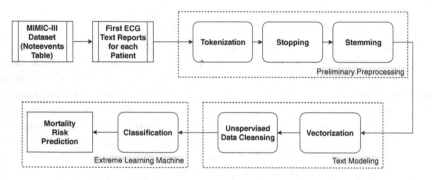

Fig. 1. Proposed methodology

2.1 Dataset and Cohort Selection

For the proposed model, unstructured text data from an open and standard dataset called MIMIC-III [9] was used. MIMIC-III (Medical Information Mart for Intensive Care III) consists of deidentified health data of 46,520 critical care patients. Clinical text records of these patients are extracted from the 'noteevents' table in the MIMIC-III dataset, from which only the ECG text reports are selected. Currently, we have considered only the first ECG report of each patient,

Table 1. ECG text Corpus statistics

(a) ECG Text Corpus Statistics

Feature	Total Number
Reports	34159
Sentences	108417
Total Words	802902
Unique Words	33748

(b) Statistics of the selected patient cohorts

Set	Total	Alive	Expired
Initial ECG Text reports	34159	30464	3695
Cluster C_1	22974	20372	2602
Cluster C_2	11185	10092	1093
Final Cohort	21465	20372	1093
Training & Test sets	10155	8068	2087
Validation set	2539	2024	515

as this is required to predict patients' mortality risk with the earliest detected condition, thereby predicting risk earlier. Next, the mortality labels of each patient are extracted from the 'patients' table in the dataset and are assigned to corresponding ECG reports of each patient. This set, now containing the first ECG text reports of 34,159 patients and the corresponding mortality labels, is used for the next phase (Details of ECG text corpus summarized in Table 1).

2.2   Preliminary Preprocessing

In the next phase, the ECG text corpus is subjected to a NLP pipeline consisting of tokenization, stopping and stemming. During tokenization, the clinical natural language text is split into smaller units called tokens. Generated tokens are filtered to remove unimportant terms (stop words) and finally, stemming is performed on the remaining tokens for suffix stripping. After the initial preprocessing, the tokens are next processed for modeling any latent clinical concepts effectively, during the Text Modeling phase.

2.3   Text Modeling

The Text Modeling phase consists of two additional levels of processing - Vectorization and Unsupervised Data Cleansing, which are discussed in detail next.

Vectorization: NLP techniques are critically important in a prediction system based on unstructured data, for generating machine processable representations of the underlying text corpus. Traditional rule and dictionary based NLP techniques, though perform well for certain applications, are not automated and require significant manual effort in tailoring them for various domains. Recent trends in ML and Deep Learning models and their usage in addition to traditional NLP techniques provide a good avenue for exploiting their performance for improved prediction. However, the effectiveness and performance of such models depend heavily on the optimized vector representations of the underlying text corpus. Several approaches have been developed for creating meaningful vector representations from text corpus, the prominent ones being Document Term Frequency vectorization and Term frequency-Inverse document frequency (Tf-Idf) Vectorization [20]. Word2Vec [15], a word embedding model, is an effective approach for generating semantic word embeddings (features) from unstructured text corpus. The generated vectors may be of several hundred dimensions, where unique terms in the text corpus are represented as a vector in the feature space such that corpus terms of similar context are closer to each other [15]. For modeling such latent concepts in the ECG text report corpus, we employed Word2Vec to generate a word embeddings matrix, which consists of the syntactic and semantic textual features obtained from the unstructured ECG corpus. The skip-gram model of Word2Vec was chosen over Continuous Bag-Of-Words (CBOW), due to its effectiveness with infrequent words and also as the order of words is important in the case of clinical reports [15]. We used a standard dimension size of 100, i.e., each ECG report is represented using a 1×100 vector, thus resulting in a final matrix of dimension 34159×100, each row representing the latent concepts in the ECG report of a specific patient.

Unsupervised Data Cleansing: The vectorized ECG text corpus data is next subjected to an additional process of data cleansing, for identifying special case data points and conflicting records. For this, K-Means Clustering was applied on the vectorized data to cluster the data into two clusters ($k = 2$, as the proposed prediction model is a two-class prediction, 'alive' and 'expired' patients) after which a significant overlap was observed in the two clusters. Cluster C_1 contained records of 20372 alive and 2602 expired patients while cluster C_2 had 10092 alive and 1093 expired patients. As a significant number of the data points representing 'alive' patients were in cluster C_1, we derived a reduced patient cohort that consists of all 'alive' patients from cluster C_1 and all 'expired' patients from cluster C_2, which were then considered for building the prediction model. The remaining patient data points exhibited anomalies due to existence of patients who might have expired due to causes not related to heart. Further examination of these special or conflicting cases revealed a requirement for considerable auxiliary analysis, therefore, we intend to consider them as part of our future work. In summary, the new patient cohort now consisted of 20372 alive and 1093 expired patients (tabulated in Table 1b).

2.4   Linguistics Driven Prediction Model

The patient cohort obtained after the data cleansing process is now considered for building the prediction model. Towards this, we designed a neural network model that is based on a fast learning architecture called Extreme Learning Machine (ELM) [7]. ELM is a single hidden layer Feedforward Neural Networks (SLFNN) where the parameters that fire the hidden layer neurons don't require tuning [7]. The hidden nodes used in ELM fire randomly and learning can be carried out without any iterative tuning. Essentially, the weight between the hidden and output layers of the neural network is the only entity that needs to be learned, thus resulting in an extremely fast learning model. Different implementations of ELMs have been used for tasks like supervised and unsupervised learning, feature learning etc., but to the best of our knowledge, ELMs have not been applied to unstructured clinical text based prediction models. In this SLFNN architecture, we set the number of nodes in the input layer to 100 as the feature vectors obtained after Word2Vec modeling are of similar dimensions. The hidden layer consists of 50 nodes and a single node is used at the output layer, to generate the predicted mortality risk of a patient. The Rectified Linear Unit (ReLU) activation function was used in the layers of the proposed ELM architecture as it is a step function and works well for binary classification. During training, the weights between the hidden and output layers are iteratively learned and optimized. Finally, the patient-specific mortality prediction is obtained at the output layer.

3 Experimental Results

For validating the proposed prediction model, an extensive benchmarking exercise was carried out. The experiments were performed using a server running Ubuntu Server OS with 56 cores of Intel Xeon processors, 128 GB RAM, 3 TB Hard Drive and two NVIDIA Tesla M40 GPUs. The patient cohort is split into training, test and validation sets (as shown in Table 1b). The vectorized feature vectors and the respective mortality labels in the training dataset are used for training the ELM model. We used 10-fold cross validation with a training to test data ratio of 75:25. Standard metrics like Accuracy, Precision, Recall, F-score and Area Under Receiver Operating Characteristic Curve (AUROC) were used for performance evaluation of the proposed model. Additionally, Matthews Correlation Coefficient (MCC) was also used as a metric, as it takes into account true positives, false positives and false negatives, therefore, regarded as a balanced measure even in presence of class imbalance [3]. We also benchmarked the performance of the proposed prediction model against well established, traditional parametric severity scoring methods. Four popular scoring systems - SAPS-II, SOFA, APS-III and OASIS, were chosen for this comparison. We implemented and generated the respective scores for each patient in the validation set. For SAPS-II, the mortality probability was calculated as per the process proposed by Le et al. [13]. In case of SOFA, the mortality prediction of each patient can be obtained by regressing the mortality on the SOFA score using a main-term logistic regression model similar to Pirracchio et al. [18], whereas for APACHE-III (APS III), it is calculated for each patient as per Knaus et al.'s method [12]. The mortality probability for each patient as per OASIS scoring system is given by the in-hospital mortality score calculation defined by Johnson et al. [8]. A classification threshold of 0.5 was considered for SAPS-II, APS-III and OASIS.

The validation patient data is fed to the trained model for prediction and its performance was compared to that of traditional scoring methods. The results are tabulated in Table 2, where, it is apparent that the proposed model achieved a significant improvement in performance over all four traditional scores. The proposed model predicted high mortality risk (label 1) correctly for most patients belonging to 'expired' class, which is a desirable outcome expected out of this

Table 2. Benchmarking ELM model against traditional severity methods SAPS-II, SOFA, OASIS and APS-III

Models	Accuracy	Precision	Recall	F-Score	AUROC	MCC
ELM *(Proposed)*	0.98	0.98	0.98	0.98	0.99	0.84
SAPS-II	0.86	0.87	0.86	0.86	0.80	0.34
SOFA	0.88	0.86	0.88	0.85	0.73	0.22
APS-III	0.89	0.86	0.89	0.86	0.79	0.26
OASIS	0.88	0.86	0.89	0.86	0.77	0.26

CDSS, which is also evident in the high precision values achieved. AUROC, F-Score and MCC are very relevant metrics for this experiment as the data exhibits class imbalance (number of patients in 'alive' class much greater those in 'expired' class; see Table 1b). MCC, which measures the correlation between the actual and predicted binary classifications, ranges between -1 and $+1$, where $+1$ represents perfect prediction, 0 indicates random prediction and -1 indicates total disagreement between actual and predicted values. The high values of F-Score and MCC for the proposed model in contrast to the others, indicates that, regardless of class imbalance in the data, the proposed model was able to achieve a good quality classification for both alive (0) and expired (1) labels. The plot of Receiver Operating Characteristic (ROC) curves generated for all models considered for comparison is shown in Fig. 2. Again, it is to be noted that the proposed model showed a substantial improvement of nearly 19% in AUROC in comparison to the best performing traditional model, SAPS-II.

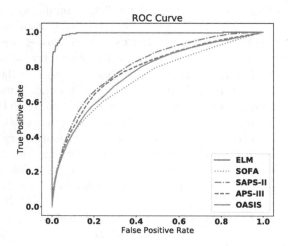

Fig. 2. Comparison of AUROC performance of the various models

4   Conclusion and Future Work

In this paper, a CDSS model for ICU mortality prediction from unstructured ECG text reports was presented. Word2Vec was used to model the unstructured text corpus to represent patient-specific clinical data. An unsupervised data cleansing process was used to handle conflicting data or special cases which represent anomalous data. A neural network architecture based on Extreme Learning Machines was used to design the proposed model. When benchmarked against popular standard severity scoring systems, the proposed model significantly outperformed them by 10–13% in terms of accuracy.

This work is part of an ongoing project with an objective of effectively using unstructured clinical text reports such as nursing/diagnostic notes and other lab test reports for improved decision-making in real-world hospital settings. Currently, our focus is using only a certain subset of the data for training and validation, however, in future, we intend to develop a prediction model that can perform well even in the presence of anomalous data. Further, we also intend to explore the suitability of deep learning architectures for unsupervised feature modeling for optimal corpus representation and improved prediction accuracy.

Acknowledgement. We gratefully acknowledge the use of the facilities at the Department of Information Technology, NITK Surathkal, funded by Govt. of India's DST-SERB Early Career Research Grant (ECR/2017/001056) to the second author.

References

1. Barak-Corren, Y., et al.: Predicting suicidal behavior from longitudinal electronic health records. Am. J. Psychiatry **174**(2), 154–162 (2016)
2. Belle, A., et al.: Big data analytics in healthcare. BioMed Res. Int. **2015**, 16 (2015)
3. Boughorbel, S., et al.: Optimal classifier for imbalanced data using matthews correlation coefficient metric. PloS one **12**(6), e0177678 (2017)
4. Calvert, J., et al.: Using EHR collected clinical variables to predict medical intensive care unit mortality. Ann. Med. Surg. **11**, 52–57 (2016)
5. Clermont, G., et al.: Predicting hospital mortality for patients in the ICU: a comparison of artificial neural networks with logistic regression models. Crit. Care Med. **29**(2), 291–296 (2001)
6. Gall, L., et al.: A simplified acute physiology score for ICU patients. Crit. Care Med. **12**(11), 975–977 (1984)
7. Huang, G., et al.: Trends in extreme learning machines: a review. Neural Netw. **61**, 32–48 (2015)
8. Johnson, A.E., et al.: A new severity of illness scale using a subset of acute physiology and chronic health evaluation data elements shows comparable predictive accuracy. Crit. Care Med. **41**(7), 1711–1718 (2013)
9. Johnson, E.W., et al.: MIMIC-III, a freely accessible critical care database. Sci. Data **3**, 1–9 (2016). 160035
10. Kim, S., et al.: A comparison of ICU mortality prediction models through the use of data mining techniques. Healthc. Inform. Res. **17**(4), 232–243 (2011)
11. Knaus, W.A., et al.: Apache-a physiologically based classification system. Crit. Care Med. **9**(8), 591–597 (1981)
12. Knaus, W.A., et al.: The APACHE III prognostic system: risk prediction of hospital mortality for critically ill hospitalized adults. Chest **100**(6), 1619–1636 (1991)
13. Le Gall, J., et al.: A new simplified acute physiology score (SAPS II) based on a European/North American multicenter study. JAMA **270**(24), 2957–2963 (1993)
14. Marafino, D., et al.: N-gram support vector machines for scalable procedure and diagnosis classification, with applications to clinical free text data from the intensive care unit. JAMIA **21**(5), 871–875 (2014)
15. Mikolov, T., Chen, K., et al.: Efficient estimation of word representations in vector space. arXiv preprint arXiv:1301.3781 (2013)
16. Nimgaonkar, A., et al.: Prediction of mortality in an Indian ICU. Intensive Care Med. **30**(2), 248–253 (2004)

17. O'malley, A.S., Grossman, J.M., et al.: Are electronic medical records helpful for care coordination? Experiences of physician practices. J. Gen. Internal Med. **25**(3), 177–185 (2010)
18. Pirracchio, R., et al.: Mortality prediction in intensive care units with the super ICU learner algorithm (SICULA): a population-based study. Lancet Respir. Med. **3**(1), 42–52 (2015)
19. Poulin, S., et al.: Predicting the risk of suicide by analyzing the text of clinical notes. PloS one **9**(1), e85733 (2014)
20. Salton, G., Buckley, C.: Term-weighting approaches in automatic text retrieval. Inf. Process. Manag. **24**(5), 513–523 (1988)
21. Vincent, J.L., et al.: The SOFA score to describe organ dysfunction/failure. Intensive Care Med. **22**(7), 707–710 (1996)
22. Williams, F., Boren, S.: The role of the electronic medical record (EMR) in care delivery development in developing countries: a systematic review. J. Innov. Health Inform. **16**(2), 139–145 (2008)
23. Yi, K., Beheshti, J.: A hidden Markov model-based text classification of medical documents. J. Inf. Sci. **35**(1), 67–81 (2009)

Multimodal Language Independent App Classification Using Images and Text

Kushal Singla[✉], Niloy Mukherjee, and Joy Bose

Samsung R&D Institute, Bangalore, India
{kushal.s,niloy.m,joy.bose}@samsung.com

Abstract. There are a number of methods for classification of mobile apps, but most of them rely on a fixed set of app categories and text descriptions associated with the apps. Often, one may need to classify apps into a different taxonomy and might have limited app usage data for the purpose. In this paper, we present an app classification system that uses object detection and recognition in images associated with apps, along with text based metadata of the apps, to generate a more accurate classification for a given app according to a given taxonomy. Our image based approach can, in principle, complement any existing text based approach for app classification. We train a fast RCNN to learn the coordinates of bounding boxes in an app image for effective object detection, as well as labels for the objects. We then use the detected objects in the app images in an ensemble with a text based system that uses a hierarchical supervised active learning pipeline based on uncertainty sampling for generating the training samples for a classifier. Using the ensemble, we are able to obtain better classification accuracy than if either of the text or image systems are used on their own.

Keywords: User modelling · App classification · Object recognition
Object detection

1 Introduction

User modelling and user interest computation, based on the user's profile and activity on mobile devices, is important to provide personalized services to users. One of the significant indicators of user interest, that is not so well studied, is the usage of mobile applications on the device. Data collected on mobile app usage is very high–dimensional, since a typical user would have multiple devices, each device having hundreds of apps and the user using these apps multiple times every day. Assigning coarse-grained categories to the apps helps to avoid the effects of the curse of dimensionality and hence can make it feasible to perform the user modelling. These app categories are typically hierarchical. For example, a flight booking app such as Skyscanner can be under the Flights category, which itself might be under another, larger, category such as Travel. Also, one app can have multiple categories. The app categories provided by the app stores such as Google Play store or the Apple app store are often labelled manually by app developers or by app store owners, and may not be accurately descriptive of the true nature of the app or context in which the app is used

© Springer International Publishing AG, part of Springer Nature 2018
M. Silberztein et al. (Eds.): NLDB 2018, LNCS 10859, pp. 135–142, 2018.
https://doi.org/10.1007/978-3-319-91947-8_14

by most users. Also, such categories may not match the categories in a taxonomy used for user interest computation, for a service seeking to provide customized recommendations to the user (Fig. 1).

Many solutions for automatic app classifications exist, but they mainly use text based data, such as metadata associated with the app, or app usage logs, as a way to classify the apps. In this paper, we argue that using app images is a better way to understand the use that a given app might be intended for. Images are independent of language, and so can be also used to classify apps with text descriptions in a different language. Also, many popular apps do not have adequate or accurate text descriptions. Our approach comprises teaching the classifier to learn to detect the objects in images as well as learn to label the objects, on the basis of optimal manual tagging and object labelling in images in the training set.

The rest of this paper is structured as follows: in the following section we survey related work in the areas of app classification. Section 3 describes our image based approach. In Sect. 4, we present the results of experiments to measure the accuracy of our method to classify apps. Section 5 concludes the paper.

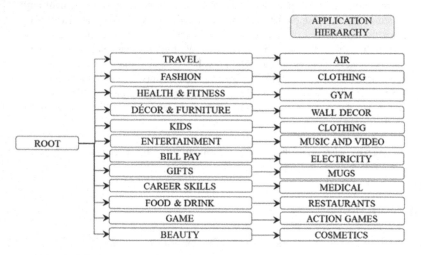

Fig. 1. Illustration of a hierarchical taxonomy for mobile applications.

2   Related Work

2.1   Related Work in the Area of Automatic App Classification

There are a number of works related to the field of automatic app classification. Zhu et al. [1, 2] have created app taxonomies based on the real world context learnt from the user's online behavior, such as URLs browsed or searches in a search engine. A similar approach was employed by Shevale [8]. Lindorfer [3] also built a system to classify apps, although their motivation was to analyze unknown apps for security considerations such as risk of malware. Padmakar [4] similarly calculated a risk score for unknown apps. Seneviratne [5] developed a system to identify spam apps using app metadata. Olabenjo [6] used a Naïve Bayes model to automatically classify apps on the Google play store based on app metadata, using existing app categories to learn categories of new apps. Radosavljevic et al. [7] used a method to classify unlabeled apps using a neural model trained on smartphone app logs.

However, as mentioned earlier, all the above models are text based. They do not use app images to increase the accuracy of the app classification. As such, they suffer from lower recall in cases where the app descriptions are short or incorrect.

2.2 Related Work in the Area of Automatic Object Detection in Images

Object detection is a computer vision technique which uses convolutional neural networks to identify the semantic objects in an image. Usilin et al. [9] discussed the application of object detection for image classification. We use a similar approach here in this paper, except that we apply the technique for app classification in order to complement, or improve the accuracy of text only based app classification methods.

3 System Overview

In this section, we describe our approach of an image based app classifier. We first explore the use of object detection for application classification.

3.1 Automatic Object Detection in Images

Our approach to object detection is based on the Fast Region based Convolutional neural network (Fast RCNN) developed by Girshick [10, 11]. We chose this model since it gave near real time speeds, taking only a few seconds for each inferencing, with reasonable accuracy of classification.

For training the Fast-RCNN model, we input images with manually labelled bounding boxes. While learning, the model generates a set of object box proposals using the selective search method to detect a large number of objects bounding box locations independent of the objects to be detected. The object proposals are fed into a single CNN which generates the object classification of each region. The output detection could cover same object with many Regions of Interest (ROIs), so we filter the output iteratively to select the ROI's with best confidence and discard the rest of them. We used the Microsoft Cognitive Toolkit (CNTK) library for image detection model training. It is observed that the apps from multiple languages might be available in a region. Therefore, building a text based app classifier for such regions is not a scalable approach, since we need to understand the language and getting sufficient apps for each language is a challenge.

3.2 Use of Object Detection in App Images for Application Classification

Therefore, we use the language agnostic object detection technique to identify the semantic objects in an image related to the app classes.

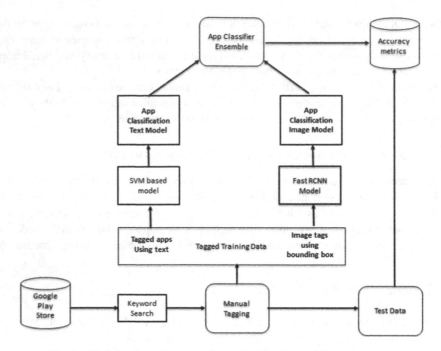

Fig. 2. High level architecture of the classifier for mobile applications.

The steps of our proposed approach are as follows:

- Manually label a number of app images belonging to a fixed number of categories to establish the ground truth.
- The labelled images are fed to the Fast-RCNN model pipeline for training.
- Build classifier based on objects detected in app images during inferencing.
- At the time of inferencing, for each app, get all the images of the app and label the objects detected in the image.
- Label the app class using a modified Borda count method (Fig. 2).

3.3   Modified Borda Count Method

In our approach, the modified Borda count method is used to label the app based on the object detection model result. Each image is seen a voter. We label each image to an app class based on the maximum count of the app class objects in the images. For the app classification, each image acts as voter and gives a vote of 1 to its app label. The class with the maximum amount of votes is assigned as the final label of the app. The algorithm to decide the final image classification label is described below.

ALGORITHM: Ranking method to determine the final app label

---

1. Input:
    A_I = Apps for inferencing where 1<=I<=n
    I_{IK} = K^{th} image for I^{th} App where 1 <=K<=m
    B_{IKJ} = J^{th} bounding box for the Kth image of the Ith App
    L_{IKJ} = Label for the Jth bounding box for the Kth image of the Ith App
2. Output:
    A_I^L = App label for the app where 1<=I<=n
    A_{IK}^L = Label of the Kth image of the app where 1<=K<=m
3. For each App A_I
4.   For each Image I_{IK}
5.   For each bounding box B_{IKJ}
6.   Fetch the label L_{IKJ}
7.   Label the image(A_{IK}^L) with the label of the maximum bounding boxes.
8.   Label(A_I^L) the app(A_I) with the label of max number of images.
9. *exit*: end procedure

3.4   Ensemble Based App Classification

Next, we build an ensemble of text and image based classifiers, which gave better performance as compared to the individual systems. The ensemble is created using the below logic:

$$CLF^T(x) = LABEL^T, CONF^T$$
$$CLF^I(x) = LABEL^I, CONF^I$$
$$CLF^{ENB}(x) = \{LABEL^I, \text{ if } LABEL^I == LABEL^T \text{ and } LABEL^I \, !=\text{"OTHERS"}$$
$$\{LABEL^T, \text{ if } LABEL^T \, !=\text{"OTHERS"}$$
$$\{LABEL^I, \text{ if } LABEL^I \, !=\text{"OTHERS"}$$
$$\{ \text{"OTHERS", otherwise}$$

Where
x: the app to be classified, $CLF^{T:}$ the text based classifier, $LABEL^{T:}$ the classification label based on the text based classifier, $CONF^{T:}$ the classification confidence based on the text based classifier, $CLF^{I:}$ the image based classifier, $LABEL^{I:}$ the classification label based on the image based classifier, $CONF^{I:}$ the classification confidence based on the image based classifier, $CLF^{ENB:}$ the ensemble based classifier.

4 Experimental Setup and Results

For our experiment, we used a dataset based on selected India region apps. Our infrastructure setup consists of two machines. One machine is used to store the application data using Elastic Search, while the second one is used for app classification.

In order to evaluate our approach, we classified the apps using (a) only the text based metadata associated with the apps, (b) using only app images and (c) using an ensemble of text and image based approaches. The following subsections describe the classification results we obtained using each of the three approaches.

4.1 Text Based App Classification Results

We built a multiclass hierarchical model to classify apps based on the textual metadata. We trained the model on 14 categories, using the one Vs rest approach for each category as follows: we manually labelled the apps as either belonging to a selected category, or not belonging to that category, and then trained the model to learn whether the app belonged to the category or not. We repeated this process for each of the 14 categories. Quality evaluation is performed using 5-fold cross-validation. This method gave encouraging results. For cross-checking, manual evaluation is performed as well. In the manual evaluation, the classifier is executed and 30 apps are chosen randomly for testing which are manually evaluated. Finally, the precision is computed for each class using this method. The result of the text only evaluation are shared in Table 2 for 4 categories.

Table 2. Comparison of classification accuracy for text, image and ensemble

App Category	Image based Classifier	Text based Classifier	Ensemble Classifier	Support
Beauty	0.65	0.72	0.76	290
Décor	0.72	0.78	0.83	126
Fashion	0.58	0.79	0.79	240
Food	0.80	0.87	0.89	454

4.2 Images Based App Classification

We first manually tagged 3048 images from 339 apps for 4 categories, along with a background class for images that do not fit into any category, to establish the ground truth. We used a test set of 339 random apps from four classes (food, décor, beauty, fashion) which are a subset of the 14 app categories used previously in the text based classification. For 4K app images taken from these 339 apps, we used the Microsoft Visual Object Tagging Tool (VoTT) to draw and manually tag the bounding boxes for each unique object in the image. We assume here that each tagged object corresponds to one unique category only. We then used these manually tagged images to train a fast RCNN model, as described previously, for the object detection. We used 10 epochs of the GPU to train our model. Then we follow the steps described in Sect. 3.2 to get the app labels using the modified Borda count method, using each image as a voter and determining the most voted app labels as label for the entire app.

4.3   Accuracy of the Text Based, Image Based Classifier and Ensemble of Text and Image Classifiers

In this section, we provide the F1 scores showing the classification accuracy of the text only, image only and ensemble models for different app categories. We chose four categories of apps (beauty, décor, fashion and food).

Table 2 and the graph in Fig. 3(a) show the F1 scores for each classifier and each app category. As we can see from the graph, the F1 scores for the text based classifier are better than the image based one, across all categories. However, the ensemble classifier performs better than both image and text in all cases. This shows that having such an ensemble model can fix some of the issues with app classification based only on the text metadata of the app.

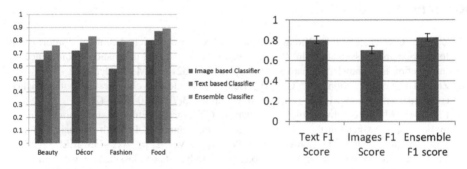

Fig. 3. (a) Classifier accuracy for different app categories. (b) Comparison of the text, image and ensemble classifier accuracy, plotted along with error bars.

The graph in Fig. 3(b) shows the weighted average of F1 scores, weighed as per the support for each app category, along with error bars for standard error, over the four categories for image, text and ensemble. Here also it is clear that the ensemble performs better than only image or only text, across apps from different categories.

4.4   Examples of Accurate App Classification Where a Text Only Model Cannot Work

After comparing the classification accuracy of the text based, image based and ensemble models, we performed a subjective test to illustrate cases where the utility of our model is visible more clearly. For this, we show some apps from the android play store having certain characteristics where a text based model alone cannot work. Table 3 shows examples of such apps. Using our ensemble approach, we could get accurate classification for all these cases of apps also.

Table 3. Examples of apps where the text only or image only classification does not work well

App characteristics	App examples
Apps having little or no text description	io.app.sigdi doorduck.com
App where confidence of the text classifier < 0.5	com.bse.cafedalalstreet
An app in a non-English language (e.g. an Indic language such as Bengali or Oriya)	com.andromo.dev293330.app466901
An app where text is right but image is wrong	com.dinein.in.android

5 Conclusion and Future Work

In this paper, we have presented an image based approach for app classification and shown it can improve the classification accuracy of a text based app classifier. In future, we will extend this model for more apps and aim to improve the accuracy.

References

1. Zhu, H., Cao, H., Chen, E., Xiong, H., Tian, J.: Exploiting enriched contextual information for mobile app classification. In: Proceedings of the 21st ACM international conference on Information and knowledge management, pp. 1617–1621. ACM, 29 Oct 2012
2. Zhu, H., Chen, E., Xiong, H., Cao, H., Tian, J.: Mobile app classification with enriched contextual information. IEEE Trans. Mob. Comput. 13(7), 1550–1563 (2014)
3. Lindorfer, M., Neugschwandtner, M., Platzer, C.: Marvin: Efficient and comprehensive mobile app classification through static and dynamic analysis. In: 39th Annual Computer Software and Applications Conference (COMPSAC), Vol. 2, pp. 422–433. IEEE (2015)
4. Lokhande, P.P., Shivaji, R.L.: A review on risk score based app classification using enriched contextual information of app context. Int. J. Comput. Sci. Inf. Technol. (IJCSIT), 5(6), 7063–7066 (2014)
5. Seneviratne, S., Seneviratne, A., Kaafar, M.A., Mahanti, A., Mohapatra, P.: Spam mobile apps: characteristics, detection, and in the wild analysis. ACM Trans. Web 11(1), 1–29 (2017). Article 4
6. Olabenjo, B.: Applying Naive Bayes Classification to Google Play Apps Categorization. Arxiv. https://arxiv.org/pdf/1608.08574.pdf
7. Radosavljevic, V., Grbovic, M.: Smartphone App categorization for interest targeting in advertising marketplace. In Proceedings of the WWW (2016)
8. Shewale, S.K., Gayakee, V.V., Ugale, P.D., Sonawane, H.D.: Personalized App service system algorithm for effective classification of mobile applications. Int. J. Eng. Tech. Res. (IJETR) 3(1), January 2015. ISSN 2321–0869
9. Usilin, S., Nikolaev, D., Postnikov, V., Schaefer, G.: Visual appearance based document image classification. In: Proceedings of 2010 IEEE ICIP (2010)
10. Girshick, R.:. Fast R-CNN. https://arxiv.org/abs/1504.08083
11. Microsoft Cognitive Toolkit Documentation. Object detection using Fast R-CNN, August 2017. https://docs.microsoft.com/en-us/cognitive-toolkit/object-detection-using-fast-r-cnn

T2S: An Encoder-Decoder Model for Topic-Based Natural Language Generation

Wenjie Ou, Chaotao Chen, and Jiangtao Ren[✉]

School of Data and Computer Science, Sun Yat-sen University, Guangzhou, China
{ouwj3,chencht3}@mail2.sysu.edu.cn, issrjt@mail.sysu.edu.cn

Abstract. Natural language generation (NLG) plays a critical role in various natural language processing (NLP) applications. And the topics provide a powerful tool to understand the natural language. We propose a novel topic-based NLG model which can generate topic coherent sentences given single topic or combination of topics. The model is an extension of the recurrent encoder-decoder framework by introducing a global topic embedding matrix. Experimental results show that our encoder can not only transform a source sentence to a representative topic distribution which can give a better interpretation of the source sentence, but also generate topic coherent and diversified sentences given different topic distribution without any text-level input.

Keywords: Natural language generation · Topic · Encoder-decoder

1 Introduction

Natural language generation (NLG) plays a critical role in various applications such as response generation in dialogue systems [17,20,21]. And the topics provide a powerful tool to understand the natural language, and make the natural language more stylize. Specifically, in this paper we study a novel NLG task called Topic-based natural language generation. Our goal is to generate topic coherent sentences given a topic distribution, which is beneficial for many real-world applications such as personalized conversation, review generation and poetry writing.

In the recurrent encoder-decoder structure (aka. Seq2Seq) [15,16], general sentence generation is usually based on sequence auto-encoder architecture [2,9]. However, these architectures cannot generate sentences when given only topic information which is more common in real applications, because they need a sentence as input. Moreover, the sentences they generated are always lack of diversity. To address these problems, we are motivated to modify current sequence auto-encoder model so that it can generate not only coherent sentences given a single topic but also diversified sentences through the combination of different topics.

© Springer International Publishing AG, part of Springer Nature 2018
M. Silberztein et al. (Eds.): NLDB 2018, LNCS 10859, pp. 143–151, 2018.
https://doi.org/10.1007/978-3-319-91947-8_15

Table 1. Examples of sentences based on topics "matlab" and "mac os" generated by T2S model trained on *StackOverflow* dataset. Topic-relevant words are in red.

Topic	Examples
1.0 matlab	matlab of vectors error
	matlab how to use regular in matrix
	matlab best way to find function in a matrix
1.0 osx	disk activity on mac os x
	sample mac os x on a windows external projects
	mac font rendering on windows in osx
0.7 matlab + 0.3 osx	looking for matlab on mac os x
	matlab in programming on mac
	an running on mac m in matlab

In our proposed Topics-to-Sentence (T2S) model, we explicitly introduce an global topic embedding matrix to represent different topics and capture the topic information in the source texts during training. The encoder and decoder are decoupled and T2S model can generate topic coherent sentences given only topic information. Moreover, it can also generate diversified sentences by designing different topic distributions. We propose to train the model in turn with a semi-supervised manner to overcome some difficult training problems.

Our contributions are summarized as follows: (1) We propose an extension of recurrent encoder-decoder model which can generate topic coherent sentences given topic distributions. (2) We discuss the challenges in training our model and propose to leverage a few topic labels to resolve the problem. (3) Results of extensive experiments demonstrate that our model can not only capture topic information in source texts, but also generate topic coherent sentences based on topic information. Table 1 illustrates some examples which is generated by our model based on a single topic or multiple topics.

2   Related Work

Recently, recurrent encoder-decoder architectures have proven to be well suited for various natural language generation tasks such as response generation [15, 16, 18–20]. Generally, both encoder and decoder are implemented by recurrent neural network (RNN), gated RNN such as Long Short-Term Memory [5] or Gated Recurrent Unit [4]. Specifically, general sentence generation usually employs sequence auto-encoder architecture [2, 9] However, such methods do not take the topic information into account so they are not suitable for topic-based natural language generation. [11] improves the performance of recurrent neural network language model by providing topic features learned by Latent Dirichlet Allocation [1]. NTM [3] combine topic model and neural network into a uniform framework. TWE [10] leverages topic information to help the training of words embedding. TA-Seq2Seq [22] employs topic model to produce topic-specific vocabulary and then generate responses in conversation based on the corresponding vocabulary. In contrast to these models, our model focuses on the task of generating topic coherent sentences given topic distributions.

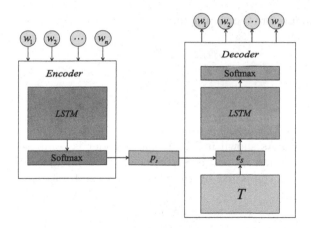

Fig. 1. Architecture of the Topic-to-Sentence model. (P_s: topic distribution; e_s: context vector; T: topic embedding matrix.)

3   Topic-Based Natural Language Generation

In this section, we introduce our Topic-to-Sentence (T2S) model which extends the conventional recurrent encoder-decoder framework for the task of topic-based natural language generation. The architecture of our T2S model is illustrated in Fig. 1.

3.1   Notation

We represent a sentence as a sequence of words $s = \{w_1, w_2, ..., w_{N_s}\}$ where N_s is the number of words in the sentence, and a sentence starts with a "<bos>" token and ends with a "<eos>" token. The word w is associated with a K-dimensional embedding $e_w = \{e_w^1, e_w^2, ..., e_w^K\}$. Let V denotes the total number of words in the vocabulary. $T \in N_T \times M$ denotes the global topic embedding matrix where N_T and M is the number and dimension of topic vectors, respectively.

3.2   Encoder

The function of the encoder in our T2S model is to transform the sentence into a topic distribution p_s. As illustrated in Fig. 1, the encoder consists of an LSTM layer and a *Softmax* layer. In the encoding process, we first input a sentence $s = \{w_1, w_2, ..., w_{N_s}\}$ into the LSTM layer and obtain the last hidden state h_{N_s} as sentence representation. Then, we add a linear layer to project sentence representation h_{N_s} into the space of N_T topics and then use a *Softmax* layer to convert real values to conditional topic distribution, which is calculated as follows.

$$h_{N_s} = LSTM_{enc}(s) \tag{1}$$

$$p_s = softmax(W_{enc}h_{N_s} + b_{ence}) \tag{2}$$

where W_{enc} and b_{enc} are the parameters of the linear layer and $softmax$ function is defined as follow.

$$softmax(x_j) = \frac{e^{x_j}}{\sum_j e^{x_j}} \tag{3}$$

3.3 Decoder

The decoder in our T2S model is used to generate topic coherent sentences given topic distributions. As shown in Fig. 1, the decoder is composed of an LSTM layer, a global topic matrix and a $Softmax$ layer. In the decoding process, the context vector e_s is computed by the product of the topic distribution p_s and the topic matrix T:

$$e_s = p_s \cdot T \tag{4}$$

where T is random initialized and need to be learned. Then the context vector is used to update the state of LSTM unit. In each time step t, the LSTM updates its hidden state as follow.

$$h_t = LSTM_{decode}(e_s, h_{t-1}, e_t) \tag{5}$$

Then h_t is used for word predicting through $Softmax$ layer as in the conventional encoder-decoder models:

$$p(w_{t+1}|e_s, w_{\leq t}) = softmax(W_{dec}h_t + b_{dec}) \tag{6}$$

where W_{dec} and b_{dec} are parameters used to compute a distribution over words. The reconstruction loss of a training sentence is defined as the negative log likelihood:

$$L_{rec} = -\sum_t log(p(w_{t+1}|e_s, w_{\leq t})) \tag{7}$$

In the generation process, we can handcraft different topic distributions p_s as the input of the decoder and generate diversified and topic coherent sentences based on the random sample method [12].

3.4 Optimization Challenges

Our model aims to learn a global topic matrix and transform a source sentence to its representative topic distribution. However, the straightforward implementation of our T2S model fails to achieve this object, because it tends to convert all source sentences to the same topic distribution. This problematic tendency in learning may be due to the LSTM decoder's sensitivity to subtle variation in the hidden states. This causes the model to initially learn to fix and ignore the topic distributions output by the encoder, and generate the target sentences only with the more easily optimized decoder. To overcome this problem, we propose to leverage a little topic labels to guide the learning of topic distributions and topic matrix. So our model is trained in a semi-supervised manner where some sentences is accompanied with their golden topic label. As for sentences with n

distinct topics, we replicate the sentence for n times but associate each instance with different topic. This ensures that all the topic labels can be utilized in the training process.

For an annotated sentence s, we further minimize its topic loss which is defined as the cross-entropy error between its gold topic distribution y_s and predicted topic distribution p_s:

$$L_{topic}(s) = - \sum_{c=1}^{N_T} y_s^c \cdot log(p_s^c) \qquad (8)$$

where y_s^c denotes the gold probability of topic c with ground truth being 1 and others being 0, p_s^c denotes the probability of topic c predicted by the encoder.

The T2S model is trained to minimize the summation of topic loss L_{topic} and reconstruction loss L_{rec} in annotated sentences and the reconstruction loss L_{rec} in unannotated sentences alternately. We use back propagation to calculate the gradients of all parameters, and update them with stochastic gradient descent.

4   Experiments

4.1   Datasets

We evaluate our proposed model on 2 datasets: *StackOverflow*[1] and *Restaurant* [14]. In the training process, only 10% training data is annotated with their topic labels. Sentences of *StackOverflow* containing only one topic, and *Restaurant* are online reviews covering multiple topics about restaurant. For more details of the datasets, please refer to their original papers.

4.2   Implementation Details

In both encoder and decoder, the dimension of hidden states in LSTM is set to 300. We initialize our word embeddings with publicly available 300-dimensional Glove vectors [13], which is trained on 840 billion tokens of Common Crawl data[2]. Words that do not exist in the pretrained Glove vectors are replaced by "<unk>" token. All the weight matrices and bias in the model are randomly initialized from uniform distribution $U(-0.01, 0.01)$. We implement our neural network model by TensorFlow[3]. We train the model with a mini-batch size of 64 examples, a maximum length of 32, and an initial learning rate of 0.001 for Adam method [6].

---

[1] Available   at   https://www.kaggle.com/c/predict-closed-questions-on-stack-overflow/download/train.zip.

[2] Pre-trained word vectors of Glove can be obtained from http://nlp.stanford.edu/projects/glove/.

[3] https://www.tensorflow.org/.

Table 2. Results of test set accuracy on *AG News*, *StackOverflow* and *TREC*. Best scores are in **bold** while results of our T2S model are in <u>underlined</u>.

Model	StackOverflow
RNN [8]	42.17
SkipVec(combine) [7]	9.74
SkipVec(bi) [7]	9.58
STC^2-LPI [23]	54.06
STC^2-LE [23]	53.93
T2S	**67.80**

Table 3. Examples of sentences and their topic distributions generated by T2S encoder on *Restaurant* data. The first block lists the results of single-topic sentences. The second block lists the results of sentences with multiple topics. The probabilities of the golden topics are in **Bold** and topic-relevant words are in red.

Sentence	service	food	anecdotes/ miscellaneous	price	ambience
the service is prompt friendly	**0.988**	0.009	0.001	0.001	0.001
i have no idea why this restaurant is overlooked	0.018	0.139	**0.823**	0.016	0.004
atmosphere is a bore	0.085	0.054	0.049	0.087	**0.726**
the place is clean and if you like soul food then this is the place to be	0.093	**0.480**	0.059	0.021	**0.346**
we enjoyed ourselves thoroughly and will be going back for the desserts	0.092	**0.486**	**0.363**	0.057	0.003
reasonably priced with very fresh sushi	0.016	**0.625**	0.071	**0.266**	0.021

4.3   Evaluation of Encoded Topic Distribution

In order to generate topic coherent sentences, the encoder in T2S is required to encode the source sentences into representative topic distributions. So we first evaluate the encoder's ability to generate meaningful topic distributions. For single-topic datasets, we use the encoded topic distribution as the result of classified task and evaluate the classification performance on the test set. The results of accuracy on *StackOverflow* summarized in Table 2. As we can see, our T2S model can achieve comparable classification performance compared with the state-of-the-art baselines including supervised methods on the *StackOverflow* dataset.

For the multi-topics *Restaurant* dataset, we illustrate some examples of sentences and their corresponding encoded topic distributions in Table 3. The probabilities of the golden topics and the words relevant to the topics are marked in bold and red, respectively. In the first block of Table 3, we can see that the T2S encoder can transform the single-topic sentences into very concentrated topic distributions. And the sentences consisting of multiple topics encoded topic distributions are consistent to their real topics. These results demonstrate that the

Table 4. Examples of sentences generated by T2S model trained on *Restaurant* dataset. Word relevant to topics are in red.

Topic	Examples
1.0 food	the portions are fresh and the oil was delicious
	the bread with the seafood is delicious
1.0 price	decent price of a nice deal
	a great choice at reasonable cost and a great deal
1.0 service	the staff and the waiters are nice
	besides the service was good
0.5 food +	great food with reasonable prices makes for dinner that can not be beat
0.5 price	excellent french food at a menu that a solid price
	decent wine at reasonable prices makes worth a great price
0.3 food +	fabulous service fantastic food and a very friendly waiter and a reasonable price
0.3 price +	delicious food at a great price but do not go here on a cold day and sit by the door
0.3 service	only would be good there they have most good thai food at a reasonable price

encoded topic distribution from T2S encoder can give a better interpretation to the source sentences.

4.4 Evaluation of Topic-Based Language Generation

In this subsection, we mainly evaluate our T2S decoder's ability to generate topic coherent sentences given topic distributions. We manually design different topic distributions p_s and input them into the decoder to generate relevant sentences. Examples produced by T2S model trained on *StackOverflow* and *Restaurant* are illustrated in Tables 1 and 4. As we can see in Table 4, T2S trained on multi-topic *Restaurant* dataset can also generate coherent reviews based on both single topic and multi-topics. For example, Even given a combination of three topics (0.33 food + 0.33 service + 0.33 price), the T2S decoder can also generate coherent reviews like "Fabulous service fantastic food and a very friendly service and a reasonable price".

These results demonstrate that our T2S model can generate more diversified and topic coherent sentences by specifically combining different topics.

5 Conclusion

In this paper, we extend an encoder-decoder models by introducing a topic matrix to the task of topic-based natural language generation, which can capture the topic information in the source texts. In the generation process, the decoder is decoupled from the encoder and we can generate topic coherent and diversified sentences by inputting different topic distributions into the decoder. We also leverage a little supervised information (i.e. topic labels) to overcome the difficulties in training process. Experiment results show that our proposed model can not only encode source sentences into representative topic distributions, but also generate topic coherent sentences given topic distributions, which can benefit many real world application such as personalized conversation and review generation.

References

1. Blei, D.M., Ng, A.Y., Jordan, M.I.: Latent dirichlet allocation. J. Mach. Learn. Res. **3**, 993–1022 (2003)
2. Bowman, S.R., Vilnis, L., Vinyals, O., Dai, A.M., Józefowicz, R., Bengio, S.: Generating sentences from a continuous space. In: Proceedings of CoNLL, pp. 10–21 (2016)
3. Cao, Z., Li, S., Liu, Y., Li, W., Ji, H.: A novel neural topic model and its supervised extension. In: Proceedings of AAAI, pp. 2210–2216 (2015)
4. Cho, K., van Merrienboer, B., Gülçehre, Ç., Bahdanau, D., Bougares, F., Schwenk, H., Bengio, Y.: Learning phrase representations using RNN encoder-decoder for statistical machine translation. In: Proceedings of EMNLP, pp. 1724–1734 (2014)
5. Hochreiter, S., Schmidhuber, J.: Long short-term memory. Neural Comput. **9**(8), 1735–1780 (1997)
6. Kingma, D.P., Ba, J.: Adam: A method for stochastic optimization. CoRR abs/1412.6980 (2014)
7. Kiros, R., Zhu, Y., Salakhutdinov, R., Zemel, R.S., Urtasun, R., Torralba, A., Fidler, S.: Skip-thought vectors. In: Proceeding of NIPS, pp. 3294–3302 (2015)
8. Lai, S., Xu, L., Liu, K., Zhao, J.: Recurrent convolutional neural networks for text classification. In: Proceedings of AAAI, pp. 2267–2273 (2015)
9. Li, J., Luong, M., Jurafsky, D.: A hierarchical neural autoencoder for paragraphs and documents. In: Proceedings of ACL, pp. 1106–1115 (2015)
10. Liu, Y., Liu, Z., Chua, T., Sun, M.: Topical word embeddings. In: Proceedings of AAAI, pp. 2418–2424 (2015)
11. Mikolov, T., Zweig, G.: Context dependent recurrent neural network language model. In: Proceedings of SLT, pp. 234–239 (2012)
12. Neubig, G.: Neural machine translation and sequence-to-sequence models: A tutorial. CoRR abs/1703.01619 (2017)
13. Pennington, J., Socher, R., Manning, C.D.: Glove: global vectors for word representation. In: Proceedings of EMNLP, pp. 1532–1543 (2014)
14. Pontiki, M., Galanis, D., Pavlopoulos, J., Papageorgiou, H., Androutsopoulos, I., Manandhar, S.: Semeval-2014 task 4: aspect based sentiment analysis. In: Proceedings of SemEval@COLING, pp. 27–35 (2014)
15. Serban, I.V., Sordoni, A., Lowe, R., Charlin, L., Pineau, J., Courville, A.C., Bengio, Y.: A hierarchical latent variable encoder-decoder model for generating dialogues. In: Proceedings of AAAI, pp. 3295–3301 (2017)
16. Shang, L., Lu, Z., Li, H.: Neural responding machine for short-text conversation. In: Proceedings of ACL, pp. 1577–1586 (2015)
17. Sordoni, A., Galley, M., Auli, M., Brockett, C., Ji, Y., Mitchell, M., Nie, J., Gao, J., Dolan, B.: A neural network approach to context-sensitive generation of conversational responses. In: Proceedings of NAACL-HLT, pp. 196–205 (2015)
18. Tran, V., Nguyen, L., Tojo, S.: Neural-based natural language generation in dialogue using RNN encoder-decoder with semantic aggregation. In: Proceedings of the 18th Annual SIGdial Meeting on Discourse and Dialogue, Saarbrücken, Germany, 15–17 August 2017. pp. 231–240 (2017)

19. Wen, T., Gasic, M., Kim, D., Mrksic, N., Su, P., Vandyke, D., Young, S.J.: Stochastic language generation in dialogue using recurrent neural networks with convolutional sentence reranking. In: Proceedings of SIGDIAL, pp. 275–284 (2015)
20. Wen, T., Gasic, M., Mrksic, N., Rojas-Barahona, L.M., Su, P., Vandyke, D., Young, S.J.: Multi-domain neural network language generation for spoken dialogue systems. In: Proceedings of NAACL-HLT, pp. 120–129 (2016)
21. Wen, T., Gasic, M., Mrksic, N., Su, P., Vandyke, D., Young, S.J.: Semantically conditioned LSTM-based natural language generation for spoken dialogue systems. In: Proceedings of EMNLP, pp. 1711–1721 (2015)
22. Xing, C., Wu, W., Wu, Y., Liu, J., Huang, Y., Zhou, M., Ma, W.: Topic aware neural response generation. In: Proceedings of AAAI, pp. 3351–3357 (2017)
23. Xu, J., Xu, B., Wang, P., Zheng, S., Tian, G., Zhao, J., Xu, B.: Self-taught convolutional neural networks for short text clustering. CoRR abs/1701.00185 (2017)

Ontology Engineering

Ontology Development Through Concept Map and Text Analytics: The Case of Automotive Safety Ontology

Zirun Qi[1(✉)] and Vijayan Sugumaran[2]

[1] J. Mack Robinson College of Business, Georgia State University,
Atlanta, GA 30302, USA
zqil@gsu.edu
[2] Center for Data Science and Big Data Analytics, Department of Decision,
and Information Sciences, Oakland University, Rochester MI 48309, USA
sugumara@oakland.edu

Abstract. Ontology development is an expensive and time-consuming process. The development of real-world organizational ontology-based knowledge management systems is still in early stages. Some existing ontologies with simple tuples and properties are not designed for domain specific requirement, or does not utilize existing knowledge from organizational database or documents. Here we propose our concept map approach to first semi-automatically create a detailed level entities/concepts as a keyword list by applying natural language processing, including word dependency and POS tagging. Then this list can be used to extract entities/concepts for the same domain. This approach is applied to automotive safety domain. The results are further mapped to existing ontology and aggregated to form a concept map. We implement our approach in KNIME with Stanford NLP parser and generate a concept map from automotive safety complaint dataset. The final results expand the existing ontology, and also bridge the gap between ontology and real-world organization ontology-based knowledge management systems.

Keywords: Ontology · Concept map · Text mining · Automotive safety
Knowledge management

1 Introduction

Knowledge Management (KM) has been an active area of research in the last three decades. Knowledge is one of the most important assets of organizations [14]. Organizational capabilities for the development, storage, sharing and effective use of knowledge have also been particularly recognized as a strategic differentiator among competing firms [16]. However, lack of mechanisms for harnessing the knowledge embedded in various documents and artifacts that are scattered throughout the organization is cited as a major limitation [18]. Ontologies have long been argued as one approach for capturing and representing domain knowledge. Several studies have been reported in the literature that tout the use of ontologies for enterprise knowledge management [19–21]. While several works on ontology learning and development have

© Springer International Publishing AG, part of Springer Nature 2018
M. Silberztein et al. (Eds.): NLDB 2018, LNCS 10859, pp. 155–166, 2018.
https://doi.org/10.1007/978-3-319-91947-8_16

been reported in the literature, there is no single methodology or tool that has emerged as a gold standard that can be used for automatically generating good quality ontologies.

Ontologies define the terminology of a domain by specifying the relevant hierarchical concepts and their relationships. They can easily include tens or hundreds of thousands of concepts and are both expensive and time-consuming to develop. Ontology creation is the process of automatically or semi-automatically constructing ontologies on the basis of textual domain descriptions. The assumption is that the domain text reflects the terminology that should go into an ontology, and that appropriate linguistic and statistical methods should be able to extract the appropriate concept candidates and their relationships from these texts. Numerous approaches for ontology creation have been proposed in the past [9–13], however, it has not been possible to fully automate the generation of ontologies. Ontology creation toolsets may generate candidate concepts and relationships, but human labor is needed to verify the suggestions and complete the ontologies.

Ontology creation is an ongoing activity with considerable effort required in different domains and industries. An ontology usually has a formal representation, with names and definition of the types, properties, and interrelationship of the entities that exist in a specific domain. Vigo et al. [3] have identified a number of pitfalls that researchers face while creating an ontology and suggest several strategies to overcome these pitfalls and designing appropriate tools for ontology authoring. While several tools exist to assist in ontology creation, Vigo et al. [3] argue that we still don't have a good understanding of the effectiveness of these tools, the kind of support that ontology creators need, and how best to create, reuse, refactor and debug ontologies. They identify several limitations for ontology creation: (a) lack of adequate tools for exploration and visualization of ontologies, (b) support for debugging and preventing errors, (c) judging the quality of ontologies, and (d) checking the correctness of ontologies.

The development of real-world organizational ontology-based knowledge management systems is still in early stages [1]. First, the ontology (such as DBpedia) is a shallow ontology with simple tuples and properties, it is not designed for domain specific requirement. Second, the large amount of information in an organization typically already exists outside of ontology, such as database, and documents. Third, ontology requires the metadata, which is a time-consuming and difficult process that organization tend to void [2]. Therefore, there is a great need for a simplified approach for knowledge mapping from an ontology to the knowledge management system [17].

A Knowledge Map or a Concept Map is a simple graphical representation that shows meaningful relationships between two or more related concepts [6]. In a knowledge map, the nodes represent concepts and the arcs represent the relationships with appropriate labels connecting the nodes. Concept maps are beginning to be used in KM systems for creating annotations and representing the knowledge captured in technical documents. They are also being used in ontology development since they have many similarities to an ontology. A concept map structurally resembles an ontology and can be used as an initial representation for developing and visualizing ontologies [4].

Most ontologies are constructed using traditional modeling approaches with teams of ontology experts and domain experts working together. Also, current ontology creation tools do not have the reliability or credibility needed in large-scale ontology engineering projects. The current ontology creation workbenches and ontology modeling approaches require the participation of ontology experts. These are modelers that are familiar with the rather cryptic syntax of modeling languages like OWL [15] and RDF [7]. The expert modelers specify ontology structures after interviewing domain experts or transform candidate lists from workbenches into legal ontology statements. The costs of developing ontologies and the competence needed to model them have effectively prevented smaller organizations or communities from developing their own ontologies.

Generating concept maps through text analytics has received increased attention and creating concept maps in a semi-automated manner is becoming feasible through several tools and templates [5]. Thus, creating a concept map from existing documents and other knowledge sources in a particular domain and using it as a starting point and transforming it into an ontology is quite appealing. Hence, the objective of this research are to develop: (a) an approach for creating concept maps from a set of domain documents, (b) transform a concept map into a corresponding ontology, and (c) demonstrate the feasibility of the approach using a case study.

The remainder of the paper is organized as follows. Section 2 provides a summary of relevant literature and Sect. 3 discusses the proposed approach. Section 3 describes the application of the proposed approach in developing an ontology for the automotive safety domain. Section 4 discusses the details of the prototype implementation and initial validation. Section 5 concludes the paper.

2   Related Work

Over the years, a number of ontology engineering methodologies have been proposed. They provide guidelines for collaborative ontology construction and maintenance, and they provide a structure for the organization of ontology development projects. Fernandez-Lopez et al. [22] tend to view ontology engineering as a specialized form of software engineering, making use of the same types of phases and the same type of modeling approaches. Methodologies such as Methontology [22], UPON [24], Diligent [25], and Stanford's Ontology Development 101 [23], all split up the process into phases that guide you through the development work.

Currently, there is no comprehensive ontology engineering methodology that reflects the state-of-the-art on automatic ontology creation. Even though these tools may have a severe effect on the construction of new ontologies and the updating of existing ones, they are mostly ignored from a methodological point of view. They may acknowledge that the tools are of value to particular tasks or stages, but the tools should not affect the overall structure of the methodologies. There are now tools that may help users in carrying out the individual steps in the ontology creation process.

Some ontology workbenches exist that support the creation of ontologies. The JATKE workbench is implemented as a plug-in to the Protégé ontology editor and helps users develop ontologies in Protégé [6]. OntoLT is another Protégé plug-in that transforms linguistically annotated entities into concepts and individuals in ontologies [2]. Text2Onto is an advanced ontology workbench that makes use of both Lucene and GATE [8] to produce many of the same ranked candidate lists [3]. Text2Onto also has some additional features for ontology maintenance and incremental ontology updating.

The OntoLearn workbench from Navigli and Velardi [35] concentrates on word sense disambiguation and makes use of the WordNet lexicon. A common aspect of all these workbenches is that they consider the ontology creation process a chain of static analysis components. The user does not have the flexibility to adapt the analysis to his or her preferences, the nature of the document collection, or aspects of the domain itself. Furthermore, it is assumed that the user is familiar with ontology editors and can afterwards refine the generated results manually.

Our proposed approach relies on generating concept maps first, which is relatively a straightforward process and use the concept map to then develop the ontology over time. This process can be automated using the concept map generation process that is widely established and accepted. Several tools and templates exist that can be used to implement the proposed approach. The following section discusses the proposed approach and provides an architecture for an environment that implements the approach.

3 Proposed Approach

We propose the following architecture to generate concept maps from processing a text corpus (Fig. 1).

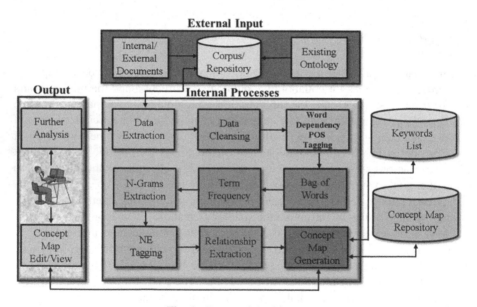

Fig. 1. Proposed Architecture

Extract the entities from the existing complaint report by using Stanford NLP package, word dependency, and POS tagger. This process will extract the domain specific relationships, classification, and properties from real-world cases.

Map the extracted entities, relationship, and properties to the existing ontologies.

3.1   The Dataset

The dataset used in this research is the National Highway Traffic Safety Administration (NHTSA) public data[1]. The NHTSA is an agency of the Executive Branch of the U.S. government, part of the Department of Transportation. The data is provided by Office of Defects Investigation (ODI) in NHTSA. The whole database dump contains about 1.5 million of vehicle safety complaints records since 1995. The data resource is consumers' complaint about the vehicle incidents. Each record includes a unique ID (ID), manufacturer' name (MFR_NAME), vehicle/equipment make (MAKE), vehicle/equipment model (MODEL), model year (YEAR), date of incident (FAIL DATE), specific component's description (COMPDESC), detailed information about consumer's vehicle (e.g., VIN number), and the content of the complaint (CDESCR). In this research, we extracted the information from the content of the complaint to construct knowledge map then mapping the results to the ontology. An example record in the dataset is shown in Table 1. Some of the columns in the dataset are omitted here for space.

Table 1. An Example Record in the NHTSA Complaint Dataset

ID	Mfr_Name	Make	Model	Year	Fail Date
1000051	Ford Motor Company	FORD	FUSION	2010	20130718
COMPDESC	VIN	CDESCR			
VEHICLE SPEED CONTROL	3FAHP0HA1AR	Vehicle keeps shutting off while driving. First happened on July 18th, second July 19th. Just started again on July 31 & continues. I was told it's my throttle.			

3.2   Automobile Ontology

We first checked the existing ontology from DBpedia ontology. There is one automobile ontology as an entity of type: Class[2]. The automobile class has several domains as the properties, which are wheelbase, bodyStyle, engine, layout, transmission, platform, fuelCapacity, and numberOfDoors. However, not all properties of a typical automobile are included. For example, the brake system, which includes the components such as brake shoes, wheel cylinder, wheel bearing, and brake fluid, etc. Like the properties already listed in DBpedia, these components are also critical to the automotive safety features.

---

[1] Dataset from NHTSA link: https://www-odi.nhtsa.dot.gov/downloads/.

[2] DBpedia automobile class link: http://dbpedia.org/ontology/Automobile.

As mentioned above in Sect. 1, DBpedia has simple tuples and properties and is not designed for a full coverage of a domain knowledge or concepts. Specifically in automotive industry, different manufacturers may have their own set of terms to represent the same property in the ontology. Moreover, from the consumer's complaint, consumers tend to use common, but not formal terms/words to describe the incidents and the components that involved in the incidents. Because of the lack of domain knowledge, consumers are less likely to use the same set of terms in their complaint. Therefore, in addition to the existing ontology, a concept map can be added to enlarge the coverage of the domain knowledge.

3.3   Concept Extracting and Mapping

Extracting. From the content of the consumers' complaints, we identified that the common pattern of the complaints is that the consumers first described the situation or the contextual information about the incidents, then the consumers used their own terms and vocabulary to describe how exact the incidents happened, and the outcomes. Fortunately, for each complaint record, there is a higher level components as one automobile property identified, such as engine, power train, electrical system, etc. However, the lower level components that involved in the incidents can only be found from the content of each complaint. And these components are usually not represented by using a formal term or followed the terminology used by manufacturer.

Table 2. The Relationships from an Example Complaint

COMPDESC	Content of Complaint	Extracted Entities/Components	
		Single Term	2-Gram
Engine	The wrench light comes on and the car loses its acceleration completely. As I pull over the side of the road and brake, the car will start shaking. The only way to fix it is to turn the car off and back on again. The wrench light will disappear after that and it'll start driving normally again. First time this happened I was on the interstate with my three year old in the car with me. One unhappy momma!!!! The code it read p2111 which I have been told it is a defect in the throttle body. Hope they are right.	Wrench Light Brake Code Throttle Body	Wrench Light Throttle Body

The first task of our approach is to extract these automotive components from the content of the complaint. We applied the Stanford NLP package to perform tokenization, word dependency, and POS tag parsing, to identify nouns as entities of components. Since the dataset has a number of different manufacturers, models, and years, to keep the components consistent, we randomly chose a single manufacturer to build a training subset of complaints, then to extract component keywords.

From about 1.5 million complaint records, we extract 10,000 records by matching the chosen manufacturer. After the POS tag parsing, all of the nouns (i.e., with POS tags as NN, NNP, NNS, NNPS) are identified and counted. There are 8,393 unique nouns, from which we manually picked the top 300 nouns (ranked by number of term frequency count descending) that related to automotive components. The component keyword list consists of these 300 nouns. Moreover, to capture the component with more than one term, we performed a 2-gram nouns phrase matching if both of the terms in the 2-gram are matched the list of 300 nouns.

Mapping. In the complaint dataset, each record has a high level component category that identified by NHTSA. We use this category as a higher level concept in the ontology to be mapped with the components extracted from the complaint content. The new relationships are created by this mapping process. Table 2 shows the relationships from an example complaint record.

From Table 2, we can see that the Engine entity as a high level component has the relationships with wrench light, throttle body, brake, and code (p2111) in this incident. These relationships are used to construct the concept map to further develop the ontology over time. We introduce our approach aggregation process to utilize the relationships in Sect. 4.

3.4   Prototype in KNIME

We implement our proposed approach in KNIME[3]. KNIME is an open source data analytic tool through its modular data pipelining concept. The prototype has the following components (Fig. 2):

- Complaint dataset loader. Import the raw dataset and remove the duplicated records.
- Preprocessor and NLP parsing. The NLP parses the complaint content for tokenization, POS tagging, and creating bag of words.
- Keyword creating. The single terms are aggregated for manual process to filter out the terms are not related to automotive component.
- Keyword loader. Load the keyword list from previous component.
- Single term/2-Gram extractor. Using the keyword list to match single keyword or 2-Gram keywords. The results can be further aggregated at different levels from the complaint dataset loader.

---

[3] For more detail about KNIME, please visit: https://www.knime.com/.

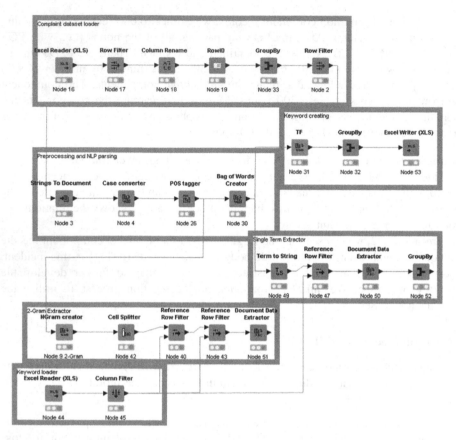

Fig. 2. The Implemented Prototype in KNIME

4 Result

Aggregation. Table 2 only shows one complaint record's result. The results from multiple complaints can be aggregated at different level. For example, the single term and 2-Gram results can be aggregated at COMPDESC (high level component) level, which can be engine, power train, etc. Other possible levels can be any columns in NHTSA dataset. Moreover, the results can be aggregated by more than one level at the same time, such as model and year. Table 3 shows the top 10 2-Grams aggregated result for different COMPDESC by descending order from 10,000 complaint records related to manufacturer Ford Motor Company. Table 4 shows the top 10 2-Grams aggregated result for different model and year by descending order from 10,000 complaint records related to manufacturer Ford Motor Company.

Table 3. 2-Gram Aggregated Result based on COMPDESC

COMPDESC	2-Gram	Count
Vehicle Speed Control	Throttle body	449
Steering	Power steering	395
Power Train	Throttle body	250
Steering	Steering wheel	229
Engine	Throttle body	168
Vehicle Speed Control	Gas pedal	142
Power Train	Wrench light	138
Engine	Engine light	131
Fuel/Propulsion System	Throttle body	111
Power Train	Engine light	100

Table 4. 2-Gram Aggregated Result based on Model and Year

Model	Year	2-Gram	Count
Fusion	2010	Throttle body	327
Escape	2010	Throttle body	247
Escape	2008	Power steering	122
Fusion	2011	Throttle body	122
Escape	2011	Throttle body	117
Fusion	2010	Wrench light	108
Escape	2009	Throttle body	94
Escape	2008	Steering wheel	76
Escape	2010	Wrench light	76
Escape	2010	Gas pedal	66

Comparing with an existing ontology on automobile industry such as DBpedia, our prototype can extract detail level entities as concepts with aggregation capability. New properties and their relationships are found from the automobile safety complaint dataset. Figure 3 shows the constructed concept map from part of the result in Table 4.

The features we may add to the knowledge management system from proposed solution are:

- New properties
- New relationships
- Product or service related information, such as model, year.
- Knowledge representation. More information in addition to simple tuples and properties in the domain specific format.

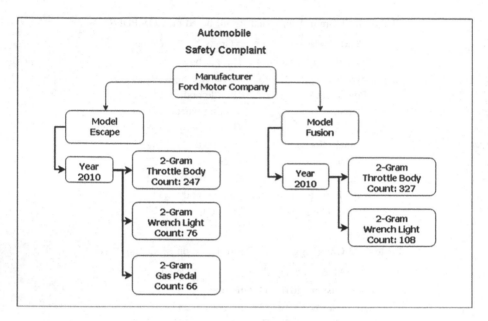

Fig. 3. Concept Map from Part of Table 4

5   Conclusion and Future Work

This paper proposes a new approach of expanding the existing ontology by creating a new concept map for domain specific needs. The knowledge sources are from public available dataset generated by consumer on automobile safety. Our approach is successfully implemented through an open source software KNIME with Stanford NLP parser to extract detail level entities as new concepts and aggregate the result at different domain specific levels. These results are used to create concept map for further analysis.

Although our prototype is applied to automotive industry in this research, we are confident that the same approach can be also used in other related domain, such as online reviews, user generated content on social media, and government reports, etc. As future work we may utilize more features from NLP parser results, and word dependency relationship (such as verb-noun) to provide more relationships between nouns, and also reasoning capability to the knowledge management systems.

References

1. Kim, S., Suh, E., Hwang, H.: Building the knowledge map: an industrial case study. J. Knowl. Manag. **7**(2), 34–45 (2003)
2. Maedche, A., Motik, B., Stojanovic, L., Studer, R., Volz, R.: Ontologies for enterprise knowledge management. IEEE Intell. Syst. **18**(2), 26–33 (2003)

3. Vigo, M., Bail, S., Jay, C., Stevens, R.: Overcoming the pitfalls of ontology authoring: Strategies and implications for tool design. Int. J. Hum. Comput. Stud. **72**, 835–845 (2014)
4. Starr, R.R., de Oliveira, J.M.: Concept maps as the first step in an ontology construction method. Inf. Syst. **38**(5), 771–783 (2013)
5. Iqbal, R., Murad, M.A.A., Mustapha, A., Sharef, N.M.: An ontology development approach using concept maps by automatic term extraction. Int. J. Inf. Commun. Technol. **10**(1), 51–65 (2017)
6. Novak, J. Canas, A.: The theory underlying concept maps and how to construct and use them. Technical Report. Institute for Human and Machine Cognition, Florida, 1-36 (2008)
7. Klyne, G. Carroll, J.J.: Resource description framework (RDF): concepts and abstract syntax (2006)
8. Cimiano, P., Völker, J.: Text2Onto. In: Montoyo, A., Muñoz, R., Métais, E. (eds.) NLDB 2005. LNCS, vol. 3513, pp. 227–238. Springer, Heidelberg (2005). https://doi.org/10.1007/11428817_21
9. Gulla, J.A., Borch, H.O., Ingvaldsen, J.E.: Ontology Learning for Search Applications. In: Meersman, R., Tari, Z. (eds.) OTM 2007. LNCS, vol. 4803, pp. 1050–1062. Springer, Heidelberg (2007). https://doi.org/10.1007/978-3-540-76848-7_69
10. Haase, P., Völker, J.: Ontology learning and reasoning - dealing with uncertainty and inconsistency. In: Proceedings of the International Semantic Web Conference. Workshop 3: Uncertainty Reasoning for the Semantic Web (ISWC-URSW'05), pp. 45–55. Springer, Berlin, Heidelberg (2005)
11. Maedche, A. Staab, S.: Semi-automatic engineering of ontologies from text. In: Proceedings of the 12th Internal Conference on Software and Knowledge Engineering, pp. 231–239. Chicago (2000)
12. Navigli, R., Velardi, P.: Learning domain ontologies from document warehouses and dedicated web sites. Comput. Linguist. **30**(2), 151–179 (2004)
13. Sabou, M., Wroe, C., Goble, C., Stuckenschmidt, H.: Learning domain ontologies for semantic web service descriptions. Web Semant. Sci. Serv. Agents World Wide Web **3**(4), 340–365 (2005)
14. Grant, R.M.: Prospering in dynamically-competitive environments: organizational capability as knowledge integration. Knowledge and Strategy, pp. 133–153 (1999)
15. Motik, B.: On the properties of metamodeling in OWL. J. Logic Comput. **17**(4), 617–637 (2007)
16. Nidumolu, S.R., Subramani, M., Aldrich, A.: Situated learning and the situated knowledge web: exploring the ground beneath knowledge management. J. Manag. Inf. Syst. **18**(1), 115–151 (2001)
17. Rus, M.: Lindvall.: knowledge management in software engineering. IEEE Softw. **19**(3), 26–38 (2002)
18. Ju, T.L.: Representing organizational memory for computer-aided utilization. J. Inf. Sci. **32**(5), 420–433 (2006)
19. Maedche, A., Motik, B., Stojanovic, L., Studer, R.: Volz, R: Ontologies for enterprise knowledge management. IEEE Intell. Syst. **18**(2), 26–33 (2003)
20. Fensel, D.: Ontology-based knowledge management. IEEE Comput. **35**(11), 56–59 (2002)
21. Chang, J., Choi, B., Lee, H.: An organizational memory for facilitating knowledge: an application to e-business architecture. Expert Syst. Appl. **26**(2), 203–215 (2004)
22. Fernández-López, M. Gómez-Pérez, A. Juristo, N.: METHONTOLOGY: from ontological art towards ontological engineering. In: AAAI-97 Spring Symposium Series, 24–26 March 1997, Stanford University, EEUU (1997)
23. Noy, N.F. McGuinness, D.L.: Ontology development 101: A Guide to Creating Your First Ontology (2001)

24. De Nicola, A., Missikoff, M., Navigli, R.: A proposal for a unified process for ontology building: UPON. In: International Conference on Database and Expert Systems Applications, pp. 655–664. Springer, Berlin, Heidelberg (2005)
25. Tempich, C., Pinto, H.S., Sure, Y., Staab, S.: An Argumentation Ontology for DIstributed, Loosely-controlled and evolvInG Engineering processes of oNTologies (DILIGENT). In: Gómez-Pérez, A., Euzenat, J. (eds.) ESWC 2005. LNCS, vol. 3532, pp. 241–256. Springer, Heidelberg (2005). https://doi.org/10.1007/11431053_17
26. Navigli, R., Velardi, P., Cucchiarelli, A., Neri, F. Cucchiarelli, R.: Extending and enriching WordNet with OntoLearn. In: Proceeding of 2nd Global WordNet Conf. (GWC), pp. 279–284 (2004)

A Fuzzy-Based Approach for Representing and Reasoning on Imprecise Time Intervals in Fuzzy-OWL 2 Ontology

Fatma Ghorbel[1,2(✉)], Fayçal Hamdi[1], Elisabeth Métais[1],
Nebrasse Ellouze[2], and Faiez Gargouri[2]

[1] CEDRIC Laboratory, Conservatoire National des Arts et Métiers (CNAM),
Paris, France
fatmaghorbel6@gmail.com,
{faycal.hamdi,metais}@cnam.fr
[2] MIRACL Laboratory, University of Sfax, Sfax, Tunisia
nebrasse.ellouze@gmail.com,
faiez.gargouri@isimsf.rnu.tn

Abstract. Representing and reasoning on imprecise temporal information is a common requirement in the field of Semantic Web. Many works exist to represent and reason on precise temporal information in OWL; however, to the best of our knowledge, none of these works is devoted to represent and reason on imprecise time intervals. To address this problem, we propose a fuzzy-based approach for representing and reasoning on imprecise time intervals in ontology. Our approach is based on fuzzy sets theory and fuzzy tools and is modeled in Fuzzy-OWL 2. The 4D-fluents approach is extended, with new fuzzy components, in order to represent imprecise time intervals and qualitative fuzzy interval relations. The Allen's interval algebra is extended in order to compare imprecise time intervals in a fuzzy gradual personalized way. Inferences are done via a set of Mamdani IF-THEN rules.

Keywords: Imprecise time interval · 4D-fluents · Allen's interval algebra
Fuzzy-OWL 2

1 Introduction

In the Semantic Web field, representing and reasoning on imprecise temporal information is a common requirement. Indeed, temporal information given by users is often imprecise. For instance, if they give the information "Alexandre was married to Nicole by 1981 to late 90" two measures of imprecision are involved. On the one hand, the information "by 1981" is imprecise in the sense that it could mean approximately from 1980 to 1982; on the other hand, the information "late 90" is imprecise in the sense that it could mean, with an increasingly possibility, from 1995 to 2000. When an event is characterized by a gradual beginning and/or ending, it is usual to represent the corresponding time span as an imprecise time interval.

In OWL, many works have been proposed to represent and reason on precise temporal information; however, to the best of our knowledge, there is no work devoted to represent and reason on imprecise time intervals.

In this paper, we propose a fuzzy-based approach for representing and reasoning on imprecise time intervals in ontology. It is based on fuzzy sets theory and fuzzy tools. It is based on Fuzzy-OWL 2 [1] which is an extension of OWL 2 that deals with fuzzy information. To represent imprecise time intervals in Fuzzy-OWL 2, we extend the 4D-fluents model [2] in two ways: (1) It is enhanced with new fuzzy components to be able to model imprecise time intervals. (2) It is enhanced with qualitative temporal expressions representing fuzzy relations between imprecise temporal intervals. To reason on imprecise time intervals, we extend Allen's work to compare imprecise time intervals in a fuzzy gradual personalized way. Our Allen's extension introduces gradual fuzzy interval relations e.g., "long before". It is personalized in the sense that it is not limited to a given number of interval relations. It is possible to determinate the level of precision that should be in a given context. For instance, the classic Allen relation "before" may be generalized in N interval relations, where "before$_{(1)}$" means "just before" and gradually the time gap between the two imprecise intervals increases until "before$_{(N)}$" which means "long before". The resulting fuzzy interval relations are inferred from the introduced imprecise time intervals using the FuzzyDL reasoner [3], via a set of Mamdani IF-THEN rules, in Fuzzy-OWL 2.

The current paper is organized as follows: Sect. 2 is devoted to present some preliminary concepts and related work in the field of temporal information representation in OWL and reasoning on time intervals. In Sect. 3, we introduce our fuzzy-based approach for representing and reasoning on imprecise time intervals. Section 4 draws conclusions and future research directions.

2   Preliminaries and Related Work

In this section, we introduce some preliminary concepts and related work in the field of temporal information representation in OWL and reasoning on time intervals.

2.1   Representing Temporal Information in OWL

Five main approaches are proposed to represent time information in OWL: Temporal Description Logics [4], Versioning [5], N-ary relations [6] and 4D-fluents [2]. All these approaches represent only crisp temporal information in OWL. Temporal Description Logics extend the standard description logics with additional temporal constructs e.g., "sometime in the future". N-ary relations approach represents an N-ary relation using an additional object. The N-ary relation is represented as two properties each related with the new object. The two objects are related to each other with an N-ary relation. Reification is "a general purpose technique for representing N-ary relations using a language such as OWL that permits only binary relations" [7]. Versioning approach is described as "the ability to handle changes in ontologies by creating and managing different variants of it" [5]. When an ontology is modified, a new version is created to represent the temporal evolution of the ontology. 4D-fluents approach represents temporal information and the evolution of the last ones in OWL. Concepts varying in time are represented as 4-dimensional objects with the 4th dimension being the temporal dimension.

Based on the present related work, we choose the 4D-fluents approach. Indeed, compared to related work, it minimizes the problem of data redundancy as the changes occur only on the temporal parts and keeping therefore the static part unchanged. It also maintains full OWL expressiveness and reasoning support [7]. We extend this approach in two ways. It is extended with new fuzzy components to represent (1) imprecise time intervals and (2) fuzzy interval relations.

2.2   Allen's Interval Algebra

In [8], Allen has proposed 13 mutually exclusive primitive relations that may hold between two precise time intervals. Their semantics is illustrated in Table 1. Let $I = [I^-, I^+]$ and $J = [J^-, J^+]$ two time intervals; where I^- (respectively J^-) is the beginning time-step of the event and I^+ (respectively J^+) is the ending.

A number of works fuzzify Allen's temporal interval relations. We classify these works into (1) works focusing on fuzzifying Allen's interval algebra to compare precise time intervals and (2) works focusing on fuzzifying Allen's interval algebra to compare imprecise time intervals.

Three approaches have been proposed to fuzzify Allen's interval algebra in order to compare two precise time intervals: [9], [10] and [11]. In [9], the authors propose fuzzy Allen relations viewed as fuzzy sets of ordinary Allen relationship taking into account a neighborhood structure, a notion originally introduced in [12]. In [10], the authors represent a time interval as a pair of possibility distributions that define the possible values of the endpoints of the crisp interval. Using possibility theory, the possibility and necessity of each of the interval relations can then be calculated. This approach also allows modeling imprecise relations such as "long before". In [11], the authors propose a fuzzy extension of Allen's work, called IA^{fuz} where degrees of preference are associated to each relation between two precise time intervals.

Four approaches have been proposed to fuzzify Allen's interval algebra to compare two imprecise time intervals: [13, 14, 15 and 16]. In [13], the authors propose a temporal model based on fuzzy sets to extend Allen relations with imprecise time intervals. The authors introduce a set of auxiliary operators on intervals and define fuzzy counterparts of these operators. The compositions of these relations are not studied by the authors. In [14], the authors propose an approach to handle some gradual temporal relations as "more or less finishes". However, this work cannot take into account gradual temporal relations such as "long before". Furthermore, many of the symmetry, reflexivity, and transitivity properties of the original temporal interval relations are lost in this approach; thus it is not suitable for temporal reasoning. In [15], the authors propose a generalization of Allen's relations with precise and imprecise time intervals. This approach allows handling classical temporal relations, as well as other imprecise relations. Interval relations are defined according to two fuzzy operators comparing two time instants: "long before" and "occurs before or at approximately the same time". In [16], the authors generalize the definitions of the 13 Allen's classic

Table 1. Allen's temporal interval relations (I:▬▬▬▬▬,J:▭▭▭▭).

Relation	Inverse	Relations between interval bounds	Illustration
Before (I, J)	After (I, J)	$I^+ < J^-$	
Meets (I, J)	Met-by (I, J)	$I^+ = J^-$	
Overlaps (I, J)	Overlapped-by (I, J)	$(I^- < J^-) \wedge (I^+ > J^-) \wedge$ $(I^+ < J^+)$	
Starts (I, J)	Started-by (I, J)	$(I^- = J^-) \wedge (I^+ < J^+)$	
During (I, J)	Contains (I, J)	$(I^- > J^-) \wedge (I^+ < J^+)$	
Ends (I, J)	Ended-by (I, J)	$(I^- > J^-) \wedge (I^+ = J^+)$	
Equal (I, J)	Equal (I, J)	$(I^- = J^-) \wedge (I^+ = J^+)$	

interval relations to make them applicable to fuzzy intervals in two ways (conjunctive and disjunctive). Gradual temporal interval relations are not taken into account.

3   Our Fuzzy-Based Approach for Representing and Reasoning on Imprecise Time Intervals in Ontology

In this section, we propose our fuzzy-based approach to represent and reason on imprecise time intervals. This approach is based on a fuzzy environment. We extend the 4D-fluents model to represent imprecise time intervals and their relationships in Fuzzy-OWL 2. To reason on imprecise time intervals, we extend the Allen's interval algebra in a fuzzy gradual personalized way. We infer the resulting fuzzy interval relations in Fuzzy-OWL 2 using a set of Mamdani IF-THEN rules.

3.1   Representing Imprecise Time Intervals and Fuzzy Qualitative Interval Relations in Fuzzy-OWL 2

We represent an imprecise time interval using fuzzy sets. We represent the imprecise beginning interval bound as a fuzzy set which has the L-function membership function and the ending interval bound as a fuzzy set which has the R-function membership function. Let $I = [I^-, I^+]$ be an imprecise time interval. We represent the binging bound I^- as a fuzzy set which has the L-function membership function ($A = I^{-(1)}$ and $B = I^{-(N)}$). We represent the ending bound I^+ as a fuzzy set which has the R-function membership function (A = $I^{+(1)}$ and B = $I^{+(N)}$). For instance, if we have the information "Alexandre was starting his PhD study in 1973 and was graduated in late 80", the beginning bound is crisp. The ending bound is imprecise and it is represented by L-function membership function (A = 1976 and B = 1980). For the rest of the paper, we use the membership functions defined in [17] and shown in Fig. 1.

The classic 4D-fluents model introduces two crisp classes "TimeSlice" and "TimeInterval" and four crisp properties "tsTimeSliceOf", "tsTimeInterval", "hasBegining" and "hasEnd". The class "TimeSlice" is the domain class for entities representing temporal parts (i.e., "time slices"). The property "tsTimeSliceOf" connects an instance of class "TimeSlice" with an entity. The property "tsTimeInterval" connects an instance of class "TimeSlice" with an instance of class "TimeInterval". The instance of class "TimeInterval" is related with two temporal instants that specify its starting and ending points using, respectively, the "hasBegining" and "hasEnd" properties. Figure 2 illustrates the use of the 4D-fluents model to represent the following example: "Alexandre was started his PhD study in 1975 and he was graduated in 1978".

We extend the original 4D-fluents model to represent imprecise time intervals in the following way. We add two fuzzy datatype properties "FuzzyHasBegining" and "FuzzyHasEnd" to the class "TimeInterval". "FuzzyHasBegining" has the L-function membership function (A = $I^{-(1)}$ and B = $I^{-(N)}$). "FuzzyHasEnd" has the R-function membership function (A = $I^{+(1)}$ and B = $I^{+(N)}$).

The 4D-fluents approach is also enhanced with qualitative temporal relations that may hold between imprecise time intervals. We introduce the "FuzzyRelationIntervals", as a fuzzy object property between two instances of the class "TimeInterval". "FuzzyRelationIntervals" represent fuzzy qualitative temporal relations. "FuzzyRelationIntervals" has the L-function membership function (A = 0 and B = 1). Figure 3 represents our extended 4D-fluents model in Fuzzy-OWL 2.

We can see in Fig. 4 an instantiation of the extended 4D-fluents model in Fuzzy-OWL2. On this example, we consider the following information: "Alexandre was married to Nicole just after he was graduated with a PhD. Alexandre was graduated with a PhD in 1980. Their marriage lasts 15 years. Alexandre was remarried to Béatrice since about 10 years and they were divorced in 2016". Let I = [I^-, I^+] and J = [J^-, J^+] be two imprecise time intervals representing, respectively, the duration of the marriage of Alexandre with Nicole and the one with Béatrice. I^- is represented with the fuzzy datatype property "FuzzyHasBegining" which has the L-function membership function (A = 1980 and B = 1983). I^+ is represented with the fuzzy datatype property "FuzzyHasEnd" which has the R-function membership function (A = 1995 and B = 1998). J^- is represented with the fuzzy datatype property "FuzzyHasBegining" which has the L-function membership function (A = 2005 and B = 2007). J^+ is represented with the crisp datatype property "HasEnd" which has the value "2016".

3.2   Representing Imprecise Time Intervals and Fuzzy Qualitative Interval Relations in Fuzzy-OWL 2

We propose a set of fuzzy gradual personalized comparators that may hold between two time instants T_1 and T_2. Based on these operators, we present our fuzzy gradual personalized extension of Allen's work to compare two imprecise time intervals.

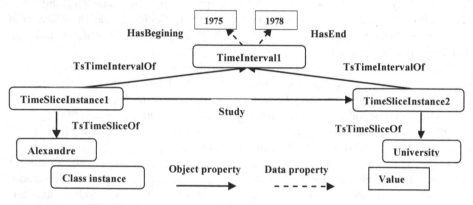

Fig. 1. R-Function, L-Function and Trapezoidal membership functions [17].

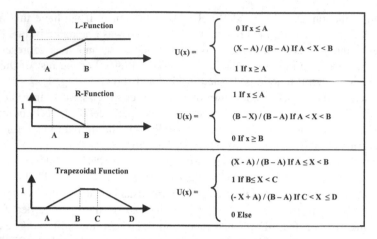

Fig. 2. An instantiation of the classic the 4D-fluents model.

Then, we infer, in Fuzzy OWL 2, the resulting temporal interval relations via a set of Mamdani IF-THEN rules using the fuzzy reasoner FuzzyDL.

We generalize the crisp time instants comparators "Follow", "Precede" and "Same", introduced in [18]. Let α and β two parameters allowing the definition of the membership function of the following comparators ($\in]0, +\infty[$); N is the number of slices; T_1 and T_2 are two time instants; we define the following comparators (illustrated in Fig. 5):

- $\{Follow_{(1)}{}^{(\alpha, \beta)} (T_1, T_2) \ldots Follow_{(N)}{}^{(\alpha, \beta)} (T_1, T_2)\}$ are a generalization of the crisp time instants relation "Follows". $Follow_{(1)}{}^{(\alpha, \beta)} (T_1, T_2)$ means that T_1 is "just after or approximately at the same time" T_2 w.r.t. (α, β) and gradually the time gap between T_1 and T_2 increases until $Follow_{(N)}{}^{(\alpha, \beta)} (T_1, T_2)$ which means that T_1 is "long after" T_2 w.r.t. (α, β). N is set by the expert domain. $\{Follow_{(1)}{}^{(\alpha, \beta)} (T_1, T_2) . \ldots Follow_{(N)}{}^{(\alpha, \beta)} (T_1, T_2)\}$ are defined as fuzzy sets. $Follow_{(1)}{}^{(\alpha, \beta)} (T_1, T_2)$ has R-Function membership function which has as parameters $A = \alpha$ and $B = (\alpha + \beta)$.

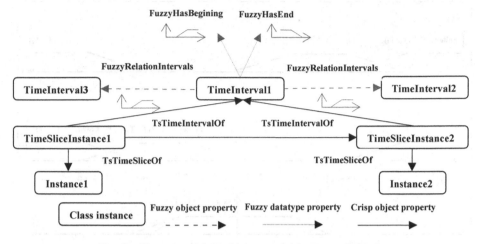

Fig. 3. The extended 4D-fluents model in Fuzzy-OWL 2.

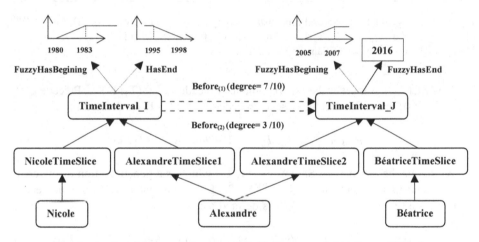

Fig. 4. An instantiation of the extended 4D-fluents model in Fuzzy-OWL 2.

All comparators $\{\text{Follow}_{(2)}{}^{(\alpha,\ \beta)}(T1, T_2) \dots \text{Follow}_{(N-1)}{}^{(\alpha,\ \beta)}(T_1, T_2)\}$ have trapezoidal membership function which has as parameters $A = ((K-1)\ \alpha)$ and $B = ((K-1)\ \alpha + (K-1)\ \beta)$, $C = (K\ \alpha + (K-1)\ \beta)$ and $D = (K\ \alpha + K\ \beta)$; where $2 \leq K \leq N-1$. $\text{Follow}_{(N)}{}^{(\alpha,\ \beta)}(T_1, T_2)$ has L-Function membership function which has as parameters $A = ((N-1)\ \alpha + (N-1)\ \beta)$ and $B = ((N-1)\ \alpha + (N-1)\ \beta)$;

- $\{\text{Precede}_{(1)}{}^{(\alpha,\ \beta)}(T_1, T_2) \dots \text{Precede}_{(N)}{}^{(\alpha,\ \beta)}(T_1, T_2)\}$ are a generalization of the crisp time instants relation "Precede". $\text{Precede}_{(1)}{}^{(\alpha,\ \beta)}(T_1, T_2)$ means that T_1 is "just before or approximately at the same time" T_2 w.r.t. (α, β) and gradually the time gap between T_1 and T_2 increases until $\text{Precede}_{(N)}{}^{(\alpha,\ \beta)}(T_1, T_2)$ which means that T_1 is "long before"

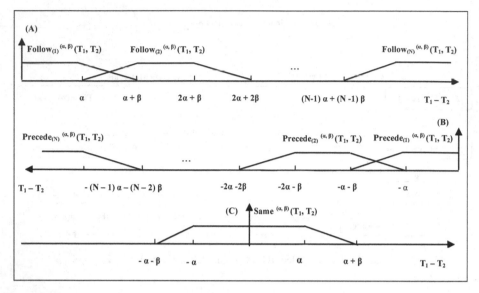

Fig. 5. Fuzzy gradual personalized time instants comparators. (A) Fuzzy sets of {Follow$_{(1)}$ $^{(\alpha, \beta)}$ (T_1, T_2) ... Follow$_{(N)}$ $^{(\alpha, \beta)}$ (T_1, T_2)}. (B) Fuzzy sets of {Precede$_{(1)}$ $^{(\alpha, \beta)}$ (T_1, T_2) ... Precede$_{(N)}$ $^{(\alpha, \beta)}$ (T_1, T_2)}. (C) Fuzzy set of Same$^{(\alpha, \beta)}$ (T_1, T_2).

T_2 w.r.t. (α, β). N is set by the expert domain. {Precede$_{(1)}$ $^{(\alpha, \beta)}$ (T_1, T_2) ... Precede$_{(N)}$ $^{(\alpha, \beta)}$ (T_1, T_2)} are defined as fuzzy sets. Precede$_{(i)}$ $^{(\alpha, \beta)}$ (T_1, T_2) is defined as:

$$Precede_{(i)}^{(\alpha, \beta)}(T_1, T_2) = 1 - Follow_{(i)}^{(\alpha, \beta)}(T_1, T_2) \tag{1}$$

- We define the comparator Same$^{(\alpha, \beta)}$ (T_1, T_2) which is a generalization of the crisp time instants relation "Same". Same$^{(\alpha, \beta)}$ (T_1, T_2) means that T_1 is "approximately at the same time" T_2 w.r.t. (α, β). It is defined as:

$$Same^{(\alpha, \beta)}(T_1, T_2) = Min(Follow_{(1)}^{(\alpha, \beta)}(T_1, T_2), Precede_{(1)}^{(\alpha, \beta)}(T_1, T_2)) \tag{2}$$

Then, we extend Allen's work to compare imprecise time intervals with a fuzzy gradual personalized view. We provide a way to model gradual, linguistic-like description of temporal interval relations. Compared to related work, our work is not limited to a given number of imprecise relations. It is possible to determinate the level of precision that should be in a given context. For instance, the classic Allen relation "before" may be generalized in N imprecise relations, where "Before$_{(1)}$ $^{(\alpha, \beta)}$ (I, J)" means that I is "just before" J w.r.t. (α, β) and gradually the time gap between I and J increases until "Before$_{(N)}$ $^{(\alpha, \beta)}$ (I, J)" which means that I is long before J w.r.t. (α, β). The definition of our fuzzy interval relations is based on the fuzzy gradual personalized time instants compactors. Let $I = [I^-, I^+]$ and $J = [J^-, J^+]$ two imprecise time intervals; where I^- has the L-function membership function ($A = I^{-(1)}$ and $B = I^{-(N)}$); I^+ is a fuzzy set which has the R-function membership function ($A = I^{+(1)}$ and $B = I^{+(N)}$); J^- is a fuzzy set which has the L-function membership function ($A = J^{-(1)}$ and $B = J^{-(N)}$); J^+ is a fuzzy set which has the R-function membership function ($A = J^{+(1)}$ and $B = J^{+(N)}$). For instance, the fuzzy interval relation "Before$_{(1)}$ $^{(\alpha,\ \beta)}$ (I, J)" is defined as:

$$\forall I^{+\,(i)} \in I^+, \forall J^{-(j)} \in J^- : Precede_{(1)}^{(\alpha,\,\beta)}(I^{+\,(i)}, J^{-(j)}) \tag{3}$$

This means that the most recent time instant of I^+ ($I^{+(N)}$) ought to proceed the oldest time instant of J^- ($J^{-(1)}$):

$$Precede_{(1)}^{(\alpha,\,\beta)}(I^{+\,(N)}, J^{-(1)}) \tag{4}$$

In the similar way, we define the others temporal interval relations, as shown in Table 2.

Finally, we have implemented our fuzzy gradual personalized extension of Allen's work in Fuzzy-OWL 2. We use the ontology editor PROTÉGÉ version 4.3 and the fuzzy reasoner FuzzyDL. We propose a set of Mamdani IF-THEN rules to infer the temporal interval relations from the introduced imprecise time intervals which are represented using the extended 4D-fluents model in Fuzzy-OWL2. For each temporal interval relation, we associate a Mamdani IF-THEN rule. For instance, the Mamdani IF-THEN rule to infer the "Overlaps$_{(1)}$ $^{(\alpha,\ \beta)}$ (I, J)" relation is the following:

(define-concrete-feature Precede$_{(1/1)}$ real) (define-concrete-feature Precede$_{(1/2)}$ real)
(define-concrete-feature Precede$_{(1/3)}$ real) (define-concrete-feature Overlaps$_{(1)}$ real)
(define-fuzzy-concept Fulfilled Right-shoulder(0,-α-β,-α,0)) (define-fuzzy-concept True
Right-shoulder(0,0,1,1))
(define-concept Rule0 (g-and (some Precede$_{(1/1)}$ Fulfilled) (some Precede$_{(1/2)}$ Fulfilled) Fulfilled)
(some Precede$_{(1/3)}$ Fulfilled) (some Overlaps$_{(1)}$ True))) //Fuzzy rule
(instance facts (= Precede$_{(1/1)}$ ($I^{-(N)}$ - $J^{-(1)}$))) (instance facts (= Precede$_{(1/2)}$ ($J^{-(N)}$ - $I^{+(1)}$)))
(instance facts (= Precede$_{(1/3)}$ ($I^{+(N)}$ - $J^{+(1)}$))) //Instantiations

---

We define three input fuzzy variables, named "Precede$_{(1/1)}$", "Precede$_{(1/2)}$" and "Precede$_{(1/3)}$", which have the same membership function than that of "Precede$_{(1)}$ $^{(\alpha,}$ $^{\beta)}$". We define one output variable "Overlaps$_{(1)}$" which has the same membership than that of the fuzzy object property "FuzzyRelationIntervals". "Precede$_{(1/1)}$", "Precede$_{(1/2)}$" and "Precede$_{(1/3)}$" are instantiated with, respectively, ($I^{-(N)} - J^{-(1)}$), ($J^{-(N)} - I^{+(1)}$) and ($I^{+(N)} - J^{+(1)}$).

4 Conclusion

In this paper, we proposed a fuzzy-based approach to represent and reason on imprecise time intervals in ontology. It is entirely based only on fuzzy environment. We extended the 4D-fluents model to represent imprecise time intervals and fuzzy interval relations in Fuzzy-OWL 2. To reason on imprecise time intervals, we extend the Allen's interval algebra in a fuzzy gradual personalized way. We infer the resulting fuzzy interval relations in Fuzzy-OWL2 using a set of Mamdani IF-THEN rules.

Table 2. Fuzzy gradual personalized temporal interval relations upon imprecise time intervals.

Relation	Inverse	Relations between bounds	Definition
Before$_{(K)}$ $^{(\alpha, \beta)}$ (I, J)	After$_{(K)}$ $^{(\alpha, \beta)}$ (I, J)	$\forall\, I^{+(i)} \in I^+, \forall\, J^{-(j)} \in J^-/$ $(I^{+(i)} < J^{-(j)})$	Proceed$_{(K)}$ $^{(\alpha, \beta)}$ $(I^{+(N)}, J^{-(1)})$
Meets $^{(\alpha, \beta)}$ (I, J)	Met-by $^{(\alpha, \beta)}$ (I, J)	$\forall\, I^{+(i)} \in I^+, \forall\, J^{-(j)} \in J^-/$ $(I^{+(i)} = J^{-(j)})$	Min(Same$^{(\alpha, \beta)}$ $(I^{+(1)}, J^{-(1)}) \wedge$ Same$^{(\alpha, \beta)}$ $(I^{+(N)}, J^{-(N)}))$
Overlaps$_{(K)}$ $^{(\alpha, \beta)}$ (I, J)	Overlapped-by$_{(K)}$ $^{(\alpha, \beta)}$ (I, J)	$\forall\, I^{-(i)} \in I^-, \forall\, I^{+(i)} \in I^+, \forall\, J^{-(j)} \in J^-, \forall\, J^{+(j)} \in J^+/$ $(I^{-(i)} < J^{-(j)}) \wedge$ $(J^{-(j)} < I^{+(i)}) \wedge$ $(I^{+(i)} < J^{+(j)})$	Min(Proceed$_{(K)}$ $^{(\alpha, \beta)}$ $(I^{-(N)}, J^{-(1)}) \wedge$ Proceed$_{(K)}$ $^{(\alpha, \beta)}$ $(J^{-(N)}, I^{+(1)}) \wedge$ Proceed$_{(K)}$ $^{(\alpha, \beta)}$ $(I^{+(N)}, J^{+(1)}))$
Starts$_{(K)}$ $^{(\alpha, \beta)}$ (I, J)	Started-by$_{(K)}$ $^{(\alpha, \beta)}$ (I, J)	$\forall\, I^{-(i)} \in I^-, \forall\, I^{+(i)} \in I^+, \forall\, J^{-(j)} \in J^-, \forall\, J^{+(j)} \in J^+/$ $(I^{-(i)} = J^{-(j)}) \wedge$ $(I^{+(i)} < J^{+(j)})$	Min(Same$^{(\alpha, \beta)}$ $(I^{-(1)}, J^{-(1)}) \wedge$ Same$^{(\alpha, \beta)}$ $(I^{-(N)}, J^{-(N)}) \wedge$ Proceed$_{(K)}$ $^{(\alpha, \beta)}$ $(I^{+(N)}, J^{+(1)}))$
During$_{(K)}$ $^{(\alpha, \beta)}$ (I, J)	Contains$_{(K)}$ $^{(\alpha, \beta)}$ (I, J)	$\forall\, I^{-(i)} \in I^-, \forall\, I^{+(i)} \in I^+, \forall\, J^{-(j)} \in J^-, \forall\, J^{+(i)} \in J^+/$ $(J^{-(j)} < I^{-(i)}) \wedge$ $(I^{+(i)} < J^{+(i)})$	Min(Proceed$_{(K)}$ $^{(\alpha, \beta)}$ $(J^{-(N)}, I^{-(1)}) \wedge$ Proceed$_{(K)}$ $^{(\alpha, \beta)}$ $(I^{+(N)}, J^{+(1)}))$
Ends$_{(K)}$ $^{(\alpha, \beta)}$ (I, J)	Ended-by$_{(K)}$ $^{(\alpha, \beta)}$ (I, J)	$\forall\, I^{-(i)} \in I^-, \forall\, I^{+(i)} \in I^+, \forall\, J^{-(j)} \in J^-, \forall\, J^{+(j)} \in J^+/$ $(I^{-(i)} < J^{-(j)}) \wedge$ $(I^{+(i)} = J^{+(j)})$	Min(Proceed$_{(K)}$ $^{(\alpha, \beta)}$ $(J^{-(N)}, I^{-(1)}) \wedge$ Same$^{(\alpha, \beta)}$ $(I^{+(1)}, J^{+(1)}) \wedge$ Same$^{(\alpha, \beta)}$ $(I^{+(N)}, J^{+(N)}))$
Equal $^{(\alpha, \beta)}$ (I, J)	Equal $^{(\alpha, \beta)}$ (I, J)	$\forall\, I^{-(i)} \in I^-, \forall\, I^{+(i)} \in I^+, \forall\, J^{+(j)} \in J^+/$ $(I^{-(i)} = J^{-(j)}) \wedge$ $(I^{+(i)} = J^{+(j)})$	Min(Same$^{(\alpha, \beta)}$ $(I^{-(1)}, J^{-(1)}) \wedge$ Same$^{(\alpha, \beta)}$ $(I^{-(N)}, J^{-(N)}) \wedge$ Same$^{(\alpha, \beta)}$ $(I^{+(1)}, J^{+(1)}) \wedge$ Same$^{(\alpha, \beta)}$ $(I^{+(N)}, J^{+(N)}))$

The works presented in this paper have been tested in two projects, having in common to manage life logging data: (1) In the VIVA[1] project, we aim to design the Captain Memo memory prosthesis [19] [20] for Alzheimer Disease patients. Among other functionalities, this prosthesis manages a knowledge base on the patient's family tree, using an OWL ontology. Imprecise inputs are especially numerous when given by an Alzheimer Disease patient. Furthermore, dates are often given in reference to other dates or events. Thus, we have been using our approach reported in this paper. One interesting point in this solution is to deal with a personalized slicing of the person's life in order to sort the different events. (2) The ANR DAPHNE project aims to allow Middle Ages specialized historians to deal with prosopographical data bases storing Middle age academic's career histories. Data come from various archives among Europe and data about a same person are very difficult to align. Representing and

---

[1] http://viva.cnam.fr/.

ordering imprecise time interval is required to redraw the careers across the different European universities who hosted the person.

Future work will be devoted to propose a crisp-based approach to represent and reason on imprecise time intervals in ontology. This approach uses only crisp standards and tools and is modeled in OWL 2.

References

1. Bobillo, F., Straccia, U.: Fuzzy ontology representation using OWL 2. Int. J. Approx. Reason. **52**(7), 1073–1094 (2011)
2. Welty, C., Fikes, R., Makarios, S.: A reusable ontology for fluents in OWL. In: Frontiers in Artificial Intelligence and Applications, pp. 226–236 (2006)
3. Bobillo, F., Straccia, U.: FuzzyDL: An expressive fuzzy description logic reasoner. In: IEEE World Congress on Computational Intelligence, pp. 923–930 (2008)
4. Artale, A., Franconi, E.: A survey of temporal extensions of description logics. In: Annals of Mathematics and Artificial Intelligence, pp. 171–210 (2000)
5. Klein, M., Fensel, D.: Ontology versioning on the semantic web. In: International Semantic Web Working Symposium (SWWS), pp. 75–91 (2001)
6. Hayes, P., Welty, C.: Defining n-ary relations on the semantic web. In: W3C Working Group Note (2006)
7. Batsakis, S., Petrakis, E.G.: Representing and reasoning over spatio-temporal information in OWL 2.0. In: Workshop on Semantic Web Applications and Perspectives (2010)
8. Allen, J.F.: Maintaining knowledge about temporal intervals. In: Communications of the ACM, pp. 832–843 (1983)
9. Guesgen, H.W., Hertzberg, J., Philpott, A.: Towards implementing fuzzy Allen relations. In: Workshop on Spatial and Temporal Reasoning, pp. 49–55 (1994)
10. Dubois, D., Prade, H.: Processing fuzzy temporal knowledge. In: IEEE Transactions on Systems, Man, and Cybernetics, pp. 729–744 (1989)
11. Badaloni, S., Giacomin, M.: The algebra IA^{fuz}: a framework for qualitative fuzzy temporal. J. Artif. Intell. **170**(10), 872–902 (2006)
12. Freksa, C.: Temporal reasoning based on semi–intervals. In: Artificial intelligence, pp. 199–227 (1992)
13. Nagypál, G., Motik, B.: A Fuzzy Model for Representing Uncertain, Subjective, and Vague Temporal Knowledge in Ontologies. In: Meersman, R., Tari, Z., Schmidt, Douglas C. (eds.) OTM 2003. LNCS, vol. 2888, pp. 906–923. Springer, Heidelberg (2003). https://doi.org/10.1007/978-3-540-39964-3_57
14. Ohlbach, H.: Relations between fuzzy time intervals. In: International Symposium on Temporal Representation and Reasoning, pp. 44–51. IEEE (2004)
15. Schockaert, S., De Cock, M., Kerre, E.E.: Fuzzifying Allen's temporal interval relations. In: IEEE Transactions on Fuzzy Systems, pp. 517–533 (2008)
16. Gammoudi, A., Hadjali, A., Ben Yaghlane, B.: Fuzz-TIME: an intelligent system for fluent managing fuzzy temporal information. Int. J. Intell. Comput. Cybern. **10**(2), 200–222 (2017)
17. Zadeh, L.: The Concept of a Linguistic Variable and its Application to Approximate Reasoning. Int. J. Inf. Sci. **9**(1), 301–357 (1975)
18. Vilain, M.B., Kautz, H.: Constraint propagation algorithms for temporal reasoning. In: Readings in qualitative reasoning about physical systems, pp. 377–382 (1986)

19. Herradi, N., Hamdi, F., Métais, E., Ghorbel, F., Soukane, A.: PersonLink: An Ontology Representing Family Relationships for the CAPTAIN MEMO Memory Prosthesis. In: Jeusfeld, Manfred A., Karlapalem, K. (eds.) ER 2015. LNCS, vol. 9382, pp. 3–13. Springer, Cham (2015). https://doi.org/10.1007/978-3-319-25747-1_1
20. Métais, E., Ghorbel, F., Herradi, N., Hamdi, F., Lammari, N., Nakache, D., Ellouze, N., Gargouri, F., Soukane, A.: Memory prosthesis. In: Non-Pharmacological Therapies in Dementia (2015)

Assessing the Impact of Single and Pairwise Slot Constraints in a Factor Graph Model for Template-Based Information Extraction

Hendrik ter Horst[1]([✉]), Matthias Hartung[1]([✉]), Roman Klinger[2],
Nicole Brazda[3], Hans Werner Müller[3], and Philipp Cimiano[1]

[1] CITEC, Bielefeld University, Bielefeld, Germany
{hterhors,mhartung,cimiano}@techfak.uni-bielefeld.de
[2] IMS, University of Stuttgart, Stuttgart, Germany
roman.klinger@ims.uni-stuttgart.de
[3] CNR and Neurology, HHU Düsseldorf, Düsseldorf, Germany
{nicole.brazda,hanswerner.mueller}@uni-duesseldorf.de

Abstract. Template-based information extraction generalizes over standard token-level binary relation extraction in the sense that it attempts to fill a complex template comprising multiple slots on the basis of information given in a text. In the approach presented in this paper, templates and possible fillers are defined by a given ontology. The information extraction task consists in filling these slots within a template with previously recognized entities or literal values. We cast the task as a structure prediction problem and propose a joint probabilistic model based on factor graphs to account for the interdependence in slot assignments. Inference is implemented as a heuristic building on Markov chain Monte Carlo sampling. As our main contribution, we investigate the impact of soft constraints modeled as single slot factors which measure preferences of individual slots for ranges of fillers, as well as pairwise slot factors modeling the compatibility between fillers of two slots. Instead of relying on expert knowledge to acquire such soft constraints, in our approach they are directly captured in the model and learned from training data. We show that both types of factors are effective in improving information extraction on a real-world data set of full-text papers from the biomedical domain. Pairwise factors are shown to particularly improve the performance of our extraction model by up to +0.43 points in precision, leading to an F_1 score of 0.90 for individual templates.

Keywords: Ontology-based information extraction
Slot filling · Probabilistic graphical models
Soft constraints · Database population

1 Introduction

Initiated by the advent of the distant supervision [13] and open information extraction paradigms [2], the last decade has seen a tendency to reduce

© Springer International Publishing AG, part of Springer Nature 2018
M. Silberztein et al. (Eds.): NLDB 2018, LNCS 10859, pp. 179–190, 2018.
https://doi.org/10.1007/978-3-319-91947-8_18

information extraction problems to relation extraction tasks. In the latter, the focus is on extracting binary entity-pair relations from text by applying various types of discriminative classification approaches. We argue that many tasks in information extraction (in particular, when being used as an upstream process for database population) go beyond the binary classification of whether a given text expresses a given relation or not, as they require the population of complex *template structures*. Such templates consist of a number of typed slots to be filled from unstructured text [6]. Following an ontology-based approach [20], we assume that the templates (including slots and the types of their potential fillers) are pre-defined in a given ontology.

We frame template-based information extraction as an instance of a structured prediction problem [17] which we model in terms of a joint probability distribution over value assignments to each of the slots in a template. Subsequently, we will refer to such templates as *schemata* in order to avoid ambiguities. Formally, a schema S consists of typed slots $(s_1, s_2, \ldots, s_n)$. The slot filling task corresponds to the maximum a posteriori estimation of a joint distribution of slot fillers given a document d

$$(s_1, s_2, \ldots, s_n) = \operatorname*{argmax}_{s_1', s_2', \ldots, s_n' \in \Phi} P(s_1 = s_1', \ldots, s_n = s_n' \mid d), \qquad (1)$$

where Φ is the set of all possible slot assignments.

Slots in a schema are interdependent, and these dependencies need to be taken into account to avoid incompatible slot assignments. A simple formulation in terms of n binary-relation extraction tasks would therefore be oversimplifying. On the contrary, measuring the dependencies between all slots would render inference and learning intractable. We therefore opt for an intermediate solution, in which we analyze as to what extent measuring *pairwise* slot dependencies helps in avoiding incompatibilities and finally to improve an information extraction model for the task.

We propose a factor graph approach to schema/template-based information extraction which incorporates factors that are explicitly designed to encode such constraints. Our main research interest is therefore to (1) understand whether such constraints can be learned from training data (to avoid the need for manual formulation by domain experts), and (2) to assess the impact of these constraints on information extraction performance.

We evaluate our information extraction model on a corpus of scientific publications reporting the outcomes of pre-clinical studies in the domain of spinal cord injury. The goal is to instantiate multiple schemata to capture the main parameters of each study. We show that both types of constraints are effective, as they enable the model to outperform a naive baseline that applies frequency-based filler selection for each slot.

2   Related Work

Template/Schema-based information extraction dates back to the MUC-4 Shared Task [18] which aimed at extracting instantiations of templates describing terrorist attacks. More recently, Haghighi et al. [7] focus on corporate acquisition events. Information extraction approaches in this line of research are commonly limited to only one or a fixed set of templates, each of them containing only a comparably small set of slots. Obviously, these assumptions pose severe restrictions to real-world application scenarios. Many tasks in the context of knowledge discovery from scientific literature [8], for instance, require a rich representation of the technical domain of interest, which commonly involves numerous templates with multiple (and possibly hierarchically embedded) slots.

Recent examples of reducing slot filling problems to relation extraction tasks are Riedel et al. [15] with a focus on knowledge base completion, Zhang et al. [21], Adel et al. [1], and Singh et al. [16] in the context of cold-start knowledge base population. While our work also addresses the cold-start problem, our domain of application requires the population of complex ontologically typed schemata. We approach this challenge using undirected probabilistic graphical models which integrate coherence constraints over pairs of slots within a schema. Similar techniques have been proposed for the more shallow problems of HMM-based sequence labeling by Chang et al. [5] and relation extraction by Lopez de Lacalle and Lapata [12]. In line with the latter approach, we aim at inducing constraint knowledge automatically from training data.

Methodologically, our work is similar to collective information extraction with undirected graphical models as proposed by Bunescu et al. [4] or Kluegl et al. [9]; however, these approaches are limited to problems of text segmentation, entity tagging and extraction of individual relations.

As the only precursor of our work towards information extraction in the spinal cord injury domain, Paassen et al. [14] address entity extraction in isolation, i. e., they aim at detecting all entities taking part in a relation, without considering the relation classification task as such.

3   Method

We frame the slot filling task as a joint inference problem in undirected probabilistic graphical models. Our model is a factor graph [11] which probabilistically measures the compatibility of a given textual document d consisting of tokenized sentences χ, a fixed set of entity annotations $\mathcal{A}$, and a to be filled ontological schema S. The schema S is automatically derived from an ontology and is described by a set of typed slots, $S = \{s_1, \ldots, s_n\}$. Let $\mathcal{C}$ denote the set of all entities from the ontology, then each slot $s_i \in S$ can be filled by a pre-defined subset of $\mathcal{C}$ called slot fillers. Further, each annotation $a \in \mathcal{A}$ describes a tuple $\langle t, c \rangle$ where $t = (t_i, \ldots, t_j) \in \chi$ is a sequence of tokens with length ≥ 1 and a corresponding filler type $c \in \mathcal{C}$.

3.1   Factorization of the Probability Distribution

We decompose the overall probability of a schema S into probability distributions over single slot and pairwise slot fillers. Each individual probability distribution

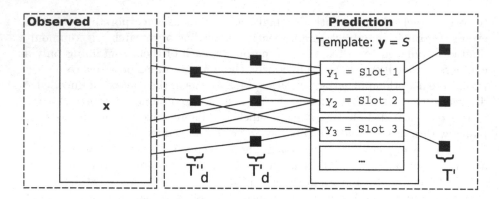

Fig. 1. Factor graph of our model for an exemplary ontological schema S.

is described through factors that measure the compatibility of single/pairwise slot assignments. An unrolled factor graph that represents our model structure is depicted in Fig. 1. The factor graph consists of different types of factors that are connected to subsets of variables of $y = \{y_0, y_1, \ldots, y_n\}$ and of $x = d = \{\chi, \mathcal{A}\}$, respectively. We distinguish three factor types by their instantiating factor graph template $\{T', T'_d, T''_d\} \in \mathcal{T}$: (i) **Single slot factors** $\Psi'(y_i) \in T'$ that are solely connected to a single slot y_i, (ii) **Single slot+text factors** $\Psi'(x, y_i) \in T'_d$ that are connected to a single slot y_i and x, (iii) **Pairwise slot+text factors** $\Psi''(x, y_i, y_j) \in T''_d$ that are connected to a pair of two slots y_i, y_j and x.

The conditional probability $P(y \mid x)$ of a slot assignment y given x is then

$$P(y|x) \quad = \quad \frac{1}{Z(x)} \prod_{y_i \in S} \left[\Psi'(y_i) \ \cdot \ \Psi'(x, y_i) \right] \prod_{y_i \in S} \prod_{y_j \in S} \left[\Psi''(x, y_i, y_j) \right], \quad (2)$$

where $Z(x)$ denotes the partition function and all factors are formulated as $\Psi(\cdot) = \exp(\langle f_T(\cdot), \theta_T \rangle)$ with sufficient statistics $f_T(\cdot)$ and parameters θ_T ($T \in \mathcal{T}$ and $\Psi \in \{\Psi', \Psi''\}$).

3.2 Inference and Learning

We perform Markov chain Monte Carlo (MCMC) sampling to approximate a posterior distribution, while sharing the factorization properties as defined by the factor graph [10]. We learn the parameters via SampleRank [19].

Ontological Sampling. The generation of proposal states in our MCMC sampling procedure follows the idea of Gibbs sampling, mainly applying atomic changes to slots. The initial state s_0 in our exploration is empty, thus $y = (\varnothing)$. A set of potential successors is generated by a proposal function changing a slot by either deleting an already assigned value or changing the value to another slot filler. The state with the highest probability s_{t+1} is chosen as successor state only if $p(s_{t+1}) > p(s_t)$. The inference procedure stops, iff $s_{t+3} = s_t$.

Objective Function. Given a predicted assignment $\boldsymbol{y}'$ of all slots in schema type $\hat{S}$ and a set $\mathcal{G}$ of instantiated schemata of type $\hat{S}$ from the gold standard, the training objective is

$$\max_{\boldsymbol{y}^* \in \mathcal{G}} F_1(\boldsymbol{y}^*, \boldsymbol{y}'), \tag{3}$$

where F_1 is the harmonic mean of precision and recall, based on the overlap of assigned slot values between $\boldsymbol{y}'$ and $\boldsymbol{y}^*$.

3.3 Factors and Constraints

At the core of our model are features that encode soft constraints to be learned from training data. In general, these constraints are intended to measure the compatibility of slot fillers within a predicted schema. Such soft constraints are designed through features that are described in the following.

Single-Slot Constraints in Template T'. We include features which measure common, acceptable fillers for single slots with numerical values. Given filler annotations $\{a_i = \langle v, c \rangle\}$ of slot y_i, the model can learn individual intervals for different types of fillers such as temperature $(-10\text{--}40)$, or weight $(200\text{--}500)$, for example. For that, we calculate the average μ and standard deviation σ for each particular slot based on the training data. For each slot s_i in schema S, a boolean feature $f_{\sigma=n}^{s_i}$ is instantiated for each $n \in \{0, \ldots, 4\}$, indicating whether the value y_i is within n standard deviations σ_{s_i} of the corresponding mean μ_{s_i}. To capture the negative counterpart, a boolean feature $f_{\sigma>n}^{s_i}$ is instantiated likewise:

$$f_{\sigma=n}^{s_i}(y_i) = \begin{cases} 1 & \text{iff } \lceil (\frac{v - \mu_{s_i}}{\sigma_{s_i}}) \rceil = n \\ 0 & \text{otherwise.} \end{cases} \qquad f_{\sigma>n}^{S_i}(y_i) = \begin{cases} 1 & \text{iff } \lceil (\frac{v - \mu_{s_i}}{\sigma_{s_i}}) \rceil > n \\ 0 & \text{otherwise.} \end{cases} \tag{4}$$

In this way, the model learns preferences over possible fillers for a given slot which effectively encode soft constraints such as "the weight of rats typically scatters around a mean of $300\,\text{g}$ by two standard deviations of $45\,\text{g}$".

Pairwise Slot Constraints in T_d''. In contrast to single-slot constraints, pairwise constraints are not limited to slots with real-valued fillers. Soft constraints on slot pairs are designed to measure the compatibility and (hidden) dependencies between two fillers, e.g., the dependency between the dosage of a medication and its applied compound, or between the gender of an animal and its weight. This is modeled in terms of their linguistic context and textual locality, as discussed in the following.

We assume that possible slot fillers may be mentioned multiple times at various positions in a text. Therefore, given a pair of slots (s_i, s_j), we define λ as an aggregation function that returns the subset of annotations $\lambda(s_i) = \{a = \langle t, c \rangle \in \mathcal{A} \mid a(c) = s_i(c)\}$. We measure the locality of two slots in the text by the minimum distance between two sentences containing annotations for the corresponding slot fillers. A bi-directional distance for two annotations is defined as $\delta(a_k, a_l) = |\text{sen}(a_k) - \text{sen}(a_l)|$ where sen denotes a function that returns the sentence index of an annotation. For each $n \in \{0, \ldots, 9\}$, a boolean feature $f_{\delta=n}$ is instantiated as:

$$f_{\delta=n}^{s_i, s_j}(y_i, y_j) = \begin{cases} 1 & \text{iff } n = \min_{a_k \in \lambda(y_i), a_l \in \lambda(y_j)} \delta(a_k, a_l) \\ 0 & \text{otherwise.} \end{cases} \tag{5}$$

To capture the linguistic context between two slot fillers y_i and y_j, we define a feature $f_{\pi_n}^{s_i}(y_i, y_j)$ that indicates whether a given $\mathcal{N}$-gram $\pi_n \in \pi$ with $1 < \mathcal{N} \leq 3$ occurs between the annotations $a_k \in \lambda(y_i)$ and $a_l \in \lambda(y_i)$ in the document.

Textual Features in T' and T'_d. Given a single slot s_i with filler y_i and the aggregated set of all corresponding annotations $\lambda(y_i)$, we instantiate three boolean features for each annotation $a \in \lambda(y_i)$ as follows.

Let $L_s(l_{y_i}, a(t))$ be the Levenshtein similarity between the ontological class label l_{y_i}, and the tokens of an annotation $a(t)$. Two boolean features $f_{bin(s_{max}) < \Delta}(y_i)$ and $f_{bin(s_{max}) \geq \Delta}(y_i)$ are computed as:

$$f_{bin(s_{max}) < \Delta}(y_i) = \begin{cases} 1 & \text{iff } b < \Delta \\ 0 & \text{otherwise.} \end{cases} \quad f_{bin(s_{max}) \geq \Delta}(y_i) = \begin{cases} 1 & \text{iff } b \geq \Delta \\ 0 & \text{otherwise.} \end{cases}, \tag{6}$$

where $b = bin(s_{max})$ is the discretization of the maximum similarity s_{max} into intervals of size 0.1, and

$$s_{max} = \max_{a \in \lambda(y_i)} L_s(l_{y_i}, a(t)) \text{ with } L_s = 1 - \frac{\text{levenshtein}(l_{y_i}, a(t))}{\max(\text{len}(l_{y_i}), \text{len}(a(t)))}. \tag{7}$$

Finally, we instantiate features $f_{\pi_k\ context}^{s_i}(y_i)$ and $f_{\pi_k\ within}^{s_i}(y_i)$, indicating whether an $\mathcal{N}$-gram π_k occurs in the context (before or after) or within any annotation of slot y_i.

4 Database Population in the Spinal Cord Injury Domain

4.1 Problem Description

We address the problem of ontology-based information extraction in a slot filling setting as a prerequisite for cold-start database population. The extraction task comprises multiple schemata of different types, each of them being provided by a domain ontology and containing multiple slots. Each slot in a schema needs to be filled either by a literal from the input document or by a class from the ontology, depending on whether it is derived from a data-type or object-type property (cf. Fig. 2).

We consider slot filling as a document-level task, i.e., entities filling the slots of a particular schema may be dispersed across the entire text. In addition, each literal or ontological category can, in principle, fill multiple slots of the appropriate type. We approach the task in a supervised machine learning approach; supervision is available at the document level in terms of fully instantiated gold schemata without direct links between slot fillers and text mentions.

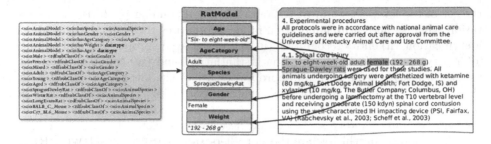

Fig. 2. Information extraction workflow: Domain concepts and associated slots are defined in a *domain ontology* (left) and transformed into *schema structures* (middle) which are automatically populated from text (right) by the *slot filling model.*

4.2   Application Context

Our work in the PSINK project[1] aims at information extraction from full-text scientific publications on pre-clinical experiments in the spinal cord injury domain. The results of the extraction process (i.e., fully instantiated schemata as shown in Fig. 2) will be made accessible in a comprehensive database in order to foster translation from pre-clinical trials into clinical therapeutic concepts bearing the potential to induce neuronal regeneration in human patients suffering from spinal cord injuries.

This information extraction task is an instance of the problem described in Sect. 4.1, with the extraction schema being derived from the specifically designed Spinal Cord Injury Ontology (cf. Sect. 4.3 below).

4.3   Ontology and Data Set

Spinal Cord Injury Ontology (SCIO). Pre-clinical trials in the spinal cord injury domain follow strict methodological patterns. Experimental protocols and the main outcomes of pre-clinical studies on spinal cord injury are formally represented in SCIO [3]. In total, the ontology contains more than 500 classes and approx. 80 properties (slots). SCIO top-level classes defining the schema types are ANIMALMODEL, INJURYMODEL, TREATMENT, INVESTIGATIONMETHOD and RESULT. Slots are either object-type properties which can be filled by a SCIO class, or data-type properties which are filled with free text. For example, Fig. 2 (left and middle part) presents the ANIMALMODEL class along with its predefined slots: *ageCategory*, *gender* and *species* are object-type properties; *age* and *weight* are data-type properties.

Annotated Data Set. The annotated data set was created by two SCI experts who annotated 25 full-text scientific papers from the SCI literature. Annotations were provided at the level of fully instantiated schemata per document, using

---

[1] http://www.psink.de.

the set of top-level classes in SCIO and their corresponding properties as annotation schema. The entire annotation process comprises three steps: (i) mention identification, (ii) entity recognition (in case of data-type properties) and linking (object-type properties), (iii) schema instantiation, and (iv) filling the slots of an instantiated schema with an appropriate entity. The latter steps are due to the fact that the cardinality of schemata of a particular type per document is unknown a priori, and multiple schemata may share individual slot fillers. The following example shows a sentence that describes two instantiations of an ANIMALMODEL schema which share the slot fillers *species* (SPRAGUEDAWLEYRAT) and *ageCategory* (ADULT): *"A total of 39 Sprague-Dawley rats were used for these experiments: adult males (285–330 g) and females (192–268 g)."*

Inter-annotator agreement at the level of fully instantiated schemata in terms of F_1 score between annotators amounts to 0.93 for ANIMALMODEL, 0.79 for INJURY, 0.77 for TREATMENT and 0.65 for INVESTIGATIONMETHOD.

5 Experiments

In the following section, we describe our experimental settings, the evaluation metrics and results. Model performances are independently reported for four SCIO schemata: ANIMALMODEL, INJURY, TREATMENT, and INVESTIGATION-METHOD (cf. Sect. 4.3). As a preprocessing step, we apply symbolic entity recognition in order to generate annotations $\mathcal{A}$. The regular expressions used are automatically generated from ontology class labels. In case of data-type properties (e.g., weight of an animal), regular expressions are manually created.

5.1 Experimental Settings

The system is evaluated in a 6-fold cross validation on the complete data set. In all experiments, we restrict the complexity of the schemata to first-order slots, i.e., ontological properties that are directly connected to their respective domain class. In the current approach, we are not aiming at predicting the correct number of instantiations per schema type. Thus, our system is restricted to fill a single schema of each type per document, even if it contains multiple instances of the same schema type (e.g., multiple TREATMENTs).

With respect to this restriction, we report the evaluation results for both (i) *Full Evaluation* (taking the actual number of gold schemata into account), and (ii) *Best Match Evaluation* (comparing the predicted schema to the best matching gold schema).

Further, we report the performance for two different models, in order to investigate the relative impact of single-slot constraints vs. pairwise slot constraints. In the *pairwise slot filling* (PSF) model, the inference and the factor graph is based on the joint assignment of slot pairs, whereas in *single slot filling* (SSF) model, all slots are independently filled.

Evaluation Metrics. We report model performances as macro precision, recall and harmonic F_1. Given a document with a set of gold schemata $\mathcal{G}$ of type

$S = \{s_0, \ldots s_n\}$ and the predicted schema p, the comparison is always based on the best assignment $g' = \text{argmax}_{g \in \mathcal{G}} F_1(p, g)$. For the computation of the overall F_1 score, we convert all ontological schemata into sets of slot-filler pairs with $p = \{s'_0 = c_j, \ldots, s'_n = c_k\}$ and $\mathcal{G} = \{g^0, \ldots, g', \ldots, g^l\} = \{(s_0^0 = c_a, \ldots, s_n^0 = c_b), \ldots, (s'_0 = c_c, \ldots, s'_n = c_d), \ldots, (s_0^l = c_e, \ldots, s_n^l = c_f)\}$. The overall F_1 score is calculated based on the two sets of p and $\mathcal{G}$. We define a true positive (tp) as a slot-filler pair that are in both p and $\mathcal{G}$, a false positive (fp) as a pair that is in p but not in $\mathcal{G}$, and a false negative (fn) as a pair that is in $\mathcal{G}$ but not in p. During the *Best Match Evaluation*, we set $\mathcal{G} = \{g'\}$.

Most Frequent Filler Baseline. We compare the performance of our models in all settings against a naive but plausible baseline. Following the intuition that important information is mentioned in a higher frequency than non-important information, a slot is always filled with the filler that has the highest annotation frequency. In the following, we refer to this procedure as Most Frequent Filler (MFF) baseline.

5.2 Results

In the following, we describe the evaluation results for all experiments. First, we compare the performance in the *Full Evaluation* vs. *Best Match Evaluation* settings. In the former setting, we expect a rather low recall due to the restriction of predicting exactly one schema per type. This leads to many false negatives, as multiple instances of the same type cannot be fully covered yet. Hence, we hypothesize a significant increase in recall in the *Best Match Evaluation* setting. By comparing the predicted schema to the best match only, we investigate whether the low recall is due to the large amount of missing schemata. If so, this would indicate that our model is able to select the correct slot fillers among a huge set of possible candidates. The performance of all models in both settings is reported in Table 1.

Full Evaluation Results. The results show a strong recall of our baseline model with a distinct lack in precision. The baseline yields the highest recall among all models and schema types except for the ANIMALMODEL (0.55 for baseline vs. 0.90 for SSF/PSF). Compared to the SSF model, we notice a considerable increase in precision in all schema types which is most pronounced in the INVESTIGATIONMETHOD (+0.64). The increase in precision for the three other schemata are between +0.24 and +0.36. Comparing the PSF to the SSF model, we observe further strong improvements in precision and slight improvements in recall. The PSF model clearly outperforms the baseline for the ANIMALMODEL with an increase in F_1 of +0.39, the INJURY +0.12, and the INVESTIGATIONMETHOD with +0.14. Despite the precision being increased by +0.46 in the TREATMENT, the baseline shows a higher F_1 score in this configuration (+0.03), due to a drop in recall by −0.10.

Table 1. Performance of Most Frequent Filler Baseline (MFF) vs. Single Slot Filler (SSF) and Pairwise Slot Filler (PSF) models in the *Full Evaluation* (full) and *Best Match* (best) setting.

		MFF			SSF			PSF		
		P	R	F_1	P	R	F_1	P	R	F_1
ANIMAL	Full	0.48	0.55	0.51	0.84	0.90	0.86	**0.91**	**0.90**	**0.90**
MODEL	Best	0.48	0.57	0.52	0.84	**1.00**	0.91	**0.91**	**1.00**	**0.95**
INJURY	Full	0.28	**0.38**	0.31	0.52	0.22	0.31	**0.77**	0.30	**0.43**
	Best	0.28	**0.43**	0.33	0.52	0.29	0.35	**0.77**	0.40	**0.50**
TREAT-	Full	0.39	**0.26**	**0.30**	0.70	0.16	0.26	**0.87**	0.16	0.27
MENT	Best	0.39	**0.74**	0.51	0.70	0.63	0.65	**0.87**	0.63	**0.73**
INVEST.	Full	0.36	**0.45**	0.36	**1.00**	0.39	0.50	**1.00**	0.39	**0.50**
METHOD	Best	0.36	0.98	0.52	**1.00**	**1.00**	**1.00**	**1.00**	**1.00**	**1.00**

Best Match Evaluation Results. In this setting, we further investigate the recall performance of our models compared to the previously discussed *Full Evaluation* results. As we only remove uncaptured schema instances from $\mathcal{G}$ (cf. Sect. 5.1), the precision remains the same. All models show an overall increase in recall for all schema types. With respect to the PSF model, we can see a strong increase in recall for INVESTIGATIONMETHOD by +0.61 and for TREATMENT by +0.47. Further, slight increases by +0.10 and +0.07 can be observed for ANIMALMODEL and INJURY, respectively. Similar observations can be made for the SSF model.

5.3   Discussion

Comparing the baseline model with the SSF model, we notice a very strong increase in precision in combination with a slight drop in recall. This positive trend in precision is continued when considering the PSF model. Further, the results show a positive impact of pairwise over single-slot constraints on recall.

The high recall of 0.90 for the ANIMALMODEL in the full evaluation is mainly due to a low number (1 to 2) of instances per schema type in each document. The fact that there is no difference in the performance of the SSF and SPF models for the INVESTIGATIONMETHOD suggests a strong slot independence, so that pairwise slot constraints do not have a big impact in this particular case. The low increase in recall between the two evaluation settings for the INJURY suggests difficulties for this schema. In contrast, the recall increase for the TREATMENT schema from 0.16 to 0.63 clearly shows that most of the errors are due to a large number of schema instances per document.

Overall, the results show that our system is often able to select the correct set of slot fillers for a schema, even from a huge set of possible schemata and their corresponding slot filler candidates.

6 Conclusions and Outlook

We have investigated the impact of single and pairwise slot constraints in a factor graph model for schema/template-based information extraction. We found that both types of constraints increase the overall performance of the slot filling model, as they are able to capture soft slot restrictions (for single slots) and (hidden) slot dependencies (for pairwise slots). We were able to show that, compared to a plausible baseline, both constraint types are effective, with pairwise constraints outperforming the single slot constraints. For future work, we plan to extend the current model by incorporating further constraints beyond the current restriction to pairwise slot dependencies, with a potential culmination in a fully joint model.

Our approach was developed in the context of the PSINK project which aims at populating a database for pre-clinical studies in the spinal cord injury domain. Our proposed approach lays the groundwork for this task by instantiating onto-logically defined schemata and filling them from unstructured text. In future work, we plan to extend our approach to more complex schemata covering the entire ontology. This raises further research questions that need to be answered, such as *How to determine the actual number of instances per schema type?* and *How to efficiently explore recursively nested properties within complex schemata?*

Acknowledgments. This work has been funded by the Federal Ministry of Education and Research (BMBF, Germany) in the PSINK project (project numbers 031L0028A/B).

References

1. Adel, H., Roth, B., Schütze, H.: Comparing convolutional neural networks to traditional models for slot filling. In: Proceedings of NAACL/HLT, pp. 828–838 (2016)
2. Banko, M., Cafarella, M., Soderland, S., Broadhead, M., Etzioni, O.: Open information extraction from the web. In: Proceedings of IJCAI, pp. 2670–2676 (2007)
3. Brazda, N., ter Horst, H., Hartung, M., Wiljes, C., Estrada, V., Klinger, R., Kuchinke, W., Müller, H.W., Cimiano, P.: SCIO: an ontology to support the formalization of pre-clinical spinal cord injury experiments. In: Proceedings of the 3rd JOWO Workshops: Ontologies and Data in the Life Sciences (2017)
4. Bunescu, R., Mooney, R.: Collective information extraction with relational markov networks. In: Proceedings of ACL, pp. 438–445 (2004)
5. Chang, M.W., Ratinov, L., Roth, D.: Structured learning with constrained conditional models. Mach. Learn. **88**(3), 399–431 (2012)
6. Freitag, D.: Machine learning for information extraction in informal domains. Mach. Learn. **39**(2–3), 169–202 (2000)
7. Haghighi, A., Klein, D.: An entity-level approach to information extraction. In: Proceedings of ACL, pp. 291–295 (2010)

8. Henry, S., McInnes, B.: Literature based discovery: models, methods, and trends. J. Biomed. Inform. **74**, 20–32 (2017)
9. Kluegl, P., Toepfer, M., Lemmerich, F., Hotho, A., Puppe, F.: Collective information extraction with context-specific consistencies. In: Flach, P.A., De Bie, T., Cristianini, N. (eds.) ECML PKDD 2012. LNCS (LNAI), vol. 7523, pp. 728–743. Springer, Heidelberg (2012). https://doi.org/10.1007/978-3-642-33460-3_52
10. Koller, D., Friedman, N.: Probabilistic Graphical Models: Principles and Techniques. MIT Press, Cambridge (2009)
11. Kschischang, F.R., Frey, B.J., Loeliger, H.A.: Factor graphs and sum product algorithm. IEEE Trans. Inf. Theory **47**(2), 498–519 (2001)
12. Lopez de Lacalle, O., Lapata, M.: Unsupervised relation extraction with general domain knowledge. In: Proceedings of EMNLP, pp. 415–425 (2013)
13. Mintz, M., Bills, S., Snow, R., Jurafsky, D.: Distant supervision for relation extraction without labeled data. In: Proceedings of ACL, pp. 1003–1011 (2009)
14. Paassen, B., Stöckel, A., Dickfelder, R., Göpfert, J.P., Brazda, N., Kirchhoffer, T., Müller, H.W., Klinger, R., Hartung, M., Cimiano, P.: Ontology-based extraction of structured information from publications on preclinical experiments for spinal cord injury treatments. In: Proceedings of the 3rd Workshop on Semantic Web and Information Extraction (SWAIE), pp. 25–32 (2014)
15. Riedel, S., Yao, L., McCallum, A., Marlin, B.M.: Relation extraction with matrix factorization and universal schemas. In: Proceedings of NAACL/HLT, pp. 74–84 (2013)
16. Singh, S., Yao, L., Belanger, D., Kobren, A., Anzaroot, S., Wick, M., Passos, A., Pandya, H., Choi, J.D., Martin, B., McCallum, A.: Universal schema for slot filling and cold start: UMass IESL at TACKBP 2013. In: Proceedings of TAC-KBP (2013)
17. Smith, N.A.: Linguistic Structure Prediction. Morgan and Claypool, San Rafael (2011)
18. Sundheim, B.M.: Overview of the fourth message understanding evaluation and conference. In: Proceedings of MUC, pp. 3–21 (1992)
19. Wick, M., Rohanimanesh, K., Culotta, A., McCallum, A.: SampleRank: learning preferences from atomic gradients. In: Proceedings of the NIPS Workshop on Advances in Ranking, pp. 1–5 (2009)
20. Wimalasuriya, D.C., Dou, D.: Ontology-based information extraction: an introduction and a survey of current approaches. J. Inf. Sci. **36**(3), 306–323 (2010)
21. Zhang, Y., Zhong, V., Chen, D., Angeli, G., Manning, C.D.: Position-aware attention and supervised data improve slot filling. In: Proceedings of EMNLP, pp. 35–45 (2017)

NLP

Processing Medical Binary Questions in Standard Arabic Using NooJ

Essia Bessaies[✉], Slim Mesfar, and Henda Ben Ghzela

Riadi Laboratory, University of Manouba, Manouba, Tunisia
{essia.bessaies,slim.mesfar}@riadi.rnu.tn,
henda.benghezala@ensi.rnu.tn

Abstract. Nowadays, the medical domain has a high volume of electronic documents. The exploitation of this large quantity of data makes the search of specific information complex and time consuming. This difficulty has prompted the development of new adapted research tools, as question-answering systems. Indeed, this type of system allows a user to ask a question in natural language and automatically identify a specific answer instead of a set of documents deemed pertinent, as is the case with search engines. For this purpose, we are developing a question answering system which is based on a linguistic approach. The use of the linguistic engine of NooJ in order to formalize the automatic recognition rules and then applying them to a dynamic corpus composed of arabic medical journalistic articles. In this paper, we present a method for analyzing medical Binary questions. The analysis of the question asked by the user by means the application of cascade of morpho-syntactic resources. The linguistic patterns (grammars) which allow us to annotate the question and the semantic features of the question of extracting the focus and topic of the question. We start with the implementation of the rules which identify and to annotate the various medical entities. The named entity recognizer (NER) is able to find references to people, places and organizations, diseases, viruses, as targets to extract the correct answer from the user. The NER is embedded in our question answering system in order to identify the answer and delimit the potential justification sequence the precision and recall show that the actual results are encouraging and could be integrated for more types of questions other than binary questions.

Keywords: Information extraction · Medical Binary questions
Arabic language · Local grammar · Named entities

1 Introduction

Nowadays, the medical domain has a high volume of electronic documents. The exploitation of this large quantity of data makes the search of specific information complex and time consuming. This complexity is especially evident when we seek a short and precise answer to a human natural language question rather than a full list of documents and web pages. In this case, the user requirement could be a Question Answering (QA) system which represents a specialized area in the field of information retrieval.

© Springer International Publishing AG, part of Springer Nature 2018
M. Silberztein et al. (Eds.): NLDB 2018, LNCS 10859, pp. 193–204, 2018.
https://doi.org/10.1007/978-3-319-91947-8_19

The goal of a QA system is to provide inexperienced users with a flexible access to information allowing them to write a query in natural language and obtain not the documents which contain the answer, but its precise answer passage from input texts. There has been a lot of research in English as well as some European language QA systems. However, Arabic QA systems could not match the pace due to some inherent difficulties with the language itself as well as due to lack of tools available to assist researchers. Therefore, the current project a tempts to design and develop the modules of an Arabic QA system.

For this purpose, the developed question answering system is based on a linguistic approach, using NooJ's linguistic engine in order to formalize the automatic recognition rules and then apply them to a dynamic corpus composed of medical journalistic articles.

In addition, we present a method for analyzing medical questions (for Binary questions). The analysis of the question asked by the user by means of the syntactic and morphological analysis. The linguistic patterns (grammars) which allow us to extract the analysis of the question and the semantic features of the question of extracting the focus and topic of the question.

In the next section, an overview of the state of art describes related works to question answering system. The Sect. 3, we describe the generic architecture of the proposed QA system. In Sect. 4, introduces our approach to annotation of medical factoid question and extraction of right answer.

2   State of Art

As explained in the introduction, QA systems for Arabic are very few. Mainly, it is due to the lack of accessibility to linguistic resources, such as freely available lexical resources, corpora and basic NLP tools (tokenizers, morphological analyzers, etc.). To our knowledge, there are only seven research works on Arabic QA systems:

- Yes/No Arabic Question Answering System [6], is a formal model for a semantic based yes/no Arabic question answering system based on paragraph retrieval. The results are based on 20 documents. It shows a positive result of about 85% when we use entire documents. Besides, it gives 88% when we use only paragraphs. The system focuses only on yes/no questions and the corpus size is relatively small (20 documents).
- QARAB [5] is an Arabic QA system that takes factoid Arabic questions and attempts to provide short answers. QARAB uses both information retrieval and natural language processing techniques.
- ArabiQA [3], which is fully oriented to the modern Arabic language, also answers factoid questions using Named Entity Recognition. However, this system is not yet completed.
- DefArabicQA [10] provides short answers to Arabic natural language questions. This system provides effective and exact answers to definition questions expressed in Arabic from Web resources. DefArabicQA identifies candidate definitions by using a set of lexical patterns, filters these candidate definitions by using heuristic

rules and ranks them by using a statistical approach. It only processes definition questions and does not include other types of question (When, How and Why).

- AQuASys [4] is composed of three modules: A question analysis module, a sentence filtering module and an answer extraction module. Special consideration has been given to improving the accuracy of the question analysis and the answer extraction scoring phases. These phases are crucial in terms of finding the correct answer. The recall rate is 97.5% and the precision rate is 66.25%.

- Idraaq [2] is an Arabic QA system is fully programmed in Java. The system also makes use of other third party components and resources. The system is designed around the three typical modules of a Question Answering system: Question analysis and classification module, Passage Retrieval (PR) module and Answer Validation (AV) module. IDRAAQ registered encouraging performances in particular with factoid questions.

- Al-Bayan [1] Question Answering system for the Quran, that takes an Arabic question as an input and retrieves semantically relevant verses as candidate passages. Then an answer extraction module extracts the answer from the retrieved verses accompanied by their Tafseer. Al-Bayan is composed of four modules: Preprocessing Operations, A question analysis module, Information Retrieval (IR) a and an answer extraction module. Evaluation results on a collected dataset show that the overall system can achieve 85% accuracy.

After this investigation, to solve the problem of question answering system, the developed question answering system is based on a linguistic approach, using NooJ's linguistic engine in order to formalize the automatic recognition rules and then apply them to a dynamic corpus composed of arabic medical journalistic articles. The named entity recognizer (NER) is embedded in our question answering system in order to identify these answers and questions associated with the extracted named entities. For this purpose, we have adapted a rules based approach to recognize Arabic named entities and right answers, using different grammars and gazetteer.

3 Architecture

From a general viewpoint, the design of a QA system (Fig. 1) must take into account four phases:

1. **Question analysis:** This module performs a morphological analysis to determine the question class (binary question). A question class helps the sys-tem to classify the question type to provide a suitable answer. This module may also identify additional semantic features of the question like the topic and the focus.
2. **Text preprocessing segmentation:** will also identify the negation status of the sentence and the style of the sentence. This module is to develop the linguistic patterns for the segmentation of sentences. These patterns are helpful in segmentation and identifying the type of sentences.
3. **Passage retrieval:** The third motivation behind the question classification task is to develop the linguistic patterns for the candidate answers. These patterns are helpful in matching in parsing and identifying the candidate answers.

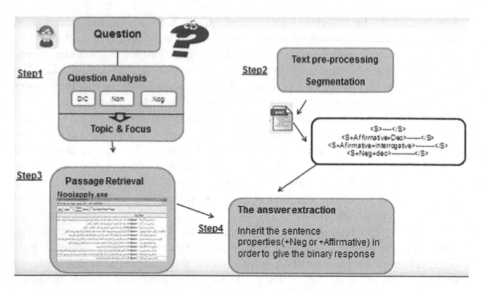

Fig. 1. Architecture of question answering system.

4. **The answer extraction:** This module selects the most accurate answers among the phrases in a given corpus. The selection is based on the question analysis. The suggested answers are then given to the user as a response to his initial natural language query. We are working on the integration of similarity scores in order to better rank the retrieved passages.

4 Our Approach

4.1 Named Entity Recognition (NER)

We think that an integration of a Named Entity Recognition (NER) module will definitely boost system performance. It is also very important to point out that an NER is required as a tool for al-most all the QA system components. Those NER systems allow extracting proper nouns as well as temporal and numeric expressions from raw text [8]. In our case, we used our own NER system especially formulated for the Arabic medical domain. We have considered six proper names categories:

- **Organization:** named corporate, governmental, or other organizational entity;
- **Location:** name of geographically defined location;
- **Person:** named person or family;
- **Viruses:** Names of medical viruses;
- **Disease:** Names of diseases, illness, sickness;
- **Treatment:** Names of Treatments;

Named Entity Recognition (NER) is a subtask of information extraction, where each proper name in the input passage such as persons, locations and numbers - is assigned a named entity tag. In our case, we used our own NER system especially formulated for the Arabic medical domain. We have considered six proper names categories: organization, location, person, viruses, diseases, and treatment (Table 1).

Table 1. Named Entity Recognition

Categories	Definitions	Examples
Organization	Names of corporations, gov. entities or ONGs	بنك آلدم = blood bank
Location	Politically or geographically defined locations	مستشفى الأطفال = Children's Hospita
Person	Names of persons or families	طبيب النساء علي طارق = Gynecologist Tariq Ali
Viruses	Names of medical viruses	فيروس الروتا = Rotavirus
Diseases	Names of diseases, illness, sickness	مرض السرطان = Cancer
Treatment	Names of treatments	علاج طبيعى = Physiotherapist

4.2 Question Analysis

The system first takes the Arabic question which is preprocessed to extract the query that will be used in the Passage retrieval module and Text preprocessing segmentation. The question is also classified to get the type of the question (binary question), and consequently the type of its expected answer, which will then be used in the Answer Extraction module.

Then, in order to look for the best answer, it gives the maximum amount of information (syntactic, semantic, distributional, etc.) from the given question, such as the expected answer the focus and topic of the question. This information will play an important role in the phase of extraction candidate answers.

- **Topic:** the topic corresponds to the subject matter of the question.
- **Focus:** the focus corresponds to the specific property of the topic that the user is looking for.

The following example shows the detailed annotation of the identified parts of a question.

198   E. Bessaies et al.

Example

Is measles a contagious disease?

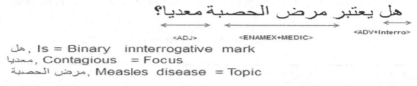

هل يعتبر مرض الحصبة معديا؟

<ADJ>   <ENAMEX+MEDIC>   <ADV+Interro>

هل, Is = Binary innterrogative mark

معديا, Contagious = Focus

مرض الحصبة, Measles disease = Topic

4.3   The Passage Retrieval Module and Text Preprocessing Segmentation

Text Preprocessing Segmentation

Sentence segmentation is the problem of dividing a string of written language into its component sentences.

In this paper, we propose a rule-based approach to Arabic texts segmentation, where segments are sentences, our approach relies on an extensive analysis of a large set of lexical cues as well as punctuation marks. Our approach relies on morphological and syntactic information using several dictionaries and orthographic rectification grammar. To this end, we use NooJ linguistic resources [8] in order to perform surface morphological and syntactic analysis. Integration of segmentation tool for Arabic texts an enhanced version of [7] (Fig. 2).

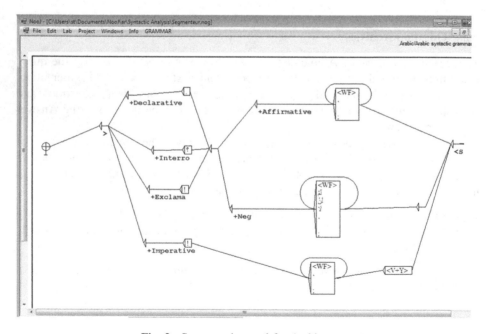

Fig. 2. Segmentation tool for Arabic texts

The segmentation tool will also identify:

- **The negation status of the sentence:** +Negative OR + Affirmative
- **The sentence style:** +Declarative, +Imperative, +Interrogative OR + Exclamative

Automatic annotation for Arabic corpora has an important role in many applications of Natural Language Processing (NLP). In this context, we are interested in the automatic annotation of Arabic corpora using transducers set implemented in NooJ platform [9]. And to achieve our aim, we must precede the annotation phase by a segmentation phase. This segmentation phase will, on the one hand, reduce the complexity of the analysis and, on the other hand, improve NooJ platform functionalities.

Also, we achieved our annotation phase by identifying different types of lexical ambiguities, and then an appropriate set of rules is proposed. These patterns are helpful in segmentation and identifying the type of sentences.

Example

هل يعتبر مرض الحصبة معدي؟ الحصبة هو مرض فيروسي حاد ومعدي يصيب الأطفال، ويسبب لهم بعض المضاعفات التي تكون خطيرة في بعض الأحيان. ويعتبر مرض الحصبة من أكثر الأمراض انتشارا في سن الطفولة بصفه خاصة، ولكنه قد يصيب الكبار أيضاً.

Step 1: After the application of the grammars on this text we notice that the concordances are as follows. The segmentation tool will also identify: the negation status of the sentence and the type of sentence (Fig. 3).

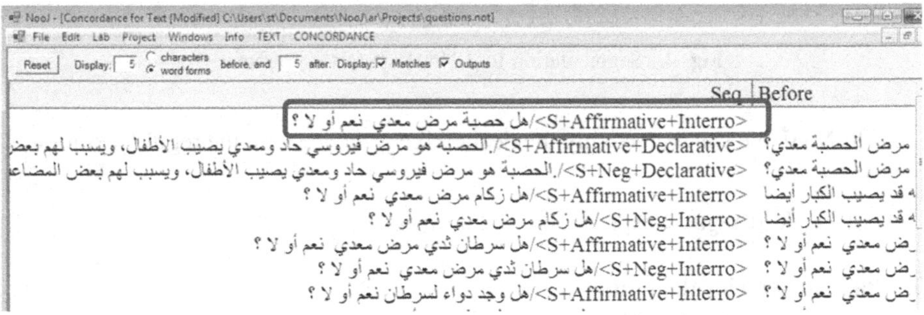

Fig. 3. Result of segmentation NooJ syntactic grammar

          ➢   هل يعتبر مرض الحصبة معدي؟ / <S+Affirmative+Interro>

These patterns of segmentation are helpful in matching in parsing and identifying the candidate answers and the negation status of the sentence, type of sentence.

Step 2: The annotation of the sentences by the recognition of the declarative sentences not the exclamatory and interrogative sentences.

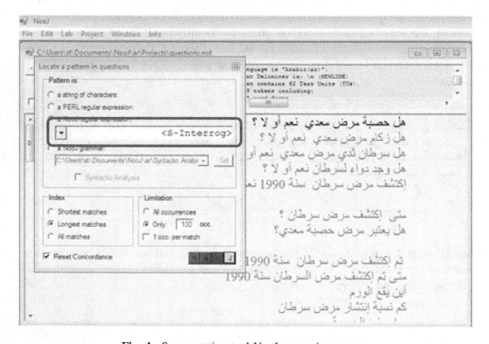

Fig. 4. Segmentation tool NooJ syntactic grammar

With NooJ We can extract only the declarative sentence with the regular expression (Fig. 4):

- Apply a NooJ regular expression: <S-Interro>

The Passage Retrieval Module

The third motivation behind the question classification task is to develop the linguistic patterns for the candidate answers. These patterns are helpful in matching in parsing and identifying the candidate answers.

Passage retrieval is typically used as the first step in current question answering systems. In particular, we show how a variety of prior language models trained on correct answer text allow us to incorporate into the retrieval step information that is often used in answer extraction (Fig. 5).

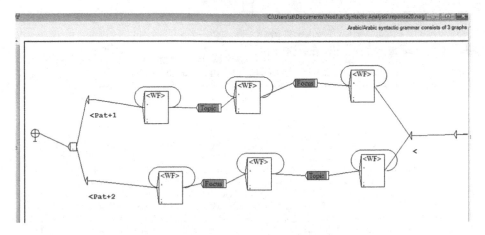

Fig. 5. Passage retrieval tool

Step 3: After analyzing the binary question and extracting the focus and the question topic and segmenting the text in order to extract the candidate responses.

5   Experiments and Results

5.1   NER

Evaluation
To evaluate our NER local grammars, we analyses our corpus to extract manually all named entities. Then, we compare the results of our system with those obtained by manual extraction. The application of our local grammar gives the following result (Table 2):

Table 2. NER grammar experiments on our corpus

Precision	Recall	F-Measure
0,90	0,82	0,88

According to these results, we have obtained an acceptable identification of named entities. Our evaluation shows F-measure of 0.88. We note that the rate of silence in the corpus is low, which is represented by the recall value 0,88 because journalistic texts of our corpus are heterogeneous and extracted from different resources (Fig. 6).

Discussion
Despite the problems described above, the used techniques seem to be adequate and display very encouraging recognition rates. Indeed, a minority of the rules may be sufficient to cover a large part of the patterns and ensure coverage. However, many other rules must be added to improve the recall.

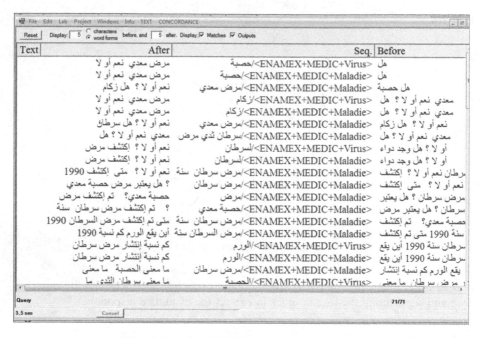

Fig. 6. Result of NER NooJ syntactic grammar

5.2   Automatic Annotation of Binary Question in Standard Arabic

Evaluation

To evaluate our automatic annotation question local grammars, we also Analyses our user's queries to extract manually the question analysis. Then, we compare the results of our system with those obtained by manual extraction. The application of our local grammar gives the following result (Table 3):

Table 3. Annotation question grammar experiments

Precision	Recall	F-Measure
0,75	0,72	0,73

According to these results, we have obtained an acceptable annotation of question.

Our evaluation shows F-measure of 0.73. We note that the rate of silence in the corpus is low, which is represented by the recall value 0.72. This is due to the fact that this assessment is mainly based on the results of the NER module (Fig. 7).

Discussion

Errors are often due to the complexity of user's queries sentences or the absence of their structure in our system In fact, the Arabic sentences are usually very long, which sets up obstacles for the question analysis. Despite the problems described above, the

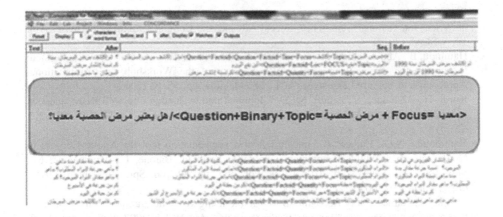

Fig. 7. Result of Annotation NooJ syntactic grammar

developed method seems to be adequate and shows very encouraging extraction rates. However, other rules must be added to improve the result.

6 Conclusion

Arabic Question Answering Systems could not match the pace due to some inherent difficulties with the language itself as well as upon to the lack of tools offered to support the researchers. The task of Question Answering can be divided into four phases; Question Analysis, Text preprocessing segmentation, Passage retrieval, and Answer extraction. Each of these phases plays crucial roles in overall performance of the Question Answering Systems.

As a future work, we work on the other phase of Question Answering Systems; that is "Answer Extraction". In Answer Extraction, we can look for such methods used in this phase including evaluation, tools, and corpus.

Finally, as a long term ambition, we intend to consider studying the processing of the "why" and "how" question types.

References

1. Abdelnasser, H., Mohamed, R., Ragab, M., Farouk, B., El-Makky, N., Torki, M.: Al-Bayan: an Arabic question answering system for the Holy Quran. In: Proceedings of the EMNLP 2014 Workshop on Arabic Natural Language Processing (ANLP) (2014)
2. Abouenour, L., Bouzoubaa, K., Rosso, P.: IDRAAQ: new Arabic question answering system based on query expansion and passage retrieval. In CLEF (2012)
3. Ben Ajiba, Y., Rosso, P., Lyhyaoui, A.: Implementation of the ArabiQA question answering system's components. In: Proceedings Workshop on Arabic Natural Language Processing, 2nd Information Communication Technologies International Symposium, ICTIS-2007, Fez, Morroco, 3–5 April 2017

4. Bekhti, S., Alharbi, M.: Aquasys: a question answering system for Arabic. In: Proceedings of WSeas International Conference. Recent Advances in Computer Engineering Series, vol. 12. WSEAS (2013)

5. Hammou, B., Abu Salam, H., Lytinen, S., Evens, M.: QARAB: a question answering system to support the Arabic language. In: Proceedings of the Workshop on Computational Approaches to Semitic Languages, ACL, Philadelphia, pp. 55–65 (2002)

6. Kurdi, H., Alkhaider, S., Alfaifi, N.: Development and evaluation of a web based question answering system for Arabic language. Comput. Sci. Inf. Technol. (CS & IT) **4**(2), 187–202 (2014)

7. Hammouda, N.G., Haddar, K.: Integration of a segmentation tool for Arabic Corpora in NooJ platform to build an automatic annotation tool. In: Barone, L., Monteleone, M., Silberztein, M. (eds.) NooJ 2016. CCIS, vol. 667, pp. 89–100. Springer, Cham (2016). https://doi.org/10.1007/978-3-319-55002-2_8

8. Mesfar, S.: Named Entity Recognition for Arabic using syntactic grammars. In: Kedad, Z., Lammari, N., Métais, E., Meziane, F., Rezgui, Y. (eds.) NLDB 2007. LNCS, vol. 4592, pp. 305–316. Springer, Heidelberg (2007). https://doi.org/10.1007/978-3-540-73351-5_27

9. Silberztein, M.: La formalisation des langues: l'approche de NooJ. ISTE, Londres (2015)

10. Trigui, O., Hadrich Belguith, L., Rosso, P.: DefArabicQA: Arabic definition question answering system. In: Proceedings of the Workshop on Language Resources and Human Language Technologies for Semitic Languages, 7th LREC, Valletta, Malta (2010)

Verbal Multi-Word Expressions in Yiddish

Chaya Liebeskind$^{(\boxtimes)}$ ⓘ and Yaakov HaCohen-Kerner ⓘ

Department of Computer Science, Jerusalem College of Technology,
Lev Academic Center, 21 Havaad Haleumi Street, P.O.B. 16031
9116001 Jerusalem, Israel
liebchaya@gmail.com, kerner@jct.ac.il

Abstract. *Verbal Multi-Word Expressions* (VMWEs) are very common in many languages. They include among other types the following types: *Verb-Particle Constructions* (VPC) (e.g. get around), *Light-Verb Constructions* (LVC) (e.g. make a decision), and idioms (ID) (e.g. break a leg). In this paper, we present a new dataset for supervised learning of VMWEs written in Yiddish. The dataset was manually collected and annotated from a web resource. It contains a set of positive examples for VMWEs and a set of non-VMWEs examples. While the dataset can be used for training supervised algorithms, the positive examples can be used as seeds in unsupervised bootstrapping algorithms. Moreover, we analyze the lexical properties of VMWEs written in Yiddish by classifying them to six categories: VPC, LVC, ID, *Inherently Pronominal Verb* (IPronV), *Inherently Prepositional Verb* (IPrepV), and other (OTH). The analysis suggests some interesting features of VMWEs for exploration. This dataset is a first step towards automatic identification of VMWEs written in Yiddish, which is important for natural language understanding, generation and translation systems.

Keywords: Multi-Word Expression (MWE)
Verbal Multi-Word Expression (VMWE) · Yiddish

1 Introduction

Multi-Word Expressions (MWEs) have been defined as "idiosyncratic interpretations that cross word boundaries (or spaces)" by Sag et al. [1]. Under this definition, there is a wide range of linguistic constructions such as compounds (*washing machine, light switch*), phrasal verbs (*grow up, take after*), and idioms (*a penny for your thoughts, cry over spilt milk*). Due to their flexible and heterogeneous nature, MWEs interpretation poses a major challenge for Natural Language Processing (NLP) systems [2]. NLP applications, such as machine translation and information extraction should identify them and deal with their lexical ambiguity when they are detected [3].

In this paper, we focus on Yiddish verbal MWEs. The VMWEs include *Verb-Particle Constructions* (VPC) (e.g. *turn on*), *Light-Verb Constructions* (LVC) (e.g. *make a mistake*), idioms (ID) (e.g. *call it a day*), *Inherently Pronominal Verb* (IPronV) (e.g. *help oneself*), *Inherently Prepositional Verb* (IPrepV) (e.g. *look for*), and other (OTH) (e.g. *drink and drive*).

© Springer International Publishing AG, part of Springer Nature 2018
M. Silberztein et al. (Eds.): NLDB 2018, LNCS 10859, pp. 205–216, 2018.
https://doi.org/10.1007/978-3-319-91947-8_20

The PARSEME (PARSing and Multi-word Expressions) COST action[1] presented a shared task on automatic identification of VMWEs and provided annotation guidelines. The guidelines distinguish different types of VMWE categories and allow us to provide a full picture of the diverse properties that Yiddish VMWEs exhibit. Although VMWEs identification has been widely investigated in English, as far as we know, our research is a first attempt to focus on VMWEs in Yiddish.

Yiddish is the historical language of the Ashkenazi Jews[2]. It is written with a fully vocalized alphabet based on the Hebrew alphabet. Yiddish originated during the 9th century [4] in Central Europe, providing the nascent Ashkenazi community with a High German-based vernacular fused with elements taken from Hebrew, Aramaic, Slavic languages, and traces of Romance languages [5][3].

The goal of our research is to manually construct a quality dataset of Yiddish VMWEs, which is a useful tool for supporting automatic extraction of additional VMWEs that are not covered by our resource. An additional contribution of the ongoing research presented in this paper is an analysis of the categories of Yiddish VMWEs. This analysis is a first necessary step for designing algorithms for automatic VMWEs extraction and identification. The dataset with the interesting features of VMWEs that the analysis suggested for exploration can be used for training supervised algorithms for automatic identification of Yiddish VMWEs. Moreover, the set of positive examples from our dataset can be used as seeds in unsupervised bootstrapping algorithms.

In this study, the basic motivation is to increase the number of quality resources and tools for NLP in Yiddish, which is regarded as a resource-poor language. We manually construct a dataset of 200 non-VMWEs and 200 VMWEs written in Yiddish divided into six categories. The main contributions of this study are the successful construction of a new dataset, which can serve as a useful tool for extraction of additional VMWEs and an analysis of six categories of VMWEs including examples of conflicting categories.

The rest of the paper is organized as follows. Section 2 describes available digital resources in Yiddish. In Sect. 3, we aim to provide the necessary background needed for the subsequent sections. Section 4 elaborates on the tools that we have constructed for learning automatic identification of Yiddish VMWEs and linguistic properties related to Yiddish VMWEs. In Sect. 5, we describe the categories of VMWEs and their distribution over the dataset's positive examples. In Sect. 6, we summarize and suggest directions for future research.

2   Digital Resources in Yiddish

Yiddish is a *resource-poor* language. Resource-poor languages lack the basic resources that are fundamental to computational linguistics, including raw resources such as large manually-tagged corpora, parallel texts, and digital dictionaries, as well as tools such as

---

[1] https://typo.uni-konstanz.de/parseme/index.php.

[2] Ashkenaz is the medieval Hebrew name for northern Europe and Germany.

[3] https://en.wikipedia.org/wiki/Yiddish.

part-of-speech taggers, stemmers, sentence parsers, and highly accurate optical character recognition programs.

In the last decades, there has been a noticeable increase in the digital resources available to the Yiddish researcher. In the following we list the main sources of Yiddish-language corpora available today.

There are a few sources of scanned Yiddish books and journals freely available online. However, they are only available as images, and so are not text searchable. Examples for this kind of sources are the National Yiddish Book Center[4] and the Index to Yiddish Periodicals[5]. The National Yiddish Book Center is a non-profit organization that has scanned over 10,000 works of Yiddish literature to rescue Yiddish books and share their content with the world. The Index to Yiddish Periodicals is a bibliographical data base of approximately 250,000 bibliographic records from about 800 distinct publications using Yiddish, which aims to record the materials published in the Yiddish press. The thematic index in particular allows the browser to reach the wealth of information published in the Yiddish press, especially about Jewish communities in Eastern Europe and abroad, political and cultural Jewish movements, institutions, Yiddish and Hebrew writers and actors, and trends in all fields of Jewish Culture. The thematic index registers materials published in the Yiddish press about general subjects as well.

OCR'ed text-searchable sources are the Yiddish books scanned by Google as part of its Google Books project and the Yiddish books and journals at hebrewbooks.org. However, since the OCR was optimized for Hebrew text and not for Yiddish, there are numerous errors.

The Penn Yiddish Corpus [6][6] is a corpus of Yiddish texts, transliterated into Latin characters, of various dialects and eras ranging from 1462 to the 1990s. The corpus contains roughly 200k word tokens and is manually annotated for part-of-speech and syntactic parsing in the Penn Treebank format.

The Corpus of Modern Yiddish (CMY)[7] is a comprehensive linguistic database of annotated texts in Yiddish. The CMY is a project currently underway at the University of Regensburg by scholars from Regensburg and Moscow. It will contain two sub-corpora. The first subcorpus, a balanced one, will contain a 10 million word forms from a large variety of texts and will cover the period from 1850 up to today. The second subcorpus will contain contemporary newspaper texts only and will be large enough to make possible large-scale statistical studies [7]. Retrieval is automatically enabled text via an internet interface that allows complex lexical morphological queries. The majority of the text in the CMY is currently from the website of the Yiddish Forward[8], a Yiddish newspaper, which has made its articles available online since 2006.

In this research, we describe the manual construction of a quality dataset of Yiddish VMWEs. For the dataset creation, we used the texts of the Yiddish Forward.

---

[4] https://archive.org/details/nationalyiddishbookcenter.

[5] http://yiddish-periodicals.huji.ac.il/.

[6] ftp://babel.ling.upenn.edu/research-material/yiddish-corpus/.

[7] http://web-corpora.net/YNC/search/.

[8] http://yiddish.forward.com.

3   Automatic Identification of Verbal MWEs

There are three main approaches to automatic identification of MWEs: (1) Statistical approaches based on frequency and co-occurrence affinity [8–10]. (2) Linguistic approaches using parsers, lexicons and language filters [11–13]; and (3) Hybrid approaches combining different methods [14–21].

Hybrid identification approaches are often based on supervised learning models whose results are evaluated by comparing automatically tagged text to reference annotations. The features used in supervised methods include scores derived from linguistic tools and statistical measures.

Lapata and Lascarides [22] presented one of the first experiments using a supervised approach. They classified noun-noun compounds, such as *income tax* and *public-relations*, into true MWEs and random co-occurrence using a C4.5 decision tree. Pecina [23] discovered collocations in Czech and German using logistic regression, linear discriminant analysis, support vector machines, and neural networks classifiers. Ramisch et al. [24] proved that the supervised approach is an effective way to combine scores, giving more weight to more discriminating features and reducing the weight of redundant ones. They showed that the supervised approach also provides a workaround for the problem of choosing a scoring method for a given data set among dozens of methods proposed in the literature. Furthermore, they demonstrated that the learned models can provide insight into features' informativeness [25]. A system that can identify Hebrew noun compounds with high accuracy, distinguishing them from non-idiomatic noun-noun constructions, was introduced by Al-Haj and Wintner [26]. Their methodology is based on careful examination of the linguistic peculiarities of the construction, followed by corpus-based approximation of these properties via a general machine learning algorithm that is fed with features based on the linguistic properties. Recently, Rondon et al. [27] proposed an iterative method for the continued discovery of new MWEs. The system requires some initial supervision to build a seed MWE lexicon and classifier, and incrementally enriches it by mining texts in the web and bootstrapping from its.

In the last decade, there is a growing interest in methods to predict the compositionality of MWEs [28–34]. Since there is a clear distributional difference between compositional and non-compositional uses of MWEs, distributional similarity methods have been used to determine whether a given MWE has non-compositional uses.

Liebeskind and HaCohen-Kerner [34] proposed a semantically motivated indicator for classifying VN-MWEs written in Hebrew and defined features that are related to various semantic spaces and combined them as features in a supervised classification framework. They empirically demonstrated that their semantic feature set yields better performance than the common linguistic and statistical feature sets and that combining semantic features contributes to the VN-MWEs identification task. This approach may be useful for poor-resource languages, such as Yiddish, where no language processing tools are available.

Considerable research has been done on automatic identification of VMWEs in many languages [34–38]. However, as far as we know, no research has been carried for Yiddish.

Furthermore, there are almost no NLP tools for Yiddish. One of the exceptions, to the best of our knowledge, is the implementation of four techniques for automatic normalization of orthographic variant Yiddish texts [39]. In this research, using a manually normalized set of 16 Yiddish documents as a training and test corpus, four techniques for automatic normalization were compared: a hand-crafted set of transformation rules, an off-the-shelf spell checker, edit distance minimization with manually set weights, and edit distance minimization with weights learned through a training set. For the given test corpus, normalization by minimization of edit distance with multi-character edit operations and learned weights was found to perform best according to the following measures: (1) proportion of correctly normalized words in a test set and (2) precision and recall in a test of information retrieval.

4 Tools for Learning Automatic Identification of Verbal MWEs Written in Yiddish

In this section, first, we describe our manually constructed labeled dataset of VMWEs and non-VMWEs. Then, we detail how a task-oriented stopword list for the task of VMWEs identification was generated.

4.1 A Dataset of Yiddish Verbal MWEs

Our dataset currently includes 200 VMWEs and 200 non-VMWEs. It was manually constructed from the website of the Yiddish Forward newspaper. We did not limit our dataset to VMWEs of a certain length. The data set includes VMWEs of one word up to 13 words. Figure 1 illustrates the distribution of the VMWEs length in our dataset.

The three most frequent lengths of these VMWEs in descending order are three, four, and two words. The average number of words is 3.45 and the average number of characters per VMWE is 18.85. Some examples for Yiddish VMWEs are given in the next section.

We conducted 2 rounds of manual categorization with 2 judges. First, the annotators identified candidate VMWEs, classified them as VMWEs or non-VMWEs, and categorized their types (see Sect. 5). Then, they discussed the conflicting annotations and agreed on a final annotation. Only candidates that were classified as non-verbal by both annotators in the first round were selected as negative examples for our dataset. Thus, a non-VMWE can be every word sequence that includes a verb and is not tagged as VMWE. About 1000 documents were used for the annotation process.

Figure 2 illustrates the distribution of the non-VMWEs length in our dataset. The three most frequent lengths of these VMWEs are three (e.g. *awisg`nicT d`m bnin*[9] - used the building), four (e.g., m` *zwkT a ciik`nwng* - looking for a drawing), and five words (e.g. *z`n`n g`blibn zii`r wwiiniq m`nTšn*- there were very few people left). The average number of words is 4.025 and the average number of characters per VMWE is 23.23.

---

[9] To facilitate readability, we use a transliteration of Hebrew using Roman characters; the letters used, in Hebrew lexicographic order, are abgdhwzxTiklmns`pcqršt.

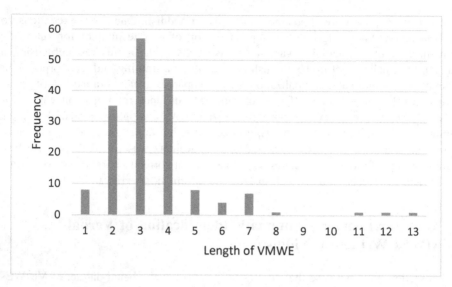

Fig. 1. The distribution of the VMWEs length in our dataset

4.2   A Task-Oriented Stopword List

Stopwords are usually the most frequent words of a corpus. Since they appear to be of little value in helping select documents matching a user need, they are often excluded from the vocabulary. In the task of VMWEs identification it is important not to omit frequent words which might be a part of a VMWE.

As far as we know, there is not any publically available list of stopwords in Yiddish. Therefore, we downloaded an English stopword list from the web. We translated each of the prepositional words and investigated its validity as a stopword for the task of VMWEs identification. For this purpose, we used a sample of positive examples from our dataset. The validation process was performed as follows: First, we searched for all the examples that include at least one Yiddish stopword. Then, we tried to omit the stopword or replace it with another propositional word. If the changed expression maintained its idiomatic meaning, then it is a stopword.

Table 1 presents six examples of three stopwords. Since the criteria were checked at the MWE level, we defined the extracted stopwords as task-oriented. Some words that would have been included in a general list of Yiddish stopwords did not meet the criteria and were not included in our list. For example, the stopword ain (in) in the MWE hak nišT ain Tšiinik (see Table 1, stopword #2, example #1) was not considered a stopword in our setting since replacing it with other prepositional words would result in a non-MWE, a sentence with its literal meaning.

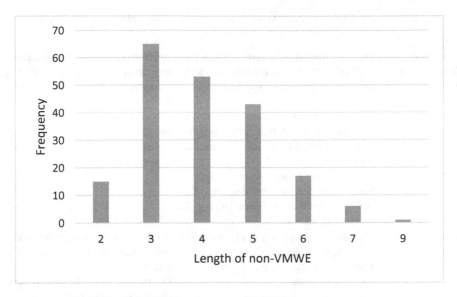

Fig. 2. The distribution of the non-VMWEs length in our dataset

Table 1. Examples of task-oriented stopwords.

#	Stopword	Example #1	Example #2
1	qiin (to)	pr`g nišT **qiin** qwšiwt – it is irrelevant (lit. do not question questions)	haq nišT **qiin** bwbq`s – do not make up what did not happen. (lit. do not throw pine cones)
2	nišT (not)	hak **nišT** ain Tšiinik – do not twist the reality (lit. do not throw in the kettle)	haq **nišT** qiin bwbq`s – do not make up what did not happen. (lit. do not throw pine cones)
3	a (a)	Zik **a**`ch g`bn – to get along (lit. to give himself an advice)	apg`Tan **a** špicl – to fool someone (lit. to do a practical joke)

5 Categories of Verbal MWEs in Yiddish

Following PARSEME shared task on automatic identification of VMWEs' pilot annotation[10], we classify Yiddish VMWEs that include at least one verb and one noun along six categories: Light Verb Construction (LVC), Verb-Particle Combinations (VPC), Idioms (ID), Inherently Pronominal Verb (IPronV), Inherently Prepositional Verb (IPrepV), and other (OTH). We note that the pilot annotation was performed with Google Spreadsheets. Later, the PARSEME shared task adopted FLAT: FoLiA Linguistic Annotation Tool[11], a web platform which allows textual annotation and

---

[10] https://typo.uni-konstanz.de/parseme/index.php/2-general/151-parseme-shared-task-pilot-annotation.

[11] http://proycon.github.io/folia/.

Table 2. The definition of the VMWEs categories with examples from our dataset.

Category	Description	Examples
Light Verb Construction (LVC)	They are formed by a verb and a noun The verb is "light" i.e. it contributes to the meaning of the whole only to a small degree The noun has one of its regular meanings and typically refers to an action or event	Ariinkapn ain a šmw`s – to cause an unplanned conversation with someone familiar (lit. to grab into a conversation) Parlwirn di hap`nwng – to despair (lit. to lose his hope)
Verb-Particle Combinations (VPC)	They are formed by a head verb and a particle Their meaning is non-compositional Notably, the change in the meaning of the verb goes significantly beyond adding the meaning of the particle	QwqT awip … wwi … – to think of (lit. look at … as …)
Idioms (ID)	They are composed of a head verb and at least one of its complements They are usually semantically totally non-compositional	Aww`q liign ain a qranqb`T – to cause suffer (lit. to put in a sick bed) D`r rwx wwiisT – he has no idea (lit. the wind knows)
Inherently Pronominal Verb (IPronV)	They are formed by a full verb combined with a clitic pronouns that refer to the subject of the verb	qwmT par – to take place in a certain date (to arrive) awnT`rg`gang`n – to expire (to go back/down)
Inherently Prepositional Verb (IPrepV)	They are combinations where the verb mandatorily requires a preposition but the meaning of the verb is more or less transparent This category includes cases where a common verb is used together with a preposition in a non-compositional sense and the original meaning of the verb might be bleached	G`pakn awip … – to fail due to (lit. to fall on) L`bT aib`r … – to live in a period that something happened (lit. to live above)
Other (OTH)	This category contains VMWEs, which do not fit to the preceding categories	D`r himl aiz d`rwwiilništ bii di `rd –meanwhile, it is possible (lit. meanwhile, the sky is not stick to the ground) A lid haTništ qiin g`bwirn Tag – not everything has an exact date (lit. the song has no birthday)

categorization of MWEs, including overlapping and discontinuous units. In Table 2, we shortly cite the description of these categories and provide examples.

In case of doubt, we allowed the judges to assign two categories for a VMWE. There were four cases of doubled annotated VMWEs, conflicting categories: LVC/OTH, LVC/VPC, VPC/OTH, and IPrepV/OTH.

The distribution of each category in our Yiddish dataset is reported in Table 3.

Table 3. The distribution of the VMWEs categories in our dataset

Category	Distribution (%)
Light Verb Construction (LVC)	37
Idioms (ID)	19
Other (OTH)	19
Inherently Prepositional Verb (IPrepV)	14
Inherently Pronominal Verb (IPronV)	8
Verb-Particle Combinations (VPC)	3

The most frequent category is LVC. The next frequently categories are the ID and OTH categories with an equal distribution of 19%. An example for the VMWE with a LVC/VPC conflict is šild`rT d`m zik`rhiiT-mwi`r (lit. protect the defensive shield) - keeps the rules. The annotators hesitated whether to remove the noun or to leave it and have two arguments in the expression. Another example for conflicting categories is the VMWE az m` haT [m`r] cw Tan aiin`r miTn cwwiTn (lit. that you have to do one with another) - there is more engagement (usually speaking) with each other, which the annotators annotated as a VPC/OTH conflict. Due to the length of the sentence, the annotators doubt if the verb should be separated from the rest of the sentence.

The Yiddish dataset is publicly available for download in uft8 format[12].

6   Summary and Future Work

In this study, we present the construction of a dataset of 200 non-VMWEs and 200 VMWEs written in Yiddish. This dataset can serve as a seed tool to extract additional VMWEs. Various statistical information was presented regarding non-VMWEs and VMWEs. Furthermore, we introduce an analysis of six categories of VMWEs including examples of conflicting categories.

We plan to investigate supervised algorithms for classifying collocations as VMWEs or non-VMWEs. Inspired by the different categories that we have discussed in the previous section we would engineer features. Since Yiddish is a poor-resource language, we plan to identify VMWEs by their semantic compositionality. This property can be modeled by co-occurrence and distributional similarity measures. Additionally, we plan to use the Yiddish VMWEs as seeds in unsupervised boot-strapping algorithms for classification of collocations. Moreover, we plan to generalize the construction process to other languages, based on our experience in Yiddish and Hebrew [40].

Acknowledgments. We would like to express our deep gratitude to Gitty Eithen, Bluma Zicherman, and Hindy Golomb, our research assistants, for carrying out the annotation process.

---

[12] http://liebeskind-chaya.blogspot.co.il/p/downloads.html.

References

1. Sag, I.A., Baldwin, T., Bond, F., Copestake, A., Flickinger, D.: Multiword expressions: a pain in the neck for NLP. In: Gelbukh, A. (ed.) CICLing 2002. LNCS, vol. 2276, pp. 1–15. Springer, Heidelberg (2002). https://doi.org/10.1007/3-540-45715-1_1
2. Biber, D., Johansson, S., Leech, G., Conrad, S., Finegan, E., Quirk, R.: Longman Grammar of Spoken and Written English. MIT Press, Cambridge (1999)
3. Fazly, A., Stevenson, S.: Distinguishing subtypes of multiword expressions using linguistically-motivated statistical measures. In: Proceedings of the Workshop on a Broader Perspective on Multiword Expressions, pp. 9–16. Association for Computational Linguistics (2007)
4. Jacobs, N.G.: Yiddish: A Linguistic Introduction. Cambridge University Press, Cambridge (2005)
5. Baumgarten, J.: Introduction to Old Yiddish Literature. Oxford University Press, Oxford (2005)
6. Santorini, B.: The Penn Yiddish Corpus. University of Pennsylvania (1997)
7. Aptroot, M., Hansen, B.: Yiddish Language Structures. vol. 52, Walter de Gruyter, Berlin (2014)
8. Dias, G., Guilloré, S., Lopes, J.G.P.: Language independent automatic acquisition of rigid multiword units from unrestricted text corpora. In: Proceedings of Conférence Traitement Automatique des Langues Naturelles (TALN) (1999)
9. Deane, P.: A nonparametric method for extraction of candidate phrasal terms. In: Proceedings of the 43rd Annual Meeting on Association for Computational Linguistics, pp. 605–613. Association for Computational Linguistics (2005)
10. Pecina, P., Schlesinger, P.: Combining association measures for collocation extraction. In: Proceedings of the COLING/ACL on Main Conference Poster Sessions, pp. 651–658. Association for Computational Linguistics (2006)
11. Bejcek, E., Stranák, P., Pecina, P.: Syntactic identification of occurrences of multiword expressions in text using a lexicon with dependency structures. In: MWE@ NAACL-HLT, pp. 106–115 (2013)
12. Green, S., de Marneffe, M.-C., Manning, C.D.: Parsing models for identifying multiword expressions. Comput. Linguist. 39, 195–227 (2013)
13. Al-Haj, H., Itai, A., Wintner, S.: Lexical representation of multiword expressions in morphologically-complex languages. Int. J. Lexicogr. 27, 130–170 (2013)
14. Baldwin, T.: Deep lexical acquisition of verb–particle constructions. Comput. Speech Lang. 19, 398–414 (2005)
15. Zhang, Y., Kordoni, V., Villavicencio, A., Idiart, M.: Automated multiword expression prediction for grammar engineering. In: Proceedings of the Workshop on Multiword Expressions: Identifying and Exploiting Underlying Properties, pp. 36–44. Association for Computational Linguistics (2006)
16. Fazly, A.: Automatic acquisition of lexical knowledge about multiword predicates. University of Toronto (2007)
17. Boulaknadel, S., Daille, B., Aboutajdine, D.: A multi-word term extraction program for Arabic language. In: LREC (2008)
18. Ramisch, C., de Medeiros Caseli, H., Villavicencio, A., Machado, A., Finatto, M.J.: A hybrid approach for multiword expression identification. In: Pardo, T.A.S., Branco, A., Klautau, A., Vieira, R., de Lima, V.L.S. (eds.) PROPOR 2010. LNCS (LNAI), vol. 6001, pp. 65–74. Springer, Heidelberg (2010). https://doi.org/10.1007/978-3-642-12320-7_9

19. Farahmand, M., Nivre, J.: Modeling the statistical idiosyncrasy of multiword expressions. In: MWE@ NAACL-HLT, pp. 34–38 (2015)
20. Sangati, F., van Cranenburgh, A.: Multiword expression identification with recurring tree fragments and association measures. In: MWE@ NAACL-HLT, pp. 10–18 (2015)
21. Mandravickaite, J., Krilavičius, T.: Identification of multiword expressions for Latvian and Lithuanian: hybrid approach. In: Proceedings of the 13th Workshop on Multiword Expressions (MWE 2017), pp. 97–101 (2017)
22. Lapata, M., Lascarides, A.: Detecting novel compounds: the role of distributional evidence. In: Proceedings of the Tenth Conference on European Chapter of the Association for Computational Linguistics, vol. 1, pp. 235–242. Association for Computational Linguistics, Stroudsburg (2003)
23. Pecina, P.: Lexical association measures and collocation extraction. Lang. Resour. Eval. **44**, 137–158 (2010)
24. Ramisch, C., Schreiner, P., Idiart, M., Villavicencio, A.: An evaluation of methods for the extraction of multiword expressions. In: Proceedings of the LREC Workshop-Towards a Shared Task for Multiword Expressions (MWE 2008), pp. 50–53 (2008)
25. Ramisch, C., Villavicencio, A., Moura, L., Idiart, M.: Picking them up and figuring them out: verb-particle constructions, noise and idiomaticity. In: Proceedings of the Twelfth Conference on Computational Natural Language Learning, pp. 49–56. Association for Computational Linguistics (2008)
26. Al-Haj, H., Wintner, S.: Identifying multi-word expressions by leveraging morphological and syntactic idiosyncrasy. In: Proceedings of the 23rd International Conference on Computational Linguistics, pp. 10–18. Association for Computational Linguistics (2010)
27. Rondon, A., de Medeiros Caseli, H., Ramisch, C.: Never-ending multiword expressions learning. In: MWE@ NAACL-HLT, pp. 45–53 (2015)
28. Katz, G., Giesbrecht, E.: Automatic identification of non-compositional multi-word expressions using latent semantic analysis. In: Proceedings of the Workshop on Multiword Expressions: Identifying and Exploiting Underlying Properties, pp. 12–19. Association for Computational Linguistics (2006)
29. Sporleder, C., Li, L.: Unsupervised recognition of literal and non-literal use of idiomatic expressions. In: Proceedings of the 12th Conference of the European Chapter of the Association for Computational Linguistics, pp. 754–762. Association for Computational Linguistics (2009)
30. Biemann, C., Giesbrecht, E.: Distributional semantics and compositionality 2011: shared task description and results. In: Proceedings of the Workshop on Distributional Semantics and Compositionality, pp. 21–28. Association for Computational Linguistics (2011)
31. Guevara, E.: Computing semantic compositionality in distributional semantics. In: Proceedings of the Ninth International Conference on Computational Semantics, pp. 135–144. Association for Computational Linguistics (2011)
32. Salehi, B., Cook, P., Baldwin, T.: A word embedding approach to predicting the compositionality of multiword expressions. In: HLT-NAACL, pp. 977–983 (2015)
33. Yazdani, M., Farahmand, M., Henderson, J.: Learning semantic composition to detect non-compositionality of multiword expressions. In: EMNLP, pp. 1733–1742 (2015)
34. Liebeskind, C., HaCohen-Kerner, Y.: Semantically motivated Hebrew verb-noun multi-word expressions identification. In: COLING, pp. 1242–1253 (2016)
35. Dandapat, S., Mitra, P., Sarkar, S.: Statistical investigation of Bengali noun-verb (NV) collocations as multi-word-expressions. In: Proceedings of Modeling and Shallow Parsing of Indian Languages, MSPIL, pp. 230–233 (2006)

36. Diab, M.T., Bhutada, P.: Verb noun construction MWE token supervised classification. In: Proceedings of the Workshop on Multiword Expressions: Identification, Interpretation, Disambiguation and Applications, pp. 17–22. Association for Computational Linguistics (2009)
37. Schneider, N., Danchik, E., Dyer, C., Smith, N.A.: Discriminative lexical semantic segmentation with gaps: running the MWE gamut. Trans. Assoc. Comput. Linguist. **2**, 193–206 (2014)
38. Todirascu, A., Navlea, M.: Aligning Verb+Noun Collocation to Improve a French-Romanian Statistical MT System. John Benjamins (2015)
39. Blum, Y.P.: Techniques for automatic normalization of orthographically variant Yiddish texts (2015)
40. Liebeskind, C., HaCohen-Kerner, Y.: A lexical resource of Hebrew verb-noun multi-word expressions. In: LREC, pp. 522–527 (2016)

Stochastic Approach to Aspect Tracking

Maoto Inoue$^{(\boxtimes)}$ and Takao Miura$^{(\boxtimes)}$

Department of Advanced Sciences, Hosei University, 3-7-2 KajinoCho,
Tokyo, Koganei 184-8584, Japan
maoto.inoue.7h@stu.hosei.ac.jp, miurat@hosei.ac.jp

Abstract. In this investigation, we discuss aspect tracking, i.e., how to identify tracking storylines of document topics. Since there happen huge amount of fragment information, it is hard to see what they mean and how they go within topics by hands. Here we attack to this kind of problems by means of stochastic models. Our basic idea is that we consider state transitions as internal structure of stories based on HMM, and we extract several storylines as aspects of topics by probabilistic likelihood. We utilize KL divergence to screen topics.

Keywords: Aspect tracking · Hidden Markov Model (HMM)
Topic detection and tracking

1 Motivation

Very often we face to an information explosion because of tremendous proliferation of internet, however they are fragmentary, unreliable and inconsistent yet the contents keep changing every second. Since most of the information are textual, we should examine which words spread over which documents with huge combination. *Topic tracking* is to detect documents (in chronological sequences) suitable for a known topic. Basically this is a kind of supervised learning using text CLAssification or clustering such as k-NN. It has been proposed to model change/transformation automatically for this purpose, but it is hard to avoid heavy computation and we need efficient, effective and sound techniques.

There have been much amount of investigation called *Topic Detection and Tracking* (TDT) proposed so far [2]. It allows us to detect an *event*, i.e., something happening at a certain time and place, and to track development of the situation. Typical examples analysis of are radio broadcasts and document articles of news papers. During initial TDT activities, they have developed many algorithms for distinction between topics of events, putting focus on specific features such as temporal natures, location and named entities. Much amount of investigation talked about *topic* detection, basically the main techniques come from unsupervised clustering, i.e., discovery of previously unidentified events classes, in a chronologically-ordered accumulation of stories. *Event tracking* aims to assign event-labels automatically when a news arrives, based on previously identified stories. Here labelling could be estimated by using k-NN techniques or classifiers.

© Springer International Publishing AG, part of Springer Nature 2018
M. Silberztein et al. (Eds.): NLDB 2018, LNCS 10859, pp. 217–229, 2018.
https://doi.org/10.1007/978-3-319-91947-8_21

Allan applies relevance feedback and some information filtering technique to these issues [1] while Carbonell discusses both decision tree and k-NN techniques [4]. Among others, Yang [9] has proposed an interesting approach to give relationship among documents and extract some transition over contents. In their idea, they apply kNN clustering to consecutive news articles based on cosine similarity of document vectors, and extract clusters from the articles' topics tracked. They also give some comparison with decision-tree and mention that this approach plays superior and improves the precision. In recent activities, they discuss feature selection such as NLP, probability/stochastic theory or domain-dependent information such as morphological aspects [8].

In this work, we discuss chronologically ordered textual information, typically news articles with timestamp but appeared over several topics. Usually every news article records one event in one topic and they constitute a meaningful story as a whole, so we might see multiple topics go simultaneously and consistently [8].

We apply Hidden Markov Model (HMM) to track aspects of topic, because we could consider state transitions as storylines and HMM generates a state transition diagram, although it requires a few training data to adjust models while others not. By using HMM, we track storylines in documents along with temporal order, here we involve forward and backward algorithm of the top-K maximum likelihood to explore better candidates. This work contributes mainly to the following 3 points:

(1) We can track aspects in documents and model storylines correctly according to each topic.
(2) We don't assume large training information in advance yet obtain aspects correctly.
(3) We also build inner-structure of aspects by which we understand easily the storylines as a whole.

The rest of the paper is organized as follows. In Sect. 2 we describe topics, aspects and tracking. In Sect. 3, we describe definition about HMM. In Sect. 4 we discuss our ideas of aspect trackings based on HMM putting attention on computational efficiencies. Section 5 contains an experimental results to see the effectiveness of this idea and we conclude this investigation in Sect. 6.

2   Topics and Aspect Tracking

In TDT, an *event* is defined as something that happens at some specific time and place, while a *topic* a seminal event or activity along with all directly related events and activities [2,9]. For example, both "Chiba stabbing incident" and "Yamaguchi arson murder" are topics which may include several news articles such as "some incident occurred" and "criminal arrested", but inherent differences as "deadly weapons" and "crime location". Certainly each "murder" topic contains a different story. A *storyline* (or just *story*) is a sequence of events over one or some topics that mentions some well-structure within by which a consistent and meaningful affair is described. In other words, relevant events constitute an *aspect* along with the storyline. *Storyline extraction* (from news sources), called *aspect tracking*, aims to extract events under a certain topic and reveal how these events are evolved over time [10]. Some technique should be provided to extract accurately relevant events and link them correctly into coherent stories, even if the news are given chronologically.

In this work, we discuss an aspect tracking issue. Note that it is different from topic tracking, since we should examine relevant events and relating them in a consistent manner. We expect to track aspects and to give possibly well-defined perspective with the help of stochastic theory or some other domain-independent techniques but not of NLP theory. The stochastic techniques to model the issues could come from Hidden Markov Model (HMM), Maximum Entropy Markov Model (MEMM) and Conditional Random Fields (CRF), but there has been no model of aspect tracking based on them proposed so far to our best knowledge.

There exist 2 big difficulties to attack to the issue of aspect tracking. First it is *no common framework* to model storylines so that it is no way to assume training data in public. For example, we can't distinguish easily aspects from others in a common topic. Second, we know there could happen *burst states* of news articles in a short period which might cause incorrect tracking aspects even if we follow single common topic.

We tackle with the issue by using HMM framework. Considering state transition as a storyline, we estimate Markov model. To do that, we construct the model with maximum likelihood during estimation steps. During the construction, we keep multiple candidate models of top-K maximum likelihood paths. We extend forward and backward approaches based on Viterbi algorithm that allow us to obtain multiple candidate models.

3 Markov Models

In this section we review some background knowledge and our motivation of the approach.

3.1 Markov Model

In probability theory, a *stochastic process* means a mathematical vehicle where random variables are associated with a set of states to model randomly changing over time, considered as a probabilistic finite state automaton (with observation output). The approach is widely used as mathematical models of systems and phenomena that appear to vary in a random manner. Common example is "speech recognition".

A *Markov Model* (MM) is one of the stochastic framework with an assumption that future states depend only on the current state but not on the events/states that occurred before it, called *Markov property*. Formally MM is defined as (N, M, A, B, Π) where $N = \{1, 2, \ldots, n\}$ a set of states, $M = \{1, 2, \ldots, m\}$ a set of output symbols, $A = \{a_{ij}, i, j \in N\}$, $0 \leq a_{ij} \leq 1$, $\Sigma_j a_{ij} = 1$, $B = \{b_{ij}(o), i, j \in N, o \in M, \Sigma_o b_{ij}(o) = 1$, and $\Pi = \{\pi_i, i \in N\}$, $\Sigma_i \pi_i = 1$.

Each a_{ij} means a transition probability from i to j, $b_{ij}(o)$ a probability of an output symbol o, $0 \leq b_{ij}(o) \leq 1$ at a transition from i to j, and π_i means an initial probability of a starting state s_i. Generally, Markov assumption enables reasoning and computation to intractable systems because we have $P(s_{t+1}|s_1 \ldots s_t) = P(s_{t+1}|s_t)$ $s_1, \ldots, s_{t+1} \in N$, and facilitates predictive modelling and probabilistic forecasting in a simple and efficient way [7].

However, MM has 2 serious deficiencies: how to obtain training data, and how to estimate Markov model A, B and Π suitable for the training data. Usually we could construct the model by examining training data, say counting state transitions and observation-output. Note there exist two kinds of *labels*, one for a state and another for class. Clearly the true problem is how we obtain many training data correctly and efficiently.

3.2   Hidden Markov Model

A *Hidden Markov Model* (HMM) is a kind of Markov Model for which the state is only partially observable [3]. Note observation-outputs are related to the states of the system, but they are typically insufficient to precisely determine the state. What *hidden* does mean is a *state* associated with observation for *best* suitable interpretation.

HMM provides us with solutions to 3 issues. Assume we have observation sequences $O = o_1 \ldots o_T, o_j \in M$. The first is that we can obtain the probability of O. In fact, once we assume a Markov Model of interests, we explore all the paths to generate O, we summarize all the probabilities of the paths. The second issue is that we can obtain the *most likely* path $S = s_1 \ldots s_T, s_j \in N$ to generate O. *Viterbi* algorithm allows us to estimate the path of the maximum likelihood efficiently using dynamic programming technique: The forward algorithm computes the probability of the sequence of observations.

The last issue is important, that is, true contribution of HMM is that we can estimate $A = (a_{ij}), i, j \in N$ and $B = (b_{ij}(o)), i, j \in N, o \in M$ for a Markov model (N, M, A, B, Π). Our state transition probabilities $P(s_j|s_i) = a_{ij}$, probabilities $P(o_k|s_i)$ of observation o_k at state s_i and initial probability $P(s_i) = \pi_i$ at start. In other words, HMM generates MM according to sequence data. Basically we apply EM algorithm to estimate MM. We maximize the likelihood of the data by iterating adjustment process through EM algorithm from scratch. HMM requires *no training data* in advance. Several well-known algorithms for Hidden Markov Models exist. *Baum Welch* (BW) algorithm estimates them efficiently from a set of (untrained) data. Let us note hat a sufficient size of "training data" helps us to start with better initialization suitable for quick convergence (of the iteration) and for avoidance of local-minimum situation. Without any training data or appropriate initialization, it might be hard to avoid these deficiencies [6].

Let us show some example of HMM behavior in a Fig. 1. Given 7 news articles after filtering (described later), we apply HMM (BW algorithm) to estimate Markov model. Here we get the initialization probability 0.9943 at a state 0, and the transitions between states. In this figure we give names to states by looking for the words corresponded to the states.

4   Aspect Tracking Using HMM

In this section, we describe a new approach to track aspects of document topics based on HMM. In our approach, we pose two assumption on aspect tracking issue.

(1) Every topic shares unified structure of a Markov model, and (2) Articles chronologically ordered within an aspect may keep distinguishable likelihood.

Our approach consists of two ideas, *extended Viterbi* algorithms and *aspect-tracking* based on the algorithms.

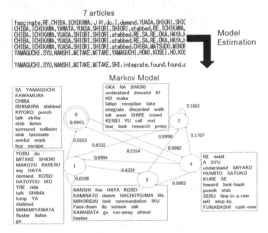

Fig. 1. Constructed Markov Model

4.1   Extending Viterbi Algorithms

Given an output sequence $O = o_1 \ldots o_T$ $(T > 0)$ along with a Markov Model (N, M, A, B, Π), we like a suitable paths $s_1 \ldots s_T$ of the maximum likelihood to estimate O. The well-known *Viterbi* algorithm provides us with the answers. We define two functions $\alpha_t(i)$ and $\beta_t(i)$, the maximum likelihood of a state i at a time t, and the state chaining of a state i at a time t of the $\alpha_t(i)$ respectively.

$$
\begin{aligned}
\alpha_0(i) &= \pi_i \\
\alpha_{t+1}(i) &= Max_j \{\alpha_t(j) \times a_{ji} \times b_{ji}(o_{t+1})\} \\
\beta_0(i) &= \phi \\
\beta_{t+1}(i) &= ArgMax_{j=1,\ldots,n} \{\alpha_t(j) \times a_{ji} \times b_{ji}(o_{t+1})\}
\end{aligned}
$$

Note π_i means a initial probability starting from i. By induction, we assume the maximum likelihood $\alpha_t(j)$ in a state j at time t. Then, to get to i at time $t+1$, we examine all the state $j \in N$ to maximize $\alpha_t(j) \times a_{ji} \times b_{ji}(o_{t+1})$, which is $\alpha_{t+1}(i)$. Our goal is to obtain $\alpha_T(s_T)$. Similarly, by induction, we assume the path of the maximum likelihood $\beta_t(j)$ in a state j at time t. Then, to get to i at time $t+1$, we examine all the state $j \in N$ to maximize $\alpha_t(j) \times a_{ji} \times b_{ji}(o_{t+1})$. Note $\beta_{t+1}(i) = j$ means we go to i at time $t+1$ from j, which maximizes the likelihood. We have $\beta_2(s_2) = s_1, \beta_3(s_3) = s_2, \ldots$, for example. Our final result is a backward chaining $\beta_T(s_T)$.

Here we extend Viterbi algorithm for two purposes of our works, backward estimation and top-K paths estimation. Given $1 \leq K \leq n$, let k be an integer of $k = 1, \ldots, K$. By *k-Viterbi* algorithm, we mean a procedure to obtain the k-th most likely sequence S of the maximum likelihood. We define two functions $\alpha_t^{(k)}(i)$ and $\beta_t^{(k)}(i)$, the k-th maximum likelihood of a state i at a time t, and the state chaining of a state i at a time t of the $\alpha_t^{(k)}(i)$ respectively.

$$\alpha_0^{(k)}(i) = \pi \ (k=1), 1 \ (k>1)$$
$$\alpha_{t+1}^{(k)}(i) = Max_{h=1,\ldots,K,j=1,\ldots,n}^{(k)}\{\alpha_t^{(h)}(j) \times a_{ji} \times b_{ji}(o_{t+1})\}$$
$$\beta_0^{(k)}(i) = \phi$$
$$\beta_{t+1}^{(k)}(i) = ArgMax_j^{(k)}\{\alpha_t^{(h)}(j) \times a_{ji} \times b_{ji}(o_{t+1})\}$$
$$j = 1,\ldots,n, h = 1,\ldots,K$$

Note that $Max^{(k)}$ is an operator of k-th maximum and $Max = Max^{(1)}$, and that $ArgMax^{(k)}$ is an `argmax` operator of k-th maximum and $ArgMax = ArgMax^{(1)}$. As previously, by induction, we assume the h-th maximum likelihood $\alpha_t^{(h)}(j)$ in a state j at time t, $h = 1,\ldots,K$. Then, to get to i at time $t+1$, we examine all the states $j \in N$ to obtain k-th maximum $\alpha_t^{(h)}(j) \times a_{ji} \times b_{ji}(o_{t+1})$, $h = 1,\ldots,K$. Our result is $\alpha_T^{(k)}(s_T)$.

As for k-th maximum path, we assume the path of the maximum likelihood $\beta_t^{(h)}(j)$ in a state j at time t, $h = 1,\ldots,K$. Then, to get to i at time $t+1$ with the likelihood of the k-th maximum, we examine all the states $j \in N$ to obtain k-th maximum $\alpha_t^{(h)}(j) \times a_{ji} \times b_{ji}(o_{t+1})$, $h = 1,\ldots,K$. Note $\beta_{t+1}^{(k)}(i) = j$ means we go to i at time $t+1$ from j but j doesn't always mean k-th, so we may have $\beta_2(s_2)^{(1)} = s_1, \beta_3^{(2)}(s_3) = s_2$. We get a backward chaining $\beta_T^{(k)}(s_T)$. It takes $O(KnT)$ since we always K possible choices at each step. Let us note that we obtain all of $\alpha_T^{(k)}(s_T)$ and $\beta_T^{(k)}(s_T)$ simultaneously, $k = 1,\ldots,K$. This extension helps us to obtain top-K candidate paths efficiently and simultaneously.

In our running example (Fig. 1), let us examine top-2 paths. An article contains words "YAMAGUCHI, SYU, NANSHI, MITAKE, SHI, HUKUMU, IU" and the first word ("YAMAGUCHI") has the output probability 0.0746 at a state 0. The cost $\alpha_1^{(}1)(s_1) = 0.074 = 0.9943 \times 0.0746$ is defined as the product with the initial state probability 0.9943. We see the next word ("SYU") only at the transition from 0 to 4 and the cost $\alpha_2 2(s_2)$ becomes 0.074 $\times$ the transition 0.9992 $\times$ the word probability.

Then we introduce *backward* algorithms. By the word *backward* Viterbi algorithm, we mean another procedure to obtain the most likely sequence S of the maximum likelihood but we obtain all the information in a backward manner. Basically there is no difference between forward and backward algorithms but some remarks remain. We define two functions $\gamma_t(i)$ and $\delta_t(i)$, the maximum likelihood starting from a state i to s_T at a time t, and the state chaining from a state i to s_T at a time t of the $\gamma_t(i)$ respectively.

$$\gamma_T(i) = 1 \ (i = s_T), 0 \ (otherwise)$$
$$\gamma_t(i) = Max_{j=1,\ldots,n}\{a_{ij} \times \gamma_{t+1}(j) \times b_{ij}(o_t)\}$$
$$\delta_T(i) = i$$
$$\delta_t(i) = ArgMax_j\{a_{ij} \times \gamma_{t+1}(j) \times b_{ij}(o_t)\}$$

By inductive assumption, we have the maximum likelihood $\gamma_{t+1}(j)$ starting from a state j to s_T at time $t+1$. Then, to get back to i at time t, we examine all the states $j \in N$ to maximize $a_{ij} \times \gamma_{t+1}(j) \times b_{ij}(o_t)$. Our final likelihood is $\gamma_1(s_1)$ in this case. Similarly, we assume the path of the maximum likelihood $\delta_{t+1}(j)$

in a state j at time $t + 1$. Then, to get back to i at time t, we examine all the states $j \in N$ to maximize $a_{ji} \times \gamma_{t+1}(j) \times b_{ji}(o_{t+1})$. Note $\delta_t(i) = j$ means we go to i at time t from j, so we have $\delta_2(s_2) = s_1, \delta_3(s_3) = s_2, \ldots$. Our result is a backward chaining $\delta_1(s_1)$. Note that it takes no additional complexity compared to Viterbi.

Finally, let us put all the results together into one procedure, i.e., k-backward Viterbi (k-$bViterbi$) algorithm, which means a backward procedure to obtain the k-th most likely sequence S of the maximum likelihood, $k = 1, \ldots, K$.

$$\gamma_T^{(k)}(i) = i \quad (i = s_T), 0 \quad (otherwise)$$
$$\gamma_t^{(k)}(i) = Max_{h=1,\ldots,K,j=1,\ldots,n}^{(k)}\{a_{ji} \times \gamma_{t+1}^{(h)}(j) \times b_{ji}(o_{t+1})\}$$
$$\delta_T^{(k)}(i) = 1 \quad (i = s_T), 0 \quad (otherwise)$$
$$\delta_t^{(k)}(i) = ArgMax_j^k\{a_{ji} \times \gamma_{t+1}^{(h)}(j) \times b_{ji}(o_{t+1})\}$$
$$j = 1, \ldots, n, h = 1, \ldots, K$$

Let us see how the backward Viterbi algorithm works. In a Fig. 1, the article of "YAMAGUCHI, SYU, NANSHI, MITAKE, SHI, HUKUMU, IU" has the last 7-th word "IU". This word arises at state 3, 2 and 1 with the probability 0.093, 0.0166 and 0.01292 respectively. The top-2 costs $C(s_7)$ can be estimated as 0.093 and 0.0166. The 6-th word ("HUKUMU") arises at the transition backwardly from 4 to 3 and from 4 to 2 so that we have the top-2 costs $C(s_6)$ 0.0108 and 6.714e04.

4.2   Tracking Aspects

Now let us describe how to track an aspect based on HMM. We assume a collection of news articles appeared in several aspects over multiple topics. That is, the articles are given in a chronologically order where all the articles in every aspect are chronologically well-ordered but interleaved.

We assume we like to track a topic of the first article. Then, we consider consecutive articles as a cluster of our topic if the first n articles D_1 are *similar* to the first $n + 1$ articles D_2 in a sense that the word distributions of the n articles are *similar* to the $n + 1$ articles but the larger articles don't satisfy this property any more. Here we say two word distributions D_1 is *similar* to D_2 if their KL-divergence $KL(D_1 \| D_2)$ keeps ϵ_1 or smaller. Note that we apply this filtering process to the articles to keep closer relevance of topics of interest.

We build a Markov Model to the articles extracted (by HMM). Using k-bViterbi algorithm on the Markov Model, we examine the next (i.e., $n + 1$-th) article, and estimate both (backward) likelihood of the most likely path containing this article and obtain the perplexity (PP) of the path[1]. To do that, we take an expect PP value in D_1, $E[D_1]$. Comparing this perplexity value to $E[D_1]$, we decide whether the article be *familiar* with the model or not; we say the article is *familiar* if the difference of the perplexity values is less than ϵ_2. If the case, we add this new comer to D_1. Otherwise, we skip the article. We consider the

---

[1] The logarithmic likelihood $H(P)$ divided by the total size of the articles on the path, and PP $= 2^{H(P)}$.

average PP in D_1 as $E[D_1]$. Finally, we take a new article and repeat the process until we examine all the candidates.

Let us illustrate the process above in a Fig. 2. We examine whether the 7-th article is familiar or not to other 6 articles discussed in a Fig. 1. Remember the Fig. 1 has been constructed by using all the 7 articles. Assuming $K = 2$, we obtain paths of the top-2 likelihood, and their PP values, PP-1 and PP-2. The right side of the Fig. 2 contains all the likelihood and the average. The difference between 7-th and the averages are 0.16145, 0.18472 respectively, so that we can't say familiar because of the threshold 0.05.

5 Experiments

5.1 Preliminaries

Let us show how well the proposed approach works.

Here we examine Yang approach [9] as a baseline where they have given relationship among documents and extracted some transition over contents. Basically they apply k-NN clustering to consecutive news

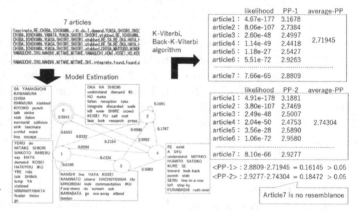

Fig. 2. Aspect tracking on Markov Model of 7 articles

articles based on cosine similarity of documents, then extract clusters from the articles topics tracked. Note the baseline assumes multiple topics and the articles come in chlonologically, which is identical to our case.

We examine News Corpus of Mainichi 2013 (in Japanese) where every article contains "topic" label but not any information about storylines. We extract two aspects (storylines), "*Chiba stabbing incident*" (abbr. Chiba) and "*Yamaguchi arson murder*" (abbr. Yama) by hands where each topic consists of 11 and 12 articles respectively since Jan., 1. We make the articles in each topic are chronologically ordered, events of multiple topics may take place in an interleaved manner at the same time, but we don't know which article belongs to which topic[2]. In other words, we will do classification as well as storyline construction at the same time. After applying morphological analysis to the articles in advance, we extract only all the proper nouns and the verbs.

We apply our top-K approach and the baseline to these articles. We take several threshold parameters in our approach, We take $K = 2$ for K-Viterbi/K-bViterbi processes. Threshold values ϵ_1 for KL divergence and ϵ_2 for PP value

---

[2] We examine the topic label only for evaluation purpose.

could be selected through preliminary experiments, here we assume $\epsilon_1 = 0.012$ and $\epsilon_2 = 0.05$. For the baseline, we assume 0.2 as the similarity as a threshold and $k = 3$ for k-NN clustering.

For evaluation, we discuss 3 criterion, *recall, precision* and *detection cost function* (DCF) C_F [5]. DCF can be calculated using "undetected rate" P_{miss} (how many articles we miss tracking, like *negative recall*) and "incorrectly detected rate" P_{false} (how many articles we track incorrectly, like *negative precision*). P_{target} means the ratio of the number of the articles in a correct topic. Then let us define $C_{Det} = C_{miss} \times P_{miss} \times P_{target} + C_{false} \times P_{false} \times (1 - P_{target})$ where $C_{miss} = 10.0, C_{false} = 1.0$. Finally let C_F be $C_{Det}/Min(C_{miss} \times P_{target}, C_{false} \times (1 - P_{target}))$. Note that the C_F value works better when the values go smaller (perfect if 0.0).

5.2 Results

Let us track a topic *"Chiba stabbing incident"* with noises of *"Yamaguchi arson murder"*, and let us illustrate the results of undetected rate, incorrect detected rate, precision and recall in a Table 1. Also in Table 2 (baseline or Ours), we show storylines extracted with (correct) labels.

As shown in the Table 1, we see the baseline approach shows 54.5% (6 articles among 11) of undetected rate and 33.3% (4/12) of incorrectly detected rate with DCF 3.45. It provides us with 9 articles extracted of which 5 shows "Chiba" incident. The storyline contains "Occurred"(Chi-1,Yama-2) → "Arrested" (Chi-4) → "Confessed" (Chi-7) → "Arrested" (Yama-6) → "Confessed" (Chi-9) → "Arrested" (Yama-10) → "Confessed" (Chi-10) → "Arrested" (Yama-11).

On the contrary, our approach shows 45.5% (5 articles among 11) of undetected rate and 16.7% (2/12) of incorrectly detected rate with DCF 2.26. We extracted 8 articles of which 6 articles concern "Chiba" incident. The storyline contains "Occurred"(Chi-1) → "Arrested" (Chi-2) → "Investigation" (Yama-4) → "Confessed" (Chi-5,9,10) → "Indictment" (Chi-11) → "Indictment" (Yama-11).

A Fig. 3 shows our Markov model constructed using 8 articles extracted. Note that we give output-words specific to individual state and state names by looking at these words by hands. Let us note that our approach provides us with "Indictment" state correctly in this storyline (which doesn't appear in baseline approach), so that *"Chiba stabbing incident"* can be modelled well without any inconsistency as a whole.

5.3 Discussion

Let us discuss the results by our approach. The recall ratios are about 50% in both approaches. From the view point of undetected rate in our approach, we take 6 articles from 11 candidates correctly and consistently along with the development of the "Chiba" story. Especially ours detects an "Indictment" state correctly.

Let us discuss why we can do that. In a Fig. 4, we show a Markov Model which is estimated by all the 11 correct articles in the "Chiba" topic and could be seen

Table 1. Evaluation results

State	Baseline Detected	Ours Detected
Occurred(1)	1	1
Arrested(3)	1	1
Confession1(1)	0	1
Victim's Personality(1)	0	0
Confession2(4)	3	2
Indictment(1)	0	1
OtherTopics(12)	4	2
Undetected Rate(%)	54.545(6/11)	45.454(5/11)
Incorrectly Detected Rate(%)	33.333(4/12)	16.666(2/12)
Norm Detection Cost	6.6107	4.3332
Precision(%)	55.555(5/9)	75(6/8)
Recall(%)	45.454(5/11)	54.545(6/11)

Table 2. Tracking by baseline and tracking by ours

Baseline		
Label-ID	Contents	State
Chi-1	Chiba stabbing incident Occurred in the front of the station in Chiba	Occurred
Yama-2	Then it was found by police investigation an arson happened	
Chi-4	A suspect was arrested unexpectedly	Arrested
Chi-7	It was found in an interview to the police that the suspect pointed	Confessed
Yama-6	a suspect was found among the mountains and accompanied to the police	
Chi-9	He deposed he reflected on his past behavior	Confessed
Yama-10	he confessed he did	
Chi-10	He deposed he sincerely repented	Confessed
Yama-11	The police decided to arrest him again on 16th day	
Ours		
Label-ID	Contents	State
Chi-1	Chiba stabbing incident Occurred in the front of the station in Chiba	Occurred
Chi-2	A suspect arrested at Kaminato harbor and ...	Arrested
Yama-4	police said they made sure the dead bodies	
Chi-5	It was found in an interview to the police that the suspect pointed	Confessed
Chi-9	He deposed he reflected on his past behavior	Confessed
Chi-10	He deposed he sincerely repented	Confessed
Chi-11	The office indicted Hayato Oka (24yo), unemployed, for murder	Indictment
Yama-12	He was charged with arson of sin	

as the *answer* model of this topic. Let us compare the two models in Figs. 3 and 4 with each other. Compared to the answer model, the one estimated by ours is closely related to the answer, because the main part of state transition works very similarly and the output words in each state appear also in the answers in the corresponded state. In fact, here is the major paths in the both models:

$$Occurred/Occurred \cdot Arrested/Arrested \cdot Confession1/$$
$$Arrested \cdot Confession2/Occurred \cdot Indictment/Occurred/$$
$$\cdots$$

We show PP values in a Table 3 using our MM where doubly underlined one means a value less than threshold.

As for incorrectly detected rate, we have 16.67(2/12) by our approach and 33.33% (4/12) by baseline. In our approach, we have decided some parameters

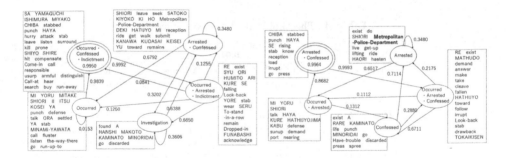

Fig. 3. Our approach Markov Model   **Fig. 4.** Answer Markov Model

Table 3. PP values by Ours

1:	1-Chiba	2-Chiba	9-Chiba	10-Chiba	11-Chiba	12-Yama			1-Yama	average	difference
PP-1	2.078527985	1.734456688	1.451737874	1.480746562	1.561471534	1.746526851			1	1.675577916	0.675577916
PP-2	2.07960942	1.735133921	1.454990313	1.48382782	1.594916409	1.748490572			1	1.682828076	0.682828076

2:	1-Chiba	2-Chiba	9-Chiba	10-Chiba	11-Chiba	12-Yama			2-Yama	average	difference
PP-1	3.1678969	2.73844672	2.499749457	2.441880777	2.542764137	2.926392123			2.881038878	2.719521686	0.161517192
PP-2	3.188159355	2.746918137	2.500719727	2.475342899	2.591735655	2.958070899			2.927718479	2.743491112	0.184227367

3:	1-Chiba	2-Chiba	9-Chiba	10-Chiba	11-Chiba	12-Yama			3-Yama	average	difference
PP-1	3.911065496	3.001982644	2.542555537	2.532693245	2.64521437	3.335302439			3.18797105	2.994802288	0.193168762
PP-2	3.915252168	3.014604627	2.56959152	2.563667345	2.707017881	3.056993687			3.194263555	2.971187871	0.223075684

4:	1-Chiba	2-Chiba	9-Chiba	10-Chiba	11-Chiba	12-Yama			3-Chiba	average	difference
PP-1	3.319597541	2.869100708	2.549796656	2.54740159	2.562376989	3.138191746			3.099702152	2.831077538	0.268624613
PP-2	3.32712179	2.869370694	2.553978382	2.595353343	2.686681913	3.163489744			3.128674972	2.865999311	0.262675661

5:	1-Chiba	2-Chiba	9-Chiba	10-Chiba	11-Chiba	12-Yama			4-Chiba	average	difference
PP-1	3.328419066	2.897115068	2.546987597	2.530027903	2.547723285	3.163541404			3.114855714	2.83563572	0.27977422
PP-2	3.333531817	2.900332244	2.565742674	2.547674148	2.585698556	3.175401785			3.13132109	2.851546871	0.27977422

6:	1-Chiba	2-Chiba	9-Chiba	10-Chiba	11-Chiba	12-Yama			4-Yama	average	difference
PP-1	3.450070263	2.836898909	2.50538	2.498866076	2.543737984	3.029135723			2.862720962	2.810681492	0.05203947
PP-2	3.453774952	2.855988822	2.528789807	2.520980706	2.65423756	3.045084659			2.873076135	2.843142751	0.090023384

7:	1-Chiba	2-Chiba	9-Chiba	10-Chiba	11-Chiba	12-Yama	4-Yama		5-Chiba	average	difference
PP-1	3.545251877	3.057047524	2.589711267	2.629067895	2.485754622	3.140404927	2.968552259		2.94895619	2.916541481	0.032414708
PP-2	3.54730834	3.058228065	2.59173959	2.631018626	2.495080913	3.147910354	2.994458192		2.949813184	2.923677726	0.026135458

8:	1-Chiba	2-Chiba	9-Chiba	10-Chiba	11-Chiba	12-Yama	4-Yama	5-Chiba	6-Yama	average	difference
PP-1	3.59499494	3.06453146	2.587402664	2.63401163	2.553265416	3.22081403	2.873628988	3.027194967	3.160321028	2.944480512	0.215840516
PP-2	3.596027564	3.06580972	2.589591597	2.636122664	2.557154826	3.224782383	2.878006946	3.027926226	3.16307107	2.945927741	0.218143329

9:	1-Chiba	2-Chiba	9-Chiba	10-Chiba	11-Chiba	12-Yama	4-Yama	5-Chiba	6-Chiba	average	difference
PP-1	3.604316139	3.06940139	2.590503148	2.632189867	2.541230447	3.205083102	2.880984103	3.014868116	5.003878613	2.942322039	2.061556574
PP-2	3.604316139	3.06940139	2.599525541	2.64087415	2.557184046	3.219905953	2.887884055	3.016447011	5.008530167	2.949442286	2.059087882

10:	1-Chiba	2-Chiba	9-Chiba	10-Chiba	11-Chiba	12-Yama	4-Yama	5-Chiba	7-Chiba	average	difference
PP-1	4.117761226	3.233030558	2.726108651	2.719426509	2.607672544	3.560545635	3.183374817	3.133962331	3.559633675	3.160235284	0.399398391
PP-2	4.120837543	3.233978271	2.732068666	2.722644069	2.695044297	3.574277253	3.189043192	3.137260478	3.563474569	3.175644221	0.387830348

11:	1-Chiba	2-Chiba	9-Chiba	10-Chiba	11-Chiba	12-Yama	4-Yama	5-Chiba	6-Yama	average	difference
PP-1	3.808552288	3.167533593	2.661204354	2.64317868	2.633367032	3.102562701	2.956608579	3.148319732	3.323764105	3.0152037	0.308560735
PP-2	3.811141453	3.168964686	2.669049557	2.656382412	2.637712242	3.123425183	2.96485655	3.149390147	3.324195449	3.022615279	0.301580171

12:	1-Chiba	2-Chiba	9-Chiba	10-Chiba	11-Chiba	12-Yama	4-Yama	5-Chiba	8-Chiba	average	difference
PP-1	4.063934391	3.214851077	2.625451219	2.641054811	2.725990365	3.594594793	3.247234557	3.107155568	3.21573573	3.152533348	0.063202383
PP-2	4.06476014	3.217576088	2.627558093	2.654669955	2.810341827	3.5997716	3.254062244	3.109137441	3.219285711	3.167234674	0.052051038

13:	1-Chiba	2-Chiba	9-Chiba	10-Chiba	11-Chiba	12-Yama	4-Yama	5-Chiba	7-Yama	average	difference
PP-1	3.791662396	3.020386772	2.687049963	2.650945341	2.661408771	3.299299277	3.143811218	3.175171929	3.597046645	3.053716958	0.543329687
PP-2	3.792634963	3.024414589	2.695467402	2.681530314	2.726862707	3.300319433	3.145343594	3.177694251	3.597433171	3.068033407	0.529399765

14:	1-Chiba	2-Chiba	9-Chiba	10-Chiba	11-Chiba	12-Yama	4-Yama	5-Chiba	8-Yama	average	difference
PP-1	4.306145434	3.304064385	2.595207982	2.689184219	2.571440481	3.444210405	3.085270947	3.240042744	3.662901489	3.154445825	0.508455664
PP-2	4.308183557	3.277557866	2.609389509	2.697987139	2.672228437	3.456156767	3.123300146	3.241177667	3.670293928	3.048247636	0.622046292

15:	1-Chiba	2-Chiba	9-Chiba	10-Chiba	11-Chiba	12-Yama	4-Yama	5-Chiba	9-Chiba	average	difference
PP-1	3.641310911	3.070952446	2.744777432	2.662032196	2.667102253	3.172548212	2.934641315	3.176271965	3.293369682	3.008704591	0.284665091
PP-2	3.647715938	2.177057995	2.781377421	2.709568291	2.7082955	3.202769174	2.954058242	3.17667293	3.298411204	2.919689436	0.378721768

16:	1-Chiba	2-Chiba	9-Chiba	10-Chiba	11-Chiba	12-Yama	4-Yama	5-Chiba	10-Yama	average	difference
PP-1	4.15354914	3.211257649	2.678730712	2.693741422	2.815669687	3.192528355	3.207793039	3.2006679	4.083552119	3.144242238	0.939309981
PP-2	4.160047788	2.234797994	2.750208316	2.698930219	2.872927645	3.198198772	3.212763361	3.206026622	4.086266696	3.04173759	1.044529106

17:	1-Chiba	2-Chiba	9-Chiba	10-Chiba	11-Chiba	12-Yama	4-Yama	5-Chiba	11-Yama	average	difference
PP-1	4.17599415	3.19078039	2.554657864	2.591940053	2.8332368	3.010507585	3.082045878	3.19194106	3.303722023	3.078887973	0.224834051
PP-2	4.177876348	2.225310428	2.619565769	2.609155534	2.873178409	3.020857092	3.082890878	3.192256801	3.319122978	2.975136407	0.34398657

Table 4. Threshold and KL-divergence

0.01			0.012		
Round	Articles	KL	Round	Articles	KL
1	Chi-1, Yama-1	0.021085951	1	Chi-1, Yama-1	0.021085951
2	Chi-1, Yama-2	0.013214638	2	Chi-1, Yama-2	0.013214638
3	Chi-1, Chi-2	0.008882086	3	Chi-1, Chi-2	0.008882086
4	Chi-1, Chi-2, Yama-3	0.020165253	4	Chi-1, Chi-2, Yama-3	0.020165253
5	Chi-1, Chi-2, Chi-3	0.012190679	5	Chi-1, Chi-2, Chi-3	0.012190679
6	Chi-1, Chi-2, Yama-4	0.013786736	6	Chi-1, Chi-2, Yama-4	0.013786736
7	Chi-1, Chi-2, Chi-5	0.012829681	7	Chi-1, Chi-2, Chi-5	0.012829681
8	Chi-1, Chi-2, Yama-5	0.014469782	8	Chi-1, Chi-2, Yama-5	0.014469782
9	Chi-1, Chi-2, Chi-6	0.013364345	9	Chi-1, Chi-2, Chi-6	0.013364345
10	Chi-1, Chi-2, Chi-7	0.01461821	10	Chi-1, Chi-2, Chi-7	0.01461821
11	Chi-1, Chi-2, Yama-6	0.017462573	11	Chi-1, Chi-2, Yama-6	0.017462573
12	Chi-1, Chi-2, Chi-8	0.012732278	12	Chi-1, Chi-2, Chi-8	0.012732278
13	Chi-1, Chi-2, Yama-7	0.017364093	13	Chi-1, Chi-2, Yama-7	0.017364093
14	Chi-1, Chi-2, Chi-9	0.010989098	14	Chi-1, Chi-2, Chi-9	0.010989098
15	Chi-1, Chi-2, Yama-8	0.015315333	15	Chi-1, Chi-2, Chi-9, Yama-8	0.014939441
16	Chi-1, Chi-2, Yama-9	0.019816968	16	Chi-1, Chi-2, Chi-9, Yama-9	0.018760831
17	Chi-1, Chi-2, Yama-10	0.016737215	17	Chi-1, Chi-2, Chi-9, Yama-10	0.01625942
18	Chi-1, Chi-2, Chi-10	0.010969145	18	Chi-1, Chi-2, Chi-9, Chi-10	0.010330202
19	Chi-1, Chi-2, Yama-11	0.016511058	19	Chi-1, Chi-2, Chi-9, Chi-10, Yama-11	0.014270153
20	Chi-1, Chi-2, Chi-11	0.011012479	20	Chi-1, Chi-2, Chi-9, Chi-10, Chi-11	0.010174216
21	Chi-1, Chi-2, Yama-12	0.014130583	21	Chi-1, Chi-2, Chi-9, Chi-10, Chi-11, Yama-12	0.011855117
Extracted Articles	Chi-1, Chi-2			Chi-1, Chi-2, Chi-9, Chi-10, Chi-11, Yama-12	

at preliminary experimental step as shown in a Table 4 where "Chi-2" means an article 2 concerns "Chiba event". This is why we have exceptional articles.

As in the table, we see difference articles extracted according to thresholds. In a case of 0.01, we have extracted the articles only in correct topic, but not in a case of 0.012.

6 Conclusion

In this investigation we have proposed a stochastic approach HMM for aspect tracking of news articles. We have extended Viterbi algorithm to model both forward and backward chaining and selected candidate models with top-K maximum likelihoods. We have also discussed the Markov Models as possible candidates for visualization of the storylines. With the help of output-words, we can expect to understand the contents easier.

Acknowledgment. We thank Prof. Wai Lam in Chinese University of Hong Kong for his helpful discussion and comments to our approach

References

1. Allan, J., Papka, R., Lavrenko, V.: On-line new event detection and tracking. In: Proceedings of the 21st ACM SIGIR Conference on Research and development in Information Retrieval, pp. 1–9 (1998)
2. Allan, J., Carbonell, J., et al.: Topic detection and tracking pilot study final report. In: Proceedings of the DARPA Broadcast News Transcription and Understanding Workshop, pp. 194–218 (1998)
3. Bilmes, J.A.: A gentle tutorial of the EM algorithm and its application to parameter estimation for Gausian mixture and hidden Markov Models. ICSI **4**, 1–15 (1998)
4. Carbonell, J.G., Yang, Y., et al.: CMU report on TDT-2: segmentation, detection and tracking, pp. 1–6. Carnegie Mellon University (1999)
5. Fiscus, J.G., Doddington, G.R.: Topic detection and tracking evaluation overview. In: Allan, J. (ed.) Topic Detection and Tracking, pp. 17–31. Kluwer Academic Publishers, Norwell (2002)
6. Inoue, M., Shirai, M., Miura, T.: Sequence classification based on active learning. In: 18th IEEE/ACIS International Conference on Software Engineering, Artificial Intelligence, Networking and Parallel Distributed Computing (SNPD) (2017)
7. Jurafsky, D., Martin, J.H.: Hidden Markov Model, Speech and Language Processing (2016)
8. Stokes, N., Carthy, J.: First story detection using a composite document representation. In: Proceedings of Human Language Technology (HLT), pp. 1–8 (2001)
9. Yang, Y., Carbonell, J., et al.: Learning approaches for detecting and tracking news events. IEEE J. Intell. Syst. **14–4**, 32–43 (1999)
10. Zhou, D., Xu, H., et al.: Unsupervised storyline extraction from news articles. In: IJCAI, pp. 3014–3020 (2016)

Transducer Cascade to Parse Arabic Corpora

Nadia Ghezaiel Hammouda[1(✉)], Roua Torjmen[2], and Kais Haddar[3]

[1] Miracl Laboratory, Institute of Computer Sciences and Communications
of Hammam Sousse, Sousse, Tunisia
Ghezaielnadia.ing@gmail.com
[2] Faculty of Economic Science and Management of Sfax, Miracl Laboratory,
University of Sfax, Sfax, Tunisia
rouatorjmen@gmail.com
[3] Faculty of Sciences of Sfax, Miracl Laboratory,
University of Sfax, Sfax, Tunisia
kais.haddar@yahoo.fr

Abstract. Arabic parsing is an important task in several NLP applications. Indeed to obtain a robust, efficient and extensible parser treating several phenomena, several issues (i.e., ambiguity and embedded structures) must be resolved. In this context, we will build an Arabic parser based on a deep linguistic study done with a new vision allowing the problem division and on a transducer cascade implemented in the NooJ linguistic platform. This parser is accomplished through our designed dictionaries, morphological grammars and transducers recognizing different sentence forms. The constructed parser is applied to two test corpora containing more than 5900 sentences with different structures. The parser outputs are XML annotated sentences. To evaluate the obtained results, we calculated the measure values of the precision, the recall and the f-measure, and compare them with those obtained by recursive transducer parser. The calculated measure values show that these results are encouraging.

Keywords: Arabic sentence · Granularity level · Lexical ambiguity
Transducer cascade · Arabic parser

1 Introduction

The parsing and annotation of Arabic corpora remain a challenge because of the richness of the Arabic language. So, the use of computers and new visions can shed new light on the study of this language. In fact, with the annotation process, syntactic information becomes clearer and this helps in several areas such as the construction of stochastic parsers, automatic translators and automatic recognizer of named entities.

Despite all the work to date, having a good syntactic representation (annotation) of Arabic texts presents a challenge. Ambiguity and non-robustness present the problems of existing works. Then, automation, optimization and disambiguation are needed. Also, the enrichment of the system of rules by the various linguistic phenomena existed increases the quality. The essential part in the annotation of a corpus is the constructed grammar. Furthermore, the tags attributed to Arabic words and phrases must be expressive to help in understanding of syntactic information. All these proposed

© Springer International Publishing AG, part of Springer Nature 2018
M. Silberztein et al. (Eds.): NLDB 2018, LNCS 10859, pp. 230–237, 2018.
https://doi.org/10.1007/978-3-319-91947-8_22

solutions are feasible through the cascade of finite transducers. A cascade is a sequential execution of a list of transducers.

In this paper, we will bring an innovative and original vision for the annotation of Arabic corpora. Therefore, the goal of this work has been the development of an automatic annotation tool for Arabic corpora. This tool is based on a symbolic approach using a cascade of finite transducers formalized in the NooJ linguistic platform [10]. The use of transducers cascade allows the reduction of the execution time for the annotation process. It facilitates the maintenance task because the modification of the rule does not require a global intervention, that is to say, it only affects the target graph. It also gives priority to certain rules and reduces ambiguities. Finite state transducer cascade has a lot of benefits that will be more detailed in the paper. But, it is very sensitive. So, we must know how to use transducers and classify them, that is to say, to assign each transducer a precise rank in order to have the right annotations and to avoid the mutual exclusion and the interaction of several linguistic phenomena.

The paper is structured in six sections. In the second section, we present the previous work on the parsing domain. In the third section, we are interested in studying the Arabic language in an innovative way that takes into account the working methodology of the transducer cascades. In the fourth section, we describe our proposed method by detailing the different phases. In the fifth section, we elaborate the experiment by testing different forms of sentences and we evaluate our transducer cascade and compare it to the recursive transducer by applying them to two corpora. This paper is closed by a conclusion and some perspectives.

2 State of the Art in Parsing Domain

Syntactic analysis is the result of a formalization of various lexical and syntactic rules. This formalization consists on the representation of various syntactic phenomena. Among the first systems created in syntactic analysis, we find Cass system (Cascaded Analysis of Syntactic Structure) for the parsing of English and German texts [1]. This system is based on a set of finite state automata applied in iterative order. In addition, the work of [9] presented the FSPar annotation system. The author proposed a cascade of transducers for the parsing of the German language. This work is based on a list of recursive transducers treating each group independently of the other forms. This cascade enriched the output text with the various syntactical features for each word. Also, in [4], the authors presented a tool for parsing Arabic text using recursive graphs. The used approach is based cascade of transducers implemented in NooJ platform. This tool is based on three steps: starting with the segmentation phase, the preprocessing phase and finishing with the annotation. Concerning the search for information, we mention the work of [7] which presented the ASRextractor system. The latter is a system for extracting and annotating semantic relations between Arabic named entities using the TEI formalism. ASRextractor is based on a transducer cascade for extraction and annotation.

Also, [2] ensured an analysis of the Arabic language, and more precisely the coordination's phenomenon. In this work, the author had developed a HPSG grammar based on a hierarchy of types in order to classify the different linguistic units for the Arabic language, essentially the coordinated forms. Besides, there are different researches linked to parsing Arabic language. In [6], the authors proposed a generator of TEI (Text Encoding Initiative) lexicons based on an Arabic word hierarchy.

Among the disambiguation's systems, we cite MADA, which is a tool providing multiple applications like POS tagging, diacritization, lemmatization, stemming and glossing. This tool is based on statistical approach by exploring SVM models (Support Vector Machines). A simple gather with the AMIRA toolkit to improves the performance of both platforms. As a result, the MADAAMIRA [8] system explored Arabic and the Egyptian language. In addition to these systems, there is the Stanford parse, which is a system, developed at Stanford University. This parser and the used pos-tagger are based essentially on the Maximum Entropy Model (MEM) called Conditional Markov Model (CMM). The tagger used by the Stanford parser was inspired from the Arabic Penn Treebank (ATB), which provided a large set of tags.

All these cited works show that parsing for the Arabic language is a difficult task and several issues remain unsolved. Full formalization is always a challenge.

3   Linguistic Study

Our linguistic study brings a new vision that that promotes the contribution of local grammars [3]. It makes it possible to extract the specificities of the Arabic language, to better understand it and to master it.

In this section, we present firstly the Arabic word hierarchy. Secondly, we mention the different lexical ambiguities. Thirdly, we discuss the importance of the granularity level in the annotation process as well as their uses in our study. Finally, we continue this study with the different types of phrases and sentences in the Arabic language.

3.1   Hierarchy of Arabic Word Types

In traditional Arabic, the word categories are reduced to three: verbs, nouns and particles. Most Arabic verbs are formed of a simple root of three consonants such as « خرج » "karaja" (exit). But, they can be formed of another root of four consonants such as « برمج » "barmaja" (program). Moreover, from a single root verb, it is possible to create seven forms of verbs, each of which carries a new meaning. Also verbs can be intransitive, transitive, double transitive and triple transitive.

The nouns may be personal pronouns as « أَنَا » "ana" (I), demonstrative pronouns as « هذا » "hedha" (this) and relative pronouns as « الَّذِي » "alladhii" (who). In addition, they can be derived from verbal roots and they can be numbers. The nouns can be also common noun as « طِفْلٌ » "tiflun" (kid) or proper noun as « أَحْمَدُ » "ahmadu" (Ahmed).

Finally, the third category concerns particles that are used to indicate the spatio-temporal setting. Among the temporal prepositions, we quote « بَعْدَ » "ba'da" (after) and « مُنْذُ » "munthu" (since) and among the spatial prepositions, we quote « أَمَامَ » "'amaama" (besides), « وَرَاءَ » "wara'a" (behind), and « قُرْبَ » "qurba" (near).

3.2   Lexical Ambiguities

The word lexical ambiguity concerns the existence of more than one meaning and category to which the word belongs. Among the language specificities that produce Arabic ambiguities, we cite the non-vocalization which means that the word has no diacritics. There is also the homograph which means that words are written in the same way, whatever the pronunciation manner. We also cite the agglutination which consists of bringing together different morphemes into a single morphological element. The lifting of different ambiguities can be done in a separate way through the lexical rule succession use. This strategy favors the establishment of local grammars.

3.3   Granularity Level of Word Features

The notion of granularity refers to the idea of showing the several levels of information of a word, that is to say, to go into details to show all its characteristics.

The word « الولد » "alwaladu" (the boy) is a common noun singular masculine and nominative. Each level shows a new feature for that noun such as its type, number, kind and semantic case. Verbs also can have a high granularity level. For example, the word « كتب » "kataba" is a transitive verb in the past with the 3rd person singular masculine. In this example, each level shows a new feature for this verb, such as its transitivity, time, gender, and number.

3.4   Different Arabic Phrases

In Arabic language, we distinguish different phrase types: verb phrase, noun phrase and prepositional phrase. Respecting certain specificities and the VSO (Verb Subject Object) order, the verb phrase is composed of a verb that can be preceded by a tool. But, some modes of verb like apocopate and subjunctive require a tool. The noun phrase has several forms like annexation compound, adjectival compound and relative clause compound. In addition, it can be simple. The noun phrase can be nominative, accusative and genitive. In fact, the noun phrases can be nominative in several cases like the subject in the verbal sentence and the topic or attribute in the nominal sentence. Also, they can be accusative in the objects of verbal sentence. They can be genitive in indirect objects that always begin with a preposition

In this paragraph, we studied the different phrases that must be taken in count during the construction of the grammar rule set. This study shows also that local grammars can reduce the complexity of an Arabic parser establishment.

3.5   ˈArabic Sentences Types

Arabic grammar distinguishes between two types of sentences. The first type is the nominal sentence; it looks like the sentences of the French and English language and begins with a topic. The second type is the verbal sentence that begins with a verb followed by a subject with the possibility of having one, two or three objects depending on the transitivity of the verb. In fact, the nominal sentence consists of two elements: a topic and an attribute. In general, the topic is a nominative noun phrase and may be a nominative noun phrase, a prepositional phrase or a noun phrase followed by a prepositional phrase. In addition, nominal sentences can be preceded either by Kana and her sisters (كان وأخواتها) either by Inna and her sisters (إن وأخواتها). These tools change the dependence of the topic and the attribute. Kana and her sisters insist that the attribute becomes accusative, that is, the noun phrase becomes accusative, but the prepositional phrase remains invariable. Unlike kana and her sisters, Inna and her sisters forced that the topic becomes accusative, i.e. the noun phrase becomes accusative. This linguistic study must be respected during the construction of the system of rules in order to overcome the ambiguities of the annotation of an Arabic corpus and to raise the aspect of granularity level.

4   Proposed Method

As we have already indicated, the proposed method is based on a previous work presented in [5]. Our new parsing method consists of transforming are cursive transducer into many elementary transducers. This transformation is not easy due to the enormous numbers of sub-graphs and especially the embedded ones. Firstly, we study the nature of each sub-graph based on the graph's complexity, the number of sub-graphs and their depth. After this study, we classify graphs by levels: from low complex one to high complex one. Secondly, we studied the relation between graphs: elementary, embedded and recursive graphs. Thirdly, we created the transducer cascade which respects the extracted graph hierarchy. In fact, the cascade executes the established transducers in an adequate ranking. This ranking is identified through multiple experimentations. In the following, we will present transducers from our proposed cascade for Arabic sentences starting from specific phrases to entire sentences.

4.1   Noun Phrase

In the transducer of Fig. 1, we can recognize all nominative noun phrases.

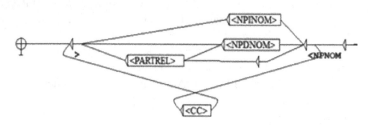

Fig. 1. Transducer for nominative noun phrase.

The transducer of Fig. 1 contains four already compiled transducers. These compiled sub-graphs generate four new annotation tags <NPINOM>, <NPDNOM>, <PARTREL> and <CC>. These annotations concern respectively indefinite nominative noun phrase (NP), definite nominative NP, relative NP and NP with conjunction. The same idea is used for recognizing accusative and genitive NPs.

4.2 Verbal Phrase

The verbal phrase is covered by the following transducer.

The recursive transducer of Fig. 2 contains two sub-graphs: VerbSTool for VP without tool and VerbTool for VP starting with a tool. The sub-graph VerbSTool uses a fine granularity level. The transitivity of the verb is controlled by using a variable called "$trans". Besides, the sub-graph VerbTool concerns verbs preceded by a tool imposing a particular verb tense. In fact, this particularity helps to the disambiguation.

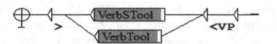

Fig. 2. Transducer for verbal phrase.

4.3 Verbal Sentence

The verbal sentence contains different rules. This sentence type starts usually with a VP succeeded by a nominative NP. In addition, depending on the transitivity of the verb, the number of accusative NP is fixed. The variable "$trans" controls the transitivity of the verbal phrase. In addition, a verbal sentence can contain one or more PP (prepositional phrase).

4.4 Nominal Sentence

The following transducer is proposed to recognize all nominal sentences forms.

The transducer of Fig. 3 enriches the tag dictionary. In fact, the proposed transducer treats the modal verb "KANA" followed by a nominative topic and an accusative attribute, the tool "INNA" followed by an accusative topic and a nominal attribute.

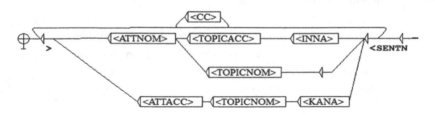

Fig. 3. Transducer for nominal sentence.

5 Experimentation and Evaluation

To experiment our transducer cascade parser on the test corpus, we have used firstly our segmentation tool presented in [4]. Secondly, we used our proper tagset inspired from Stanford's tagset, a set of morphological grammars (113 infected verb patterns, 10 broken plural patterns and one agglutination transducer) and dictionaries that are already exist in the NooJ linguistic platform (24732 nouns, 10375 verbs and 1234 particles). However, the resources that exist in NooJ are not complete and sufficient. For this raison, we have added two other dictionaries. The first dictionary named "verbeintr.nod" contains 91 entries. The second dictionary named "prenom.nod" for proper nouns contains 105 entries. Thus, all these resources have the same priority except the "verbeintr.nod"; it has a high priority of a single level "H1" compared to others. Thirdly, the proposed cascade executes 23 transducers including 100 sub-graphs. The transducers that are edited in NooJ linguistic platform called in a fixed ranking. Fourthly, we tested our parser on two corpora. The first corpus is extracted from the Arabic Treebank (ATB) containing 836 sentences and the second is extracted from Arabic stories containing 5900 sentences. Concerning the ATB corpus, we deleted the sentence annotations in order to obtain a raw corpus or the test. In addition, the number of words in the sentences of the two corpora varies between 4 and 83 words. These sentences have different forms of verb phrases (with and without tools) with different tenses and modes. Also they have several noun phrase structures.

For the evaluation, we used the known metrics: recall, precision and f-measure. We obtain the following results for the ATB corpus presented in Table 1.

Table 1. Summarizing the metrics obtained for ATB corpus.

	ATB corpus	Recall	Precision	F-measure
Cascade parser	836 sentences	0.9	0.94	0.91

The major advantage of the new parser compared with Stanford is the great reduction of the execution time (parsing 836 sentences in 59.2 s). But, there is no improvement in the measure values. Concerning the stories corpus, we obtained the following evaluation illustrated in Table 2.

Table 2. Summarizing the metrics obtained for stories corpus.

	Stories corpus	Recall	Precision	F-measure
Cascade parser	5900 sentences	0.74	0.82	0.77
Recursive parser	5900 sentences	0.6	0.7	0.62

The measure values of Table 3 show the efficiency of the cascade parser compared to the recursive parser. But some parsing problems are detected especially in complex and embedded structures. To solve these problems, we must add some constraints. Also to improve the measure values, we must add other syntactical rules concerning untreated linguistic phenomena.

6 Conclusion

In the present paper, we developed an automatic parsing tool for Arabic corpora in NooJ linguistic platform. This tool that contains a segmentation phase is accomplished through our cascade of transducers and dictionaries, and is based on a deep linguistic study. In addition, we have shown the efficiency of transducer cascade versus the recursive transducer. Thus, the evaluation is performed on a set of sentences belonging to two corpora. The results obtained are ambitious and show that our parser can treat efficiently different sentence forms. As perspectives, we will increase the coverage of our designed dictionaries. We will also improve our parser by adding other syntactic rules recognizing frozen forms of sentences. These rules will be specified by finite transducers.

References

1. Abney, S.: Partial parsing via finite-state cascades. Nat. Lang. Eng. **2**(4), 337–344 (1996)
2. Boukedi, S., Haddar, K.: HPSG grammar for Arabic coordination experimented with LKB system. In: Proceedings of the Twenty-Seventh International Florida Artificial Intelligence Research Society Conference, FLAIRS 2014, Pensacola Beach, Florida, 21–23 May 2014, pp. 166–169 (2014)
3. Ghezaiel, N., Haddar, K.: Parsing Arabic nominal sentences with transducers to annotate corpora. Computación y Sistemas, **21**(4), 647–656 (2017). Advances in Human Language Technologies (Guest Editor: A. Gelbukh)
4. Hammouda, N.G., Haddar, K.: Integration of a segmentation tool for Arabic corpora in NooJ platform to build an automatic annotation tool. In: Barone, L., Monteleone, M., Silberztein, M. (eds.) NooJ 2016. CCIS, vol. 667, pp. 89–100. Springer, Cham (2016). https://doi.org/10.1007/978-3-319-55002-2_8
5. Hammouda, N.G., Haddar, K.: Arabic NooJ parser: nominal sentence case. In: Mbarki, S., Mourchid, M., Silberztein, M. (eds.) NooJ 2017. CCIS, vol. 811, pp. 69–80. Springer, Cham (2018). https://doi.org/10.1007/978-3-319-73420-0_6
6. Maamouri, M., Bies, A., Buckwalter, T., Mekki, W.: The Penn Arabic Treebank: building a large-scale annotated Arabic corpus. In: NEMLAR Conference on Arabic Language Resources and Tools, vol. 27, pp. 466–467 (2004)
7. Mesmia, F.B., Zid, F., Haddar, K., Maurel, D.: ASRextractor: a tool extracting semantic relations between Arabic named entities. In: 3rd International Conference on Arabic Computational Linguistics, ACLing 2017, 5–6 November 2017, Dubai (2017)
8. Pasha, A., Al-Badrashiny, M., Diab, M.T., El Kholy, A., Eskander, R., Habash, N., Roth, R.: MADAMIRA: a fast, comprehensive tool for morphological analysis and disambiguation of Arabic. In: Proceedings of LREC, Reykjavik, vol. 14, pp. 1094–1101 (2014)
9. Schiehlen, M.: A cascaded finite-state parser for German. In: Proceedings of EACL 2003, vol. 2, pp. 163–166 (2003)
10. Silberztein, M.: A new linguistic engine for NooJ: parsing context-sensitive grammars with finite-state machines. In: Mbarki, S., Mourchid, M., Silberztein, M. (eds.) NooJ 2017. CCIS, vol. 811, pp. 240–250. Springer, Cham (2018). https://doi.org/10.1007/978-3-319-73420-0_20

Multi-Word Expressions Annotations Effect in Document Classification Task

Dhekra Najar[✉], Slim Mesfar, and Henda Ben Ghezela

RIADI, University of Manouba, Manouba, Tunisia
Dhekra.najar@gmail.com,
hhbg.hhbg@gmail.com, mesfarslim@yahoo.fr

Abstract. Document classification is a necessary task for most Natural Language Processing tools since it classifies documents content in a helpful and meaningful way. The main concern in this paper is to investigate the impact of using multi-words for text representation on the performances of text classification task. Two text classification strategies are proposed to observe the robustness of each of them. First, we will deal with the literature review of existing linguistic resources in Arabic language. Secondly, we will present a classification method that is based on domain candidate simple terms. These terms are automatically extracted from multiple specialized corpora depending on their appearance frequency. Then, we will present a detailed description of a classification method based on multi-word expressions dictionary. CompounDic, an Arabic multi-word expressions dictionary, will be used to automatically annotate multi-word expressions and variations in text. Finally, we carried out a series of experiments on classifying specialized text based on simple words and multi-word expressions for comparison purposes. Our experiments show that the use of multi-word expressions annotations enhances the text classification results.

Keywords: Text classification · Topic identification · Multi-Word Expressions
Natural language processing · NooJ · Arabic language · MWEs variation

1 Introduction

Multi-word expressions (MWEs) have received a wide attention in natural language processing domain (NLP). Handling MWEs as the single terms in the linguistic analysis to increase the performance of NLP tasks, such us machine translation and text classification, has been a controversial subject of various researches. This paper focuses on investigating the effectiveness of using multi-words expressions annotations for text representation on the performances of text classification.

We present a semantic document classification approach based on language analysis. Two text classification strategies are proposed to observe the robustness of each of them. Firstly, we will apply a classification method that is based on domain candidate simple terms. Then, we will apply a classification method that is based on multi-word expressions annotations. The rest of this paper is organized as follows: Sect. 2 presents preliminaries of this paper, including the basic techniques and concepts that will be used to implement our approach. The concept of document classification and MWEs

M. Silberztein et al. (Eds.): NLDB 2018, LNCS 10859, pp. 238–246, 2018.
https://doi.org/10.1007/978-3-319-91947-8_23

are introduced in this section. Section 3 describes the both proposed strategies for comparison purpose. The corpora, experimental results and methods evaluation are specified in Sect. 4. Section 5 concludes the whole paper.

2 Preliminaries

2.1 Document Classification

Text classification is known under a number of synonyms such as document/text categorization/routing and topic identification. It can be defined as content-based assignment of one or more predefined categories (topics) to documents for purposes of easier information retrieval and better organization. This task is necessary for most Natural Language Processing tools since it classifies documents content in a meaningful way. The task of text classification includes two kinds of properties of indexing term: semantic quality and statistical quality [1]. Semantic quality is related to a term's meaning, i.e., to how much extent the index term can describe text content. Statistical quality is related to the discriminative (resolving) power of the index term to identify the category of a document in which the term occurs. Document classification is often used to subsume two types of analyses: document categorization and document clustering. The distinction is that categorization is a form of supervised approach and clustering an unsupervised approach of grouping textual objects. Document classification task involve main steps [2]: document pre-processing, feature extraction/ selection, Model selection and training and testing the classifier. We are mainly concerned with feature extraction using simple words and MWEs. Authors in [3] used a statistic method to investigate the impact of using multi-word expressions in several document classification environments. They have significantly better retrieval effectiveness using multi-words. In the study of [4], authors explain the importance of integrating MWEs analysis in the preprocessing stages in order to obtain viable linguistic analysis using a pure linguistic approach. They also show how the grammar is responsible for detecting MWEs with morphosyntactic variations. They obviously reduce ambiguities and give a noticeable degree of certitude to the analysis and machine translation outputs.

2.2 Multi-Word Expressions

Multiword expressions are defined as idiosyncratic interpretations that cross word boundaries or spaces [5]. This means that multi-Word Expressions are groups that work together as units to express a specific meaning. They can be formed by combining two or more words together. The diverse analyzers, based on morphological aspect, are not able to recognize multiword expressions. They usually separate MWEs into single terms which can affect the semantic representation of texts.

In Arabic language, we still have a lack of linguistic resources comparing to Latin languages. We cite in the table below some Arabic linguistic resources of MWE:

For example, authors in [6] have followed a hybrid method to Multi-word term extraction system for Arabic language. The system takes into account several types of morphosyntactic variations. They were able to annotate an environmental corpus. In total they compiled a list of 65,000 MWTs. Authors in [7] presented a pattern-matching algorithm to construct a list of 4,209 Arabic MWEs, collected from various dictionaries, grouped based on their syntactic type. They have automatically annotated a large corpus of Arabic text. The authors considered some variations and found 481,131 MWE instances.

3   Approach

3.1   Resources

3.1.1   CompounDic: Arabic Multi-Word Expressions Dictionary

In previous work, we have semi-automatically built CompounDic [8], an Arabic 2 units MWEs thematic lexicon. For this purpose, we have taken advantage of NooJ's linguistic engine strength in order to create this large coverage terminological MWEs dictionary for Modern Standard Arabic language CompounDic.

CompounDic is an Arabic MWEs dictionary that lists many entries, divided into more than 20 domains. It lists only MWEs in their base form. With regard to syntactic and morphological flexibility, the lexicon covers 2 types of MWEs: Fixed MWEs (no variation allowed) and semi-fixed MWEs (variation in their structural pattern).

CompounDic contains 36960 entries classified into more than 20 semantic domains. All the entries of CompounDic are manually set in the base form: "indefinite singular form". Then, all the listed MWEs are voweled manually so that NooJ would be able to recognize unvoweled, semi-voweled as well as fully voweled MWEs. The manual vocalization is an extremely important step since it allows to vowel entries depending on their semantic information since we can find a word that has different way of vocalization and different meanings. This helps reducing linguistic ambiguities in Arabic texts. The final manual step is classifying the MWEs according to 2 criteria: the grammatical composition (N1 N2), (N1 ADJ). The entries of CompounDic are classified into more than 20 domains.

Every entry in CompounDic is stored with information about its structure, number of units and domain. To give an example from the technical domain in our lexicon:

إِنْعِدَام الإِتِّزَان,N+Structure=N1_N2+CMPD+Units=2+Domain=Technical

Fixed MWEs always occur in exactly the same structure and can be easily recognized by a lexicon. However, most MWEs allow different types of variations.

3.1.2   Simple Terms Dictionary

«El-DicAr» is an Electronic Dictionary for Arabic language [9]. El-DicAr is a dictionary of lemmas containing the vocabulary of the Arabic language. This dictionary contains 10,500 verbs, 6,204 names and 36 pronouns. Moreover, it associates lemmas with their inflectional, morphological and syntactic-semantic paradigms in order to produce more than 3 million inflectional forms. For the following parts of this work, we will use the Electronic Dictionary for Arabic "El-DicAr" as the basis to build our simple terms dictionary using NooJ's morphological analyzer. The dictionary is created based on a collection of 3000 potential candidate simple terms. Firstly, terms are automatically extracted from multiple specialized corpora depending on their appearance frequency. They are originally enriched with the distributional description (22 domains inherited from specialized corpus). The extracted terms are automatically voweled, set in their base form, and classified based on their semantic category. Secondly, we have automatically enriched the extracted candidate terms in El-DicAr with their semantic description (Domain = Technical).

Example: In El-DicAr, we have enriched the derivational paradigms of the candidate term (جولة, inning) with the distributional information "+Domain":

جَوْلَة,N+FLX=Flexion1+DRV=NomDRV:NomDuel3+DRV=FemininPluriel9 +**Sport**

Moreover, the dictionary takes into account the morphosyntactic variations using the inflectional, morphological and syntactic-semantic paradigms of "El-DicAr".

3.2   Methods

Two text classification strategies are proposed to observe the robustness of each of them. Firstly, we present a classification method that is based on the dictionary of domain candidate simple terms. Secondly, we present a classification method based on the dictionary of MWEs "CompounDic.

3.2.1   Candidate Single Terms Method

In order to annotate candidate terms in texts with their domain, we implement a simple syntactic local grammar (using NooJ[1] platform) composed of 22 sub graphs. Each sub graph concerns a domain. However in NooJ "simple words and multi-words units are processed in a unified way: they are stored in the same dictionaries, their inflectional and derivational morphology is formalized with the same tools and their annotations are undistinguishable from those of simple words" [10] (Fig. 1).

The main graph, as shown below, indicates the constraint (Domain=#$var $Domain). This constraint allows inheriting the semantic information (Domain) from the recognized term to annotate the matching sequence in text.

We carried a series of experiments to classify specialized text based on simple terms in order to test the robustness of this method. We launched linguistic analysis on heterogeneous articles from Internet. As an example, we present a part of annotations in a document as shown in Fig. 2 below. These annotations were generated by our system by applying the lexical grammars.

---

[1] See www.nooj4nlp.net.

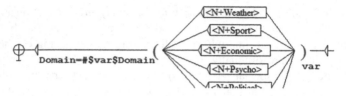

Fig. 1. Local grammar of classification

Example: Document 1 (Weather)

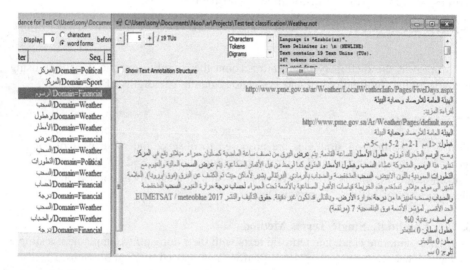

Fig. 2. Generated annotations of candidate simple terms in document 1

3.2.2   MWEs Method

Example 1: Document 1 (Weather)

In order to annotate MWEs in test corpus, we implement a second local grammar. This grammar allows inheriting the semantic information of all expressions with the paradigm: <N+CMPD+Domain>. We present a part of annotations that were generated after analyzing the same document of the previous example (Fig. 3).

4   Results and Discussion

4.1   Results

Illustrating the concordances, our experiments show that the use of multi-word expressions annotations enhances the text classification results more than the candidate simple terms. Multi-word expressions have a great importance in topics identification

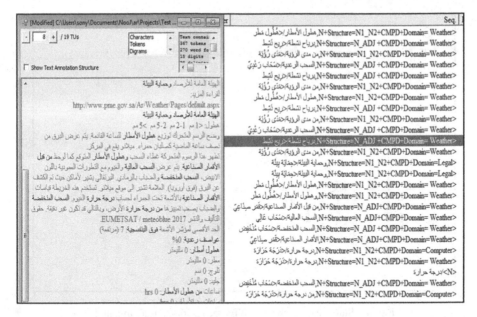

Fig. 3. Generated annotations of MWEs in document 1

since they constitute domain relevant terms. As we can see in concordances, in most cases the annotations of MWE domains are correct.

We manually exported the concordances of MWEs method with NooJ and process them to generate statistical graphs related to the semantic information (domains) of a document (textual article). In the first document, the system has successfully classified the document into "Weather" domain. This graph shows the number and the allocation of MWEs in the document that was recognized by our system per domain.

MWEs method has also classified most of the articles successfully. The Fig. 4 above shows the recall and precision obtained by testing the both strategies.

The results, as seen in Table 1, indicate that we have reached better quality results of recognition and classification using MWEs method. The results of MWEs method in term of precision and recall are better than candidate simple terms method.

There is some silence (non recognized candidate terms and MWEs) due to some missing entries in our dictionary but we judge our results as precise.

Several obstacles make the document classification in Arabic language complicated such as high inflectional nature, morphological ambiguity… These specificities of Arabic language represent the most challenging problems for Arabic NLP researchers.

4.2 Discussion

We have been faced with some specific cases like in document 2 that is specialized in psychological domain. The system (MWEs method) has classified this document as Psychological document (Fig. 5).

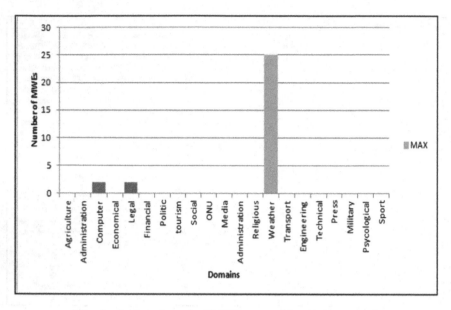

Fig. 4. Statistical graph of MWEs recognition in document 1

Table 1. Results

	Precision	Recall
Candidate simple terms method	0.71	0.65
MWEs method	0.90	0.73

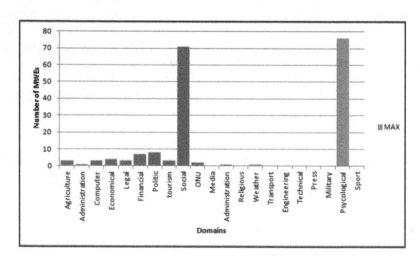

Fig. 5. Statistical graph of MWEs recognition in document 2

According to Statistical analysis, we notice that the text contains many Social MWEs as well. This can be explained by the fact that the text is talking about the psychological illness «social phobia».

We follow the same steps as previously detailed. The system has classified document 3 as economical document. However, domains like Legal, Financial and Political were also high rated. This leads us to the conclusion that there are some domains that are related to each other just like: (Politic, press, media); (Industry, technical); (Economic, politic, financial). In these cases we risk to have multiple domains highly rated. In order to solve this difficulty, we are studying the ability of using a lexical ontology in order to relate these domains (concepts). Considering the results of this comparing study, we have concluded that handling MWEs improve the performance of classification tasks. The analysis and recognition of MWEs must be integrated in the preprocessing stages of text analysis if we want to obtain better classification results with NLP tools.

5 Conclusion

This paper focuses on investigating the effectiveness of using multi-words expressions annotations for text representation on the performances of text classification. Two text classification strategies were proposed to observe the robustness of each of them. Firstly, a classification method that is based on domain candidate simple terms was presented. Then, a classification method that is based on Multi-Word Expressions was presented. Our research has shown that handling MWEs in the linguistic analysis is more efficient to increase the performance of NLP tasks.

References

1. Lewis, D.D.: Text representation for intelligent text retrieval: a classification oriented view. In: Text-Based Intelligent Systems: Current Research and Practice in Information Extraction and Retrieval, pp. 179–197 (1992)
2. Dalal, M.K., Zaveri, M.A.: Automatic text classification: a technical review. Int. J. Comput. Appl. **28**(2), 37–40 (2011)
3. Papka, R., Allan, J.: Document classification using multiword features. In: Proceedings of the Seventh International Conference on Information and Knowledge Management, pp. 124–131. ACM (1998)
4. Attia, M.A.: accommodating multiword expressions in an Arabic LFG grammar. In: Salakoski, T., Ginter, F., Pyysalo, S., Pahikkala, T. (eds.) FinTAL 2006. LNCS (LNAI), vol. 4139, pp. 87–98. Springer, Heidelberg (2006). https://doi.org/10.1007/11816508_11
5. Sag, I.A., Baldwin, T., Bond, F., Copestake, A., Flickinger, D.: Multiword expressions: a pain in the neck for NLP. In: Gelbukh, A. (ed.) CICLing 2002. LNCS, vol. 2276, pp. 1–15. Springer, Heidelberg (2002). https://doi.org/10.1007/3-540-45715-1_1
6. Boulaknadel, S., Daille, B., Aboutajdin, D.: A multi-word term extraction program for Arabic language. In: LREC (2008)
7. Hawwari, A., Bar, K.., Diab, M.: Building an Arabic multiword expressions repository. In: Proceedings of the 50th ACL, pp. 24–29 (2012)

8. Najar, D., Mesfar, S., Ghezela, H.B.: A large terminological dictionary of Arabic compound words. In: Okrut, T., Hetsevich, Y., Silberztein, M., Stánislavenka, H. (eds.) NooJ 2015. CCIS, vol. 607, pp. 16–28. Springer, Cham (2016). https://doi.org/10.1007/978-3-319-42471-2_2

9. Mesfar, S.: Analyse morpho-syntaxique automatique et reconnaissance des entités nommées en arabe standard. Doctoral Dissertation, Besançon (2008)

10. Silberztein, M.: The Formalisation of Natural Languages: The NooJ Approach, p. 346. Wiley, Hoboken (2016)

RDST: A Rule-Based Decision Support Tool

Sondes Dardour$^{(\boxtimes)}$ and Héla Fehri$^{(\boxtimes)}$

MIRACL Laboratory, University of Sfax, Sfax, Tunisia
dardour.sondes@yahoo.com, hela.fehri@yahoo.fr

Abstract. With the unfriendly wellbeing impacts of antibiotics and chemical drugs, medical herbalism has been a resurgence of interest in last years. Thus, medicinal plants are capable of treating disease and improving wellbeing, frequently without any significant side effects. This paper presents a Rule-based decision support tool aimed at helping users to identify accurate medicinal plants according to their symptoms taking into account the contraindications of each plant. This tool is based on IF-THEN rules, dictionaries and transducers. It permits the identification of the accurate medicinal plants, the recognition of medicinal plant properties and it incorporates user feedback to refine its results. Dictionaries and transducers are implemented in NooJ linguistic platform and applied in JAVA application with the command-line program noojapply. Experimentations of the Rule-based decision support tool show interesting results. Performance is satisfactory since our tool could act as a consultant. Furthermore, the functionality can be extended to other medicinal plants in the aim to treat the whole body health system.

Keywords: Medicinal plants · Symptoms · Recognition · Transducer
Dictionary · Decision

1 Introduction

The medical herbalism, also referred to as botanical medicine or phytotherapy, is being used for treating different human (or animal) health conditions from the very beginning of human existence. Historically, several plant species are preferably considered as a best treatment to help people feel better or cope with anxiety, depression, cold, fever, skin diseases, etc. In fact, plants have been used since ancient times to heal and cure diseases and improve health and wellbeing of organisms [1]. Recently, it has been confirmed by World Health Organization (WHO) that 80% of people worldwide rely chiefly on herbal medicine to meet their primary healthcare needs [2].

The increasing for self-medication by patients is attributed to medicinal plants. However, instances of self-medication also occur, occasionally resulting in complications and negative impact on patient health, because of wrong medicinal plant administration. Therefore, in the aim of identifying the accurate medicinal plant, a rule-based decision support tool that supports patients to minimize or to prevent further recurrence of this practice, specifically the mistaken use of plants is crucial.

Natural language is a necessary entity for medical consultations to gather information in order to understand disease and associated symptoms and to provide

© Springer International Publishing AG, part of Springer Nature 2018
M. Silberztein et al. (Eds.): NLDB 2018, LNCS 10859, pp. 247–255, 2018.
https://doi.org/10.1007/978-3-319-91947-8_24

necessary medicinal plants. To make machine understand such kind of natural languages, Natural Language Processing (NLP) is used.

In this context, our objective is to propose using Rule-based approach, a decision-support tool. This tool allows the identification of medicinal plants that better treat patients' diseases and associated symptoms and allows the recognition of medicinal plant properties. Our RDST (Rule-based Decision-Support Tool) is based on IF-THEN rules, dictionaries and transducers. Indeed, finite state transducers (FST) can play an important role in solving different decision situations and provide relevant solutions to alleviate the analysis process. Finite state transducers are finite state machines [3], which define relations between two sets of strings by means of a transformation of one string into another. Furthermore, the FST can produce some outputs and this property determines the way transducers are being used in NLP. They can be visually presented by graphs; this makes them convenient for human use, too.

The remainder of this paper is structured as follows. Section 2 presents the literature review. Our approach is described in Sect. 3. Section 4 deals with the experimentation carried out to evaluate the RDST efficiency. Finally, we conclude about the main contributions and propose further perspectives.

2   Literature Review

The concept of a named entity (NE) was first introduced at the Message Understanding Conference (MUC) [4]. NE defines some informative words, such as plant names, disease names, as information units in text. Recognition of named entity is aimed to further extract information by identifying the key ideas or concepts. Research on NEs revolves around two complementary axes: the first is concerned with the identification and typing of NEs while the second with translation of NEs. Approaches to NER include linguistic, statistic, and hybrid approaches.

At the best of our knowledge, there are two works on NEs in herbalism among which we cite the work of [5]. In this work, the authors proposed a decision-support tool based on the rule-based approach. This tool allows the identification of medicinal plants. The process of identifying medicinal plants is based on two steps: collecting criteria for identifying syntactic patterns and transformation of these patterns to transducers. The evaluation of this tool gives satisfactory results. However, the major problem lies in the modeling of transducers, which is illegible and difficult to understand. Indeed, when adding criteria or medicinal plants, it is necessary to construct another transducer.

We can cite also the work described in [6]. This work allows the identification of medicinal plants names from French-Arabic parallel corpora. The proposed method mainly involves two steps. The first is concerned with the construction of corpus, which is formed by several texts composed of the multilingual encyclopedia Wikipedia, and the second step is based on a morphological analysis followed by a syntactic analysis.

In the medical domain, linguistic and terminological resources were developed by National Library of Medicine [7] to solve problems such as medical NER. The most used terminology resource is the Unified Medical Language System (UMLS) which is maintained as a support for integration of biomedical textual annotations scattered in distinct databanks. The UMLS is used by [8] to recognize diseases. Other works are dealing with the recognition tasks [9, 10].

3 Proposed Approach

The challenges discussed in the first section make clear the need for rule-based approach to deal with identification of medicinal plants. In this section, we describe the RDST, our approach proposed to perform the identification task. In our proposal, the RDST is based on three phases as illustrated in Fig. 1: the identification of resources, the construction of resources, and user feedback. In the following, we detail these phases.

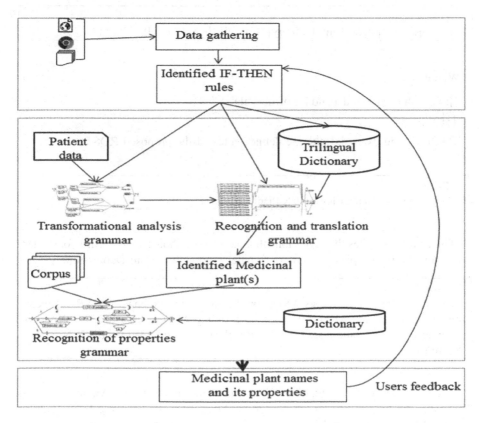

Fig. 1. Architecture of RDST

3.1   Identification of Resources

This phase is composed of two steps: data gathering and rules identification to identify medicinal plants.

Data Gathering. Although medicinal plants are widely used to cure safely, however, they can potentially cause side effects especially in breastfeeding and pregnancy. Therefore, we gathered data about medicinal plants through studying several online sources including doctossimo[1], creapharma[2], pharmacognosy[3], and search results returned by the search engine Google. This study provides knowledge about medicinal plants, contraindications and the use of each medicinal plant to ensure the safety practice.

Rules Identification. This step is based on IF-THEN rules. According to our study, we identify 396. Each rule takes the following format:

Rule format
IF [Age] [Sex] Symptom[+][Allergy] [Chronic diseases] THEN Plant[+]

Where:

[]: Optional criteria depending of the rule
+:1 or more

The following examples cite the defined rules of the proposed RDST.

Rule 1
IF Symptom= Depression THEN Plant= Birdsfoot

The rule 1 describes the conditions that lead to the choice of the "Birdsfoot" plant. This plant can treat a patient (male or female) who suffers from Depression.

Rule 2
IF Age= More than 12 years AND Sex= Female (NOT Pregnant AND Breastfeeding) AND Symptom= Abundant menstrual bleeding AND Allergy= NOT Allergy to Artemisia AND Chronic disease= NOT Renal or Hepatic impairment THEN Plant= Artemisia

The rule 2 describes the conditions that lead to the choice of the "Artemisia" plants. This latter can treat a woman whose age is above 12 years suffering from abundant menstrual bleeding on condition she did not pregnant and breastfeeding.

---

[1] http://www.doctissimo.fr/html/sante/phytotherapie/plante-medicinale/guide-phyto.htm.

[2] https://www.creapharma.ch/pl-plantes.htm#horizontalTab2.

[3] https://www.medicinalplants-pharmacognosy.com/.

3.2   Construction of Resources

This phase is based on four steps: the identification of accurate medicinal plant, the recognition of medicinal plant properties, the construction of the necessary dictionaries, and the construction of user feedback module.

Identification of Medicinal Plant. In this step, the IF-THEN rules were transformed into transducers. Nevertheless, to identify such medicinal plants, some challenges must be overcome. In fact, the number of diseases is very important. Moreover, the symptoms related to some diseases appear to be similar. In this case, it will be difficult to model the IF_THEN rules. For this, it is relevant to classify symptoms for better readability. Indeed, for each combination of symptoms of such disease, we define the corresponding class according to the International Classification of Diseases (ICD) defined by WHO.

Using these classes as our departure point, we propose a modeling composed of two grammars: the transformational analysis grammar and the recognition and translation of medicinal plants grammar.

Transformational Analysis. In order to identify the appropriate medicinal plant among those that exist in our dictionary, we must respect the same codes used in this dictionary. Thus, we used transformational analysis of patient data. Indeed, the submitted patient data (age, sex, symptoms, chronic diseases and allergies) represent an input of the transducer of transformational analysis as shown in Fig. 2.

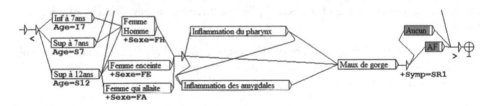

Fig. 2. Extract of the sub-graph "SR1"

The transducer of Fig. 2 describes all the paths of criteria that a patient can introduce them: symptoms and contraindications (age, sex, allergies, and chronic diseases). This transducer takes into consideration all allergies and chronic diseases which cannot be caused by the medicinal plant to be proposed. The graph in Fig. 3 shows an example of allergies and chronic diseases.

Recognition and Translation of Medicinal Plants. Transformational analysis provides a bridge between the submitted patient data and the identification of the appropriate medicinal plant. Indeed, the next application consists in finding out the medicinal plant that has the same properties as those generated in the transformational analysis. However, we cannot directly extract the right medicinal plant from the dictionary. Therefore, we have concatenated the annotation of transformational analysis with the

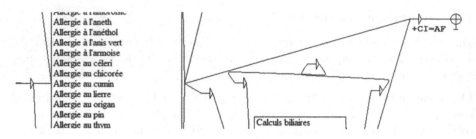

Fig. 3. Extract of the sub-graph "AF"

Fig. 4. Extract of the sub-graph "Recognition and translation"

list of medicinal plants treated in our dictionary. This concatenation represents the input of the recognition and translation transducer. This later is illustrated in Fig. 4.

As shown in Fig. 4, this transducer allows the recognition of the appropriate medicinal plant according to the annotation of transformational analysis. The output of this transducer is the name of medicinal plant in three languages: French, Arabic, and English.

Recognition of Medicinal Properties. For each identified medicinal plant, medicinal properties are recognized from corpora: medicinal plant family, harvest period, and posology as illustrated in Fig. 5.

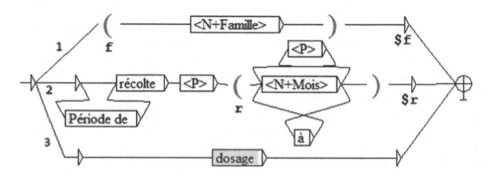

Fig. 5. Recognition of medicinal properties

Construction of Dictionaries. For our approach, we construct two dictionaries. One contains the different medicinal plants name. This dictionary contains the lemma, its category, age, sex, symptom-class, chronic diseases and allergy. Therefore, in the aim to guide the identification process, the transducers of Fig. 4 will use all these properties. The second dictionary contains trigger words such as usage interne (internal use), decoction (décoction).

3.3 User Feedback

This phase is one of the most important issues in our RDST. It includes the user feedback about the effectiveness of medicinal plants based on his/her own experience. This feedback is critical to both creating a user-friendly tool and improving performance early.

In the ulterior interactions with the RDST, the user feedback about the medicinal plants is exploited to collect implicitly information and therefore to update, add or remove criteria. Indeed, decision-making concerning a medicinal plant is carried out as follows:

> IF Number of voters said Not Effective> Number of voters said Effective THEN Add OR Update OR Remove criterion

4 Experimentation and Evaluation

In the aim investigate the feasibility of the rule-based approach, we implemented a prototype and applied it to health care decision-making. In this prototype, linguistic resources are built with the linguistic platform NooJ [11], and applied with the command-line program noojapply. We conduct a set of experimentations to evaluate the RDST efficiency. Table 1 shows the data used in the experimentation.

Table 1. Experimental data

	Total number
Texts in corpora	82
IF-THEN rules	396
Medicinal plants	75
Symptoms	209
Allergies	23
Chronic diseases	30

As a matter of fact, the prototype calls the transducers with certain logic. Moreover, the choice of passing the transducers is not random. First, the patient data are registered in the following order: age followed by sex followed by symptoms followed by chronic diseases and allergies. Then, the prototype must annotate the submitted patient data (Fig. 2). As an output, we get an annotated file.

Afterward, we move to the recognition and the translation of the recommended medicinal plants according to the obtained annotation using our dictionary and the transducer of Fig. 4. As an output, we get the identified medicinal plant(s) in Arabic, French, and English. According to the identified medicinal plant, the prototype recognizes properties from corpora (Fig. 5). Finally, we finish the identification process by displayed the French, English, and Arabic medicinal plant name; family, harvest period, posology and photo.

To evaluate the effectiveness of our RDST, we used more than 32 real users over a period of time. Each of these users has a healthcare problem and had manipulated our tool for a while.

The obtained results give 89% of well-identified medicinal plants and recognized properties. The results are satisfactory. Furthermore, the proposed RDST takes into consideration illogical situations such as a man suffering from irregular menstrual bleeding, or a pregnant woman who suffering from testicular pain. In this case, our RDST confirms validity of submitted symptoms by filtering the symptoms according to the genre.

The architecture proposed for our RDST based on dictionaries and transducers is very flexible and is able to incorporate new rules without major modification. However, the main limitation of the experimental result is the necessity of intervention of a physician in certain diagnosis. Let's note that our RDST does not treat certain diagnosis to ensure the well-being of users such as the case of a pregnant woman who suffers from abdominal pain, hypotension or hypertension.

5   Conclusion

In this work, we have proposed a RDST that provides an effective solution to help users identify medicinal plants that better treat their diseases and associated symptoms.

From the experimental results, we show that our proposed RDST achieves satisfactory results.

In the future, we plan to enrich our RDST by adding other medicinal plants to treat larger set of symptoms. We can also study synonymy and add a criterion named region to distinguish the names of medicinal plants according to the region. Furthermore, we plan to automatize the decision-making task by using machine learning algorithms.

References

1. Arvind Kumar, S.: Medicinal plants: future source of new drugs. Int. J. Herb. Med. 4(4), 59–64 (2016)
2. Hishe, M., Asfaw, Z., Giday, M.: Review on value chain analysis of medicinal Plants and the associated challenges. J. Med. Plants Stud. 4(3), 45–55 (2016)
3. Pajić, V.: Putting encyclopaedia knowledge into structural form: finite state transducers approach. J. Integr. Bioinform. (JIB) 8(2), 115–129 (2011)

4. Grishman, R., Sundheim, B.: Message understanding conference - 6: a brief history. In: Proceedings of the 16th Conference on Computational Linguistics (COLING-96), Copenhagen (1996)
5. Fehri, H., Seideh, M.A.F., Dardour, S.: A decision-support tool of medicinal plants using NooJ platform. In: Barone, L., Monteleone, M., Silberztein, M. (eds.) NooJ 2016. CCIS, vol. 667, pp. 246–257. Springer, Cham (2016). https://doi.org/10.1007/978-3-319-55002-2_21
6. Seideh, M.A.F., Fehri, H., Haddar, K.: Named entity recognition from Arabic-French herbalism parallel corpora. In: Okrut, T., Hetsevich, Y., Silberztein, M., Stanislavenka, H. (eds.) NooJ 2015. CCIS, vol. 607, pp. 191–201. Springer, Cham (2016). https://doi.org/10. 1007/978-3-319-42471-2_17
7. NLM: UMLS Knowledge Sources, 1997. U.S. Dept of Health and Human Services, 8ème edn. (1997)
8. Huang, Z., Hu, X.: Disease named entity recognition by machine learning using semantic type of metathesaurus. Int. J. Mach. Learn. Comput. 3(6), 494–498 (2013)
9. Embarek, M., Ferret, O.: Learning patterns for building resources about semantic relations in the medical domain (2008)
10. Abinaya, N., Kumar, M., Soman, K.: Randomized kernel approach for named entity recognition in Tamil. Indian J. Sci. Technol. 8(24) (2015). https://doi.org/10.17485/ijst/ 2015/v8i24/85350
11. Silberztein, M.: Formalizing Natural Languages: The NooJ Approach. London, Cognitive Science Series. Wiley-ISTE, UK. ISBN 9781848219021 (2016)

Mention Clustering to Improve Portuguese Semantic Coreference Resolution

Evandro Fonseca[1(✉)], Aline Vanin[2], and Renata Vieira[1]

[1] Pontifícia Universidade Católica do Rio Grande do Sul, Porto Alegre, Brazil
evandro.fonseca@acad.pucrs.br, renata.vieira@pucrs.br
[2] Universidade Federal de Ciências da Saúde de Porto Alegre,
Porto Alegre, Rio Grande do Sul, Brazil
aline.vanin@ymail.com

Abstract. This paper evaluates the impact that different clustering techniques may have on grouping referential mentions on rule-based coreference resolution systems. As a result, we show that our approach outperforms commonly applied methods.

Keywords: Coreference resolution · Semantics · Portuguese

1 Introduction

In this paper, we evaluate mention clustering methods (understood here as methods to group mentions) for rule-based coreference resolution models. Coreference resolution basically consists on identifying different references to the same entity in a text. For example, in the sentence: *"The opinion is from the agronomist Miguel Guerra, of University of Santa Catarina. Guerra has participated..."*, the noun phrases [*the agronomist Miguel Guerra, of University of Santa Catarina*] and [*Guerra*] are considered coreferent; in other words, they belong to the same coreference chain.

The coreference task has received a great deal of attention from the computational linguistics community. There is a variety of models that solve coreferences for one or multiple languages [2,5,9,11,14–16]. Most of these models are based on machine learning approaches. However, rule-based approaches are also competitive for the coreference resolution task, specially for languages with fewer resources that lack annotated corpus to train consistent coreference models.

In a previous shared task for coreference [12], Lee et al. presented a winning system that was purely based on rules [10]. Lately, similar approaches solving coreferences for other languages have been developed. Examples are [3,4,6] dealing with Portuguese, Spanish and Galician. However, few or no attention is given for the mention clustering process in these rule-based models. In this paper, we claim that through more elaborated methods, it is possible to improve

M. Silberztein et al. (Eds.): NLDB 2018, LNCS 10859, pp. 256–263, 2018.
https://doi.org/10.1007/978-3-319-91947-8_25

the performance of the systems. These methods make use of a discourse structure representation.

This paper is organized as follows: Sect. 2 presents the mention clustering problem; Sect. 3 presents related work; in Sect. 4, we describe our proposed clustering methods; in Sect. 5, we describe our experiments; in Sect. 6, we discuss results; and, finally in Sect. 7, conclusions and the future works are presented.

2   Mention Clustering Problem

Coreference resolution is a non-trivial task and it involves two main processing levels: (i) Mention Detection, which consists in determining which tokens belong to an entity mention, and (ii) Classification, which consists in classifying a given pair of mentions as coreferent or not. Based on the coreferent pairs, sets of entities are formed, which is called the clustering process. It consists in creating partitions with mentions that refer to the same entities.

Currently, the most used method to link mentions and generate coreference chains (in rule-based models) consists in applying a sequence of deterministic rules considering string similarity and noun phrase modifiers (i.e. pre- and post-modifier), as in [3,4,6,10]. Their approaches link a mention m_x to its antecedents if some rule is satisfied. However, in some cases we must decide whether a mention belongs to one or another partition. Consider the following example:

(a) "... *informed by [the Governor of São Paulo$_1$], [Geraldo Alckmin$_2$] at the last meeting. [The governor$_3$] says that ... However, [José Ivo Sartori$_4$] said that it will not change ... [The governor$_5$]...*"

In this sentence we have two distinct named entities: "Geraldo Alckmin" and "José Ivo Sartori" (both governors). Although there is an exact match among the noun phrases [The governor$_3$] and [The governor$_5$] they refer each to a different entity. Coreference models will probably produce a coreference link between these nouns based on their string similarity. To deal with that, it is possible to use a more elaborated discourse representation for determining whether a mention m_x belongs to a chain C_x or C_y.

When we consider semantic relations in the resolution process, this problem increases. This occurs because semantic knowledge, considering synonymy and hyponymy relations, is likely to introduce spurious links, as in:

(b) "*[Earth$_1$] is [an astro$_2$] and is in a distance of 149.600.000 km of [the Sun$_3$] ... the universe is widely and, when we consider the size of [the Sun$_4$], that is, [the star$_5$] ...*"

In example (b), there is a hyponymy relation between [Earth$_1$] and [an astro$_2$] and a synonymy relation between [an astro$_2$] and [star$_4$]; however, it is possible to see that there are two coreference chains: $C_1 = \{[Earth_1], [an astro_2]\}$ and $C_2 = \{[the Sun_3], [the Sun_4], [the star_5]\}$. Considering usual clustering methods (as the one proposed in [10]), the output will be a single coreference chain, containing all the mentions. In semantic approaches, cases like that are frequent and decrease considerably the precision of models.

3   Related Work

The mention clustering problem is a well-known issue in coreference resolution. There are two popular methods: Entity Mention and Mention Ranking. In the first, proposed by Cardie et al. [1], each noun phrase in a document is represented as a vector of attribute-value pairs. Given the feature vector for each noun phrase, the clustering algorithm coordinates the classification of partitions into equivalence classes. In his method, each noun phrase is compared to all preceding noun phrases. If the distance between two noun phrases is less than the Φ, then Φ is the cluster criteria. The authors also consider some features, like partial matching and head matching between the partitions and the candidate mention.

The second, proposed by Rahman et al. [13], uses a ranked clustering algorithm. Their model consists in establishing a rank based in distance between the antecedent and the candidate mention. Basically, assuming that I_k is a set of training instances, created for an anaphoric mention m_k, the rank for (m_j, m_k) in I_k is the rank of m_j among antecedent candidates, which is 2 if m_j is the antecedent closest to m_k or 1 otherwise. In other words, the closest antecedent receives a higher ranking in relation to other mentions. Besides these more popular approaches, there are also latent trees, as proposed in [2].

Fernandes et al. [2] propose an approach based on latent coreference trees. They argue that it may provide a powerful auxiliary structure for solving the mention clustering problem. In their approach, for each document, a set of trees and sub-trees is generated. Each sub-tree represents a mention and its referential mentions. To decide whether a mention belongs to a specific chain, they use a trained perceptron, which prioritizes the max score to generate the predicted clusters. This and other cluster methods are compared in [11] where the authors show that ranking approaches presented better results.

Although the mention clustering problem is well explored by machine learning approaches, for rule-based models it is at an initial stage. In 2013, [10] proposed a rule-based model for the coreference resolution using deterministic rules. The same model architecture is also used by [3,4,6]. It consists in linking one mention to its antecedent mentions, when at least one rule/sieve is satisfied. In this paper, we called this clustering method as baseline.

In this paper, we propose an approach that takes into account discourse structure. We use CORP[1] as case study, the most recent rule-based model available for Portuguese. In our approach, besides refering to discourse structure, we combine a set of attributes and features, like the distance among noun phrases, similar to Mention Ranking approach [13], feature vectors as in [1] and weight for links, like [2,11].

---

[1] http://www.inf.pucrs.br/linatural/wordpress/index.php/recursos-e-ferramentas/corp-coreference-resolution-for-portuguese/.

4   Clustering Methods

In this section we explain the cluster methods we adopted in our study. We will see that each method explores specific features.

4.1   Baseline

We consider as baseline the method that is commonly used by rule-based models, like [3,4,6,10]. It consists in linking one mention with its antecedents when at least one rule/sieve is satisfied, using a single iteration.

4.2   Proposed Clustering Method

Our proposed method is inspired on Heim's work [8] and it consists on exploring discourse representation. We assume that any mention is new in discourse if it does not have a link with one or more antecedents. Thus, whenever a mention does not have a coreference relation, a new chain is generated. When there is relation with more than one chain, a clustering criteria (see Sect. 4.3) will be adopted for the decision to which unique chain it will be linked.

As shown in Algorithm 1 we use an ordered list M, containing all mentions from input document. Note that each mention may have a coreference relation with one or more chains. Thus, we store the Chain_id in a vector S (just whether M_0 has some coreference relation with C_i). The next step is responsible to group the current mention (M_0) in an existing Chain C_k or to create a new coreference chain.

Algorithm 1. Clustering Algorithm

1: **while** (size of $M > 0$) **do**
2:     int j ← 0;
3:     int[ ] S;
4:     **for** each $i \in C$ **do**
5:         **if** M_0 has relation with C_i **then**
6:             $S_j \leftarrow C_i$
7:             $j \leftarrow j + +$
8:         **end if**
9:     **end for**
10:     **if** $j > 0$ **then**
11:         int $k \leftarrow Clustering\_Criteria(M_0, S, C)$
12:         $C_k \leftarrow M_0$
13:     **else**
14:         $C \leftarrow New\_Chain(M_0)$
15:     **end if**
16:     $M \leftarrow Remove(M, 0)$
17: **end while**

4.3   Clustering Criteria

Closest Chain: The Closest Chain criteria consists on, for each mention M_0, explore the set C, aiming to find the shortest distance between M_0 and the mentions from C_x. The chain whose shortest distance is found receives M_0. This distance is based on the noun phrases position in the text.

Chain Weight: Each M_0 explores C, aiming to find the chain with higher weight. Our system has thirteen rules. For each satisfied rule, we sum +1. Thus, if, for example, in C_x, there are two coreference mentions to M_0 (M_a and M_b), containing three and two satisfied rules respectively, the chain weight will be five.

Mention Weight: It selects C_x, which contains the greater amount of coreference mentions related to M_0 (each coreferent mention adds 1 to the mention weight).

Mention + Chain Weight: This criteria considers the sum of the two previous criteria.

F-Score Weight: This weight is calculated using the sum of the weights of all satisfied rules between M_0 and the mentions from C_x. To determine the weights for the rules, we use the CoNLL metric values (F-score), presented in [3]. These values were calculated by applying each of the rules of the system separately. This method selects C_x which WC_x with the higher weight.

5   Experiments

In our experiments, we evaluate and compare the proposed method with the baseline. To evaluate that, we use six known evaluation metrics described in [3]. In Table 1, we can see the results of each clustering method (including our five variations). According to the results, our five proposed methods have presented better precision, except for $CEAF_e$. However, for this case all methods have presented better recalls. Analyzing specifically the CoNLL metric, we can see that all clustering criteria outperformed the baseline. In special, "Chain" weight has presented a better result, outperforming the baseline in 2.6%.

The experiment of Table 1 was performed using the Summ-it++, due to its annotation quality. However, we perform experiments involving Summ-it++ and Corref-PT, in order to compare the impact of our method together with the use of semantic relations may have. We highlight that Corref-PT has 8 times as many semantic relations between its mentions than Summ-it++.

In Table 2 we include an analysis considering the inclusion of rules dealing with semantic relations. It is possible to see that our proposed model outperforms the baseline when we consider semantic relations. Basically, our proposed model

Table 1. Clustering methods evaluation

Method	Criteria	MUC			B³			CEAF$_m$			CEAF$_e$			BLANC			CONLL
		P	R	F	P	R	F	P	R	F	P	R	F	P	R	F	F
Baseline	-	40.6	55.0	46.7	35.9	52.4	42.6	43.0	54.3	48.0	47.1	50.2	48.6	60.4	54.9	54.1	46.0
Our	Closest chain	42.3	48.0	45.0	42.5	46.6	44.4	46.1	55.0	50.2	43.1	56.7	49.0	67.7	53.9	51.3	46.1
	Chain	45.1	52.1	48.3	43.8	49.5	46.5	48.0	57.0	52.1	45.7	57.4	50.9	68.2	54.7	53.0	48.6
	Mention	44.3	51.1	47.5	42.7	48.6	45.5	47.6	56.0	51.5	45.3	55.7	50.0	66.7	54.5	52.7	47.7
	Chain + Mention	45.0	52.0	48.2	43.6	49.3	46.3	48.0	56.8	52.0	45.7	56.9	50.7	68.0	54.7	53.0	48.4
	F-Score wgt	45.0	52.0	48.2	43.9	49.5	46.5	52.3	57.3	52.3	45.3	57.8	50.8	68.6	54.6	52.8	48.5

Table 2. Comparative analysis among our proposed method and baseline, using two corpora (Summ-it++ and Corref-PT)

Corpus	Method	Semantics	MUC			B³			CEAF$_m$			CEAF$_e$			BLANC			CoNLL
			P	A	F	P	A	F	P	A	F	P	A	F	P	A	F	F
Summ-it++	Baseline	X	56.8	49.7	53.0	55.3	46.3	50.4	58.1	54.1	56.0	53.4	56.2	54.8	73.2	57.5	58.6	52.7
		✓	40.6	55.0	46.7	35.9	52.4	42.6	43.0	54.3	48.0	47.1	50.2	48.6	60.4	54.9	54.1	46.0
	Chain	X	58.8	44.4	50.6	59.3	41.7	49.0	60.5	50.6	55.1	53.7	54.2	54.0	76.7	58.6	60.3	51.2
		✓	45.1	52.1	48.3	43.8	49.5	46.5	48.0	57.0	52.1	45.7	57.4	50.9	68.2	54.7	53.0	48.6
Corref-PT	Baseline	X	55.4	49.5	52.3	49.3	42.4	45.6	52.2	48.7	50.4	46.9	48.8	47.8	65.4	54.2	51.9	48.6
		✓	44.2	52.2	47.9	35.8	45.8	40.2	41.5	46.5	43.9	46.1	43.9	44.9	59.3	55.2	54.0	44.3
	Chain	X	64.2	47.8	54.8	61.2	40.5	48.7	59.9	49.0	53.9	50.2	51.0	50.6	79.8	56.1	55.5	51.4
		✓	54.9	50.2	52.5	51.8	43.6	47.3	52.9	51.6	52.3	46.2	52.8	49.3	73.3	53.8	50.2	49.7

outperforms the baseline method in 2.6% for CoNLL metric with Summ-it++ corpus evaluation, and 5.4% for Corref-PT corpus evaluation. If we compare the precision of MUC metric for Corref-PT, our proposed clustering method outperforms the baseline method in 10.7%; when we do not include semantics, our clustering method outperforms the baseline model in 8.8%. This improvement is discussed in Sect. 6.

6 Comparative Analysis

Here we present a comparative analysis of the behavior of the baseline and our proposed method. Example (c) shows a text fragment, where noun phrases are highlighted.

(c) "... appropriate biotechnoloy to the growth of [the country$_1$]. Guerra has cited the micropropagation of vegetables (production of seeds in laboratory, made to avoid diseases and select [healthy vegetables$_2$]) as an example of low-cost biotechnology. With that, the production of strawberries in [the south$_3$] of [the country$_4$] has increased from 3.2kg to 60kg per hectare. For the agronomist, [Brazil$_5$] must seek the development of transgenics that try to improve the conditions of local agriculture, such as the cultivation of [plants$_6$] with the capability of capture certain elements on [the land$_7$]. The president of [Embrapa$_8$] ([Brazilian Agricultural Research Corporation$_9$]), Alberto Portugal, pointed out that [the company$_1$0] seeks solutions to the problems of national agriculture..."

In this text fragment we observed two specific coreference chains:

Gold chains

C1 *[the country$_1$], [the country$_4$], [Brazil$_5$]*
C2 *[Embrapa$_8$], [Brazilian Agricultural Research Corporation$_9$], [the company$_{10}$]*

Baseline method chains

C1 *[the country$_1$], [healthy vegetables$_2$], [the south$_3$], [the country$_4$], [Brazil$_5$], [plants$_6$], [the land$_7$], [Embrapa$_8$], [Brazilian Agricultural Research Corporation$_9$], [the company$_{10}$]*
C2 *has no elements*

Our method chains

C1 *[the country$_1$], [the country$_4$], [the land$_7$]*
C2 *[Embrapa$_8$], [Brazilian Agricultural Research Corporation$_9$], [the company$_1$0]*

Analyzing results of the baseline, we can see that C2 was not recognized because there is at least one rule satisfied, which links these two chains. As example of that, in our database [7] there are triples indicating the following relations: [land, country, synonymy], [plant, vegetable, synonymy], [plant, company, hyponymy]. Note also that there is an ambiguity problem here and the baseline method does not consider the context like our proposed method. Semantic relations like these may introduce many spurious links. This example shows the importance of considering the discourse representation. It is possible to note that our method loses the noun phrase [Brazil$_5$] and links the noun [the land$_7$] using two synonymy connections.

7   Conclusion

In this paper we evaluate mention clustering alternatives for the coreference resolution task. We have shown that by considering discourse structure representation we improve significantly the coreference resolution task, and we argue that this is specially relevant when considering semantic relations for the links generation. As future work, we intend to test the impact of our method for other languages, and to compare it directly with other solutions to the coreference task, including machine-learning approaches.

Acknowledgments. The authors acknowledge the financial support of CNPq, CAPES and Fapergs.

References

1. Cardie, C., Wagstaff, K.: Noun phrase coreference as clustering. In: Proceedings of the Joint SIGDAT Conference on Empirical Methods in Natural Language Processing and Very Large Corpora, pp. 82–89 (1999)
2. Fernandes, E.R., dos Santos, C.N., Milidiú, R.L.: Latent trees for coreference resolution. Comput. Linguist. **4**, 801–835 (2014)
3. Fonseca, E.B., Sesti, V., Antonitsch, A., Vanin, A.A., Vieira, R.: CORP - Uma abordagem baseada em regras e conhecimento semântico para a resolução de correferências. Linguamatica **9**(1), 3–18 (2017)
4. Fonseca, E.B., Vieira, R., Vanin, A.: Adapting an entity centric model for portuguese coreference resolution. In: Proceedings of the 10th Annual Conference on Language Resources and Evaluation, Portorož, Slovenia (2016)
5. Fonseca, E., Vieira, R., Vanin, A.: Improving coreference resolution with semantic knowledge. In: Silva, J., Ribeiro, R., Quaresma, P., Adami, A., Branco, A. (eds.) PROPOR 2016. LNCS (LNAI), vol. 9727, pp. 213–224. Springer, Cham (2016). https://doi.org/10.1007/978-3-319-41552-9_21
6. Garcia, M., Gamallo, P.: An entity-centric coreference resolution system for person entities with rich linguistic information. In: Proceedings of 25th International Conference on Computational Linguistics, COLING, Dublin, Ireland, pp. 741–752 (2014)
7. Oliveira, H.G.: Onto. PT: Towards the automatic construction of a lexical ontology for Portuguese. Ph.D. thesis, University of Coimbra/FST (2012)
8. Heim, I.: File change semantics and the familiarity theory of definiteness. In: Rexach, J.G. (ed.) Semantics Critical Concepts in Linguistics, pp. 108–135. Routledge (1983)
9. Kaumanns, F.D.: Assessment and analysis of the applicability of recurrent neural networks to natural language understanding with a focus on the problem of coreference resolution. Ph.D. thesis, Ludwig Maximilian University of Munich, Germany (2016)
10. Lee, H., Chang, A., Peirsman, Y., Chambers, N., Surdeanu, M., Jurafsky, D.: Deterministic coreference resolution based on entity-centric, precision-ranked rules. Comput. Linguist. **39**, 885–916 (2013)
11. Martschat, S., Strube, M.: Latent structures for coreference resolution. Trans. Assoc. Comput. Linguist. **3**, 405–418 (2015)
12. Pradhan, S., Ramshaw, L., Marcus, M., Palmer, M., Weischedel, R., Xue, N.: Conll-2011 shared task: modeling unrestricted coreference in ontonotes. In: Proceedings of the Fifteenth Conference on Computational Natural Language Learning: Shared Task, pp. 1–27. Association for Computational Linguistics (2011)
13. Rahman, A., Ng, V.: Narrowing the modeling gap: a cluster-ranking approach to coreference resolution. J. Artif. Intell. Res. **40**, 469–521 (2011)
14. Rahman, A., Ng, V.: Resolving complex cases of definite pronouns: the winograd schema challenge. In: Proceedings of the 2012 Joint Conference on Empirical Methods in Natural Language Processing and Computational Natural Language Learning, pp. 777–789. Association for Computational Linguistics (2012)
15. Silva, J.F.D.: Resolução de correferência em múltiplos documentos utilizando aprendizado não supervisionado. Dissertação de Mestrado, Universidade de São Paulo (2011)
16. Soon, W.M., Ng, H.T., Lim, C.Y.: A machine learning approach to coreference resolution of noun phrases. Comput. Linguist. **27**(4), 521–544 (2001)

Resource Creation for Training and Testing of Normalisation Systems for Konkani-English Code-Mixed Social Media Text

Akshata Phadte[(✉)]

Department of Computer Science and Technology, Goa University,
Taleigao Plateau, Goa, India
akshataph07@gmail.com

Abstract. Code-Mixing is the mixing of two or more languages or language varieties in speech. Apart from the inherent linguistic complexity, the analysis of code-mixed content poses complex challenges owing to the presence of spelling variations and non-adherence to a formal grammar. However, for any downstream Natural Language Processing task, tools that are able to process and analyze code-mixed social media data are required. Currently there is a lack of publicly available resources for code-mixed Konkani-English social media data, while the amount of such text is increasing everyday. The lack of a standard dataset to evaluate these systems makes it difficult to make any meaningful comparisons of their relative accuracies.

In this paper, we describe the methodology for the creation of a normalisation dataset for Konkani-English Code-Mixed Social Media Text (CMST). We believe that this dataset will prove useful not only for the evaluation and training of normalisation systems but also help in the linguistic analysis of the process of normalisation Indian languages from native scripts to Roman. Normalisation refers to the process of writing the text of one language using the script of another language whereby the sound of the text is preserved as far as possible [3].

Keywords: Code-mixing · Social media text · Normalisation
Natural Language Processing

1 Introduction

Multilingual speakers tend to exhibit code-mixing and code-switching in their use of language on social media platforms. Code-Mixing is the embedding of linguistic units such as phrases, words or morphemes of one language into an utterance of another language whereas code-switching refers to the co-occurrence of speech extracts belonging to two different grammatical systems [2]. Here we use code-mixing to refer to both the scenarios. Konkani-English bilingual speakers produce huge amounts of CMST. [12] noted that the complexity in analyzing

© Springer International Publishing AG, part of Springer Nature 2018
M. Silberztein et al. (Eds.): NLDB 2018, LNCS 10859, pp. 264–271, 2018.
https://doi.org/10.1007/978-3-319-91947-8_26

CMST stems from non-adherence to a formal grammar, spelling variations, lack of annotated data and inherent conversational nature of the text. Therefore, there is a need to create datasets and Natural Language Processing (NLP) tools for Code-Mixed social media text as traditional tools are ill-equipped for it. Taking a step in this direction, we describe the Normalisation task built during this study.

2 Related Work

Code-mixing being a relatively newer phenomena has gained attention of researchers only in the past two decades. Word normalisation of code-mixed data has been extensively studied, but there is little work done on the Konkani-English, Hindi-English language pair. One of the earliest works on code-Mixing for facebook data was done by [5] and showed that Facebook users tend to mainly use inter-sentential switching over intra-sentential, and report that 45% of the switching was instigated by real lexical needs, 40% was used for talking about a particular topic, and 5% for content clarification. Owing to massive growth of SMS and social media content, text normalisation systems have gained attention where the focus is on conversion of these tokens into standard dictionary words. The first Chinese monolingual chat corpus was released by [13]. They also introduced a word normalisation model, which was a hybrid of the Source Channel Model and phonetic mapping model.

A commonly accepted research methodology is treating normalisation as a noisy channel problem. [15] explain a supervised noisy channel framework using HMMs for SMS normalisation. This work was then extended by [16] to create an unsupervised noisy channel approach using probabilistic models for common abbreviation types and choosing the English word with the highest probability after combining the models. [17] adopted the noisy-channel framework for normalisation of microtext and proved that it is an effective method for performing normalisation. For Hindi normalisation, they used the system built by [18] but they did not normalise English text as they used the [19] Twitter POS Tagger in the next step, which does not require normalised data. [11] worked on a complete pipeline for shallow parsing. They performed tokenisation, language identification, normalisation, POS tagging and finally, shallow parsing for code-mixed Hindi-English social media text.

3 Data Preparation

In order to create the dataset, 20 multilingual speakers who were fluent in Konkani and English were selected. They were divided into 5 Watsapp chat groups and asked to converse on topics related to daily life. Due to the usage of Watsapp as the underlying crowd sourcing engine, the data generated was highly conversational and had reasonable amount of social-media lingo. From their interactions, 7000 code-mixed sentences were selected. The names of the participants are anonymised. In addition to these dataset we had used dataset prepared by [1]. These works discuss the dataset preparation and dataset statistics of code-mixing of Konkani-English as well as its linguistic nature. For the normalisation task of Konkani-English language we extracted the code-mixed corpus which was discussed in [1] as well as dataset created by 20 multilingual speakers on watsapp chat.

3.1 Data Statistics

The size of the original data was 12088 code-mixed sentences (7000 code mixed sentences of watsapp chat and 5088 code-mixed sentences used by [1]). Total of 1,18,118 tokens contains in 12088 code mixed sentences. Of these tokens, 74,118 (62.74%) are Konkani words which are in Roman script, 27,764 (23.50%) are English words. 16,236 (13.74%) are acronym, slag words, hindi words etc. which are marked as 'Rest'.

3.2 Dataset Annotation Guidelines

The creation of this linguistic resource involved Language identification, Normalisation layer. In the following paragraphs, we describe the annotation guidelines for these tasks in detail. Manual Annotation was done on the following layer:

1. **Language Identification**
   Similar to [6], we will be treating language identification as a three class ('kn', 'en', 'rest') classification problem. Every word was given a tag out of three - en, kn and rest to mark its language. Words that a bilingual speaker could identify as belonging to either Konkani or English were marked as 'kn' or 'en', respectively. The label 'rest' was given to symbols, emoticons, punctuation, named entities, acronyms, foreign words. The label 'rest' was created in order to accommodate words that did not strictly belong to any language, described below:
   1 Symbols, emoticons and punctuation
   2 **Named Entities**: Named Entities are language independent in most cases. For instance, 'Jack' would be represented by equivalent characters in Konkani and English.
   3 **Acronyms**: This includes SMS acronyms such as 'LOL', and established contractions such as 'USA'. Acronyms are very interesting linguistic units, and play an important role in social media text. They represent not just entities but also phrases and reactions. We wanted to keep their analysis separate from the rest of the language; and hence they were categorised as 'rest' in our dataset.
   4 **Foreign Words**: A word borrowed from a language except Konkani and English has been treated as 'rest' as well. This does not include commonly borrowed Hindi words in Konkani; they are treated as a part of Konkani language.
   5 **Sub-lexical code-mixing**: Any word with word-level code-mixing has been classified as 'rest', since it represents a more complex morphology.

2. **Normalisation**

Words with language tag 'kn' in Roman script were labeled with their standard form in the native script of Konkani Devanagari, i.e. a back-transliteration will be perform. Words with language tag 'en' were labeled with their standard spelling. Words with language tag 'rest' were kept as they are.

Following are some case-specific guidelines.

1 In case a token consists of two words (due to an error in typing the space), the tokens are separated and written in their original script. For instance, 'whatis' would be normalised to 'what is', with the language ID as English.

2 In cases where multiple spellings of a word are considered acceptable, we have allowed both spelling variations to exist as the standard spellings. For instance, in 'color' and 'colour', 'dialogue' and 'dialog', both spellings are valid.

3 Contractions such as 'don't' and 'who's' have been left undisturbed. The dataset thus contains both variations - 'don't' and 'do not', depending on the original chat text.

4 Konkani has evolved through the past decades, and often we see variations in spelling of a single word. We observed the variation patterns and choose the standard spellings.

4   Experiments and Results

In this section, we explain the base system used for the Normalisation system with new dataset discussed in Sect. 3, along with improvements that increased the accuracy of our system. Here, we repeated the experiments performed by [1] and re-created part of their system, which will be referred to as our base system. The original system first tokenizes an utterance into words. Then, a language identification module classifies each word as Konkani, English or Rest. Based on the language assigned, the Normalisation module runs the Konkani or English normalisers.

1. **Base System Approaches:-**

In the base system, the authors have used the frequencies of the ILCI[1] corpus to create a Konkani dictionary. users were given some data files of ILCI corpus and were asked to transcribe these sentences in Roman script. This process was adopted instead of look and type interface, to avoid the influence of the native spelling of the word that the visual presence of the original word might have on the transliteration. This ensured that the users used the transliteration scheme that came to them most naturally. The purpose of this dictionary is to create a large mapping of normalised-unnormalised pairs, enabling a Phrase Based Machine Translation System to learn the transformations required for normalisation. On observing the low-frequency words in

---

[1] Indian Language Corpora Initiative corpus.

the dictionary, we found a lot of misspelled words, and words from other languages merely transliterated in Konkani. These were pruned to create a more accurate Konkani dictionary, totaling to 1,25,000 words. The base system uses following approaches

(a) **Konkani Transliterator and Normalizer (Normalizer)**

We use CMU Part of Speech tagger[2] on English words which reported an accuracy of 65.39%, it normalizes English words as a primary step. We used Python-Irtrans[3] developed by IIIT-Hyderabad for transliteration of Konkani words from Roman to Devanagari. We ran the konkani words on transliteration system in order to normalize it. This tool is used to convert roman into Konkani script i.e. Python-Irtrans which reported an accuracy of 60.09%.

(b) **Noisy Channel Framework**

For transliterating the detected Romanized Konkani words and for noisy English words, we built A Two Layer Normalizer was built for both Konkani and English. The message is processed using the following techniques described in following sections.

1. Compression: In Social Media platform, while chatting, users most of the time express their emotions/mood by stressing over a few characters in the word. *For example*, usage of words are *thanksss, sryy, gooooood* which corresponds the person being obliged, needy, apologetic, emotional, amazed, etc. As we know, it is unlikely for an English word to contain the same character consecutively for three or more times hence, we compress all the repeated windows of character length greater than two, to two characters. Each window now contains two characters of the same alphabet in cases of repetition. Let n be the number of windows, obtained from the previous step. Since average length of English word [20] is approximately 4.9, we apply brute force search over 2n possibilities to select a valid dictionary word. If none of the combinations form a valid English word, the compressed form is used for normalization.

Table 1 contains sanitized sample output from our compression module for further processing.

Table 1. Sample output of compression module

Input Sentence	Output Sentence
I am so gooood! tuuu kaashe asa...	I am so good! tu kashe asa...

2. Normalizer: Text Message Normalization is the process of translating ad-hoc abbreviations, typographical errors, phonetic substitution and ungrammatical structures used in text messaging (SMS and Chatting) to plain English. Use of such language (often referred as Chatting Language) induces noise which poses additional processing challenges.

---

[2] http://www.cs.cmu.edu/ark/.
[3] https://github.com/irshadbhat/indic-trans.

While dictionary lookup based methods[4] are popular for Normalization, they can not make use of context and domain knowledge. For example, *yr* can have multiple translations like *year, your.*

We tackle this by building our normalization system based on the Phrase Based Machine Translation System (PB-SMT), that learns normalization patterns from a large number of training examples. We use Moses [8], a statistical machine translation system that allows training of translation models. PB-SMT is a machine translation model; therefore, we adapted the PB-SMT model to the transliteration task by translating characters rather than words as in character-level translation. For character alignment, we used GIZA++ implementation of the IBM word alignment model. To suit the PB-SMT model to the transliteration task, we do not use the phrase reordering model. The target language model is built on the target side of the parallel data with Kneser-Ney [10] smoothing using the IRSTLM tool [9]. In a bid to simulate syllable level transliteration we also built a Normalization model by breaking the English and Konkani words to chunks of consecutive characters and trained the transliteration system on this chunked data. Training process requires a Language Model of the target language and a parallel corpora containing aligned un-normalized and normalized word pairs. For English and Konkani word Normalization, our language model consists of 50,156 English unnormalized and normalized words taken from the web, 25195 Konkani words taken from Indian Language Corpora Initiative (ILCI) Corpus and manually transliterated. Following are Parallel corpora was used:-

i. Lexical normalization dataset released by [7] which consists of 41,118 pair of un-normalized and normalized words/phrases.

ii. Dictionary of Internet slang words was extracted from http://www. noslang.com.

iii. We developed wordlists for English-Konkani language pairs using (ILCI) corpus. The wordlists contained few overlapping words.

The accuracy for the Konkani Normalizer using noisy channel framework was 78.21%, and for the English Normalizer was 74.81%.

2. Normalization Noisy Channel Framework

To fix this, a normalization module that performs language-specific transformations, yielding the correct spelling for a given word was built. Two language specific normalizers, one for Konkani and other for English/Rest, had two sub-normalizers each, as described below. Both sub-normalizers generated normalized candidates which were then ranked, as explained later in this subsection.

(a) Noisy Channel Model

A generative model was trained to produce noisy (unnormalized) tokens from a given normalized word. Using the models confidence score and the probability of the normalized word in the background corpus, n-best normalizations were chosen. First, we obtained character alignments between

---

[4] http://www.lingo2word.com.

noisy Konkani words in Roman script (Kr) to normalized Konkani words-format(Kw) using GIZA++ on developed wordlists for English-Konkani language pairs using (ILCI) corpus of konkani word pairs of the form (kw-kr) [14].

$$Kw = argmax_{Kwi}p(Kwi|Kr)$$

$$Kw = argmax_{Kwi}p(Kwi|Kr)$$

where p(Kwi) is the probability of word Kwi in the background corpus. Next, a CRF classifier was trained over these alignments, enabling it to convert a character sequence from Roman to Devanagari using learnt letter transformations. Using this model, noisy Kr words were created for Kw words obtained from a Brahmi-Net dataset [4]. Finally, using the formula below, we computed the most probable Kw for a given Kr. A similar approach was used for English text normalization, using the English normalization pairs from [7] for the noisy channel framework, and Aspell[5] as the spell-checker. Words with language tag 'rest' were left unprocessed. The accuracy for the Konkani Normalizer was 81.25%, and for the English Normalizer was 79.98.

5   Conclusion and Future Work

In this paper, we have focused on the process of creating and annotating a much needed dataset for code-mixed Konkani-English sentences in the social media context. We have used an existing language identification system, and improved a normalisation system. We are actively working on enhancing the normalisation module by taking the context of words into account. In the future we would also like to evaluate how adding more language classes, particularly for named entities and acronyms influences the overall accuracy of our system. We intend to use this dataset to build tools for code-mixed data like POS taggers, morph-analysers, chunkers and parsers.

References

1. Phadte, A., Thakkar, G.: Towards normalising Konkani-English code-mixed social media text. In: Proceedings of the 14th International Conference on Natural Language Processing (ICON-2017), pp. 85–94 (2017)
2. Gumperz, J.J.: ZIB Discourse Strategies. Cambridge University Press, Cambridge (1982)
3. Knight, K., Graehl, J.: Machine Transliteration. Comput. Linguist. **24**(4), 599–612 (1998)
4. Kunchukuttan, A. Puduppully, R. Bhattacharyya, P.: Brahmi-Net: a transliteration and script conversion system for languages of the Indian subcontinent. In: Proceedings of the 2015 Conference of the North American Chapter of the Association for Computational Linguistics: Demonstrations, pp. 81–85 (2015)

---

[5] http://aspell.net/.

5. Hidayat, T.: An analysis of code switching used by facebookers (a case study in a social network site). Sekolah Tinggi Keguruan dan Ilmu Pendidikan (STKIP) Siliwangi Bandung (2012)
6. Barman, U., Das, A., Wagner, J., Foster, J.: Code mixing: a challenge for language identification in the language of social media. In: EMNLP 2014, p. 13 (2014)
7. Han, B., Baldwin, T.: Lexical normalisation of short text messages: Makn sens a# twitter. In: Proceedings of the 49th Annual Meeting of the Association for Computational Linguistics: Human Language Technologies, vol. 1, pp. 368–378. Association for Computational Linguistics (2011)
8. Koehn, P., Hoang, H., Birch, A., Callison-Burch, C., Federico, M., Bertoldi, N., Cowan, B., Shen, W., Moran, C., Zens, R., et al.: Moses: open source toolkit for statistical machine translation. In: Proceedings of the 45th Annual Meeting of the ACL on Interactive Poster and Demonstration Sessions, pp. 177–180. Association for Computational Linguistics (2007)
9. Federico, M., Bertoldi, N., Cettolo, M.: IRSTLM: an open source toolkit for handling large scale language models. In: Interspeech, pp. 1618–1621 (2008)
10. Kneser, R., Ney, H.: Improved backing-off for m-gram language modeling. In: 1995 International Conference on Acoustics, Speech, and Signal Processing, ICASSP-1995, vol. 1, pp. 181–184. IEEE (1995)
11. Sharma, A., Gupta, S., Motlani, R., Bansal, P., Srivastava, M., Mamidi, R., Sharma, D.M.: Shallow parsing pipeline for Hindi-English code-mixed social media text (2016)
12. Vyas, Y., Gella, S., Sharma, J., Bali, K., Choudhury, M.: Pos tagging of English-Hindi code-mixed social media content. In: EMNLP, vol. 14, PP. 974–979 (2014)
13. Wong, K.-F., Xia, Y.: Normalization of Chinese chat language. Lang. Resour. Eval. **42**(2), 219–242 (2008)
14. Gupta, K., Choudhury, M., Bali, K.: Mining Hindi-English transliteration pairs from online Hindi lyrics. In: LREC, pp. 2459–2465 (2012)
15. Choudhury, M., Bali, K., Dasgupta, T., Basu, A.: Resource creation for training and testing of transliteration systems for Indian languages. In: LREC (2010)
16. Paul Cook and Suzanne Stevenson.: An unsupervised model for text message normalization. In: Proceedings of the Workshop on Computational Approaches to Linguistic Creativity, pp. 71–78. Association for Computational Linguistics (2009)
17. Xue, Z., Yin, D., Davison, B.D.: Normalizing microtext. In: Workshops at the Twenty-Fifth AAAI Conference on Artificial Intelligence (2011)
18. Gella, S., Sharma, J., Bali, K.: Query word labeling and back transliteration for Indian languages: shared task system description. In: FIRE Working Notes (2013)
19. Owoputi, O., O'Connor, B., Dyer, C., Gimpel, K., Schneider, N., Smith, N.A.: Improved part-of-speech tagging for online conversational text with word clusters. In: Association for Computational Linguistics (2013)
20. Mayzner, M.S., Tresselt, M.E.: Tables of single-letter and digram frequency counts for various word-length and letter-position combinations. Psychonomic monograph supplements (1965)

A TF-IDF and Co-occurrence Based Approach for Events Extraction from Arabic News Corpus

Amina Chouigui[1,2]($\boxtimes$), Oussama Ben Khiroun[1,2], and Bilel Elayeb[1,3]

[1] RIADI Research Laboratory, ENSI, Manouba University, Manouba, Tunisia
aminachouigui@gmail.com, oussama.ben.khiroun@gmail.com,
Bilel.Elayeb@riadi.rnu.tn
[2] National Engineering School of Sousse, Sousse University, Sousse, Tunisia
[3] Emirates College of Technology, Abu Dhabi, United Arab Emirates

Abstract. Event extraction is a common task for different applications such as text summarization and information retrieval. We propose, in this work, a TF-IDF based approach for extracting keywords from Arabic news articles' titles. These keywords will serve to extract the main events for each month using a Part-of-Speech (POS) co-occurrence based approach. The precision values are computed by corresponding the extracted events with another news website. Results show that the approach performance depends on categories and performs well for domain specific ones such as economy.

Keywords: Event extraction · Arabic language · Terms weighting Morphological analysis

1 Introduction

Event extraction is the task of automatically extracting information about events from unstructured and/or semi-structured machine-readable documents. This problem is often resolved by using Natural Language Processing (NLP) techniques. Previous researches on event extraction have focused on the extraction from news texts, blogs and tweets in English. However, the research for event extraction from Arabic texts is limited.

The Arabic language is known as a difficult language for NLP tasks. Indeed, Arabic includes sophisticated grammar rules and grammatical irregularities [1–3]. So, a simple Arabic word can present morphological, syntactical or semantic ambiguity and can be interpreted with different meanings. We take as example the word "ذهب" which has two meanings (respectively (he went) and (gold)) depending on the diacritic marks [4]. We mention also the problem of agglutination for the Arabic language. We present an example of the agglutinated Arabic word: "وجهة" (destination) and "و جهة" (and a side) [5].

This paper is organized as follows: We present in Sect. 2 some related work. Then, Sect. 3 details our proposed event extraction approach based on keywords selection. The experimentation and data analysis are presented in Sect. 4. Finally, we summarize our work and present the future work.

2 Related Work

Three main approaches of event extraction are identified in the literature [6]. First, we distinguish data-driven approaches which aim to convert data to knowledge through the usage of statistics, machine learning, linear algebra, etc. Second, we distinguish expert knowledge-driven methods which extract knowledge through representation and exploitation of expert knowledge, usually by means of pattern-based approaches. Finally, the hybrid event extraction approaches combine knowledge and data-driven methods.

In the work of Naughton et al. [7], authors extracted events from 219 news texts about the Iraq war that are present in 46 different sources. The collection is manually annotated by attributing two different labels for relevant and non-relevant events. Besides, the event sentences are clustered by two distance metrics methods. The first method is about the regularities in the sequential structure of events in a document. The second distance metric uses the weighting TF-IDF-like to attribute frequency to each word.

Zhou et al. [8] extracted events from tweets using an unsupervised framework. The process of extraction consists of filtering, extraction and categorization. They filtered the tweets by keeping only event-related ones. Zhou et al. used an unsupervised Bayesian model to extract event-related keywords. They evaluated over 60 millions tweets published in December 2010. The event extraction's precision achieved 70.49%. The extracted events are also categorized, according to their type of labels, into coherence groups.

AL-Smadi and Qawasmeh [9] extracted event trigger, event time and event type information from Arabic Tweets. They presented a knowledge-based approach to extract the events and evaluated the approach on 1 000 Arabic tweets about different type of events. The proposed approach uses an unsupervised rule-based technique and disambiguates the event-related named entities. The event type identification has the high accuracy with a value of 97.7%, then the event time extraction accuracy with 87.5% and finally the event trigger extraction accuracy with 75.9%.

In general, data-driven methods require many data and little domain knowledge and expertise. Despite the little data required in knowledge-based event extraction approaches, domain knowledge and expertise is needed [6].

3 The Proposed Events Extraction Process

We used an Arabic news texts corpus, for the event extraction task. We consider that an increased use of specific words could be related to a specific event. In general, the main news texts contain heterogeneous sentences that may or not be related to the event. Meanwhile, the titles are represented by only one sentence that describes the article new. So we choose to extract the keywords presented by the most repeated words from titles. Afterward, we extract co-occurring named entities with previous filtered keywords as presented in Fig. 1.

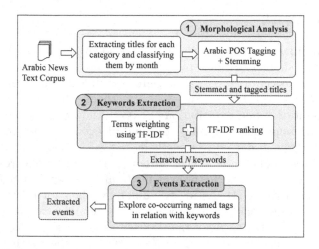

Fig. 1. Event extraction process from news texts

3.1   Morphological Analysis

After extracting and classifying titles by month, we use a Part-of-Speech (POS) tagger to annotate news titles with the suitable grammatical tag for each word. In order to do this preprocessing task, we used the Arabic POS tagger OpenNLP[1]. We present in Fig. 2 an example of tokenized and tagged title with OpenNLP tagger.

ترامب يقيد الهجرة من هذه الدول العربية

(Trump restricts immigration from these Arab countries)

ترامب B-PER_يقيد V_الهجرة D+SFN_من P_هذه DM_الدول D+PIN_العربية D+AJ

Fig. 2. Example of POS tagged title with OpenNLP tool

Recognized as an important preprocessing task, word stemming serves to remove the prefixes and suffixes from a word in order to get the same stem for different grammatically related words [11]. We take as example the words "مؤسّسات" (institutions) and "مؤسّسة" (institution) that return the same stem

---

[1] http://www.arabicnlp.pro/.

"مؤسّس" (founder). Different stemmers are available for the Arabic language in the literature [11]. We chose the Light Stemmer [12] which is a simple and freely available tool that removes prefixes and/or suffixes without dealing with infixes, patterns or irregular plurals. So, it removes the affixes without any prior knowledge in linguistic rules.

3.2 Keywords Extraction

Depending on the nature of the used corpus, the events are often described by named entities. Thus, we kept only the words having nouns relative POS tags from the tagged and stemmed titles. We consider the following set of the tags that refer to nouns in the OpenNLP POS tagger: N for a noun, NE for a named entity, NM for a number, PER for a person, ORG for an organization and LOC for a location.

Afterwards, we calculate the TF-IDF score for the extracted nouns [13]. The TF-IDF measure is often used as a weighting factor and for stop-words filtering in information retrieval and text mining fields. In order to identify the most important keywords for each month, we adapted the TF-IDF measure for a given title as follows:

$$TF * IDF(t_i, m_j) = ntf_i \times \log(\frac{n}{n_i}) \tag{1}$$

where:

- $TF * IDF(t_i, m_j)$ is the weight of the word t_i in the corresponding title for the month m_j;
- ntf_i represents the normalized frequency of a word t_i in the corresponding title;
- n is the number of titles in a month;
- n_i is the number of titles in which the word t_i occurs at least once.

Therefore, the nouns are sorted according to their computed TF-IDF weights for each month. We take the top N nouns as keywords related to events. Then, we keep only titles that contain one of these keywords in the event extraction process.

3.3 POS Co-occurrence Based Event Extraction

The main idea of using co-occurrence is to consider that two words are correlated word pairs if they co-occur in the same document or in the same sentence. Starting from this statement, we consider that co-occurring nouns may be semantically related in the case of event extraction process. So, we use the following rules to define the word co-occurrence: To extract the events from the titles, we seek the keyword in a title. Then we search for co-occurring nouns, based on POS tags, from the right and the left of the corresponding keyword. We present in Fig. 3 an example explaining the event extraction process.

4   Experimentation and Data Analysis

4.1   Test Collection

We use the ANT Corpus v1.1[2], containing news articles from January to June 2017 and classified by domain [10]. ANT Corpus texts are collected from the Tunisian web radio site Jawhara FM. Since the number of articles existing in some categories is small, we choose only 6 categories from all the 9 categories for the event extraction task. The total number of documents in these categories is 9 479 articles.

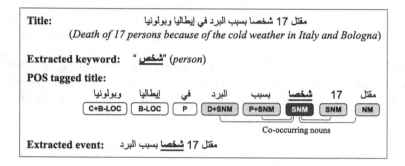

Fig. 3. Event extraction based on window co-occurring named tags

4.2   Extracting Keywords

When using the global keyword extraction process all over the 6 months, we notice that some important keywords can be omitted. In fact, using a wide time range when extracting keywords penalizes some important events that are specific to a limited period.

Table 1 presents some examples of keywords that are ranked over the 20th position when using a 6 months time range. Meanwhile, the same keywords are among the top five TF-IDF ranked terms for some months.

In order to select the most important keywords, we define a threshold to choose N nouns for each month according to their TF-IDF values. The number of extracted terms (NET) is defined as follows:

$$NET(c_i, m_j) = nt_{ij} - mt_{ij} \qquad (2)$$

where:

- nt_{ij} represents the number of all extracted terms for the category c_i in the month m_j;
- mt_{ij} represents the number of extracted terms that appear in less than five titles in the category c_i and month m_j.

---

[2] https://antcorpus.github.io/.

Table 1. Comparison examples of extracted keywords ranks through semester vs monthly time window

Term	Category	TF-IDF rank for 6 months	TF-IDF rank for a specific month
"ماكرون" (Macron)	InternationalNews	23	1st on May
"ارهاب" (terrorism)	LocalNews	14	4th on January
"فساد" (corruption)	LocalNews	22	4th on May

We compute the mean threshold for the number of keyword, that will be considered in the event extraction process, by applying normalization factor, as follows:

$$MeanThreshold = \sum_{i \in categories} \sum_{j \in months} NET(c_i, m_j)/(nC * nM) \qquad (3)$$

where:

- nC is the number of categories;
- nM is the number of months.

Meanwhile TF-IDF is a statistical measure, using it in the keyword extraction process have semantic and temporal relevance. As examples, on May which was the presidential election period in France, the extracted term from internationalNews "ماكرون" (Macron) has the highest value of TF-IDF. The term "ترامب" (Trump) also has the best TF-IDF value on January. Besides, for the economical news in Tunisia, we notice the appearance of the term "دينار" (Dinar: Tunisian currency) with the highest TF-IDF value on April because of its value drop.

4.3 Experimental Results

Since the used corpus has no relevance judgment information for the event extraction task, we search the extracted events on another Tunisian news website. So, we consider a relevant event if there is a publication about it on the Mosaïque FM[3] news radio in the same month.

We applied two different scenarios for computing the precision value as presented in Table 2. On the one hand, we consider all extracted events for precision computation. On the other hand, we calculate the precision for events that appear in at least two titles and we neglect named entities that appear in one title only.

Results show that the extracted events from the source website should be filtered by considering only the set of titles appearing at least 2 times per month. The Fig. 4 presents the detailed precision values for each month.

---

[3] https://www.mosaiquefm.net/.

Table 2. Precision values and number of extracted events per category for all events and for events appearing in at least 2 titles

Category	Number of extracted events		Precision	
	All events	Events in 2 titles or more	All events	Events in 2 titles or more
Economy	229	18	0.372	0.976
InternationalNews	532	107	0.257	0.843
LocalNews	1439	196	0.245	0.683
Politic	337	38	0.409	0.871
Society	696	73	0.146	0.690
Sport	1237	188	0.165	0.483

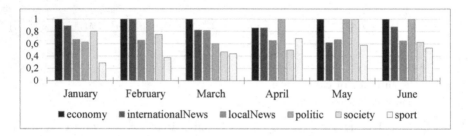

Fig. 4. Events' extraction precision for each month

By focusing on these results, the economy category has the highest precision value since it has a fixed jargon. The reduced number of events in this category, as presented in Table 2, can have a positive impact on the precision value. The same reason can explain the high precision values for the politic category. The sport category has the lowest precision value. This low value could be explained by the semantic richness of this category. Table 3 presents some examples of the same events, in sport category, that were not written the same way in the referred news website for relevance judgment. In the first presented example, »مالك الجزيري« (Malik Al Jaziri) is a Tunisian tennis player. So adding or removing the word "تنس" (tennis) does not really have an impact on the context of the corresponding event. In the second example, "مقابلات" (matches) is semantically the same as "مباريات". The last example represents the same number written in letters in the extracted event but in numbers in Mosaïque FM article. Thus, in order to enhance the performance of the proposed approach, a semantic reformulation task should be integrated with the events extraction process.

Table 3. Examples of same events written differently

Extracted events (from ANT Corpus)	Events on Mosaïque FM website	Reformulation type
دورة دبي: مالك الجزيري (Dubai course: Malik Al Jaziri)	دورة دبي للتنس : مالك الجزيري (Dubai Tennis Tournament: Malek Al Jaziri)	Title expansion
كأس تونس: برنامج مقابلات الدور ربع (Tunisia cup: quarter round matches program)	كأس تونس: برنامج مباريات الدور ربع (Tunisia cup: quarter round matches program)	Word reformulation
الدور السادس عشر (the sixteenth round)	الدور 16 (the sixteenth round)	Number/Letter representation

5   Conclusion and Future Work

We presented in this paper a statistical approach for event extraction task. We calculated the terms weighting for each noun in the news article's title using the TF-IDF measure. Then, we extracted the keywords by considering a computed threshold value. For each extracted keyword, we defined the corresponding event using a POS co-occurrence rule. In order to experiment the proposed approach, we matched the extracted events to another news website source. Results show that the proposed approach performs well by considering events that appear in two titles at least. The results depend also on categories and perform well for domain specific ones such as economy. As a perspective for the current work, we propose to combine knowledge based resources such as semantic networks and ontologies in the event extraction process. Titles reformulation would help also to identify semantic related events through machine learning algorithms.

References

1. Darwish, K., Magdy, W.: Arabic information retrieval. Found. Trends Inf. Retr. **7**(4), 239–342 (2014)
2. Elayeb, B., Bounhas, I.: Arabic cross-language information retrieval: a review. ACM Trans. Asian Low-Resour. Lang. Inf. Process **15**(3), 18:1–18:44 (2016)
3. Elayeb, B.: Arabic word sense disambiguation: a review. Art. Int. Rev. **50**, 1–58 (2018)
4. Bounhas, I., Elayeb, B., Evrard, F., Slimani, Y.: Organizing contextual knowledge for arabic text disambiguation and terminology extraction. Knowl. Organ. **38**(6), 473–490 (2011)
5. Habash, N., Rambow, O., Roth, R.: MADA+TOKAN: a toolkit for arabic tokenization, diacritization, morphological disambiguation, pos tagging, stemming and lemmatization. In: Proceedings of MEDAR'2009, pp. 102–109 (2009b)
6. Hogenboom, F., Frasincar, F., Kaymak, U., De Jong, F., Caron, E.: A Survey of event extraction methods from text for decision support systems. Decision Support Systems **85(C)**, 12–22 (2016)
7. Naughton, M., Kushmerick, N., Carthy, J.: Event extraction from heterogeneous news sources. In: Proceedings of AAAI, pp. 1–6 (2006)

8. Zhou, D., Chen, L., He, Y.: An unsupervised framework of exploring events on twitter: filtering, extraction and categorization. In: Proceedings of AAAI'2015, pp. 2468–2474 (2015)

9. AL-Smadi, M., Qawasmeh, O.: Knowledge-based approach for event extraction from Arabic tweets. Int. J. Adv. Comput. Sci. Appl. **7**(6), 483–490 (2016)

10. Chouigui, A., Ben Khiroun, O., Elayeb, B.: ANT corpus : an Arabic news text collection for textual classification. In: Proceedings of AICCSA'2017, pp. 135–142 (2017)

11. Dahab, M.Y., Ibrahim, A.A., Al-Mutawa, R.: A comparative study on Arabic stemmers. Int. J. Comput. Appl. **125**(8), 38–47 (2015)

12. Larkey, L.S., Ballesteros, L., Connell, M.E.: Light stemming for Arabic information retrieval. Arabic Computational Morphology, pp. 221–243. Springer, Dordrecht (2007)

13. Chowdhury, A., Aljlayl, M., Jensen, E.C., Beitzel, S.M., Grossmanand, D.A., Frieder, O.: Linear combinations based on document structure and varied stemming for Arabic retrieval. In: Proceedings of TREC'2002, pp. 1–12 (2002)

English Text Parsing by Means of Error Correcting Automaton

Oleksandr Marchenko[1($\boxtimes$)], Anatoly Anisimov[1], Igor Zavadskyi[1],
and Egor Melnikov[2]

[1] Taras Shevchenko National University of Kyiv, Kyiv, Ukraine
omarchenko@univ.kiev.ua
[2] P1:k ltd, London, UK

Abstract. The article considers developing an effective flexible model for describing syntactic structures of natural language. The model of an augmented transition network in the automaton form is chosen as a basis. This automaton performs the sentence analysis algorithm using forward error detection and backward error correction passes. The automaton finds an optimal variant of error corrections using a technique similar to the Viterbi decoding algorithm for error correction convolution codes. As a result, an effective tool for natural language parsing is developed.

Keywords: Grammatical error correction
Augmented transition network · Grammatical automaton
Syntactic analysis

1 Introduction

Automatic syntactic analysis and grammaticality judgement on the sentence structure in furtherance of grammatical error correction is a vital task of Natural Language Processing. Moreover, it is a trending IT topic, and hence the software market of Grammatical Error Correction (GEC) services is growing rapidly.

The sentence syntactic structure validation consists of analyzing an input sequence of words of a given sentence and matching it to a grammatical sentence model. Complete models for a given language are generated by specific rules. Traditionally, those are described by generative grammars. In case the sentence structure is incorrect, an error correction algorithm is applied. It is able to modify grammatical features of a word, its part of speech or word order, etc.

A processor that checks the grammaticality of a sentence can be implemented as a standard syntactic top-down or bottom-up parser. If the parse tree is built successfully then correctness of the sentence is approved. This sort of a parser is entirely sufficient to verify the correctness of the sentence structure. However, the search for specific errors and their correction requires implementation of a special technique. A development of an original efficient systematic method for these purposes based on ideas from the transmission error correcting codes is the subject of the presented article.

© Springer International Publishing AG, part of Springer Nature 2018
M. Silberztein et al. (Eds.): NLDB 2018, LNCS 10859, pp. 281–289, 2018.
https://doi.org/10.1007/978-3-319-91947-8_28

2   Related Work

In early works pattern matching was proposed for implementing error correction systems for different purposes [1,2]. String editing and pattern matching guided by a set of rules were later proposed [3]. The rule-based approach has inherent clarity, it can be augmented both by pattern matching and grammar rules. Man-made grammar rules [4–6] were considered as more efficient in comparison to complex models [1] especially when one has to consider errors that cannot be covered with the dictionary approach. Taking into account the way language evolves, it seems to be impossible to create a single fixed set of rules whatever elaborated it may be. Moreover, the ways a human-being makes mistakes are also evolved, making it impossible to cover all errors with a finite number of rules. As for the subset of rules to cover the most of errors, there is no efficient way to select a good subset without adapting it to certain conditions.

For the aforementioned reasons, for error correction systems the machine learning approach was proposed. Several systems applied convolutional neural networks (CNNs) for grammatical error detection and correction [7]. These works were followed by attempts to combine neural networks with machine translation (MT) for the purpose of grammar error correction [8]. The neural network (NN) MT-based projects implement the encoder-decoder architecture. The motivation for NNMT is based on the fact that NN allows for correcting previously unseen errors by far more efficiently as compared to statistical MT. The main drawback of NNMT-based systems is the implicit way of data representation that leads to the impossibility to clarify the algorithm behavior on the base of obtained results. Thus, there is the need to create a more transparent but not less powerful and flexible model for building a GEC system.

3   The Main Idea

The issue we are considering has similarities with other tasks of computational linguistics, for example, with the problem of automatic spell checking. Spell checkers use a dictionary as a static list of correct words of a particular language and a correction algorithm. In case some input sequence of characters, i.e. a token does not match any dictionary word, dynamic programming algorithms are performed. In order to find the most probable correction for the input word, they use counting edit distances between the misspelled word and words in the dictionary. There is an obvious analogy in the description of these problems and, consequently, in approaches to their solution.

It is worth noting that these two problems require fundamentally different ways for assigning sets of correct sequences. For a spell checker, the static list of words of a language (and their forms) is needed, while a grammar checker requires a dynamic and recursive model which is able to generate the entire set of all acceptable sequences that correspond to the syntactic rules and principles of a given language. Traditionally, rules are defined in the same way in which they are presented in a generative grammar. Complex and comprehensive verification of the sentence correctness requires not only part-of-speech tagging but also it strongly depends on extracting all possible grammatical features in order to support matching of word forms, verb times, ect. Sometimes the decisive factor is the presence of concrete lexemes in the sequence of sentence words. Therefore, there is a need to develop a lexicalized grammar in which, in addition to the grammar rules coordinating parts of speech, the syntactic compatibility of particularly important lexemes would be described.

As a result of our research, it was decided to represent, analyze and generate all admissible sequences of English words using the augmented transition network proposed by Woods [9]. In a transition from word to word of the analyzed sentence the automaton changes its states. Thus, its states are a set of Q positions that correspond to words or parts of speech in different syntactic constructions. Being in some state q_i, the automaton reads a next word and, depending on its lexical-grammatical value, goes to the next state q_j. A transition function is represented as $F : Q \times (T \cup N) \to Q$, where N is the set of part of speech tags, and T is the set of words in the language with prescribed lexical-grammatical characteristics.

If we consider the aforementioned problems in general terms, we can draw an analogy with the problems of antinoise coding and error correction in communication channels. This allows us to implement the approaches used to solve such problems in communication area. For example, the Viterbi method is used to decode convolutional codes [10]. This method allows us finding a sequence of signals with a minimum metrics, which is a measure of the closeness of this sequence to a sequence that could be generated by an encoder. The behavior of an encoder and a decoder can be described using a finite state machine that generates signals while performing transitions from state to state. Assigning metrics to transitions of the network, we specified an analogy between automata used in convolutional codes and the Woods augmented transition network. If the morphilogical characteristics of a word under processing transition correspond to the parameters of the transition itself, then the metrics is zero, otherwise it equals to some positive number. In our method, metrics are summarized following the run transactions from state to state. Among all states, a subset Q_{fin} of final states is defined. The automaton comes into a final state with zero metrics only in the case of processing a grammatically correct sequence representing the complete syntactic structure of a sentence. If it does not stop in a final state or stops in a final state, but with a positive metrics, this indicates an error. To correct errors, we adapt the aforementioned Viterbi method using the ideas of dynamic programming.

4   Grammatical Automaton: General Structure

A grammatical automaton is a set of states and transitions between them. A state S is a 2-tuple $\{N, I\}$, consisting of the name N of the state and a set of additional attributes I, which we call an internal state. These attributes could be changed during transitions and, thus, the internal state can be considered as a certain cumulative characteristic of the path that was passed from the initial state to the current one. Among the attributes of internal states the most important are:

- A **path marker**, which is a text string that is formed as a concatenation of markers that were marked along the path of previous transitions. The purpose of these markers is examined in more details hereinafter.
- A **metrics**, which is a number indicating the deviation of the current path from a standard path corresponding to some correct grammar construction.
- A **stack of grammatical features**, which is a list of pairs (gender, number) corresponding to some tokens. For example, the transition to a noun or a pronoun in direct case pushes their features into the stack.

The automaton is intended to parse sentences. While parsing each token, a set of transitions is produced, i.e. the next-state function $\{S, L\} \to S$ is defined, and automaton is non-deterministic. Here by S we denote the set of states of the automaton, and L denotes the set of tokens of the language. We perceive a token as a set of grammatical features, and, in special cases, we consider it as a word itself. When we process the transition between states A and B, the internal state B is created by default as a copy of the internal state A. However, the values of its attributes may also be changed by an operation attributed to the transition. A list of the most common types of such changes and operations is given as follows.

- A path marker is changed if the transition requires appending a substring to it. The transition may also contain a command to remove all occurrences of a substring in the path marker or to completely clear it.
- A metrics is non-decreasing; during transitions, it either increases or does not change. An increment can either be directly specified in the parameters of a transition (this is the so-called **penalty jump**) or calculated automatically as a measure of discrepancy between the grammatical features of a token and the parameters of the transition.
- The stack of grammatical features is changing according to the commands indicated in Table 1. In general, when tokens that are characterized by number and/or person (for example, pronouns or verbs in the present tense), are analyzed, their attributes are added to the stack. The analysis of other tokens does not change the stack.

Additionally, special conditions may be applied for a transition. In case these conditions are violated, the transition will be blocked. Most common conditions are the equality of a pair of token grammatical features and a pair of values from the top of the grammatical feature stack and also the presence or absence of certain substrings in the path marker.

These conditions ensure the implementation of rules that determine the relationship between grammatical features of tokens in different parts of a sentence. The conditions of the first type ensure the agreement of the noun or pronoun form with the form of a verb in the present tense (number, as well as the person for the "pronoun plus be" construction), and the agreement of the noun number with the determiner (for example, the rule prohibiting the use of the article "a" with plural nouns). Conditions of the second type ensure the implementation of more complex rules, such as "if in the independent clause of a complex sentence the verb is in the past, it should be in the past in the dependent clause also".

In order to simplify the representation of the grammatical finite-state machine, empty transitions are also used; they do not imply the analysis of any tokens. For example, if after states A and B, the sets of subsequent acceptable grammatical constructions are the same, one can create an empty transition from state A to state B, instead of duplicating the set of transitions. In fact, internal states are also means for simplification of the machine structure.

Penalty Transitions. Penalty transitions attributed with non-zero metrics are a powerful mechanism for correcting typical errors. Such transitions correspond to frequent irregular grammatical constructions and may also contain correct grammar features to correct mistakes.

The error correction algorithm looks for a path from the initial state to a final one with a minimal metrics in the grammatical machine graph, which corresponds to the sequence of sentence lexemes. If the grammatical structure of a sentence cannot be considered as correct, the path with zero metrics will not be found, however, another path can be found that contains penalty transitions, and hence this gives information how to correct errors.

In case a wrong token should be replaced by another one, a penalty transition is created along with the normal one. For example, the common mistake is to use an adjective instead of the adverb in constructions such as "I run quickly". Then, in a machine that recognizes grammatically correct sentences, there will be a transition that corresponds to the adverb after the verb, along with which it is necessary to create a transition for the adjective which stands after the verb. Thus, the incorrect sentence "I run quick" will not match any path with zero metrics in the graph of the grammatical machine, but will match the path with the metrics attributed to the penalty transition from the state "after the verb" to the "adjective" token. If this metrics is small enough, it will be the minimal possible metrics for this sentence, and the path containing this penalty transition will include the information that the adjective should be replaced by the corresponding adverb.

If you need to add a token to correct an error, a penalty transition is created. For it there is no "parallel" correct transition. If a token needs to be deleted, an empty penalty transition is created.

A Stack of Grammatical Features. The transition from state A to state B returns a set of states $B_1, \ldots, B_N$, in which the external state is the same and is determined by the grammatical automaton, while the internal states are generated according to the rules described in Table 1.

As seen, the **push** operation leads to cloning the state to N others, containing all possible combinations of number and person. In turn, this leads to the branching possible ways to analyze the sentence. The **push** operation is used for the words that determine the grammar form of other words, e.g. pronoun "I" determines the form of the verb in the sentence "I have a pen". And if the

Table 1. Operations with the stack of grammatical features

Action/Command	State A	The way the internal states $B_1, \ldots, B_N$ are formed
Push	If the top of the stack contains an empty state	The top of stack A is replaced by all possible internal states
	Otherwise	All possible internal states are pushed into stack A
New		An empty state is pushed into stack A
Reset		Stack is cleared except of its bottom
None		Creates one state B, in which stack of A is copied unchanged

grammar form of this "other word" is disagreed with one defined in the branch, this branch must be cut. The need for such "check for cut" is indicated by the **check mark** on the transition. In other words, it means that the grammatical features from the top of the stack should match the corresponding grammatical features of some tokens.

Sentence Analysis Algorithm. The sentence analysis consists of two phases, which, by analogy with the Viterbi algorithm for convolutional codes, can be called forward and backward passes. During a forward pass, a graph of all possible transitions corresponding to the sentence with a metrics not exceeding a certain value is constructed. During the backward pass, the final state with the minimum metrics and the path to it is determined. The transitions in this path contain the information about the error correction. The backward pass construction algorithm is trivial enough, whereas a forward pass can be implemented as a breadth-first search, or as a series of depth-first searches with a growing metrics limit. The advantage of breadth-first search is its ability to find all possible corrections, however, it takes significantly more time and space.

5   Experiments

There are several well-known special corpora for evaluating GEC systems: The National University of Singapore Corpus of Learner English (NUCLE)[11], The Cambridge Learner Corpus (CLC), HOO-2012 [12], and the CoNLL-2013 shared task ect.

The augmented transition automaton developing is still goes on, and due to the incompleteness of syntactic structure description in our model, the aforementioned corpora do not meet the requirements that provide an objective evaluation of the developed system. The first condition is that only the syntactic structures already described in our model should be used in the corpus texts. The second condition is that texts must contain only types of errors supported by the model. To evaluate the effectiveness of the proposed model for finding and correcting grammatical errors there was a need for the construction of a special text corpus with marked up errors.

The Reuters-21578 News Corps and a collection of articles from English-language Wikipedia (approximately 8,000 randomly selected articles) were used as the basis for creating the corpus. The sentences of texts were successively processed by the developed augmented transition automaton. If the automaton successfully worked out the sentences and stopped in a final state with zero sum, such sentences were included into the corpus, otherwise sentences were rejected. Consequently, we collected statistics on the relative coverage of English language syntactic structures by the developed automaton. In the case of news texts Reuters-21578, the automaton showed coverage of 66.4% (66.4% of corpus sentences were correctly handled by the automaton). In the case of the collection of English-language Wikipedia articles, the automaton showed a more modest result - 62.7%. In this way, the corpus of texts without errors was generated.

At the second stage, errors were made in texts using a specially designed utility. In texts, the program generated the following errors with a uniform distribution:

1. the lack of words belonging to the certain class in a sentence (by removal);
2. the presence of an extra word (by insertion);
3. incorrect order of words (by permutation or several permutations of adjacent words);
4. changing the verb form (time, person, number) for violation of the concordance with the form of a noun;
5. changing the form of nouns (number) for violation of the concordance with the form of a verb;
6. changing the form of nouns for violation of the concordance with the determinant;
7. replacement of the adjective with the adverb and vice versa etc.

In addition, some typical mistakes were made manually, such as the selection of the adjective form ("good"/"best"/"better"), time inconsistency for verbs in compound sentences, form inconsistency for homogeneous members in a sentence, the form of a predicate with homogeneous subjects, etc.

As a result, three versions of the corpus with errors were generated: the corpus of texts with error density - 3 errors for a sequence of 100 words (Corp A); the corpus of texts with error density - 6 errors for a sequence of 100 words (Corp B); the corpus of texts with error density - 9 errors for a sequence of 100 words (Corp C).

Simultaneously with the insertion of errors in texts, the program writes data about them in the format (place of error, type of error) in the special log file. Next, a group of linguists checked and approved the mistakes made in cases where the errors of corresponding types were generated correctly. Due to the developed utility, it was possible to form the representative text corpus in three versions with different error density.

After that the developed augmented transition automaton handled three created corpora. Before processing the created test corpora were partitioned into pieces (each of them contains approximately 300 000 words). In the automatic mode, statistics of error identification and correction were collected. Also, the most popular system Grammarly was tested on the received corpora. The comparative statistics are given in Table 2. The table presents average estimates of Precision, Recall and F1 that were measured on partitions of the corpora. In Precision, Recall and F-measure the true positive is considered as a number of properly identified and corrected errors.

Table 2. Experimental results

	Our system					Grammarly				
	Total number of mistakes	Corrected mistakes	Improper fixes	Precision	Recall	F1-score	Corrected mistakes	Improper fixes	Precision	Recall	F1-score
Corp A	8904	3461	675	0.84	0.39	0.53	2948	341	0.89	0.33	0.48
Corp B	18007	6177	3120	0.66	0.34	0.452	5423	727	0.88	0.30	0.449
Corp C	26014	6135	5839	0.51	0.24	0.32	6730	1536	0.81	0.258	0.39

As shown, the developed system overcomes Grammarly in Recall-score and F1-score when it processes texts with low and middle density of aforementioned type errors, although it loses in Precision-score. This is expectable due to different underlying approaches: we try to find a whole correct sentence in a certain sense close to a processed one, while Grammarly apparently localizes errors. Thus, our system is able to identify and to correct more mistakes, however, the probability of finding some correct sentence which distorts the meaning of the initial one, is also higher. For texts with high density of mistakes, Precision-score and Recall-score become lower because of increasing the quantity of variants for fixing errors. When we deal with texts that contain reasonably acceptable quantity of mistakes the F-score of our system demonstrates a confident advantage in comparison with the state-of-the-art system Grammarly.

6   Conclusion

The paper describes a new method for parsing natural language texts, and grammatical error identification and correction. The model of an augmented transition network in the automaton form is chosen as a basis. This automaton performs the sentence analysis algorithm using forward error detection and backward error correction passes. The automaton finds an optimal variant of error corrections using a technique similar to the Viterbi decoding algorithm for error correction convolutional codes. Conducting experiments confirmed high efficiency and correctness of the developed model.

References

1. Szanser, A.J.: Automatic error-correction in natural languages. Inf. Storage Retr. **5**(4), 169–174 (1970)
2. Riseman, E.M., Hanson, A.R.: A contextual postprocessing system for error correction using binary n-grams. IEEE Trans. Comput. **C–23**(5), 480–493 (1974)
3. MacDonald, N.H., Frase, L.T., Gingrich, P.S., Keenan, S.A.: The Writer's workbench: computer aids for text analysis. IEEE Trans. Commun. **30**(1), 105–110 (1982)
4. Jensen, K., Heidorn, G.E., Miller, L.A., Ravin, Y.: Parse fitting and prose fixing: getting a hold on ill-formedness. Comput. Linguist. Spec. Issue Ill-formed Input **9**(3–4), 147–160 (1983)
5. Richardson, S., Braden-Harder, L.: The Experience of Developing a Large-Scale Natural Language Processing System: Critique. In: Jensen, K., Heidorn, G.E., Richardson, S.D. (eds.) Natural Language Processing: The PLNLP Approach. The Kluwer International Series in Engineering and Computer Science (Natural Language Processing and Machine Translation), pp. 77–89. Springer, Boston (1993). https://doi.org/10.1007/978-1-4615-3170-8_7
6. Adriaens, G.: Simplified English grammar and style correction in an MT framework: the LRE SECC project. Aslib Proc. **47**(3), 73–82 (1995)
7. Sun, C., Jin, X., Lin, L., Zhao, Y., Wang, X.: Convolutional neural networks for correcting english article errors. In: Li, J., Ji, H., Zhao, D., Feng, Y. (eds.) NLPCC 2015. LNCS (LNAI), vol. 9362, pp. 102–110. Springer, Cham (2015). https://doi.org/10.1007/978-3-319-25207-0_9
8. Yuan, Z. and Briscoe, T.: Grammatical error correction using neural machine translation. In: Proceedings of NAACL-HLT 2016, pp. 380–386 (2016)
9. Woods, W.: Transition network grammars for natural language analysis. Commun. ACM **13**(10), 591–606 (1970)
10. Feldman, J., Abou-Faycal, I., Frigo, M.: A fast maximum-likelihood decoder for convolutional codes. In: Vehicular Technology Conference, pp. 371–375 (2002)
11. Dahlmeier, D., Ng, H.T., Wu, S.M.: Building a large annotated corpus of learner english: the NUS Corpus of Learner English. In: Proceedings of the 8th Workshop on Innovative Use of NLP for Building Educational Applications, pp. 22–31 (2013)
12. Dale, R., Kilgarriff, A.: Helping our own: the HOO 2011 pilot shared task. In: Proceedings of the 13th European Workshop on Natural Language Generation, pp. 242–249 (2011)

Annotating Relations Between Named Entities with Crowdsourcing

Sandra Collovini[(✉)], Bolivar Pereira, Henrique D. P. dos Santos,
and Renata Vieira

Pontifícia Universidade Católica do Rio Grande do Sul,
Porto Alegre, Rio Grande do Sul, Brazil
{sandra.abreu,bolivar.pereira,henrique.santos.003}@acad.pucrs.br,
renata.vieira@pucrs.br

Abstract. In this paper, we describe how the CrowdFlower platform was used to build an annotated corpus for Relation Extraction. The obtained data provides information on the relations between named entities in Portuguese texts.

Keywords: Crowdsourcing · Semantic relations annotation
Portuguese

1 Introduction

The task of extracting relations is usually given as a problem in which sentences that have already been annotated with entity mentions are additionally annotated with relations between pairs of those mentioned entities. In general, the performance of this task is measured against gold datasets, such as ACE 2008 RDC[1] for English, and HAREM[2] for Portuguese. These gold collections were created through a manual annotation process based on guidelines, which took extensive time and effort to be developed by experts.

To tackle this problem, Crowdsourcing platforms became an alternative, broadly used by researchers [11], to build annotated corpus. Amazon Mechanical Turk and Crowdflower are some of the tools that could be used to handle the annotation task. Those systems are able to publish a task to hundreds of annotators spread through the globe, in several countries [10], to solve problems manually. In this work, we describe an annotation task using Crowdsourcing, which consists of annotating any type of semantic relations between pairs of Named Entities (NE) in sentences of Portuguese text, based on the proposed annotation instructions.

This work is organized as follows: In Sect. 2, we present the background. The task description is detailed in Sect. 3. In Sect. 4, we discuss the results. One of the contributions of this work is briefly presented in Sect. 5. Finally, Sect. 6 presents the conclusions and future works.

---

[1] http://projects.ldc.upenn.edu/ace.
[2] http://www.linguateca.pt/harem.

© Springer International Publishing AG, part of Springer Nature 2018
M. Silberztein et al. (Eds.): NLDB 2018, LNCS 10859, pp. 290–297, 2018.
https://doi.org/10.1007/978-3-319-91947-8_29

2   Background

Relation Extraction (RE) is the task of identifying and classifying the semantic relations between entities from natural language texts, such as placement (Organization, Place) or affiliation (Person, Organization). It focuses on extracting structured relations from unstructured sources using different approaches. Thus, depending on the application and on the resources available, the RE task can be studied for different settings. Many approaches to RE use supervised machine learning, but these methods require human-annotated training datasets that may be unavailable [12].

For Portuguese, there are few annotated data for RE compared to other languages such as English [1]. One of the obstacles to create high quality annotated data is the lack of detailed guidelines for executing manual annotation of relations. Unfortunately, for works in Portuguese, it is not possible to use resources and databases developed for English. Also, a major difficulty is the availability of experts to perform the annotation.

One way to address these problems is to apply Crowdsourcing[3] annotation to RE. The basic workflow of Crowdsourcing in all platforms is similar. First, a requester (a human or computer) creates a task for workers to complete and posts that task on a platform of its choice. Generally, a requester also specifies certain characteristics a worker must meet to perform the task. Next, workers find the task, complete it, and return the results to the requester. If the results meet a requester's approval criteria, it compensates the worker.

Many NLP tasks have successfully used Crowdsourcing approaches, such as Named Entities [9], Relation Extraction [12], Ontology [14], Text Classification, Sentiment Analysis [13], Topic Detection [8], among others. However, these works treat languages other than Portuguese. The present work has the challenge of applying a Crowdsourcing platform to build annotated corpus quickly and efficiently from Portuguese texts, aiming at the extraction of semantic relations between named entities. There are online platforms that enable crowdsourcing, such as Amazon's Mechanical Turk (MTurk)[4] and CrowdFlower[5].

3   Task Description

In this work, we describe an annotation task of semantic relations between named entities of Summ-it++ corpus [3], using the CrowdFlower platform. Therefore, annotation instructions were developed to serve as a guide for annotators, as well as test questions to evaluate the annotators' knowledge for the proposed task. The input data and the construction/design of the job in CrowdFlower platform are presented below.

---

[3] Crowdsourcing has been defined as the act of taking a job traditionally performed by a designated agent and outsourcing it to an undefined, generally large group of people in the form of an open call.

[4] http://www.mturk.com/mturk.

[5] http://www.crowdflower.com.

3.1   Annotation Instructions

We provided the workers with annotation instructions containing an overview of the annotation task. In this task, the workers should check whether or not there is an explicit relation located between named entities of Organization, Person or Place in a sentence. If it occurs, the workers should identify all the words that describe the relation.

Table 1 shows the example (1), where a placement relation occurs between "Marfinite" (Organization) and "Brasil" (Place), identified by the elements "fica em" (verb + preposition). The relations identified in the sentence are represented as a triple: (NE1, relation, NE2), in the case of this example we have the triple (Marfinite, fica em, Brasil). If a relation does not occur, one must inform the incident, as in example (2) of Table 1 which shows no relation between "Turquia" (Place) and "Pentágono" (Organization). In general, the annotators should follow these instructions:

- Annotate only the words that occur between the pair of named entities in the sentence (see example (1) in Table 1);
- Annotate the smallest number of elements required to describe the relation, as in example (3) of Table 1, where "abre perspectivas em" (open perspectives in) is sufficient to express the relation between "Marfinite" (Organization) and "Brasil" (Place).

We highlight the difficulty to determine which elements between named entities are in fact part of the relation [7]. Thus, our guidelines were described as clearly as possible. The list of elements that form a relation and illustrative examples are presented below.

- Relations must be delimited/considered up to the preposition, if it occurs. However they are dismembered in preposition ("de", "em") plus article ("o", "a"), and the article should not be included (see example (4) in Table 1);
- Relations composed by nouns, such as nouns expressing titles/jobs (see example (5) in Table 1);
- Relations composed by verbs (predicates of the sentence), as in example (6) of Table 1;
- Relations composed only by preposition, as in example (7) of Table 1;
- Relations formed by punctuation such as: parentheses, dashes, commas etc. (see example (8) in Table 1);

There are elements that should not be included in the relation, such as adjectives: "exercício", "excelente" (see example (9) in Table 1); and pronouns: "seu", "sua" (see example (10) in Table 1).

3.2   Dataset

In order to accomplish the task, we used the Summ-it++ corpus [3], which originated from the Summ-it corpus [5]. Summ-it is one of the first corpora of the Portuguese Language gathering annotations from various levels of discourse.

Table 1. Input examples.

Examples
(1) A **Marfinite** fica em o **Brasil**. (**Marfinite** is located in **Brasil**.)   Relation: fica em (located in)   Triple: (Marfinite, fica em, Brasil)
(2) Os aparelhos regressaram à base na **Turquia**, acrescenta o comunicado do **Pentágono**. (The equipment returned to the base in **Turquia**, added the statement of **Pentágono**.)   Relation: no relation
(3) A **Marfinite** abre perspectivas de negócios através de novos distribuidores em o **Brasil**. (A **Marfinite** opens business perspectives through new distributors in **Brasil**).   Relation: abre perspectivas em (opens perspectives in)   Triple: (Marfinite, abre perspectivas em, Brasil)
(4) **Ronaldo_Goldone** continua atuando em as atividades de o **Niterói_Rugby**. (**Ronaldo_Goldone** continues to work in the activities of **Niterói_Rugby**.)   Relation: atuando em (to work in)   Triple: (Ronaldo_Goldone, atuando em, Niterói_Rugby)
(5) **Hugo_Doménech**, professor de a **Universidade_Jaume_de_Castellón**. (**Hugo_Doménech**, teacher of **Universidade_Jaume_de_Castellón**.)   Relation: professor de (teacher of)   Triple: (Hugo_Doménech, professor de, Universidade_Jaume_de_Castellón)
(6) Em 1956, **Amílcar_Cabral** criou o **Partido_Africano**. (In 1956, **Amílcar_Cabral** created the **Partido_Africano**.)   Relation: criou (created)   Triple: (Amílcar_Cabral, criou, Partido_Africano)
(7) **António_Fontes** de a **AIPAN**. (**António_Fontes** of the **AIPAN**.)   Relation: de (of)   Triple: (António_Fontes, de, AIPAN)
(8) A **USP** (**Universidade_de_São_Paulo**) aprovou a iniciativa dos alunos. (**USP** (**University_of_São_Paulo**) approved the students' initiative.)   Relation: (Triple: (USP, (, Universidade_de_São_Paulo)
(9) O **Presidente** em exercício de o **Conselho**. (The current **Presidente** of the **Conselho**.)   Relation: de (of)   Triple: (Presidente, de, Conselho)
(10) A **Legião_da_Boa_Vontade** comemora o aniversário da sua implantação em **Portugal**. (**Legião_da_Boa_Vontade** celebrates the birthday of its establishment in **Portugal**.)   Relation: implantação em (establishment in)   Triple: (Legião_da_Boa_Vontade, implantação em, Portugal)

Summ-it++ consists of fifty journalistic texts from the Science section of the Folha de São Paulo newspaper and has the following annotations: morphosyntactic annotation provided by CoGrOO[6]; coreference annotation from original Summ-it [5]; named entities recognition by system NERP-CRF [2]; and annotation of some relations between pairs of named entities (Organization and Person or Place) by system RelP [7]. These last annotations (named entities and relations) were automatically annotated and manually revised by humans.

---

[6] http://cogroo.sourceforge.net/.

3.3 Pre-annotation

In the context of this work, we selected 3 representative texts from the Summit++ corpus to be annotated manually by 3 annotators (students of Computer Science). These 3 texts resulted in 43 selected sentences, that is, sentences containing the pair of named entities (Organization and Person or Place). To measure the agreement of annotation between the 3 annotators, we calculated the kappa coefficient [4] which reached the result "K = 0.625" - "substantial agreement", resulting in the annotation of 26 sentences containing relations between named entities in focus and 17 without a relation. These gold sentences served as test questions for the training/evaluation of annotators in the CrowdFlower platform, presented in Sect. 3.4.

3.4 Annotation with CrowdFlower

The CrowdFlower platform was used to perform our annotation task, which involved 50 texts from Summ-it++. In order to develop the job, it was necessary to customize the task for application in the platform. First, the texts were pre-processed to serve as input to the workers, and only the sentences composed by the pair of named entities in focus were selected. However, each input sentence can have only one pair of entities (parameters). Because of that, if a sentence had more than one pair, the number of times it was selected equals it's number of pairs. This process of sentence selection produced an output of 243 sentences/input question.

In addition, we selected 10 gold sentences for test questions taken from 3 texts of Summ-it++ corpus (see Sect. 3.3). Along the course of the task, we turned 9 regular questions that had 100% consensus among the workers into test questions, resulting in a total of 19 test questions.

To avoid mistakes by the annotator and ensure a quality annotation, we develop some features at CrowdFlower Design Step. Using Javascript/jQuery functions and CML (CrowdFlower Makeup Language), we display the text to the annotator and he is able to click at each word he believe that belong to the relation between the entities. Only the words between the entities are clickable, avoiding annotator mistakes. Those words and its indexes are used in the responses to compute agreement and build the final corpus.

For the application of the job, the first step was to present the task to the workers. Figure 1(a) provides a test question of the task as presented to a worker. We can verify that the named entities involved are highlighted in bold, indicating which parameters are being considered and which part should be analyzed. The worker must click on the words that compose the relation between these highlighted pairs. If there is no relation, the worker must select the option: "Sentença sem relação" (no relation). There is also the option "Corrigir" (Correct) to correct the annotation, if necessary. Figure 1(b) also shows the annotation resulting from the task performed by the worker, where a affiliation relation occurs between "Cassius Vinicius Stevani" (Person) and "USP" (Organization), identified by the elements "químico de" (chemist of).

(a)

Segundo **Cassius_Vinicius_Stevani,** químico de a **USP**, que coordena os estudos, é possível que o material recolhido por o menos_dez espécies novas.

Relação selecionada: (required)

[   ]

ID dos tokens da relação: (required)

[   ]

☐ Sentença sem relação

[Corrigir]

Tripla:

(b)

Segundo **Cassius_Vinicius_Stevani,** **químico de** a **USP**, que coordena os estudos, é possível que o material recolhido por o menos_dez espécies novas.

Relação selecionada: (required)

[químico de   ]

ID dos tokens da relação: (required)

[3,4   ]

☐ Sentença sem relação

[Corrigir]

Tripla:

(Cassius_Vinicius_Stevani, químico de, USP)

Fig. 1. Example task presented to a worker (a), before and (b) after the annotation.

Each input question was annotated by three workers, who should be Brazilian, Portuguese speakers and Level 2 qualified - "Higher Quality: Smaller group of more experienced, higher accuracy contributors". The 19 test questions were annotated by at least three workers and interspersed with other input questions (243 sentences). If any annotator has a wrong test question, he is immediately notified of it. If one's accuracy on the test questions falls below 70%, they are rejected and all of their annotations are discarded [13]. However, the worker may comment the test question that has been reported. A total of 55 annotators participated in the CrowdFlower platform job, each one receiving 2 cents per annotation.

4   Results and Discussion

In this Section, we present the result of our task on CrowdFlower platform, which considered the judgments per sentence of each worker. Table 2 shows the agreement between annotators. In the first column, we illustrate the inter-annotator agreement, and in the second column the number of sentences annotated. We can see that 69 sentences reached an agreement between 2 annotators and 167 sentences resulted in an agreement between 3 annotators, in a total of 236 annotated sentences. After, the cases with relations between named entity pairs and the unrelated cases, which obtained agreement between 2 and 3 annotators are quantified in Table 2, respectively. We obtained a total of 127 relations annotated among the named entity pairs in focus and 109 cases containing no relation.

We measure the quality/reliability of the annotations by the Kappa coefficient, evaluating whether there is a relation or not in each input sentence. For this, 52 input sentences were selected, the biggest subset annotated by the same three workers, resulting in a kappa coefficient "K = 0.815", considered "almost perfect agreement".

Table 2. Results of the annotation task. ORG-ORG (relation between Organizations), ORG-PER (relation between Organization and Persons), ORG-PLC (relation between Organization and Places), NO-REL (no relation between the named entities)

Annotation	Sentences	ORG-ORG	ORG-PER	ORG-PLC	NO-REL
Agreement - 2 annotators	69	8	14	7	40
Agreement - 3 annotators	167	25	40	33	69
Total	236	**33**	**54**	**40**	109

5   Portuguese Data Enrichment

The resulting data from the annotation task (see Table 2) will be used to enrich an annotated corpus for RE from Portuguese described in [6,7]. The authors present a subset of the Golden Collections from the two HAREM[7] conferences, to which they added manual annotation of relations expressed between particular named entities (ORG, PER or PLC), in a total of 516 relation instances. We unified both corpus and added annotations made in Summ-it ++, totalizing 752 relation instances.

6   Conclusion

In this work we show how to build a task for annotation of semantic relations between named entities (ORG, PER or PLC) for Portuguese language. Also, we make available a corpus to train and evaluate algorithms for RE in Portuguese texts. This corpus is the starting point to enrich the NLP community that research the field of RE. All the content of this work is available in PUCRS-PLN Group website: http://www.inf.pucrs.br/linatural/re-annotation.

As future work, we intend to consider other categories of named entities and relations between them in the annotation task. We also intend to apply and evaluate different supervised machine learning techniques for RE to Portuguese using the resulting corpus.

Acknowledgments. We thank the CNPQ, CAPES and FAPERGS for their financial support.

References

1. Abreu, S.C., Bonamigo, T.L., Vieira, R.: A review on relation extraction with an eye on Portuguese. J. Braz. Comput. Soc. **19**(4), 553–571 (2013)
2. Amaral, D.O.F.d., Vieira, R.: NERP-CRF: uma ferramenta para o reconhecimento de entidades nomeadas por meio de conditional random fields. Linguamática **6**, 41–49 (2014)

---

[7] https://www.linguateca.pt/harem.

3. Antonitsch, A., Figueira, A., Amaral, D., Fonseca, E., Vieira, R., Collovini, S.: Summ-it++: an enriched version of the Summ-it corpus. In: Proceedings of 10th Edition of the Language Resources and Evaluation Conference (LREC 2016) (2016)
4. Carletta, J.: Assessing agreement on classification tasks: the kappa statistic. Comput. Linguist. **22**(2), 249–254 (1996)
5. Collovini, S., Carbonel, T.I., Fuchs, J.T., Coelho, J.C., Rino, L., Vieira R.: Summ-it: Um corpus anotado com informações discursivas visando a sumarização automática. In: Proceedings of V Workshop em Tecnologia da Informação e da Linguagem Humana, Rio de Janeiro, RJ, Brasil, pp. 1605–1614 (2007)
6. Collovini, S., Machado, G., Vieira, R.: Extracting and structuring open relations from Portuguese text. In: Silva, J., Ribeiro, R., Quaresma, P., Adami, A., Branco, A. (eds.) PROPOR 2016. LNCS (LNAI), vol. 9727, pp. 153–164. Springer, Cham (2016). https://doi.org/10.1007/978-3-319-41552-9_16
7. de Abreu, S.C., Vieira, R.: ReIP: Portuguese open relation extraction. Knowl. Organ. **44**(3), 163–177 (2017)
8. dos Santos, H.D.P., Woloszyn, V., Vieira, R.: Portuguese personal story analysis and detection in blogs. In: 2017 IEEE/WIC/ACM International Conference on Web Intelligence (WI), Leipzig, Germany, August 2017
9. Finin, T., Murnane, W., Karandikar, A., Keller, N., Martineau, J., Dredze, M.: Annotating named entities in twitter data with crowdsourcing. In: Proceedings of the NAACL HLT 2010 Workshop on Creating Speech and Language Data with Amazon's Mechanical Turk, CSLDAMT 2010, Stroudsburg, PA, USA, pp. 80–88. Association for Computational Linguistics (2010)
10. Kittur, A., Chi, E.H., Suh, B.: Crowdsourcing user studies with mechanical turk. In: Proceedings of the SIGCHI Conference on Human Factors in Computing Systems, pp. 453–456. ACM (2008)
11. Kucherbaev, P., Daniel, F., Tranquillini, S., Marchese, M.: Crowdsourcing processes: a survey of approaches and opportunities. IEEE Int. Comput. **20**(2), 50–56 (2016)
12. Liu, A., Soderland, S., Bragg, J., Lin, C.H., Ling, X., Weld, D.S.: Effective crowd annotation for relation extraction. In: NAACLHLT 2016, The 2016 Conference of the North American Chapter of the Association for Computational Linguistics: Human Language Technologies, San Diego California, USA, pp. 897–906 (2016)
13. Mohammad, S.M., Bravo-Marquez, F.: WASSA-2017 shared task on emotion intensity. In: Proceedings of the Workshop on Computational Approaches to Subjectivity, Sentiment and Social Media Analysis (WASSA), Copenhagen, Denmark (2017)
14. Mortensen, J.M., Musen, M.A., Noy, N.F.: Crowdsourcing the verification of relationships in biomedical ontologies. AMIA Annu. Symp. Proc. **2013**, 1020–1029 (2013)

Automatic Detection of Negated Findings with NooJ: First Results

Walter Koza[1(✉)], Mirian Muñoz[1], Natalia Rivas[1], Ninoska Godoy[1],
Darío Filippo[2], Viviana Cotik[3], Vanesa Stricker[3],
and Ricardo Martínez[4]

[1] Proyecto FONDECyT 1171033, Pontificia Universidad Católica de
Valparaíso, Avenue El Bosque 1290, 2530388 Viña del Mar, Chile
walter.koza@pucv.cl, m.araceli.m.a@gmail.com,
Natalia.rivas.f@gmail.com, ninoshka.godoy@gmail.com
[2] Hospital Pediátrico Garrahan,
Combate de los Pozos 1881, C1245AAM Buenos Aires, Argentina
dfilippo@gmail.com
[3] Departamento de Computación, FCEN, Pabellón I, Ciudad Universitaria,
C1428EGA Buenos Aires, Argentina
{vcotik,vstricker}@dc.uba.ar
[4] Universidad Diego Portales, Vergara 210, 8370067 Santiago, Chile
ricardomartinezg@gmail.com

Abstract. The objective of this study is to develop a methodology for the automatic detection of negated findings in radiological reports which takes into account semantic and syntactic descriptions, as well as morphological and syntactic analysis rules. In order to achieve this goal, a series of rules for processing lexical and syntactic information was elaborated. This required development of an electronic dictionary of medical terminology and computerized grammar. Computational framework was carried out with NooJ, a free software developed by Silberztein, which has various utilities for treating natural language. Results show that the detection of negated findings improves if lexical-grammatical information is added.

Keywords: Negated finding · Syntactic rules · NooJ

1 Introduction

The objective of this study is to develop a methodology for the automatic detection of negated findings in radiological reports which takes into account semantic and syntactic descriptions, as well as morphological and syntactic analysis rules. For this goal, a series of rules for processing lexical and syntactic information was elaborated. This required development of an electronic dictionary of medical terminology and computerized grammar. Pertinent information for the assembly of the specialized dictionary was extracted from the ontology SNOMED CT [2] and a medical dictionary [3]. Likewise, a general language dictionary was also included [4]. Lexicon-Grammar (LG), proposed by Gross [5], was used to set up the database, which allowed an exhaustive description of the argument structure of predicates projected by lexical units. Computational framework was carried out with NooJ, a free software developed by Silberztein [6], which has various utilities for treating natural language, such as morphological and syntactic grammar, as well as dictionaries. This method was compared with NegEx [1] and the results show that the detection of negated findings improves if lexical-grammatical information is added.

2 Method

The method involved two stages: (i) the elaboration of an electronic dictionary; (ii) a formalization of the syntactic structures which organize finding negations.

A compilation of an electronic dictionary that would contemplate lexical units involved in the conformation of negated findings was made. It was considered: (i) morphosyntactic information; (ii) flexional variations, and (iii) semantic information (Table 1).

Table 1. Grammatical and lexical units declared in electronic dictionaries

Units	Category		Example
Grammatical	Negation particles		'no', 'ni', 'ningún', 'sin'
	Determinants		'el', 'la', 'un', 'una'
	Clitic pronouns		'se', 'le', 'lo'
	Auxiliary verbs		'haber', 'poder', 'ser'
	Coordinating conjunctions		'y', 'o'
Lexical	Observational verbs		'observar', 'examinar', 'ver'
	Nouns	Findings	'meteorismo', 'dilatación'
		Anatomical zones	'bazo', 'riñón', 'útero'
		Zones	'zona', 'vía'
		Studies	'radiografía', 'ecografía'
		Other	'madre', 'importancia'
	Adjectives	Findings	'dilatado', 'nodular'
		Anatomical zones	'testicular', 'ventricular'
		Zones	'izquierdo', 'derecho'
		Studies	'radiográfico', 'ecográfico'
		Other	'importante'

The subsequent step was description and, later, the formalization of syntactic structures which contained negated findings. Modeling and the subsequent implementation in machines focused on structures formed by any of the following constructs:

- **[Negation adverb ('no') + 'se' + Observational V (OV)] + N_Finding (NHall) + Anatomic Zone (Zanat):** 'no se detectó dilatación en la vía urinaria'
- **N_Zana + [prep 'sin' + NHall]:** 'vía urinaria sin alteraciones'.

For the negated structures formalization the Lexicon-Grammar (LG) [5] was used. LG allows an exhaustive description of argument structures and transformational possibilities. In this way it is possible to elaborate a computerized grammar which contemplates all possible combinations. Noun phrases for findings (SNHall), are composed of a substantive lexical head, corresponding to the finding category and can be modified by an adjective phrase (SA) and a prepositional phrase (SP), corresponding or not with a Zanat. Rules for regrouping were established for the recognition of various lexical syntaxes involved (Fig. 1):

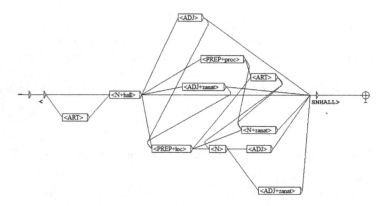

Fig. 1. Syntactic grammar for SNHALL recognition

However, sentences may contain more than one finding in coordination (two findings) or in enumeration (more than two findings). Between each finding it was necessary to contemplate the presence of a comma and the Spanish coordinators 'y' (or 'e') (Fig. 2).

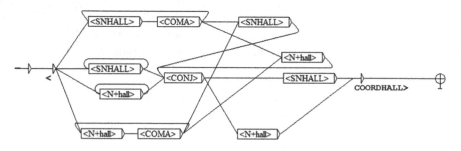

Fig. 2. Syntactic grammar for the recognition of coordinated findings

Finally, grammar was established for negated sentence recognition (Fig. 3).

One of the advantages of considering syntactic structure is a guaranteed high percentage of accuracy. Furthermore, several cases in which errata occur are fixed. To do this, grammar derived from grammatical category deduction were used:

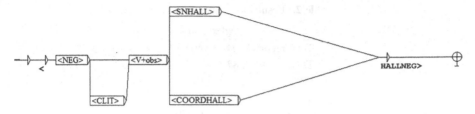

Fig. 3. Syntactic grammar for negated finding recognition

This type of grammar showed in Fig. 4 obeys Spanish language requirements. For example, after a negation followed by a clitic pronoun, the only possible word is a finite verb.

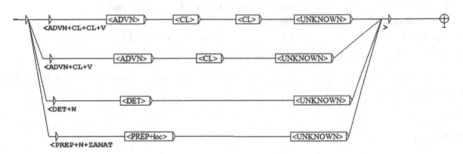

Fig. 4. Productive grammar for recognition of terms not included in the electronic dictionary (fragments)

3   Results

In order to verify NooJ potential in this type of task, a comparison with NegEx was carried out. A total of 535 sentences containing findings were extracted from a corpus of radiological reports, of which 404 contained negations (0.93 of sentences were negated). This comparison was necessary due to preprocessing required by NegEx in previous labeling of findings. Subsequently, results for both algorithms were analyzed, obtaining the following percentages (Table 2):

The NooJ algorithm are slightly higher. Although, three advantages of NooJ over NegEx can be mentioned:

- It does not require prior labeling of findings, but rather it is done simulta- neously and in a single stage, along with detecting negations.
- It allows detecting findings not included in the dictionary (or findings with spelling or typing errors) by deduction from the syntactic context.
- It can identify coordinated findings.

Table 2. Results for negated findings

	NegEx	NooJ
Total reports	404	404
TP	387	384
FP	5	0
Precision	0.987	1
Recall	0.957	0.950
F1	0.968	0.974

4 Future Works

Future works will be organized according to the following axes:

- Continued formalization of syntactic structures of negated findings.
- Enrichment of specialty dictionary.
- Creating more complex rules for deduction of grammatical categories.
- Testing this methodology in a corpus of diverse specialties.

Accordingly, we expect to achieve greater results and propose effective tools for this type of task.

References

1. Chapman, W., Bridewell, W., Hanbury, P., Cooper, G.F., Buchanan, B.: A simple algorithm for identifying negated findings and diseases in discharge summaries. J. Biomed. Inf. **34**(5), 301–310 (2001)
2. https://mlds.ihtsdotools.org/
3. RANM: Diccionario de términos médicos. Editorial Médica Panamericana, Buenos Aires (2012)
4. RAE: Diccionario de la lengua Española. Santillana, Bogotá (2014)
5. Gross, M.: Méthodes en syntaxe. Hermann, París (1975)
6. Silberztein, M.: Formalizing Natural Languages: The NooJ approach. ISTE, Londres (2016)

Part-of-Speech Tagger
for Konkani-English Code-Mixed
Social Media Text

Akshata Phadte$^{(\boxtimes)}$ and Radhiya Arsekar$^{(\boxtimes)}$

Department of Computer Science and Technology, Goa University,
Taleigao Plateau, Goa, India
akshataph07@gmail.com, radhiya.arsekar@gmail.com

Abstract. In this paper, we propose efficient and less resource-intensive strategies for Konkani-English code-mixed social media text. which witnesses several challenges as compared to tagging general normal text. Part-of-Speech Tagging is a primary and an important step for many Natural Language Processing Applications. This paper reports work on annotating code-mixed Konkani-English data collected from social media site Facebook, which consists of more than four thousands posts from Facebook and developed automatic Part-of-Speech Taggers for this corpus. Part-of-Speech tagging is considered as a classification problem and we use different classifiers such as CRFs, SVM with different combinations of features.

Keywords: Code-mixing · Social media text · Part-of-Speech tagging

1 Introduction

India is a land of many languages. People on social media often use more than one language to express themselves. But the problem starts with Multilingual speakers tend to exhibit code-mixing and code-switching in their use of language on social media platforms. Code-mixing refers to the mixing of two or more languages or language varieties in speech. Code-mixing occurs due to various reasons. According to a work by [1], the major reasons for Code-Mixing:- 45% Real lexical needs, 40% Talking about a particular topic and 5% For content clarification. [3] noted that the complexity in analyzing code-mixed social media text (CMST) stems from non adherence to a formal grammar, spelling variations, lack of annotated data, inherent conversational nature of the text and of-course, code-mixing. Therefore, there is a need to create datasets and Natural Language Processing tools for code-mixed social media text as traditional tools are ill-equipped for it. Taking a step in this direction, we present our work on building a POS tagger for Konkani-English code-mixed data collected from social media site Facebook.

M. Silberztein et al. (Eds.): NLDB 2018, LNCS 10859, pp. 303–307, 2018.
https://doi.org/10.1007/978-3-319-91947-8_31

2   Related Work

Code-mixing being a relatively newer phenomena has gained attention of researchers only in the past two decades. POS taggers on monolingual data give an accuracy of about 97.3% for English text [6]. They are often seen as sequence labeling problems and have used the context based information in the form of lexical and sub-lexical characteristics of neighboring words. But in code-mixed setting, the context information can be in a different language which makes the understanding difficult. [3] reported challenges in processing Hindi-English CMST and performed initial experiments on POS tagging. Their POS tagger accuracy fell by 14% to 65% without using gold language labels and normalization. Thus, language identification and normalization are critical for POS tagging [3]. [7] also built a POS tagger for Hindi-English CMST using Random Forests on 2,583 utterances with gold language labels and achieved an accuracy of 79.8%.

[8] further improved this POS tagger, increasing the accuracy to 93%. [11] worked on a complete pipeline for shallow parsing and performed tokenisation, language identification, normalisation, POS tagging and finally, shallow parsing and achieved accuracy of 83.4% for code-mixed Hindi-English social media text.

3   Data Preparation

Significant studies and dataset of the code-mixing phenomenon can be found in [2]. These works discuss the dataset preparation and dataset statistics of code-mixing of Konkani-English as well as its linguistic nature. For the POS tagging of Konkani-English language we extracted the code-mixed corpus which was discussed in [2]. We then manually tagged them by their language, normalisation form and by their POS tags.

3.1   Dataset Annotation Guidelines

The creation of this linguistic resource involved Language identification, Normalisation and POS tagger layer. The following paragraphs describe the annotation guidelines for these tasks in detail.

1. **Language Identification:** Every word was given a tag out of three - en, kn and rest to mark its language. Words that a bilingual speaker could identify as belonging to either Konkani or English were marked as 'kn' or 'en', respectively. The label 'rest' was given to symbols, emoticons, punctuation, named entities, acronyms and foreign words.
2. **Normalisation:** Words with language tag 'kn' in Roman script were labeled with their standard form in the native script of Konkani Devanagari, i.e. a back-transliteration was performed. Words with language tag 'en' were labeled with their standard spelling. Words with language tag 'rest' were kept as they are.

3. **Part-of-Speech Tagging:** The universal Part-of-speech tagset [9] was used to label the POS of each word as this tagset is applicable to both English and Konkani words, and it contained a level of coarseness that suited our goals. The following case-specific guidelines were also observed:
    1. Sub-lexical code-mixed words were annotated based on their context, since POS is a function of a word in a given context.
    2. Words embedded in a sentence of another language were tagged as per context of the matrix language, irrespective of the POS tag of the word in its original language.

4 Experiments and Results

Here, we repeated the experiments performed by [2] and added new part to system. The original system first tokenizes an utterance into words. Then, a language identification module classifies each word as Konkani, English or Rest. Based on the language assigned, the Normalisation module runs the Konkani or English normalisers. In this section, we explain the POS Tagging system used after the Normalisation system.

4.1 Part-of-Speech Tagging System

Understanding the Part-of-Speech POS tagging, which provides a basic level of syntactic analysis for a given word or sentence. It was modeled as a sequence labeling task using CRFs [10] and SVM following paper. The feature set comprised of:

1. **Basic Word Features:** Word based features such as affixes, context and the word itself.
2. **LANG:** Language label of the token, obtained from the Language Identification system. This can have the values - 'en', 'kn' or 'rest'.
3. **NORM:** Lexical features extracted from the normalised form of the word. These include linguistic features such as bound and free morphemes, suffixes, prefixes.
4. **TPOS:** Output of Twitter POS tagger [8] for the given word.
5. **KPOS:** Output of Konkani POS tagger[1] for the given word.

To obtain the Konkani POS tag output, the output from the normalisation module was used by transliterating Romanised Konkani words into WX-notations. Konkani POS tags were obtained using the Cdac Konkani POS tagger http:// kbcs.in/tools.html. This POS Tagger is trained on WX-notation, thus English and 'Rest' words were transliterated to WX-notation. These transliterations along with the Konkani normalised data was sent to the POS tagger and final POS tag was obtained. The features ablation for the POS Tagger are shown in Sect. 4.1. Each feature was added only if it showed a positive increase in the system accuracy. Table 1 presents the obtained results.

---

[1] http://kbcs.in/tools.html.

Table 1. Token level POS Tagger Accuracy

Features	CRF Accuracy (%)	SVM Accuracy (%)
BASELINE	85.53	85.73
+LANG	86.53	87.53
+NORM	87.72	88.53
+TPOS	90.59	91.53
+KPOS	91.47	92.53

5 Conclusion and Future Work

In this Paper, we have focused on building first step of shallow parser for Konkani-English code-mixed social data. Through this paper we present our efforts at attempting various statistical methods for POS tagging of code-mixed social media data. We have attempted to build Part-of-speech tagger for this language pair, which we hope would result in better data-mining and sentiment analysis across the Indian subcontinent. We also create a standard dataset of 5088 code-mixed Konkani-English sentences for building supervised models of shallow parsing on this data which we consider as our immediate future work. In the future, we intend to continue creating more annotated code-mixed social media data. We intend to use this dataset to build tools for code-mixed data like morph analysers, chunkers and parsers.

References

1. Hidayat, T.: An analysis of code switching used by Facebookers (a case study in a social network site). Sekolah Tinggi Keguruan dan Ilmu Pendidikan (STKIP) Siliwangi Bandung (2012)
2. Phadte, A., Thakkar, G.: Towards normalising Konkani-English code-mixed social media text. In: Proceedings of the 14th International Conference on Natural Language Processing (ICON-2017), pp. 85–94 (2017)
3. Vyas, Y., Gella, S., Sharma, J., Bali, K., Choudhury, M.: Pos tagging of English-Hindi code-mixed social media content. In: EMNLP, vol. 14, pp. 974–979 (2014)
4. Dey, A. Fung, P.: A Hindi-English code-switching corpus. In: LREC, pp. 2410–2413 (2014)
5. Bali, K., Vyas, Y., Sharma, J., Choudhury, M.: I am borrowing ya mixing? an analysis of English-Hindi code-mixing in Facebook. In: Proceedings of the First Workshop on Computational Approaches to code Switching, EMNLP 2014 (2014)
6. Rao, D., Yarowsky, D.: Part of speech tagging and shallow parsing of Indian languages. Shallow Parsing for South Asian Languages (2007)
7. Jamatia, A., Gambäck, B., Das, A.: Part-of-speech tagging for code-mixed English-Hindi Twitter and Facebook chat messages. In: RANLP, pp. 239–248 (2015)
8. Owoputi, O., O'Connor, B., Dyer, C., Gimpel, K., Schneider, N., Smith, N.A.: Improved part-of-speech tagging for online conversational text with word clusters. Association for Computational Linguistics (2013)

9. Petrov, S., Das, D., McDonald, R.: A universal part-of-speech tagset (2011). arXiv preprint. arXiv:1104.2086
10. Lafferty, J., McCallum, A., Pereira, F.C.N.: Conditional random fields: probabilistic models for segmenting and labeling sequence data. In: Proceedings of the Eighteenth International Conference on Machine Learning, ICML, vol. 1, pp. 282–289 (2001)
11. Sharma, A., Gupta, S., Motlani, R., Bansal, P., Srivastava, M., Mamidi, R., Sharma, D.M.: Shallow parsing pipeline for hindi-english code-mixed social media text (2016)

Word2vec for Arabic Word Sense Disambiguation

Rim Laatar[(⊠)], Chafik Aloulou, and Lamia Hadrich Belghuith

ANLP Research Group, MIR@CL Lab, Faculty of Economics and Management,
University of Sfax, Sfax, Tunisia
rimlaatar@yahoo.fr, {chafik.aloulou,
l.belguith}@fsegs.rnu.tn

Abstract. Word embedding, where words are represented as vectors in a continuous space, has recently attracted much attention in natural language processing tasks due to their ability to capture semantic and syntactic relations between words from a huge amount of text. In this work, we will focus on how word embedding can be used in Arabic word sense disambiguation (WSD).

Keywords: Word sense disambiguation · Word embedding · Word2vec
Arabic · NLP

1 Introduction

In the last few years, many researchers became interested in representing the language words as vectors in a multidimensional space. As a result, many NLP applications benefited from these representations.

One of the most used techniques for building word representation in vector space are the models proposed by Mikolov et al. [1]: continuous bag of word (CBOW) and Skip gram models.

Although most of the work which represents words as vectors focuses on the English language, many researchers have recently built word embedding model for the Arabic language.

When dealing with Arabic, removing diacritics is the first step to build the word embedding model.

Within the context of word disambiguation in Arabic, the different pronunciations of one word may have different meanings. In fact, these small signs added to letters let readers distinguish between similar words. However, dropping diacritics from some words may lead to many lexical ambiguities.

Our objective is therefore to build a word embedding model without omitting diacritics from words. This paper describes the various steps followed to create a word embedding model using both CBOW and Skip gram techniques. We will analyze the role of the different training parameters in the WSD performance.

The rest of this paper is divided as follows. Section 2 will deal with the construction of word embeddings model for Arabic. In Sect. 3, we will give a detailed account of the proposed method to disambiguate Arabic words. In Sect. 4, we will

© Springer International Publishing AG, part of Springer Nature 2018
M. Silberztein et al. (Eds.): NLDB 2018, LNCS 10859, pp. 308–311, 2018.
https://doi.org/10.1007/978-3-319-91947-8_32

present the experimental results and finally in Sect. 5, we will draw a conclusion and suggest some future work ideas.

2 Arabic Word Embeddings

Our objective is to map the Arabic words to real-number vectors without omitting their diacritics and evaluate word representations using skip gram and CBOW models in order to choose the best architecture to generate better word embedding model for Arabic Word Sense Disambiguation.

In this work, we have utilized The Historical Arabic Dictionary Corpus [2]. We have also added to this corpus about 200 texts extracted from Arabic Wiki Source.

The processing steps for cleaning and normalizing the corpus are as follows: Remove punctuation, and non Arabic words. Our word vectors were trained using the Skip-gram and CBOW models.

3 Methodology

In this section, we are going to describe our method to disambiguate Arabic words using word embeddings. This method is made up of two stages. The first one consists in representing the context of the word to be disambiguated together with its different definitions with the help of vectors. The second one is about the calculation of the similarities between the context vector and each definition vectors.

3.1 Context and Sense Representation

The context vector of an ambiguous word is made up of a concatenation of vectors of the words surrounding a target word. The different meanings of the ambiguous word are extracted from Alwaseet Arabic dictionary[1].

Now that the definition of the word to be disambiguated is obtained, we will also represent the sense as a vector by concatenating the word vectors which constitute the definition of the ambiguous word.

The main problem we faced is that we sometimes find some definition vectors which are null and can meanwhile reflect the final result. This could be explained by the fact that the lexicon contains some discretized and undiscretized words. In fact, we may find two words with the same non diacritics letters and different diacritics that have the same meaning. To overcome this problem, we chose to apply the Implication Relation Algorithm (IRA). IRA is an algorithm that aims at comparing two words with the same non-diacritics letters and decides whether they are the same or not [3]. The conclusion is that if the definition vector is null, we are obliged therefore to compare the words which make the definition of the ambiguous word with the vocabulary words and with the same non diacritics letters and different diacritics using IRA algorithm.

---

[1] Awaseet Dictionary, The Arabic Union Council in Cairo, 2004.

According to the sum of words obtained, we can change the word of a null vector with the word that has the nearest vector of the ambiguous word.

3.2 Similarity Using Cosine Distance

To attribute for each ambiguous word its appropriate sense, we choose the sense with the closest semantic similarity to its local context. To measure the similarity between context vector and sense vectors, we use a cosine distance metric.

4 Experimental Results

The experiences we are presenting in this article are based on the analysis of the role of the different training parameters of embedding can play on the WSD performance.

For the disambiguation of the Arabic language, we need to have an Arabic-Arabic dictionary which contains the different meanings of the ambiguous words. For this purpose, we have opted for «Alwaseet» dictionary. To evaluate the proposed method, OSAC corpus [4] has been used.

We have tested about 100 ambiguous words. For each ambiguous word, we have evaluated 100 contexts extracted from OSAC corpus. We have used Word2vec[2] toolkit to learn vectors.

The following table shows results comparing the skip gram and CBOW models as well as the best configuration chosen to run our evaluations (Table 1).

The skip-gram vectors show better results than CBOW in Arabic Word sense disambiguation.

Figures 1, 2 and 3 shows the impact of different training parameters on WSD performance using skip gram architecture.

The result shows that using Skip gram architecture by combining a 10 window-size with a 150 vector dimension and 5 negative examples produced the best model among all tested parameters.

Experiments also show that employing the IRA algorithm permits to ameliorate the precision from 51, 52 to **52, 32**. It is concluded then that using the IRA algorithm can help us avoid having a definition with a null vector and thus rise the probability to find the appropriate sense of the ambiguous word.

5 Conclusion

In this study we have shown how word embedding can be used to solve word sense disambiguation for the Arabic language. We have tested two popular word representation techniques, skip gram and CBOW. Skip gram obtained the best result for Arabic word sense disambiguation. In addition to that, the combination of a 10 window-size with a 150 vector dimension and 5 negative examples produced the best results among all the tested parameter configurations.

---

[2] http://code.google.com/archive/p/word2vec/.

Table 1. WSD performance using Skip gram and CBOW models

Models	Dimensionality	Window size	Negative	WSD precision
Skip gram	150	10	5	51,52
CBOW	200	5	10	50,25

Fig. 1. Dimentionality   **Fig. 2.** Window size

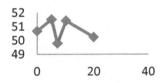

Fig. 3. Negative examples

To disambiguate Arabic words, the proposed method consists of measuring the semantic relation between the context of use of the ambiguous word and its sense definitions. The use of The Historical Arabic Dictionary Corpus remains open for future work. In fact, in addition to disambiguate words in Modern Standard Arabic, we can focus on disambiguating words in classical Arabic and study the evolution of the sense of a word throughout history.

References

1. Mikolov, T., Sutskever, I., Chen, K., Corrado, G.S., Dean, J.: Distributed representations of words and phrases and their compositionality. In: Advances in neural information processing systems (2013)
2. Al-Said, A.B., Medea-García, L.: The historical arabic dictionary corpus and its suitability for a grammaticalization approach. In: 5th International Conference in Linguistics (2014)
3. Jarrar, M., Zaraket, F.A., Asia, R.: Diacritic-aware arabic word matching. Int. Arab J. Inf. Technol (2016)
4. Saad, K.M., Ashour, W.: OSAC: open source arabic corpora. In: International Conference on Electrical and Computer Systems (2010)

Text Similarities and Plagiarism Detection

HYPLAG: Hybrid Arabic Text Plagiarism Detection System

Bilal Ghanem[1(⊠)], Labib Arafeh[2], Paolo Rosso[3],
and Fernando Sánchez-Vega[4]

[1] Arab American University, Jenin, Palestine
bi.ghanem@outlook.com
[2] Arab Open University, Ramallah, Palestine
larafeh@staff.alquds.edu
[3] Universitat Politècnica de València, Valencia, Spain
prosso@dsic.upv.es
[4] Instituto Nacional de Astrofísica, Óptica y Electrónica, Puebla, Mexico
fer.callotl@ccc.inaoep.mx

Abstract. Plagiarism is specifically defined as literary theft of paragraphs or sentences from unreferenced source. This unauthorized behavior is a real problem that targets scientific research scope. This paper proposes a Hybrid Arabic Plagiarism Detection System (HYPLAG). The HYPLAG approach combines corpus-based and knowledge-based approaches by utilizing an Arabic semantic resource (Arabic WordNet). A preliminary study on texts from undergraduate students was conducted to understand their behavior and the patterns used in plagiarism. The results of the study show that students apply different techniques to plagiarized sentences, also it shows changes in sentence's components (verbs, nouns, and adjectives). HYPLAG was evaluated on the ExAraPlagDet-2015 dataset against several other approaches that participated in the AraPlagDet PAN@FIRE shared task on Extrinsic Arabic plagiarism detection obtaining a higher performance (F-score 89% vs. 84% obtained by the best performing system at AraPlagDet) with less computational time.

Keywords: Plagiarism detection · Arabic language · Arabic WordNet

1 Introduction

With the advancement of Internet, search engines and e-publishing tools, plagiarism detection is becoming more and more important. Worldwide, a huge number of researchers have published their researches in different journals, conferences and digital libraries. Plagiarism detection is an complicated process, especially in the Arabic language since it is complex language morphologically and there is no specific criterion for plagiarism cases, which made it more difficult task for systems to be enough reliable and sensitive to detect plagiarized cases. Many statistics[1,2,3] have emphasized that plagiarism

---

[1] http://www.plagiarism.org/resources/facts-and-stats/; accessed on October 2016.
[2] http://www.checkforplagiarism.net/cyber-plagiarism; accessed on October 2016.
[3] https://infogr.am/Plagiarism-606324; accessed on October 2016.

© Springer International Publishing AG, part of Springer Nature 2018
M. Silberztein et al. (Eds.): NLDB 2018, LNCS 10859, pp. 315–323, 2018.
https://doi.org/10.1007/978-3-319-91947-8_33

is a real problem and it poses risk on literary content. A set of plagiarism detection tools have been proposed, based on different approaches for a variety of languages. However, fewer systems have been proposed for the Arabic language due to its complexity and rich morphological with complex linguistic structure. In this paper, we will address extrinsic plagiarism detection. The rest of the paper is structured as follows. Section 2, related work, focusing on Arabic plagiarism detection. The HYPLAG is described in Sect. 3. Section 4 presents the experiments and discusses the results. Finally, in Sect. 5 we draw some conclusions.

2 Related Work

Different systems have been proposed in the Arabic domain. Generally, most of these proposed systems tried to tackle the problem using fingerprinting and Tf-Idf techniques. An overview for the proposed systems in literature are summarized in Table 1. In the RDI method, an extrinsic plagiarism detection approach was proposed [1] for an Arabic language plagiarism detection shared task "ExAraPlagDet-2015". It attains the first three positions with a value of 92% precision and value of 84% recall. The proposed approach consists of three base models, document retrieval, alignment model and filtering model. None of the proposed systems has employed a knowledge-based measure to infer the plagiarized cases semantically. Based on that, we will combine knowledge-based with corpus-based information to investigate its affectivity on Arabic plagiarism cases.

3 The Hybrid Arabic Plagiarism Detection System

3.1 Pre-preprocessing and Indexing

In the first phase, each input sentence is processed to remove diacritics, non-Arabic letters, all words that contain numbers and words that are composed of one letter to reduce the error rate. Then, named entities were extracted to be excluded from the next process which is stemming[4]. We stemmed the input words to carry out the similarity at word level. Based on current comparative studies, we have chosen the Motaz stemmer [10] to be used in our approach. Finally, POS tagging was applied to tag the main sentence components that we have used in the comparison process. In our approach, we used Farasa POS tagger [11].

3.2 System Architecture

HYPLAG is comprised of several components (see Fig. 1), and in each component a set of methods are applied. The proposed approach has been designed to detect all

---

[4] We used Farasa NER tool for named entity recognition, http://qatsdemo.cloudapp.net/farasa/; accessed on December 2016.

Table 1. Overview of related systems

System	Approach	Processing level	Knowledge resources	Knowledge-based measures	Dataset	Performance	
						Precision	Recall
ADP [8]	Fingerprinting	Chunk-based	No	No	ExAraPlagDet-2015	83%	53%
SFF [9]	Fingerprinting and fuzzy set	Statement-based	Yes	No	Own	90%	85%
APag [10]	Fingerprinting	All	Yes	No	Own	93%	100%
Iqtebas [4]	Fingerprinting with windowing	Chunk-based	No	No	Own	99%	94%
ZPlag [5]	Fingerprinting with windowing	Chunk-based	Yes	No	Own	97%	94%
FPDA [3]	Feed windowed query to Google SE	Chunk-based	No	No	Google engine	–	–
RDI [2]	Idf weights with windowing	Chunk-based	No	No	ExAraPlagDet-2015	92%	84%
PDSA [6]	Tf-Idf matrix with singular value decomposition algorithm	Document-based	Yes	No	Own	97%	
ETFIDF [7]	Tf-Idf	Statement-based	Yes	No	Own	90%	92%

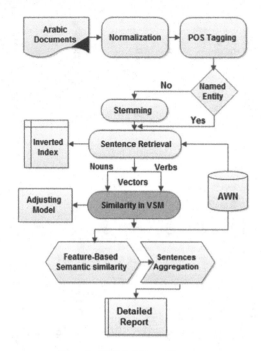

Fig. 1. HYPLAG architecture

types of plagiarism which include copy & paste, paraphrasing, and synonym substitution.

3.2.1   Sentence Ranking

The HYPLAG approach is based on search engine structure. This component mainly depends on a ranked query technique to retrieve target sentence. An inverted index is constructed using source documents that were split to length of n-word sentences. For each input sentence, using a POS tagger, sentence components were extracted and their synonyms were retrieved from Arabic WordNet (AWN)[5]. Then, each term was sent to the ranking component where the score of those sentences that contain this term will be increased. So, after passing all terms with their synonyms, the sentence that is the most related to the terms will have the highest rank. In our implementation, the highest ranked sentence should contain at least (t - threshold) terms, where t is the number of input terms without their synonyms. The threshold value has been set empirically.

3.2.2   Term Frequency × Inverse Document Frequency Weighting

This technique is widely used in NLP domains and confirmed its success and effectiveness. For each input and retrieved sentence, a two vectors (nouns & verbs vectors) are constructed. Using Td-Idf terms weights, we measured the similarity between the

---

[5] http://globalwordnet.org/arabic-wordnet/; accessed on November 2016.

two verbs vectors using the cosine similarity measure. The same process is applied for nouns vectors and one score value is resulted by applying weighted average.

After producing the final score, we will have one of three cases:

- If (final_score < min_threshold) then discard.
- If (final_score > max_threshold) then plagiarized sentence is detected.
- If (final_score > min_threshold & final_score < max_threshould) then verbs & nouns vectors are passed to Feature-based semantic similarity.

During the comparison process both sentences are adjusted in the adjusting model by K-overlapped terms once from previous sentence, and once from the next to get best match score. The Tf-Idf technique has detect both basic methods of plagiarism (copy & paste and paraphrasing), but it can't cover synonyms replacement. Therefore, we have coupled it with Feature-based semantic measure.

3.2.3   Feature-Based Semantic Similarity

Many knowledge-based semantic similarity measures have been proposed in the literature and they are used in different domains. These measures can be grouped mainly to three main classes: Path length based measures, information content based measures and feature based measures [12]. In our approach we have employed feature based measure which is based on the assumption that each concept is described by a set of words. Therefore, the more common features two concepts have, the more similar the concepts are [13]. One of the main measures is Tversky's one [14], but the problem with it is that the more unique features a concept presents the lower similarity is. Therefore, we have adopted a ratio model of Tversky's formula [15]:

$$sim_{tvr-ratio}(c_1, c_2) = \frac{F(\Psi(c_1) \cap \Psi(c_2))}{\beta F(\Psi(c_1) \backslash \Psi(c_2)) + \gamma F(\Psi(c_2) \backslash \Psi(c_2)) + F(\Psi(c_1) \cap \Psi(c_2))} \quad (1)$$

Where c_i is a concept, $\Psi(C_i)$ is a feature set of the concept c_i (The synsets of a concept), B and Υ are values between 0–1 and have been set experimentally.

From the previous step, we have received two vectors for each sentence (nouns and verbs vectors for the input sentence, and another two vectors for the retrieved one). For each vector we build a matrix to represent it with its synonyms. The matrix size for each vector depends on the number of terms in its correspondent vector. The similarity measuring process of the same component matrices are started by reading the first row (vector) of the first matrix with the first row of the second one and passing them to Tversky's measure. Finally, we computed the average for the generated results. After the comparison process for both couples of matrices achieved, the weighted average of results has computed.

4   Evaluation

The performance of our approach was tested on the corpus of Extrinsic Arabic plagiarism detection shared task 2015. Our approach has achieved a better performance. Since we used different sentence components in our system, we conducted a preliminary study on a texts form a sample of students to investigate the behavior of components from real samples. In the following section, we will describe our study.

4.1   Preliminary Study

A brief study on undergraduate students before system implementation has been conducted to study student's plagiarism behavior in changing sentence components. Our study was conducted in a course entitled "Fundaments of Research Methods" offered by the Arab American University, Palestine. We have chosen this course for different factors: The nature of this course is diversified which contains students from different studying levels (1–5 years), also the diversity of specialties is high since this course contains students from different faculties in the university. So, we have prepared a questionnaire by adding a long sentence, and we have ensured that it has diversity in components to allow all possibilities of changes in the sentence. The sample consists of 59 students, 4 students have not changed anything (not participated). For the rest 55 students, we have detected 34 verbs replacement cases, 15 nouns replacement cases and 5 cases of adjectives replacement. Also, we have detected a 29 students have paraphrased sentences. Based on components synonym replacement results, we have developed our approach. In ranked query component, we have used verbs, nouns and adjectives to retrieve the most relevant where in sentence similarity we used just verbs and nouns without adjectives since it has been replaced in a fewer number comparing to other components, also to reduce the computation cost in the comparison process.

4.2   Dataset

ExAraPlagDet shared task has provides two datasets[6], one for training and the other for the final testing. There are two types of plagiarism cases in the datasets, artificial (created automatically) and simulated (created manually). For the first type they used phrase shuffling and word shuffling as obfuscation strategies and for the second one, synonym substitution and paraphrasing are conducted. A statistical details of the corpuses is shown in [16].

4.3   Performance Measure

Using different threshold values, we have performed different runs to achieve the best result in precision and recall. On the training corpus, our approach has achieved 96% of precision and 96% for recall. The training corpus consists of 4 main documents: no-plagiarism, no-obfuscation (copy & paste), artificial-obfuscation and simulated-obfuscation. We have measured our approach on each independently and it achieved the mentioned results (see Fig. 2) for each dataset type. For the second dataset (testing corpus), HYPLAG has achieved 92% as precision value and 87% value of recall. To measure the effectiveness of coupling feature-based semantic measure with TF-IDF technique, we have disabled feature-based model and re-measured the

---

[6] http://misc-umc.org/AraPlagDet/?i=1; accessed on September 2016.

performance over the simulated-obfuscation type since it represents manual synonym substitution cases and paraphrasing which are the most related to real cases. As we see in Fig. 2, the precision and recall for simulated-obfuscation type is 93% and 89%, and after disabling feature-based model the precision value is 83% and the value of recall is 87%. Coupling both techniques has improved the results clearly, and we can recognize the improvements by running the approach on more complicated synonym substitution cases. In our approach we have built two vectors for each sentence, one for nouns and the other for verbs to carry out the comparison process on each component independently. This way may decrease the system performance since that in paraphrased plagiarism some verbs could be converted to nouns and vice versa, to mislead the detection system. Therefore, we conducted another run over the simulated-obfuscation subset to investigate the validity of it, combining both verbs and nouns for each sentence into one vector. HYPLAG has achieved 65% value of precision and 75% for recall, which shows clearly that comparing each component independently is better than combining both of them. This reveals to us that, the paraphrasing behavior that replaces these components with each other is not prevalent in plagiarism of Arabic text.

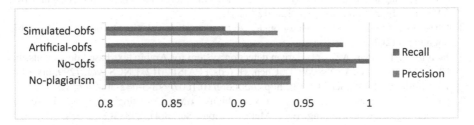

Fig. 2. Performance results on training corpus

4.4 Comparative Results

Based on the results of ExAraPlagDet shared task, RDI has achieved the best performance with 85% value of precision and 83% value of recall. Using the F-score measure and based on execution times that summarized in Table 2, we can conclude that HYPLAG clearly overcomes RDI system in detection accuracy on both corpora and in execution time period.

Table 2. Competition against ExAraPlagDet participants

Rank	Approaches	Precision	Recall	F-score	Execution time (hours)
–	HYPLAG	92%	87%	89%	39.98
ExAraPlagDet participants					
1	RDI_2	85%	83%	84%	44
2	RDI_3	85%	75%	80%	41
3	RDI_1	80%	78%	79%	43.79
...					
8	Palkovskii_2	56%	59%	58%	–

5 Conclusion

In this paper, we have proposed a Hybrid Arabic plagiarism detection system that couples both corpus-based and knowledge-based approaches to overcome main plagiarism scenarios. The proposed system is based on making use of sentence components such as verbs, nouns and adjectives to detect plagiarism cases. This approach has been adopted based on a preliminary study on a sample of university students to study their behavior on plagiarism. HYPLAG shows during experimental results that outperform RDI that showed to be the best performing system at PAN@ Fire 2015.

Acknowledgment. The work of Paolo Rosso was funded by the SomEMBED TIN2015-71147-C2-1-P MINECO research project.

References

1. Magooda, A., Mahgoub, A.Y., Rashwan, M., Fayek, M.B., Raafat, H.: RDI system for extrinsic plagiarism detection (RDI_RED), working notes for PANAraPlagDet at FIRE 2015. In: FIRE Workshops, pp. 126–128 (2015)
2. Khan, I.H., Siddiqui, M.A., Mansoor, K.: A framework for plagiarism detection in Arabic documents (2015)
3. Jadalla, A., Elnagar, A.: A plagiarism detection system for Arabic text-based documents. In: Chau, M., Wang, G.A., Yue, W.T., Chen, H. (eds.) PAISI 2012. LNCS, vol. 7299, pp. 145–153. Springer, Heidelberg (2012). https://doi.org/10.1007/978-3-642-30428-6_12
4. Farahat, F.F., Asem, A.S., Zaher, M.A., Fahiem, A.M.: Detecting plagiarism in Arabic E-Learning using text mining. Br. J. Math. Comput. Sci. **8**(4), 298–308 (2015)
5. Hussein, A.S.: A plagiarism detection system for Arabic documents. In: Filev, D., Jabłkowski, J., Kacprzyk, J., Krawczak, M., Popchev, I., Rutkowski, L., Sgurev, V., Sotirova, E., Szynkarczyk, P., Zadrozny, S. (eds.) Intelligent Systems'2014. AISC, vol. 323, pp. 541–552. Springer, Cham (2015). https://doi.org/10.1007/978-3-319-11310-4_47
6. Yousef, A.A., Aziz, M.J.: Enhanced Tf-Idf weighting scheme for plagiarism detection model for Arabic language. Aust. J. Basic Appl. Sci. **9**(23), 90–96 (2015)
7. Alzahrani, S.: Arabic plagiarism detection using word correlation in N-Grams with K-overlapping approach, working notes for PAN-AraPlagDet at FIRE 2015. In: FIRE Workshops (2015)
8. Alzahrani, S., Salim, N.: Statement-based fuzzy-set information retrieval versus fingerprints matching for plagiarism detection in Arabic documents. In: 5th Postgraduate Annual Research Seminar (PARS 2009), pp. 267–268 (2009)
9. Menai, M.E.B.: Detection of plagiarism in Arabic documents. Int. J. Inf. Technol. Comput. Sci. (IJITCS) **4**(10), 80 (2012)
10. Saad, M.K., Ashour, W.: Arabic morphological tools for text mining. In: 6th ArchEng International Symposiums, EEECS 2010, The 6th International Symposium on Electrical and Electronics Engineering and Computer Science, p. 19. European University of Lefke, Cyprus (2010)
11. Zhang, Y., Li, C., Barzilay, R., Darwish, K.: Randomized greedy inference for joint segmentation, POS tagging and dependency parsing. In: HLT-NAACL, pp. 42–52 (2015)
12. Sánchez, D., Batet, M., Isern, D., Valls, A.: Ontology-based semantic similarity: a new feature-based approach. Expert Syst. Appl. **39**(9), 7718–7728 (2012)

13. Meng, L., Huang, R., Gu, J.: A review of semantic similarity measures in WordNet. Int. J. Hybrid Inf. Technol. **6**(1), 1–12 (2013)
14. Tversky, A.: Features of similarity. Psychol. Rev. **84**(4), 327 (1977)
15. Pirró, G., Euzenat, J.: A feature and information theoretic framework for semantic similarity and relatedness. In: Patel-Schneider, P.F., Pan, Y., Hitzler, P., Mika, P., Zhang, L., Pan, J.Z., Horrocks, I., Glimm, B. (eds.) ISWC 2010. LNCS, vol. 6496, pp. 615–630. Springer, Heidelberg (2010). https://doi.org/10.1007/978-3-642-17746-0_39
16. Bensalem, I., Boukhalfa, I., Rosso, P., Abouenour, L., Darwish, K., Chikhi, S.: Overview of the AraPlagDet PAN@ FIRE2015 shared task on Arabic plagiarism detection. In: FIRE Workshops, pp. 111–122 (2015)

On the Semantic Similarity of Disease Mentions in MEDLINE® and Twitter

Camilo Thorne(✉) and Roman Klinger

Institut für Maschinelle Sprachverarbeitung (IMS), University of Stuttgart, Stuttgart,
Germany
{camilo.thorne,roman.klinger}@ims.uni-stuttgart.de

Abstract. Social media mining is becoming an important technique to
track the spread of infectious diseases and to understand specific needs
of people affected by a medical condition. A common approach is to
select a variety of synonyms for a disease derived from scientific literature to then retrieve social media posts for subsequent analysis. With this
paper, we question the underlying assumption that user-generated text
always makes use of such names, or assigns them the same meaning as
in scientific literature. We analyze the most frequently used concepts in
MEDLINE® for semantic similarity to Twitter use and compare their normalized entropy and cosine similarities based on a simple distributional
model. We find that diseases are referred to in semantically different
ways in both corpora, a difference that increases in inverse proportion
to the frequency of the synonym, and of the commonness of the disease
or condition. These results imply that, when sampling social media for
disease-related micro-blogs, query expressions must be carefully chosen,
and even more so for rarily mentioned diseases or conditions.

Keywords: Social media mining · Twitter · MEDLINE®
Disease names

1 Introduction

Named entity recognition (NER) is a well-established task in biomedical information extraction. It covers a wide variety of entity classes that can be tackled,
for instance: gene and protein names [16], chemical names [7], drug names [6],
or disease names [3]. Diseases are specifically interesting for a number of healthcare information extraction tasks related to, *e.g.*, pharmacovigilance [8,12,14,17],
where social media corpora need to be processed. A key goal in pharmacovigilance
is to detect if a disease or condition is spawned by a particular drug or medication,
by tracking the way it is mentioned in social media over time. Another important
task is to determine which are the most salient diseases and disease names. One
key assumption is that diseases are referred to, used and crucially *meant* in social
media – hence, by laymen – in a manner similar to scientific literature. Hence, it is
on the one hand sufficient to apply entity recognition models and methods trained

© Springer International Publishing AG, part of Springer Nature 2018
M. Silberztein et al. (Eds.): NLDB 2018, LNCS 10859, pp. 324–332, 2018.
https://doi.org/10.1007/978-3-319-91947-8_34

on scientific text, and on the other hand to reuse scientific terminology to query for biomedical-related microblogs such as tweets [8].

Furthermore, it has been observed that language in social media is more metaphorical, more prone to typos and newly coined words than more formal texts [15]. For instance in Twitter, people variously refer to schizophrenia as "schizo", "derangement", "bipolar disorder", "delusional disorder", and use "schizophrenia" in a metaphorical manner (unrelated to health) as in "business schizophrenia".

In this paper, we hypothesize by contrast that authors in social media (here, Twitter) use disease names in a manner different from scientific literature. This hypothesis has important consequences, that, to the best of our knowledge, until now have not been fully studied by biomedical social media mining literature: It implies that only some biomedical terms are useful for retrieving biomedical microblogs. It also implies that such terms need not necessarily be canonical terms, but rather less technical – though salient – synonyms.

To test our hypothesis, we seek to answer the following research question: *do the most frequent diseases used in* MEDLINE® *correspond in meaning and variety of use to those discussed in Twitter?* To this end, we study and compare the meaning of normalized, canonical disease mentions (henceforth: *concepts*) and of their different surface-level realizations (henceforth: *synonyms*), in large samples of Twitter (approx. 150M words) and MEDLINE® abstracts (approx. 1B words). To model and compare the meaning of disease concepts and synonyms we resort to *distributional semantics* [5]. The main tenet of distributional semantics (the so-called "distributional hypothesis") is that the meaning of a linguistic expression can be characterized or approximated by the company it keeps in corpora, *i.e.*, by the words with which it co-occurs. Such words are called the distributional *context* of a word or phrase: disease concepts and synonyms in the present study. Contexts thereafter induce word distributions and vector space representations for concepts and synonyms.

2   Methods and Data Collection

Data Collection. We build our experiment on top of MEDLINE®1 (a large bibliographic database covering abstracts of biomedical papers since the early 1950's) and Twitter. A central element of our work is DNorm [8], a disease named entity recognizer and normalizer to MeSH and OMIM. The Medical Subject Headings (MeSH) terminology [9] and Online Mendelian Inheritance in Man (OMIM) [4] are controlled vocabularies of canonical, unique disease names organized into taxonomies of categories. We first apply DNorm to the last ten years of MEDLINE® (Jan. 2007 to Dec. 2017). We then focus our study to those concepts which appear to be of highest relevance in MEDLINE® and search for postings in Twitter using the 20 most frequent synonyms associated to the most frequent 100 MeSH concepts (with the official Twitter API, between Dec. 2017 and Mar. 2018). Detailed corpus statistics are shown in Table 1.

---

[1] https://www.nlm.nih.gov/bsd/pmresources.html.

Table 1. Statistics of the samples studied. "Concepts" refer to MeSH *canonical* names.

Corpus	#Tokens	Units	Concepts	Synonyms	Time span
MEDLINE®	1,037,482,692	5,374,700 abstracts	8,386	2,190,522	Jan. 2007–Dec. 2017
Twitter	145,793,358	7,193,077 tweets	4,908	201,712	Dec. 2017–Mar. 2018

Named Entity Recognition and Frequency. We observe disease synonyms d_s and concepts d_c (normalized MeSH identifiers) and (2) identify their span within each unit of text (resp., a tweet or an abstract). For each d_m, for $m \in \{s, c\}$, *frequency* is measured. This is done in order to rank diseases by frequency and (1) compute the Jaccard (set) similarity at frequency rank $r \leq k$ for the topmost k concepts in MEDLINE® and Twitter respectively: $\mathrm{sim}_{\mathrm{jacc}}(M_r, T_r) = |M_r \cap T_r|/|M_r \cup T_r|$. We also (2) test for concept frequency correlation across both corpora.

Specificity. We represent each d_m as a *distributional context* $\boldsymbol{D}_m$ – the bag or multiset of words with which d_m co-occurs in corpus C – by counting in a window of five words before and after the mention. Distributional contexts allow to quantify the semantic *specificity* (conversely, *ambiguity*) of d_m in each corpus via *normalized entropy*:

$$H_n(d_m) = -\sum_{\boldsymbol{D}_m(w) > 0} \frac{P(w) \cdot \log_2 P(w)}{\log_2 \boldsymbol{D}_m} \tag{1}$$

where $P(w)$ is estimated via relative frequencies, and $\boldsymbol{D}_m = \sum\{\boldsymbol{D}_m(w) \mid \boldsymbol{D}_m(w) > 0\}$ is the *size* of $\boldsymbol{D}_m$. We use normalized entropy to be able to compute comparable measures (scaled to $[0, 1]$) for each d_m independent from context sizes. The higher $H_n(d_m)$, the more ambiguous d_m [10]. Thereafter we test (1) if concept frequency correlates with normalized entropy within each corpus, and (2) if normalized entropy in Twitter correlates with normalized entropy in MEDLINE®. Hence, contexts of co-occurring words can be seen as discrete word distributions. Entropy thus measures the dispersion of this induced distribution.

Similarity. We exploit distributional contexts to build a distributional model (vector space) $\mathbb{R}^{|W|}$, where each d_m is represented as a $|W|$-dimensional *distributional vector* $\boldsymbol{d}_m$ of log-scaled co-occurrence counts over the vocabulary W of MEDLINE® and Twitter. We rely on [2] to build the model. Classical count models are more appropriate in the context of this study than more state-of-the-art methods such as Word2Vec [11] or GloVE [13] word embeddings, which encode words directly into real-valued vectors and make it more difficult to compute entropies – that rely on discrete co-occurence counts [1]. Once we have learned the model from the distributional contexts, we compute cosine similarity for (1) the topmost k matching concepts in MEDLINE® and Twitter, *i.e.*, the d_cs in $M_k \cap T_k$. Additionally, we (2) study their *similarity spread*: we group their 20 most frequent synonyms to compute and compare their average similarities.

MeSH Hierarchy. MeSH IDs – disease concepts d_c – can be organized into a hierarchy H – a tree – of disease categories[2], from more general (root) to more specific (leaves). As an additional experiment, the relationship between concept normalized entropy in MEDLINE® and Twitter, and its position in H is studied. Highly ambiguous, high-entropy concepts should be located closer to the root node of the category hierarchy and have a low depth. To this end, we measure the depth of each concept d_c occurring in either Twitter or MEDLINE® in H, and test if this measure correlates to their normalized entropy in either corpus.

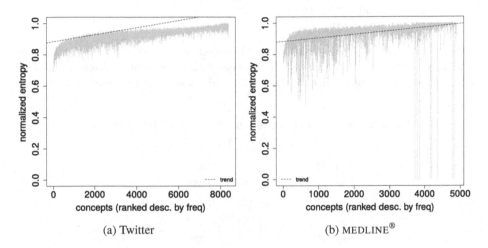

(a) Twitter   (b) MEDLINE®

Fig. 1. Normalized entropies ranked by frequency in descending order. For each concept, we take as distributional context the union of those of all its synonyms.

3 Results

3.1 Experiments

Concept Specificity Analysis. In this experiment, we analyze if disease concepts mentioned in Tweets are more specific than concepts mentioned in MEDLINE®. Results are summarized in Fig. 1. We observe a difference in concept normalized entropy distribution among both corpora. As Fig. 1 shows, normalized entropy in Twitter is on average higher than in MEDLINE®: 0.95 vs. 0.92, the difference being statistically significant (t test with $p < 0.01$). Also, larger variations in normalized entropy can be seen in Twitter.

In both cases we observe a statistically significant negative correlation (-0.69 for MEDLINE® and -0.53 for Twitter, for the Kendall τ test for rank-sensitive correlation) between normalized entropy and frequency within each corpus.

---

[2] https://www.nlm.nih.gov/mesh/intro_trees.html.

In Twitter moreover, disease concepts with both high normalized entropy and very low frequency can be observed, such as *Susceptibility of Obesity* (OMIM UID 602025, point at frequency rank 4609 in Twitter) or *Pseudopseudohypoparathyroidism* (MeSH UID D011556, point at frequency rank 3713 in Twitter).

This seems due to the fact that frequent MEDLINE® concepts tend to be referred to with a smaller number of salient semantically specific synonyms, whereas tweets are overall more ambiguous. A cross-corpus correlation analysis (Kendall τ test) of both normalized entropy and frequency shows no significant correlation between normalized entropy or frequency distributions for concepts between Twitter and MEDLINE®.

Concept Similarity Analysis. In this experiment, we study the semantic similarity of MEDLINE® and Twitter, by computing the cosine similarity of the vector representations of diseases concepts and their most frequent synonyms.

Table 2. Jaccard and cosine similarity of top 100 concepts.

Frequency rank	Jaccard	Cosine (avg.)
Top 20	0.212	0.365
Top 40	0.356	0.358
Top 60	0.446	0.345
Top 80	0.553	0.342
Top 100	0.550	0.338

This is a key aspect of our analysis, in which we endeavor this time to determine and quantify the semantic (di)similarity of diseases in MEDLINE® and Twitter, with standard and robust distributional semantic techniques. The results are summarized by Tables 2, 3 and 4, and by Fig. 2. The results show that, while there is actually a large overlap between the topmost 100 disease concepts mentioned both in Twitter and MEDLINE® (55% overlap/Jaccard similarity, see Table 2), distributional meanings differ considerably, yielding instead a 34% average cosine similarity for the same 100 concepts.

Interestingly (see Tables 3 and 4) common diseases such as *Hepatitis C* obtain a comparatively high cosine score, whereas very rare diseases such as *Behcet Syndrome* (a rare blood vessel chronic inflammation) obtain very low scores, as do diseases for which only ambiguous names exist, such as *BMD* (which may stand for *Becker muscular dystrophy*, or for bone mineral density, a mere symptom).

Figure 2b visualizes how similarity varies across matching concepts, ranked decreasingly by their frequency in MEDLINE®. As the reader can observe in the figures, cross-corpus similarity has a slight tendency to decrease with frequency ($\tau = 0.18$, $p < 0.01$).

This seems to be due to differences in language use in both corpora: (1) different synonyms are associated to each concept, (2) they are assigned different meanings with (3) a higher semantic variability for Twitter compared to MEDLINE® – a fact consistent with our specificity analysis. Observations (2) and (3) are substantiated by Fig. 2a that visualizes the average similarity among the (topmost 20) synonyms of each concept in each corpus, and shows that they are much higher in MEDLINE® than in Twitter. It is also possible to observe that concepts with above average intra-concept similarity in one corpus, exhibit an above average similarity in the other corpus as well. In fact, one can observe a statistically significant positive correlation ($\tau = 0.37$, $p < 0.01$).

Hierarchy Analysis. In this experiment, we study if specificity of disease concepts in Twitter and MEDLINE® decreases the higher the concepts are located in the MeSH disease category tree. We observe a slight, though statistically significant negative correlation, between normalized entropy and the depth of d_c in $\boldsymbol{H}$, *i.e.*, the lower the depth, the higher the specificity of d_c ($\tau = -0.129$ ($p < 0.01$) for MEDLINE®, -0.051 ($p < 0.01$) for Twitter). This can be interpreted as meaning that the distributional ambiguity as measured by $H_n(d_c)$ overlaps with the semantic generality of a disease concept d_c, while remaining largely distinct.

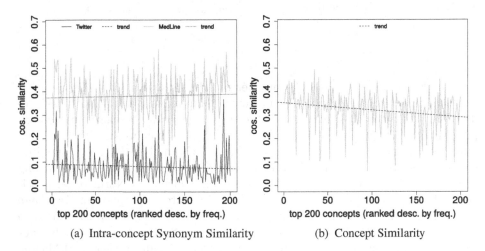

(a) Intra-concept Synonym Similarity   (b) Concept Similarity

Fig. 2. Similarity of concepts and synonyms, ranked by frequency in descending order. Tables 5 and 6 zoom into at concepts at rank 121 and 176 respectively.

3.2   Synonym Analysis

Our results are further substantiated by a qualitative analysis of concepts d_c and their synonyms d_s in both corpora, illustrated by Tables 5 and 6. We observe that higher similarity tend to coincide with a higher number of shared synonyms: in general the top two or three most frequent d_s for each d_c, more or less coincide across both corpora, but diverge afterwards the more dissimilar d_c in between Twitter and MEDLINE®. In Tables 5 and 6 we outline the main synonyms of an above average similarity disease concept, *Multiple Myeloma* (MeSH ID D009101, 0.39 similarity), and a below average similarity concept, *Angelman Syndrome* (MeSH ID D017204, 0.17 similarity). As the reader can see in the tables, while two out of three of the topmost synonyms of *Multiple Myeloma* coincide, the same only holds for one synonym for *Angelman Syndrome*.

Table 3. 7 most similar MeSH concepts.

MeSH ID	Similarity	Canonical name
D006526	0.496	Hepatitis C
D005910	0.463	Glioma
D003920	0.459	Diabetes mellitus
D006521	0.453	Chronic hepatitis
D000860	0.451	Hypoxia
D003327	0.446	Coronary disease
D015658	0.445	HIV infections
...	...	...

Table 4. 6 least similar MeSH concepts.

MeSH ID	Similarity	Canonical name
...	...	...
D015458	0.170	T cell leukemia
D002547	0.155	Cerebral palsy
C536528	0.122	Van der Woude syndrome
C535984	0.116	Congenital bilateral aplasia of vas deferens
D029461	0.109	Sialic acid storage disease
C537666	0.109	BMD

We also observe that a large number of false positives (false disease entities) are detected for Twitter. For instance, "Happiness", and "happiness" are detected in Twitter as synonyms of *Angelman Syndrome*, but also (in the 38th position) the catchphrase "AS IF IT". This likely explains the low synonym pairwise similarity observed in Fig. 2, left, and the higher average normalized entropy observed in Twitter. The reason for this behavior is likely the strong bias of disease NER and normalization systems such as DNorm towards scientific terminology. Indeed, "happiness" may indicate a symptom of *Angelman Syndrome*, that is a severe genetic disorder affecting children and associated to frequent smiling. But in Table 6, a large share of those synonyms actually refer to emotions or wishes (as in "Happy birthday to our bright dancer (...) May he find happiness (...)").

Table 5. Top 3 and bottom 3 synonyms of *Multiple Myeloma* in MEDLINE® and Twitter.

	Synonym	Entropy	Freq.
MEDLINE®	Myeloma	0.807	10,706
	Multiple myeloma	0.830	4,559
	AL	0.867	3,684
	Extramedullary myeloma	0.936	15
	Myeloma tumors	0.956	16
	Lymphoma	0.944	16
Twitter	Myeloma	0.868	1,787
	Multiple myeloma	0.832	525
	Myeloma	0.911	389
	Myelomas	1.000	5
	Myeloma diagnosis	0.989	4
	Gamida	0.914	4

Table 6. Top 3 and bottom 3 synonyms of *Angelman Syndrome* in MEDLINE® and Twitter.

	Synonym	Entropy	Freq.
MEDLINE®	AS	0.813	24,585
	AS-OCT	0.872	615
	Angelman syndrome	0.856	422
	AS-PC	0.948	8
	AS-AIH	0.932	8
	AS infection	0.931	9
Twitter	AS	0.927	598
	Happiness	0.901	483
	Happiness	0.850	135
	Militer AS	0.976	3
	AS A CHILD	0.968	3
	Angelman syndrome	0.947	3

4   Conclusions

We have carried out an extensive distributional analysis of the semantics of disease concepts and their synonyms in both social media (Twitter) and biomedical literature (MEDLINE®). To this end, we have measured and compared the normalized entropy – the distributional ambiguity – and distributional similarity of disease concepts observed in both corpora. Our analysis shows low distributional similarity among both corpora, coupled with a higher ambiguity in Twitter compared to MEDLINE®.

Our preliminary qualitative analysis shows that standard disease recognition methods such as DNorm result in high numbers of false positives, due to the larger use of catchphrases and metaphorical expressions in Twitter.

Future work will focus on the development of methods which can separate such non-disease name mentions from actual disease mentions in social media, and that help in identifying tweets substantially similar to scientific texts. Ultimately, our goal is to build upon such methods to design techniques that identify and match disease-centric relations across both social media and MEDLINE®.

Acknowledgments. This work was supported by a grant from the Ministry of Science, Research and Arts of Baden-Württemberg to Roman Klinger.

References

1. Baroni, M., Dinu, G., Kruszewski, G.: Don't count, predict! A systematic comparison of context-counting vs. context-predicting semantic vectors. In: Proceedings of ACL 2014 (2014)
2. Dinu, G., Pham, N.T., Baroni, M.: DISSECT - DIStributional SEmantics composition toolkit. In: Proceedings of ACL 2013 (2013)
3. Doğan, R.I., Leaman, R., Lu, Z.: NCBI disease corpus: a resource for disease name recognition and concept normalization. J. Biomed. Inform. **47**, 1–10 (2014)

4. Hamosh, A., Scott, A.F., Amberger, J.S., Bocchini, C.A., McKusick, V.A.: Online Mendelian Inheritance in Man (OMIM), a knowledgebase of human genes and genetic disorders. Nucl. Acids Res. **33**, 514–517 (2005)
5. Harris, Z.: Distributional structure. Word **10**(23), 146–162 (1954)
6. He, L., Yang, Z., Lin, H., Li, Y.: Drug name recognition in biomedical texts: a machine-learning-based method. Drug Discov. Today **19**(5), 610–617 (2014)
7. Klinger, R., Kolářik, C., Fluck, J., Hofmann-Apitius, M., Friedrich, C.M.: Detection of IUPAC and IUPAC-like chemical names. Bioinformatics **24**(13), 268–276 (2008)
8. Leaman, R., Islamaj Doğan, R., Lu, Z.: DNorm: disease name normalization with pairwise learning to rank. Bioinformatics **29**(22), 2909–2917 (2013)
9. Lipscomb, C.E.: Medical subject headings (MeSH). Bull. Med. Lib. Assoc. **88**(3), 265–266 (2000)
10. Melamed, I.D.: Measuring semantic entropy. In: Proceedings of the SIGLEX Workshop on Tagging Text with Lexical Semantics (1997)
11. Mikolov, T., Chen, K., Corrado, G., Dean, J.: Efficient estimation of word representations in vector space. CoRR abs/1301.3781 (2013)
12. Nikfarjam, A., Sarker, A., O'Connor, K., Ginn, R.E., Gonzalez, G.: Pharmacovigilance from social media: mining adverse drug reaction mentions using sequence labeling with word embedding cluster features. JAMIA **22**(3), 671–681 (2015)
13. Pennington, J., Socher, R., Manning, C.D.: Glove: global vectors for word representation. In: Proceedings of EMNLP 2014 (2014)
14. Sarker, A., O'Connor, K., Ginn, R., Scotch, M., Smith, K., Malone, D., Gonzalez, G.: Social media mining for toxicovigilance: automatic monitoring of prescription medication abuse from Twitter. Drug Saf. **39**(3), 231–240 (2016)
15. Seargeant, P., Tagg, C. (eds.): The Language of Social Media. Palgrave Macmillan, London (2014)
16. Wei, C.H., Kao, H.Y., Lu, Z.: GNormPlus: an integrative approach for tagging genes, gene families, and protein domains. BioMed Res. Int. **2015** (2015) (2015). ID 918710
17. Yang, C.C., Yang, H., Jiang, L., Zhang, M.: Social media mining for drug safety signal detection. In: Proceedings of the 2012 International Workshop on Smart Health and Wellbeing (SHB 2012) (2012)

Gemedoc: A Text Similarity Annotation Platform

Jacques Fize[1,2]([✉]), Mathieu Roche[1,2], and Maguelonne Teisseire[1]

[1] TETIS, Univ. Montpellier, APT, Cirad, CNRS, Irstea, Montpellier, France
jacques.fize@cirad.fr
[2] Cirad, TETIS, Montpellier, France

Abstract. We present GEMEDOC, a platform for text similarity annotation based on the spatial and the thematic dimension. To this end, a two-step annotation protocol was designed to assess the similarity between two documents: (1) identification of salient features according to the two analysis dimensions; (2) similarity assessment according to a 4-degree scale. Ultimately, the labeled data retrieved from different corpora could be used as benchmark for text-mining applications.

Keywords: Web platform · Text matching · Geo-similarity

1 Introduction

The past decade has witnessed a significant growth of the available data which are now called *Big Data*. Big Data generally refers to the 3 V's – Volume, Velocity, and Variety [1]. Volume concerns the data storage, velocity relates to the update frequency, and variety refers to the heterogeneity of data (*e.g. newspapers, images, sensor data, etc.*). While volume and velocity are well-known problems, variety still remains a serious challenge for the scientific communities [2].

In this work, we focus on matching heterogeneous textual data. In this context, a common representation of textual data is needed for evaluating similarity between documents. To address this issue, we compare heterogeneous data by exploiting different dimensions of text, *i.e.*, *thematic*, *spatiality* and ultimately *temporality*. This kind of approaches could contribute to enhance specific applications such as epidemic monitoring [3] where thematic and spatial indicators need to be combined. In this context, the identification of similar behaviors is based on the thematic (*e.g. diseases, symptoms, ...*) and the spatial information (*e.g. epicenter location, spread of the disease, ...*). In order to evaluate the impact of each dimension for document matching, the use of corpora labeled on thematic and spatial information is required.

Unfortunately, to the best of our knowledge, there are no corpora currently available that give similarity indicators according to these two dimensions. To annotate document similarity within a corpus, tools such as SEMILAR[1] or

---

[1] http://www.semanticsimilarity.org/.

© Springer International Publishing AG, part of Springer Nature 2018
M. Silberztein et al. (Eds.): NLDB 2018, LNCS 10859, pp. 333–336, 2018.
https://doi.org/10.1007/978-3-319-91947-8_35

STSANNO[2] have been developed. SEMILAR is a semantic similarity framework which integrates state-of-the-art methods for similarity annotation, called SEMILAT. STSANNO was implemented for sentence similarity annotation, particularly for minor language such as Serbian language [4].

In this article, we present the GEMEDOC platform for text similarity based on thematic and spatial dimensions. We propose a two-step annotation protocol: (1) identification of salient features according to the two analysis dimensions; and (2) similarity assessment on a 4-degree scale.

2   Assessing Similarity Between Documents

The assessment of similarity between two documents based on specified dimensions is a difficult task due to the annotation reliability, the rating scale, etc. Furthermore, the annotation between documents within a corpus can be tedious.

In order to support users during the annotation process, we defined a two-step protocol as described in the next subsection.

2.1   Gemedoc Annotation Protocol

The annotation protocol is divided in two steps (see Fig. 1). First, the user conducts a corpus analysis by extracting salient features from each text. Second, the user associates two similarity degrees (*i.e. one for each dimension*) between documents. Different document similarities based on thematic and spatial features are proposed.

In the next subsection, using a text sample, we introduce the given indicators for each dimension. Then, we outline the 4-degree similarity scale used in GEMEDOC (see Sect. 2.2).

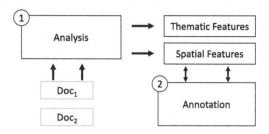

Fig. 1. Annotation process associated with GEMEDOC

---

[2] https://vukbatanovic.github.io/STSAnno/.

2.2 Identification of Salient Features

Thematic Analysis. The thematic information is highlighted by salient terms in the texts. In our tool, these important terms are extracted with specific text-mining process [5]. In Fig. 2, some words may be highlighted to identify themes, such as *humanitarian help, hardship, family,* and *immigration.*

Spatial Analysis. The spatial features provide another point-of-view of documents (*e.g. for trajectory matching*). The proposed spatial entity extraction is described in [6]. In the example of Fig. 2, spatial features such as the narrator's location (*Idomeni*) or spatial entities (*Europe, Macedonia, etc.*) are extracted.

The winding **road** across the wheat fields near the Greek village of *Idomeni* is **full** of **people** carrying large bags on their shoulders, **babies** in their arms and putting one step in front of the other. The **stream** of **humanity** continues day and night but not an average of 150 a day, (and only Syrians and the Iraqis who are **lucky** enough to have a **passport** or **ID card** from their **home** country) can continue the **journey** out of this place and across the **border** into the Former Yugoslav Republic of *Macedonia* (FYROM) and onwards to western and northern *Europe.* Few are leaving but more, many more keep coming, only to end up getting **stranded** in what is becoming **unsustainable humanitarian** situation.

Fig. 2. Testimony from a humanitarian non-governmental organization (i.e. *MSF*) with *spatial entities* and **thematic indicators**

Similarity Annotation. We defined a 4-degree similarity scale to assess the similarity between two documents and for each dimension (*i.e. thematic and spatial*):

- **Don't know:** You cannot evaluate the similarity between the documents
- **Different:** Both documents are totally different
- **Similar:** Both documents share a large set of similarities with few differences
- **Very similar:** Both documents are highly similar, apart from a few tiny differences.

3   Gemedoc

In this context, the GEMEDOC[3] tool enables us to label inter-document similarity. GEMEDOC is a web application using the Flask web framework.

GEMEDOC proposes a corpus annotation interface, along with features to display statistical indicators (*i.e Fleiss' and Cohen's Kappa, etc.*) on the annotation quality. Corpora and users' annotations are currently saved directly on the application server. Figure 3 shows the annotation interface.

---

[3] GEMEDOC is available at http://gemedoc.jacquesfize.com/about.

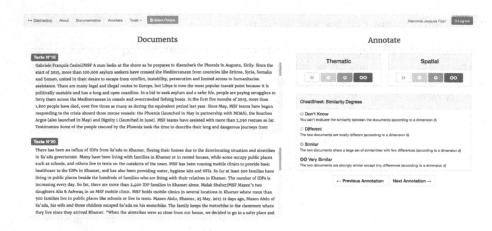

Fig. 3. GEMEDOC annotation interface

4   Conclusion

We present the GEMEDOC tool, a text similarity annotation platform. This system is based on a two dimension annotation (i.e. *thematic* and *spatiality*) and a 4-degree similarity scale. We plan to use this platform to build a corpus to evaluate text matching methods on different topics (*e.g. epidemiology studies, trajectory analysis*, and so forth).

References

1. Russom, P., et al.: Big data analytics. TDWI best practices report, fourth quarter, vol. 19, no. 4, pp. 1–34 (2011)
2. Mao, R., Xu, H., Wu, W., Li, J., Li, Y., Lu, M.: Overcoming the challenge of variety: big data abstraction, the next evolution of data management for aal communication systems. IEEE Commun. Mag. **53**(1), 42–47 (2015)
3. Arsevska, E., Roche, M., Lancelot, R., Hendrikx, P., Dufour, B.: Exploiting textual source information for epidemiosurveillance. In: Closs, S., Studer, R., Garoufallou, E., Sicilia, M.-A. (eds.) MTSR 2014. CCIS, vol. 478, pp. 359–361. Springer, Cham (2014). https://doi.org/10.1007/978-3-319-13674-5_33
4. Batanović, V., Cvetanović, M., Nikolić, B.: Fine-grained semantic textual similarity for Serbian. In: Proceedings of the 11th International Conference on Language Resources and Evaluation (2018)
5. Arsevska, E., Roche, M., Hendrikx, P., Chavernac, D., Falala, S., Lancelot, R., Dufour, B.: Identification of terms for detecting early signals of emerging infectious disease outbreaks on the web. Comput. Electron. Agric. **123**, 104–115 (2016)
6. Fize, J., Shrivastava, G.: Geodict: an integrated gazetteer. In: Proceedings of Language, Ontology, Terminology and Knowledge Structures Workshop (LOTKS 2017). Association for Computational Linguistics (2017)

Text Classification

Addressing Unseen Word Problem in Text Classification

Promod Yenigalla[(✉)], Sibsambhu Kar, Chirag Singh, Ajay Nagar,
and Gaurav Mathur

Samsung R&D Institute India, Bangalore, India
{promod.y,c.singh,ajay.nagar,gaurav.m4}@samsung.com,
sibsambhukar@gmail.com

Abstract. Word based Deep Neural Network (DNN) approach of text classification suffers performance issues due to limited set of vocabulary words.
Character based Convolutional Neural Network models (CNN) was proposed by
the researchers to address the issue. But, character based models do not inherently capture the sequential relationship of words in texts. Hence, there is scope
of further improvement by addressing unseen word problem through character
model while maintaining the sequential context through word based model. In
this work, we propose methods to combine both character and word based
models for efficient text classification. The methods are compared with some of
the benchmark datasets and state-of-the art results.

Keywords: Text classification · Word embedding · Character embedding
Multi-channel CNN

1 Introduction

Text based deep learning models require vector representation of text (usually a
numeric vector, called embedded vectors) at the very first layer of model. Different
kinds of text embedding have been proposed, such as word embedding [1, 2], character
embedding [3], document embedding [4] etc. in past few years. The embedded vectors
are formed using a large text corpus [2]. As a result, the positional attributes of the
word or character in embedded space is indicative of semantic relationship with other
words or character as available in the corpus. The type of embedding to be used in a
given problem depends on the type of task and the network architecture.

1.1 Word Embedding

Word embedding gained popularity through its use in language modelling [5], text
classification [1] and sentiment analysis [6] in recent past. For a network model, word
embedding can be initialized with random values or pre-trained vectors [2]. Mikolov
et al. [2] prepared pre-trained embedding of words based on a large text corpus. Randomly initialized embedding is updated during model training process and pre-trained
embedding can be trained further during actual model training or can be made static (no
update of embedding during training). Usually, pre-trained word embedding based

© Springer International Publishing AG, part of Springer Nature 2018
M. Silberztein et al. (Eds.): NLDB 2018, LNCS 10859, pp. 339–351, 2018.
https://doi.org/10.1007/978-3-319-91947-8_36

models converge faster compared to randomly initialized models and also provide embedding of words that are unseen in training data but may appear in test data.

Performance of word based models is limited by the dictionary that we include to build the model. Majority of the spoken or written languages corpus contain millions of words though only fraction of that is used frequently. A large size vocabulary increases both model size and training time. Moreover, the words which appear less frequently in the corpus do not learn well during training. As a result, for all word-based deep learning models a subset of words are considered as dictionary. Any word beyond the dictionary is considered as unseen and such words are either tagged as unknown or removed from the input text during training as well as testing. In case of text classification, one or more unseen word in the text may increase misclassification. Lower the vocabulary size, higher the chance of misclassification.

1.2 Character Embedding

To address above drawbacks in word-based deep network models, researchers applied character-based models in various applications [3–5]. Character embedding resolves the problem of out-of-vocabulary words as any word can be formed using set of characters. Moreover, the size of character-based models is much less as compared to word-based models due to limited number of characters forming the language corpus. Hence, the character based CNN models are much faster (training time of each epoch) than word based models though the number of epoch for convergence is much less for word based models. For character based CNN models the context or relationship between multiple characters and words are captured by convolution filters or kernels. Higher the kernel size, better is the context captured. But capturing longer context is difficult in character models as compared to word based CNN models.

Le and Mikolov [7] proposed embedding of documents through paragraph vectors to avoid word-order weakness of bag-of-words and applied for text and sentiment classification. Dai et al. [4] explored the capability of paragraph vectors for document modelling. Such document or paragraph vectors may be considered as enhanced word embedding that captures the sequence of information as a part of embedding. Wieting et al. [8] proposed character n-gram embedding (CHARGRAM) for sentence similarity study and part-of-speech tagging. Recently, a combination of word and character embedding gave relatively better performance in various NLP problems [9, 10].

1.3 Interdependency of Network Models and Type of Text Embedding

CNN and Recurrent Neural Network (RNN) provide complementary information [11] for NLP problems. The basic structure of CNN does not inherently support sequential learning but is good for extracting multi-resolution features. As a result, for text analysis, CNN finds most informative n-grams (set of word or characters) features present in the text. On the other hand, basic structure of RNN models (both GRU and LSTM) is suitable for storing past information and hence can learn sequence of texts better. So, different forms of RNN were used for language modeling [12] or sentiment analysis [13] related research. But, with time, different versions of CNNs and RNNs were developed with various types of embedding which outperformed the conventional models. For example, Kim *et al.* [5] proposed a character based CNN language model which provide much better performance than conventional LSTM version. Thus, there is no clear distinction between RNN and CNN in terms of their applicability in NLP problems though there is a preference in the type of embedding. We find both character and word embedding for CNN models whereas word embedding is usually preferred for RNN models.

CNN models with character embedding learn the most informative set of consecutive characters (character n-grams). For example, if a filter of length 3 is used for convolution followed by max-pooling, the filter will extract most-informative character-set of length 3 (3-gram character) present in the input text. We use multiple such filters to learn different inputs. But such filters cannot find any relationship between the n-grams at different position of text. Similarly, word based CNN models find the set of most informative word n-grams through convolutions and max pooling. For a fixed convolution kernel, the character n-grams span a smaller length of text as compared to word based CNN models. Increasing kernel size increases the model training time though it depends on length of embedding too. Character embedding uses much smaller vectors as compared to word embedding due to limited set of characters.

Recurrent neural network models work mostly with words. Kim 2015 *et al.* [5] used a combination of CNN and RNN in which the character embedding is input to the CNN part and the output of CNN layer act as input to the RNN part. In such models, the CNN component captures the n-gram features of text and RNN takes care of sequence context of such features in the text. As a result, these models are able to address the unseen word issues. In this work, we analyze a number of existing state-of-the-art methods and propose different models to address the unseen word issue through combined character and word embedding.

1.4 Prior Work on Text Classification Using Neural Networks

The Advances in computing and deep neural networks have helped to achieve significant enhancements in NLP. Kim [1] proposed a word based CNN model for text classification which uses Word Embedding and a parallel Convolution based architecture and provided benchmark accuracy on different datasets. Wang *et al.* [14] used clustering of word vectors followed by a CNN to classify short texts. Later, Zhang *et al.* [3] proposed a 9 layer character based CNN model, which addresses the unseen word problem in Word CNN's models, but it needs huge training corpus and takes longer time to converge with a small increase in accuracy. Recently, a very deep character CNN network with 29 CNN Layers was proposed by Conneau *et al.* [15] to show that the performance increases with depth. A detailed study on using shallow word based CNN or deeper character based CNN was published by Jonhson *et al.* [16].

With respect to using RNN for text classification, Johnson and Zhang [17] suggested a supervised and semi-supervised learning for text classification using LSTM. Tang *et al.* [13] proposed a gated RNN method for text classification in which the model learns sentence representation with CNN or LSTM and the semantic relation is encoded with gated RNN. Zhou *et al.* [18] proposed a combined CNN and LSTM model, where CNN captures high level phrase representation and the LSTM layer to understand overall sentence semantics. Wang *et al.* [9] proposed a combination of character and word model where the Word CNN is similar to Kim [1] architecture and Char CNN similar to Zhang *et al.* [3] model. In this work, we propose multiple combined word and character based CNN models.

Main contribution of the paper:

– New CNN architecture for addressing the issues with word embedding.
– Experimental observation and guidelines to work with combined word and character model for different CNN architectures.

2 Proposed Models

2.1 Model-1: CNN with Word Embedding

The model (see Fig. 1) includes word embedding, followed by convolution layer with different kernels. The features generated in the convolution layers are flattened and fed to the fully connected (FC) layer. Finally, a softmax layer is used for classification.

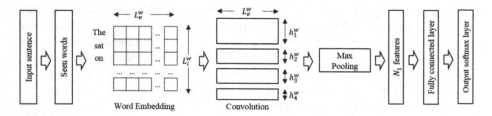

Fig. 1. *Model-1*: CNN model with word embedding (L_e^w = *Word embedding length*, L_i^w = *Input sequence length*, h_i^w = *Word filter height*, i = 1, 2, 3, 4)

We consider pre-trained word embedding of length L_e^w (300 in this work) based on Google word2vec [2, 19]. The input data length (L_i^w) for the models depends on text statistics available in the dataset as shown in Table 1. If the input sentence length differs from L_i^w, the text sample is padded or truncated to match L_i^w.

The model uses 4 convolution kernel of dimension $h_i^w \times L_e^w$, i = 1, 2, 3, 4. A kernel of size 4 × 300 means the filter is spanned over a length of 4 words. These filters are applied in parallel and the coefficients are tuned as a part of training. Number of filters of each type is 100 and max-pooling is used to extract the feature for FC layer. One FC layer is used in this work, though multiple layers may provide better results.

2.2 Model-2: Multi-channel CNN Model with Word (Seen Words) and Character Embedding (Unseen Words)

In this model (Fig. 2), all the seen words (words that appear more than a specified number of times in the dataset) are fed to the word CNN module and remaining words (not part of dictionary, i.e. unseen words) are sent to character CNN module. Though

Table 1. Word and character length of different datasets and input data length (L_i^w) for word and character CNN models

Dataset	Number of words in text			Number of characters in text		
	Mean	Max	L_i^w	Mean	Max	L_i^c
MR	20	56	56	80	256	256
AGNews	7	24	24	27	90	90
TREC	11	37	37	45	197	197
Subj	23	122	56	110	681	256
SST2	19	55	55	87	256	256
CR	19	105	56	79	533	256
MPQA	3	36	36	15	226	226

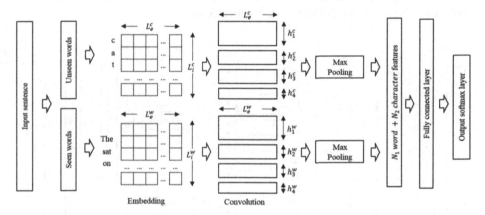

Fig. 2. *Model-2*: CNN model with both word (seen words) and character embedding (unseen words)

the sequence of words is not maintained if one or more unseen words are removed, the presence of those unseen word/words in character module helps increasing the classification accuracy. The model consists of independent word and character embedding, followed by separate convolution layers. The features generated in the convolution layers are fed together to the fully connected layer. Thus, the features from both word and character embedding are trained together in the FC layer. The word CNN part of the model is same as *Model-1* which is explained in the previous section.

Character embedding dictionary consists of lower case letters, digits and special symbols. Each character is represented as 100 dimension vector. We used both random and pre-trained character embedding in this work. The input character length (L_i^c) for the models depends on text statistics as shown in Table 1. Since we pass only unseen words in character model, the effective input length is small (as only limited number of unseen words). If the input character length differs from L_i^c, the text sample is padded or truncated to match L_i^c. We use maximum length of unseen text as L_i^c.

We use 4 convolution kernel of dimension $h_i^c \times L_e^c$, $i = 1, 2, 3, 4$. There are 100 filters of each dimension and max-pooling is used to extract the feature for FC layer. Like word model, only one FC layer is used in this work. Both the character and word level convolution outputs are fed together to the FC layer so that the model learns both character n-gram features or sub-word information (from character CNN model) and word n-gram like features (from word CNN model) together and prioritize as per output classes. We use dropout at FC layer to prevent over-fitting of the model (due to usage of both character and word model with multiple filters or kernels).

2.3   Model-3: CNN Model with Word (Seen Words) and Character (All Words)

In *Model-3* (see Fig. 3), all the seen words are fed to the word CNN module and all (both seen and unseen) words are sent to character CNN module. The remaining part of the architecture remain same as *Model-2*. The character module takes entire text as input and hence L_i^c is much larger as compared to *Model-2*.

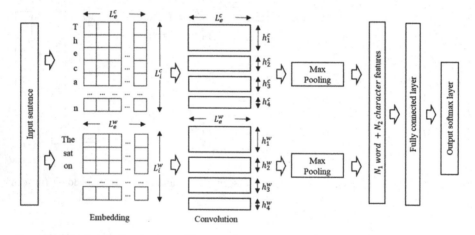

Fig. 3. *Model-3*: CNN model with both word (seen words) and character (all words)

2.4   Model-4: Word-Character Interleaved CNN Model (Word Embedding for Seen Words and Character Embedding for Unseen Words)

In *model-4*, our objective is to maintain the sequence of words by embedding character sets of unseen words in between seen words as shown in Fig. 4.

Each character of the unseen word is represented as vector of same length as words (L_e^w) and is added to the word vocabulary. The vectors are initialized with random values and were trained during model training. The remaining part of the model remain same as *Model-1*.

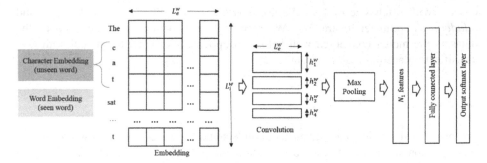

Fig. 4. *Model-4*: Word-character interleaved CNN model (word embedding for seen words and character embedding for unseen words)

3 Datasets and Experiment Detail

We consider 7 benchmarked datasets [1] to evaluate our proposed models. All the datasets are taken from a GitHub source [20]. Some basic word and character statistics of the datasets are presented in Table 1.

Movie Review: This dataset contains 10662 movie review comments and has two classes: positive and negative. Each text consists of one single sentence. In the current work, we split 80% of the texts as the training set and the 20% as test set. We present both 5 fold cross validation result as well as best accuracy among all combinations.

TREC: This is a 6-class classification problem related to question answering. We consider 5452 number for training and 500 to test the model.

AG News: The dataset is AG news corpus [9] where the articles are divided into 4 categories. The dataset has two parts: title and description, but we only considered the title for text classification. The dataset contains 120000 training and 7600 test samples.

Subj: This is a binary (subjective and objective) classification problem. There are 10000 sentences among which 80% used for training and 20% for testing. We present both 5 fold cross validation result as well as best accuracy among all combinations.

CR: This is a customer review dataset with positive and negative classes. The Dataset has 3771 sentences and we used 80% for the training and remaining 20% for testing. We use 5 fold cross validation result as well as best accuracy among all combinations.

MPQA: This dataset has 10606 sentences with two classes. We use 80% of the data for model training and remaining 20% for testing. A 5 fold cross validation result as well as best accuracy among all combinations are computed for comparison.

SST-2: Stanford Sentiment Treebank dataset with two classes: positive and negative reviews. There are 76961 phrases and 6919 sentences for training and 1820 sentences for testing.

We used LeNet CNN architecture for implementing the Multi-channel parallel-net CNN in Theano. For all the datasets, training is carried out using mini batch gradient descent with batch size of 100. We use *negative log-likelihood* as loss function and *adadelta* as optimizer for training. We used 50 epoch for training, though for all the datasets the model converged well below 50 epochs. We select the model with best classification accuracy on test data.

4 Evaluation and Discussion

We performed a series of experiments to understand the effect of unseen words on text classification accuracy with multiple standard datasets. The experimental results for all the models for different datasets are presented and discussed in the following sections. The probable reason of better or inferior performance of different experiments is also discussed. Such experimental approach may help researchers to decide models in different types of text classification problems.

We worked with variable size word dictionary for different datasets to present the relative changes in accuracy with difference in count of seen and unseen words (Table 2). Since, the total vocabulary size for different data set are different, the vocabulary size is defined in terms of minimum word frequency for each dataset. Minimum word frequency 3 indicates that all the words that appear less than 3 times in the dataset (combined training and test data) are marked as unseen or unknown word. It is observed that with increased minimum word frequency for seen word, the number of unseen words in the data increases which causes decrease in classification accuracy.

4.1 Evaluation of Model-1: CNN Model with Word Embedding

CNN with word embedding is a known technique for text classification [1]. The accuracy of the word only CNN Model decreases with decrease in vocabulary size with increased word frequency threshold. We tried to improve the accuracy in the subsequent models by passing the unseen words to a Char CNN in different ways.

CNN model size increases with higher seen or dictionary words and hence plays an important role in case of embedded applications. Hence, to work with a large dataset using word embedding based CNN models, we need to decide the final model based on both accuracy and model size. Moreover, training time (in terms of number of epochs for convergence) is also important in deciding the model and vocabulary size.

4.2 Evaluation of Model-2: CNN Model with Both Word (Seen Words) and Character Embedding (Unseen Words)

A dictionary of seen words was prepared based on the frequency of words used in the dataset and remaining unseen words were sent to character model. We vary the dictionary size to check how the accuracy of classification varies with count of seen words. We used pre-trained word embedding and allowed the embedding to learn as part of model training process. For character embedding, both pre-trained character embedding and randomly initialized (Xavier Initialization [21]) character embedding are examined. The results for both are presented in Table 2. There is no significant difference observed in accuracy on these two versions. But, it is observed that the

Table 2. Accuracy of different models with varying minimum word frequency and % of seen words (minimum word frequency is used to decide seen and unseen words)

Methods	Datasets													
	MR					AG News				TREC				
	1	3 (40%)	5 (25%)	7 (19%)	9 (15%)	1	5 (40%)	9 (27%)	13 (21%)	1	3 (26%)	5 (15%)	7 (10%)	9 (8%)
Model 1: Word CNN	82.8	81.8	80.3	79.1	78.4	87.1	86.8	86.2	86.0	93.6	93.2	92.0	90.6	89.0
Model 2: Random Char	82.8	81.9	81.1	80.1	78.7	88.2	87.5	87.3	86.6	93.6	93.6	93.0	90.6	89.2
Model 2: Pre-train Char	82.8	81.6	80.9	80.2	79.1	88.2	87.8	87.7	87.2	93.6	93.8	92.0	90.4	88.8
Model 3: Random Char	82.9	82.0	80.4	79.9	79.1	88.7	87.5	86.8	86.8	94	93.2	92.5	91.4	89.8
Model 3: Pre-train Char	83.3	81.3	80.8	79.8	78.9	88.6	87.7	87.4	87.0	94.2	93.4	92.2	90.2	88.0
Model 4: Interleaved	82.8	82.6	80.4	80.4	78.9	88.2	87.6	87.4	87.3	93.6	92.6	91.2	91.0	89.4

accuracy decreases with more unseen words though the accuracy is higher compared to word-only models. The higher accuracy of *Model-2* as compared to *Model-1* is due to the inclusion of unseen words using character model. This model takes care of unseen word problem and also tries to establish a semantic relationship between the words, as the features extracted from Word Convolution Layer and Character Convolution Layer are together trained in Fully Connected Layer. But the continuity or the sequencing between the seen and unseen words which is very critical for generating sematic relationship is missing. We partially address the problem in *Model-3* by including both seen and unseen words as input to both word and character models.

4.3 Evaluation of Model-3: CNN Model with Both Word (Seen Words) and Character Embedding (All Words)

In this model, the seen words were sent to word based model with pre-trained embedding whereas both seen and unseen words were sent to character model. Like model-2, we computed results for both pre-trained character embedding and randomly initialized character embedding. The accuracy of the model for different datasets with varying dictionary size is presented in Table 2. It can be seen that for multiple datasets this model has the best accuracy compared to other models especially when the complete dataset is considered as seen. The reason behind the high accuracy can be due to the sub-word context information learnt by character network and word n-grams context information learnt by the Word network which complements each other. One disadvantage of this model compared to *Model-2* and *Model-4* is higher training time and memory consumption as we train the whole text using a char CNN.

4.4   Evaluation of Model-4: Word-Character Interleaved CNN Model (Word Embedding for Seen Words and Character Embedding for Unseen Words)

As discussed before, the objective of this model is to maintain the word sequence by embedding character sets of unseen words in between seen words. The accuracy of the model is presented in Table 2. The accuracy for different thresholds of seen words, the result is in the same range as highest accuracy. Though we have not achieved the highest accuracy in this model, the model has potential to perform better for large datasets. The reason for low accuracy might be due to inadequate training with less unseen data. For all the above models the accuracy decreases with increase in word frequency threshold for seen data. But, with different combined models we are able to compensate the difference significantly.

4.5   Performance Comparison of State-of-the-Art Techniques

In this sub-section, we compare the performance of our method with various deep learning based state-of-the-art text classification methods reported in recent past. Seven benchmark datasets were considered for comparison as described in Sect. 2. The comparison of accuracy with state of art techniques is presented in Table 3.

Table 3. Comparison of accuracy (%) of the proposed method on benchmark datasets

Methods	Datasets						
	MR	AG News	TREC	SUBJ	CR	MPQA	SST-2
Model-3: Random Char	Max: 82.9 CV: 81.7	**88.7**	94	Max: **94.3** CV: 93.3	Max: **87.9** CV: **85.7**	Max: 91.3 CV: 90	89
Model-3: Pre-train Char	Max: 83.3 CV: 81.8	88.6	**94.2**	Max: **94.3** CV: 93.3	Max: 87.1 CV: 85.5	Max: **91.4** CV: **90**	88.2
Word CNN [1]	81.5	86.1	93.6	93.4	85	89.6	88.1
CharCNN [3]	77.0	78.3	76	-	-	-	-
WCCNN [9]	**83.8**	85.6	91.2	-	-	-	-
KPCN [9]	83.3	88.4	93.5	-	-	-	-
RNN [22]	-	-	-	94.1	-	-	87.9
BiLSTM-CRF [6]	82.3	-	-	-	85.4	-	88.3
F-Dropout [1]	79.1	-	-	93.6	81.9	86.3	-

It can be seen that for 3 datasets (AG News, TREC and SST-2) where the dataset has a fixed train and test data we achieved the best accuracy. For CR and MPQA datasets where we need to apply 80:20 for train/test ratio we achieved the best result with Cross Validation and much higher accuracy with the best train/test set. For SUBJ we achieved the best accuracy for the best train/test sample and for MR Dataset we achieved better results compared to all models except one. We present both the maximum and cross-validation results for data where the test datasets are not fixed.

To conclude, Model-3 Random Char model gave state of art results on 3 datasets out of seven considered and Model-3 Pre-trained Char model gave the best results on 3 other datasets. Not all benchmark datasets were balanced, 3 datasets namely CR, TREC and MPQA datasets were imbalanced with respect to class representation. So, we present Precision, Recall and F1 Score for these datasets in addition to accuracy as shown in Table 4.

Table 4. Performance metrics for imbalanced datasets

Dataset	Accuracy (%)	Precision	Recall	F1-Score
TREC	94.2	0.94	0.94	0.94
CR	87.9	0.88	0.88	0.88
MPQA	91.4	0.91	0.91	0.91

4.6 Discussions

This is an experimental work and in this sub-section we share some of our observations while working with various datasets.

Hyper-parameter Tuning: During experiments on various models, we tried different hyper-parameter tuning to achieve best results. Convolution filters for character was finalized as {2, 3, 4, 5} whereas for word module it was finalized as {2, 4, 6, 9} through multiple iterations and experiments. Selection of the number of convolution filters is fixed to 100 after several trials in the range of 50 to 200. We also tried multiple activation functions for convolution layer and FC layer like *Sigmoid, ReLU, TanH, SoftReLU* etc. We got best results for *SoftReLU* for Convolution Layer and *ReLU* for FC Layer. For output layer we always used *Softmax*. We have also tried various dropouts at FC layer and achieved best accuracy with 25% dropout.

Effect of Pre-trained and Randomly Initialized Word and Character Embedding: In this work, we use pre-trained word embedding using Word2Vec whereas character embedding is of two types: randomly initialized and pre-trained on Google 1 billion dataset. As mentioned in Tables 2 and 3, we see a small improvement in the accuracy using pre-trained char embedding, but not consistent across different experiments.

Unseen with Respect to Word2Vec as Compared to Working Dataset: We performed another experiment with architecture discussed in *Model-2*. In this model, a word not present in Word2Vec was considered as unseen and sent to character CNN and remaining all words were sent to word CNN part of the model. We see a maximum accuracy of 83.6% on MR Dataset with this approach. Similar technique was also employed with *Model-4* where we got a maximum accuracy of 82.9% on MR data.

Input Length of CNN Layers and Training Time: :Input word length (L_i^w) or character length (L_i^c) affect the training time due to change in number of convolutions. In *Model-2*, character-CNN takes only unseen words, which makes L_i^c to be lesser by

70–75% compared to *Model-3* which takes the all the characters of sentence as input. The training time of *Model-2* is almost equal to *Model-1* (only Word CNN), though *Model-3* achieves better accuracy. For *Model-4*, dimension of both word and character embedding are same (as word embedding). Hence, more unseen words cause higher training time. In our work, on average, *Model-2* takes approximately 10 min per epoch whereas *Model-3* and *Model-4* takes about 17–20 min per epoch.

5 Conclusions

Multiple CNN architectures are proposed to address the unseen word problem of word based models. The models show enhanced text classification accuracy for benchmark datasets. Relative merits of each model are discussed keeping various aspects in mind such as time of training, model size and classification accuracy. This might be helpful to the research community to decide architecture for a given problem. The proposed models have limited number of layers to make it faster. Though the proposed architecture is able to exceed the benchmark results, increased number of layers may enhance the accuracy further.

To extend the accuracy further we are working on adding part-of-speech (POS) embedding to character based embedding of unknown words. POS embedding is a small size embedding indicating the position of word in a vector space spanned by various POS. Character or sub-word embedding of unseen words does not capture the syntactic relationship which might be corrected by POS embedding.

References

1. Kim, Y.: Convolutional neural networks for sentence classification. In: Proceedings of EMNLP 2014 Conference, pp. 1746–1751 (2014)
2. Mikolov, T., Chen, K., Corrado, G., Dean, J.: Efficient estimation of word representations in vector space. In: Proceedings of Workshop at ICLR (2013)
3. Zhang, X., Zhao, J., LeCun, Y.: Character-level convolutional networks for text classification. In: Proceedings of INTERSPEECH 2015, pp. 3057–3061 (2015)
4. Dai, A.M., Olah, C., Le, Q.V.: Document embedding with paragraph vectors. arXiv:1507. 07998v1 [cs.CL], 29 July 2015
5. Kim, Y., Jernite, Y., Sontag, D., Rush, A.M: Character aware neural language models. arXiv:1508.06615v4 [cs.CL], 1 December 2015
6. Chen, T., Xu, R., He, Y., Wang, X.: Improving sentiment analysis via sentence type classification using BiLSTM-CRF and CNN. Expert Syst. Appl. **72**, 221–230 (2017)
7. Le, Q., Mikolov, T.: Distributed representations of sentences and documents. In: Proceedings of the 31st ICML, Beijing, China, vol. 32, JMLR: W&CP (2014)
8. Wieting, J., Bansal, M., Gimpel, K., Livescu, K.: CHARAGRAM: Embedding Words and Sentences via Character n-grams. arXiv:1607.02789v1 [cs.CL], 10 July 2016
9. Wang, J., Wang, Z., Zhang, D., Yan, J.: Combining knowledge with deep convolutional neural networks for short text classification. In: Proceedings of IJCAI (2017)

10. Liang, D., Xu, W., Zhao, Y.: Combining word-level and character-level representations for relation classification of informal text: In: Proceedings of the 2nd Workshop on Representation Learning for NLP, Vancouver, Canada, pp. 43–47, 3 August 2017
11. Yiny, W., Kanny, K., Yuz, M., Schutze, H.: Comparative Study of CNN and RNN for Natural Language Processing. arXiv:1702.01923v1 [cs.CL], 7 February 2017
12. Mikolov, T., Karafiat, M., Burget, L., Cernoky, J.H., Khundanpur, S.: Recurrent neural network based language model. In: Proceedings of Interspeech (2010)
13. Tang, D., Qin, B., Liu, T.: Document modeling with gated recurrent neural network for sentiment classification. In: Proceedings of EMNLP 2015, pp. 1422–1432 (2015)
14. Wang, P., Xu, J., Xu, B., Liu, C., Zhang, H., Wang, F., Hao, H.: Semantic clustering and convolutional neural network for short text categorization. In: Proceedings ACL, pp. 352–357 (2015)
15. Conneau, A., Schwenk, H., Barrault, L., Lecun, Y.: Very deep convolutional networks for text classification. In: Proceedings of the 15th Conference of the European Chapter of the ACL, vol. 1, Long Papers, pp. 1107–1116 (2017)
16. Johnson, R., Zhang, T.: Convolutional neural networks for text categorization: Shallow word-level vs. deep character-level (2016). arXiv preprint: arXiv:1609.00718
17. Johnson, R. Zhang, T.: Supervised and semi-supervised text categorization using LSTM for region embeddings. In: Proceedings of the 33rd ICML, New York, USA (2016)
18. Zhou, C., Sun, C., Liu, Z., Lau, F.C.M.: A C-LSTM Neural Network for Text Classification. https://arxiv.org/pdf/1511.08630
19. Mikolov, T., Sutskever, I., Chen, K., Corrado, G., Dean, J.: Distributed Representations of Words and Phrases and their Compositionality. arXiv:1310.4546
20. Dataset source. https://github.com/AcademiaSinicaNLPLab/sentiment_dataset, 2 January 2018
21. Glorot, X., Bengio, Y.: Understanding the difficulty of training deep feedforward neural networks. In: AISTATS, vol. 9, pp. 249–256 (2010)
22. Liu, P., Qiu, X., Huang, X.: Recurrent Neural Network for Text Classification with Multi-Task Learning. arXiv:1605.05101v1 [cs.CL], 17 May 2016

United We Stand: Using Multiple Strategies for Topic Labeling

Antoine Gourru[1](✉), Julien Velcin[1], Mathieu Roche[2,5], Christophe Gravier[3], and Pascal Poncelet[4]

[1] Université de Lyon, Lyon 2, ERIC EA 3083, Lyon, France
{antoine.gourru,julien.velcin}@univ-lyon2.fr
[2] TETIS, Univ. Montpellier, APT, Cirad, CNRS, Irstea, Montpellier, France
mathieu.roche@teledetection.fr
[3] Université Jean Monnet, Laboratoire Hubert Curien UMR CNRS 5516,
Saint-Étienne, France
christophe.gravier@univ-st-etienne.fr
[4] Univ. Montpellier, LIRMM, Montpellier, France
pascal.poncelet@lirmm.fr
[5] Cirad, TETIS, Montpellier, France
mathieu.roche@cirad.fr

Abstract. Topic labeling aims at providing a sound, possibly multi-words, label that depicts a topic drawn from a topic model. This is of the utmost practical interest in order to quickly grasp a topic informational content – the usual ranked list of words that maximizes a topic presents limitations for this task. In this paper, we introduce three new unsupervised n-gram topic labelers that achieve comparable results than the existing unsupervised topic labelers but following different assumptions. We demonstrate that combining topic labelers - even only two - makes it possible to target a 64% improvement with respect to single topic labeler approaches and therefore opens research in that direction. Finally, we introduce a fourth topic labeler that extracts representative sentences, using Dirichlet smoothing to add contextual information. This sentence-based labeler provides strong surrogate candidates when n-gram topic labelers fall short on providing relevant labels, leading up to 94% topic covering.

Keywords: Topic model · Topic labeling · Topic coherence

1 Introduction

With the ever-growing information flow, we are facing unprecedented difficulties to quickly grasp the informational content on large streams of textual documents. This happens in our daily life when we are browsing social media or syndicated contents, but also in professional processes for which the stakes are to categorize and understand large chunks of documents in an unsupervised setting. Topic models are a solution to tackle this issue. They have been successfully applied to

M. Silberztein et al. (Eds.): NLDB 2018, LNCS 10859, pp. 352–363, 2018.
https://doi.org/10.1007/978-3-319-91947-8_37

many situations, including historical record analysis, machine translation, sentiment analysis, to name a few [1]. While topic models make it possible to build sound representations of the different main *topics* discussed in the corpus they are draw from, they hardly provide human-friendly outputs. As a consequence, gaining a quick understanding of topics content is a cumbersome task. Interactive topic visualization [2] does help in this matter, though they require an in-depth and time-consuming cognitive effort to grasp each topic informational content. This is the compelling rationale for the advent of topic models based on n-grams [3] and topic labeling techniques [4,5]. Such techniques aim at providing a topic label, which means a single or multi-word title that is relevant to represent the informational content being discussed in the documents falling into that topic.

With the existing topic labeling techniques that we will discuss in Sect. 2, the popular approach for topic labeling is to rely on a single topic labeling technique. This implies a single measure to rank label candidates – therefore assuming the hypothesis that a single measure exists that befits any kind of topic regardless their number, their informational content type, their size, etc. While a single strategy may be overall satisfactory, reaching about 2.0 on average on a 0–3 Likert scale in a unsupervised setting [6], it also leaves many topics with very poor labels (0 or 1 on such scale).

In this paper, we investigate how to improve topic labeling. In this context, a new multi-strategy method is proposed with two main functions.

First, we propose in Sect. 3 three new n-gram topic labeling techniques, called M-Order, T-Order, and document-based labelers. M-Order and T-Order both leverage the odds for label candidates to be generated by the other topics as a background distribution as a penalty, while T-order also demotes label candidates with high score when they are nested in another significant candidate. Document-based labeler investigates the possibility that the best label may be found in a very few number of documents that are central to the category. We discuss how each labeler performs but also how they complement each other by showing that it is possible to reach a 64% improvement when using at least 2 labelers.

The second function of our multi-strategy approach (see Sect. 4) consists in surrogating labels using sentence information retrieval and showing that they provide a complementary approach for some topics that cannot find a proper fit with n-gram labels. Section 5 concludes and provides further lines of research.

2   Related Work

Topic labeling aims at finding a good label or title to provide a better understanding of what constitutes the homogeneity of a given topic [4]. Several measures have been proposed to associate either a term or a phrase that roughly summarizes the top-K words associated to a given topic. More recently, [6] explored new solutions relying on letter trigram vectors and word embeddings.

Both information-based topic labeling strategies [4,7] and those based on external semantic resources [5,8] only provide a satisfactory labeling efficiency in a subset of topics in a corpus. On a 0–3 Likert scale, [5] achieves 2.03 at best while [6] reaches 2.13 at best – and less than 2 in most of the cases. It does not mean that every topic finds a good match (score 2), but that a set of topics achieves 3 whereas another set achieves 1 or even 0. This has been confirmed in our own experiments that show standard deviations from 0.74 to 0.88 (see Table 2). We demonstrate in this paper that in an unsupervised setting, without the need of external semantic resources or user supervision, it is nonetheless possible to achieve results that let most of the topics be associated to a good label.

Another option is to work with representative sentences, which has been successfully integrated into topic-modeling oriented applications [9]. However, their usefulness in how the topic is understood has not been evaluated yet. In particular, in this paper we evaluate to which extent the sentences can complement the n-gram labels.

Other work investigated the use of other modalities for topic labeling, such as images. It turns out that previous work proved that labels based on n-grams showed the best performances when evaluated by human beings [10]. This explains why we focus on n-gram labeling here, with an additional interest in the possibility to enhance topic understanding with sentences as a surrogate.

3   Topic Labeling Based on n-Grams

First, our multi-strategy approach proposes several labels (i.e., phrases) by using different statistical ranking measures. In what follows, we score a candidate term t that is a sequence of p consecutive words, also called n-grams: $t = (w_1, w_2 \ldots w_p)$. We consider these candidates as possible labels for a given topic using different new proposed measures. The probability $p(w/z)$ of word w given topic z and the probability $p(z/d)$ of topic z given document d are given by the topic modeling algorithm (here, LDA).

3.1   T-Order, M-Order and Document-Based Labelers

M-Order Labeler. Our first contribution aims at improving the 0-order measure [4] that is computed by $\sum_{i=1}^{p} \log \frac{p(w_i|z)}{p(z)}$. Instead of normalizing by the marginal probability, we use the odds for the candidate to be generated by the other topics as a background distribution. With $p(w_i|z)$ the probability for the topic z to generate the ith word of t, we define a first score of relevance *M-order* as follows:

$$M\text{-}order(t, z) = \sum_{i=1}^{p} \log \frac{p(w_i|z)}{\frac{1}{|Z|-1} \sum_{z' \neq z} p(w_i|z')}$$

with Z the set of extracted topics. The denominator penalizes the candidates that are also likely to be generated by the other topics than topic z.

T-Order Labeler. To introduce the notion of "termhood" [11], we define that a term t is a short term if it is nested in a longer term t' that has a bigger value for some base measure of termhood (e.g., c-value). For example, in a computer science corpus, "Gibbs" would be a short term, because it is usually nested in "Gibbs Sampling" that has a higher termhood. In this case, the term t can be ignored. This post-processing method is akin to the "completeness" measure usage in [7]. Finally, the score is divided by the length *len* of the candidate. We can now define our new measure:

$$T\text{-}order(t, z) = \begin{cases} 0 & \text{if } t \text{ is a short term} \\ \frac{1}{len(t)} \cdot M\text{-}order(t, z) & \text{else} \end{cases}$$

The score adds a notion of distance maximization between topics to 0-order, and dividing by the term length prevents from favoring long terms too much. The use of a termhood base measure prevents the labels to be terms that are not semantically relevant.

Document-Based Labeler. In some cases, the best label can be found in a very few number of documents that are central to the category. We define our second new measure by averaging the importance of the set of documents featuring a given term:

$$Doc - Based(t, z) = \Big(\prod_{d \in D_t} imp_z(d) \Big)^{\frac{1}{|D_t|}}$$

where D_t is the set of documents in which the term t can be found and $imp_z(d)$ stands for the importance of document d in z. We decided to estimate $imp_z(d)$ in two ways. First, $doc\text{-}based_u(d, z)$ is based on $p(z/d)$ with a natural bias towards short documents. The second measure $doc\text{-}based_n(d, z)$ is based on $p(d/z) \propto p(z/d) * p(d)$ for a given z. We decided to approximate $p(d)$ by the ratio between the length of d and the total length of the corpus. The rationale of the new measure is therefore to find terms very specific to the topic although they exhibit moderate topic covering.

3.2 Evaluation

Methodology. We experiment on two case studies with the LDA model [12] and a fixed number of topics $k = 100$. The maximum number of iteration is set to 2000 and the hyperparameters α and β are automatically tuned as explained in [13]. We may do other choices, such as using other topic learning algorithms or setting k to other values. However, we think that the following experimental design is sufficient for supporting the claim of this paper.

Datasets. Topic models are drawn from the following datasets:

- **Sc-art:** a set of 18 465 scientific abstracts gathered by [14] over a period of 16 years. Many contributions in topic visualization and labeling techniques have been applied on similar datasets, on which n-grams seem to provide very relevant labels.

- **News-US:** a set of 12 067 news we gathered automatically from the Huffington Post RSS feeds (US version). This set spans a period of almost 3 months (from June the 20th until Sept. the 8th, 2016). Monitoring news feed is another case study of the utmost practical interest.

After learning topic models, we perform standard post-processing that is removing the topics that present a poor homogeneity. Actually, topics of poor quality would bias our experiments since no labeler will be capable of finding a satisfactory solution. To this end, we compute the Normalized Pointwise Mutual Information (NPMI) of the top 10 words, and we eventually remove the topics with a negative score, as it is shown in [15] that this score is correlated with human measure of the topic coherence. Therefore, we keep 145 topics (on 200 initial topics) and every annotator had 48 or 49 tasks to complete (2 annotators × 145 topics/6). Each task corresponds to the evaluation of two types of elements: (i) evaluation of candidate labels (i.e. words and/or phrases), (ii) evaluation of representative sentences.

As in previous works, the evaluation consists in measuring how well an n-gram candidate labels the topic on a Likert scale. A scale of five points is used in [7] and four in [6,16]. Our four points scale is presented on Table 1; it was constructed so that we can easily compare our results to the literature. We called six computer scientists as human annotators. The annotation task aims at evaluating the three main n-gram labels provided by the different labelers for a given tuple (dataset, topic). For any given tuple to annotate candidates were ranked randomly and the annotators were blind to the kind of labeler which generated each label. Each annotation was given to two annotators in order to calculate an agreement score.

Table 1. Our likert scale

Score	Description
3	Yes, perfectly
2.a	Yes, but it is too broad
2.b	Yes, but it is too precise
1	It is related, but not relevant
0	No, it is unrelated

For a given annotation task, we provided five documents that maximize $p(d|z)$, three documents that maximize $p(z|d)$, plus the thirty top words with their associated probabilities, as in [16].

The three most highly ranked labels were evaluated, either they have been computed by the basic measures of [4] or by our own measures presented in Sect. 3.1. The 0-order was computed with both uniform and frequency-based background distribution. The T-order was computed using the LIDF-value from [17] as a termhood measure, which seems to provide better results than the other

Table 2. Average score for the top-3 labels proposed on a Likert scale from 0 (unrelated) to 3 (perfect). σ details the average standard deviations for the two datasets.

Top-3	News-US	Sc-Art	All			
Max-Score	2.23	2.40	2.33	σ	**Too broad**	**Too precise**
T-order	1.27	1.24	1.26	0.81	13%	15%
M-order	1.25	1.20	1.22	0.8	13%	14%
$doc\text{-}based_n$	0.98	1.12	1.05	0.74	4%	16%
$doc\text{-}based_u$	1.03	1.17	1.10	0.75	4%	17%
1-order	1.07	1.31	1.19	0.84	8%	16%
$0\text{-}order_{uniform}$	1.10	1.63	1.36	0.82	7%	24%
$0\text{-}order_{frequence}$	1.18	1.23	1.20	0.88	9%	17%

termhood scoring measures after a preliminary screening. We choose not to limit the labeling candidate to be a bigram set, but to keep any length for the labels. The candidate generation was performed using the BioTex API[1] [17].

Results. Table 2 shows the average score (from 0 to 3) of the top-3 results of the labeling systems. Max-Score is the upper bound, which corresponds to selecting the score of the system that achieves the highest score for every topic. As it was presented earlier, the annotator was able to give some qualitative information when rating with 2: "it is too broad" or "it is too precise". The distribution by labeler is provided in Table 2. The kappa shows a fair agreement (0.34, 0.49, 0.51) that allows us to look for an automated labeling recommender system.

One first observation is that the simple $0\text{-}order_{uniform}$ is clearly better than the others for the Sc-Art dataset (1.63 against 1.24), but the T-order measure outperforms it for the News-US dataset (1.27 against 1.1). As it is well illustrated in Table 2, labeling systems are not always good (maximum 1.36 on average), but there is (almost) always a labeling system that is able to provide a good label (2.33/3 on average). This means that we can expect an improvement of about 64% in the labeling task. Tables 3 and 4 give a more concrete illustration of that idea on a selected set of topics.

An important result is that with 90% of the evaluated topic, a good label (meaning rated 2 or 3) is found. We name this amount as the *covering* of the labeling strategy. When we reduce the labels to those produced by the T-order and the $0\text{-}order_{uniform}$ only, there is a good label 83% of the time. A labeler alone can only achieve a good score at best for 63% ($0\text{-}order_{uniform}$) and 62% (*T-order*) of all topics.

Discussion. The presented results mean that even with a very small set of proposed labels, one can access the inner semantic content of a given topic. In the case of the two datasets we experiment on, we only need six labels (meaning,

---
[1] http://tubo.lirmm.fr/biotex/.

Table 3. Examples of topics learned on our datasets

Topic 1 (News-US)	Topic 2 (Sc-Art)	Topic 3 (News-US)	Topic 4 (Sc-Art)
EU	Detection	Mental	User
Brexit	Event	Health	Web
Britain	Events	Depression	Users
European	System	Illness	Filtering
Leave	Detecting	Suicide	Profiles
Vote	False	Anxiety	Collaborative
British	Detect	Disorder	Usage
London	Intrusion	Care	Preference
Minister	Vehicle	Social	System
Referendum	Anomaly	Bell	Site

Table 4. The words in bold where rated 3, the others 1. We see that for some topics the 0-order is able to find a good label whereas it is the T-order for other topics.

	Topic 1	Topic 2
T-order	**Brexit**	Intrusion
0-order	British prime minister david cameron	**Intrusion detection systems**
	Topic 3	Topic 4
T-order	**Bipolar disorder**	Preference
0-order	National suicide prevention lifeline	**User preference**

three labels produced by two labelers, if there is no overlap). On average, among these six labels, 15% are rated as unrelated, 46% as related, 29% as good label and 9% as perfect label, 1% have been rated "I don't know".

If we consider 2 and 3 as good labeling scores, the precision is about 38% (29% + 9%), even though the other answers are not totally wrong. For instance, the topic 3 presented in Table 3 gets the following labels (0-order followed by T-order labels): "bipolar disorder, depression, anxiety, national suicide prevention lifeline, mental health disorder, mental health care".

The presented results can be thought as over-optimistic: they need further experiment on other various datasets (e.g., book series or blog posts) and we know that within the labels given to the users there is still unrelated/non relevant items (in our case, about 15% of the proposed labels). This is the reason why we need to find advanced strategies to (a) improve the quality of the recommended labels, (b) complete the labels when the n-gram based approach is not sufficient to fully capture the inner semantics.

Regarding (a), we can naturally think at integrating other features than the relation between topic-word probabilities and the candidate label. For instance, we can add features related to the topic (importance of the topic, skewness of

Table 5. Performance of the labeling systems, meaning the percent of a least one good label (rated 2 or 3) in the top-3 labels

Systems	Performance
Max-Score	90
T-order	62
M-order	60
$doc\text{-}based_n$	46
$doc\text{-}based_u$	51
1-order	53
$0\text{-}order_{uniform}$	63
$0\text{-}order_{frequence}$	55
T-order + $0\text{-}order_{uniform}$	83

the word distribution, etc.) or to the dataset (for instance, longer n-grams can be favored in scientific datasets). We can also follow the work of Lau et al. [5] in leveraging more supervision in the labeling process.

Regarding (b), it seems that for some cases, the n-gram labelers can never achieve a satisfactory output. This is the reason why we propose, in the next section, to leverage information retrieval techniques to find relevant *sentences*.

4   Topic-Relevant Sentence Extraction

In the second function of our multi-strategy approach, different representative sentences are proposed in order to identify the semantic content of the topics.

4.1   Rationale

Would no n-gram labeler yields an acceptable label candidate, looking for representative sentences from the corpus is another solution to label a topic. For example, the labels returned by all the n-gram based systems for the two topics in Table 6 were rated low (the maximum score being 0 for the first and 1 for the second). But we can find sentences that were well rated, as shown in Table 7.

4.2   Sentence Extraction Solution

With this fourth new labeling technique, we assume that an information retrieval procedure can be used to post-process the top documents (considered as the "context") and look for representative sentences. The top documents, i.e. the documents that maximize $p(d/z)$, are split into sentences. We propose to use a Dirichlet smoothing to add contextual information.

Table 6. Example of two topics badly labeled by n-grams.

Topic 5 (News-US)	Topic 6 (News-US)
Photo	Facebook
Posted	Media
2016	Social
PDT	Online
Jul	App
39	Internet
Instagram	Video
Aug	Google
Jun	Users
34	Site

Table 7. Two extracted sentences that can help the user capturing the meaning of topics 5 and 6 given in Table 6 (words occurring in top words are highlighted in bold).

Topic	Example of sentence returned by our systems
5	A **photo posted** by Laura Izumikawa Choi (@lauraiz) on **Jun** 17, 2016 at 11:05 am **PDT**
6	So 'follow' or 'Like' them on **social media** sites like Twitter, **Facebook**, LinkedIn, **Google** + and Pinterest

We define β, the context distribution of a document collection, by:

$$\beta_w = \frac{c(w, C)}{\sum_{w \in V} c(w, C)} \tag{1}$$

where $c(w, C)$ counts the frequency of word w in the context C. With μ as a positive real number, we obtain the following language model:

$$\theta_w^x = \frac{c(w, x) + \mu \beta_w}{len(x) + \mu} \tag{2}$$

where $c(w, x)$ stands for the frequency of word w in the candidate sentence x. We can then compute different distance measures between the sentence vector representation and the topic. We choose to compute a negative Kullback-Leibler distance, like [4], and a simple cosine similarity. If $\mu = 0$, the θ_w^x calculated is a simple TF representation of the sentence. The greater μ is, the more importance we give to the context (the top documents). Our model is parameterized by: β (more precisely, the number of top documents $|\beta|$ we choose to keep) and μ (the amount of context we want to take into account).

4.3   Evaluation

We experiment with the same models and datasets than the n-gram evaluation of the previous section. We choose to ask the following question: "Does the sentence give a clear understanding of the topic content?". Then, the rater could choose between "yes", "no", or "don't know". The systems are presented in Table 8. We choose to compare our systems with random sentences, extracted from documents that do not maximize $p(d/z)$. We call **Rand** this system based on random sentences.

Table 8. Evaluated systems with different parameters' values.

Name	Similarity	μ	$|\beta|$
$COS10$	Cosine	0	10
$COS15$	Cosine	0	15
$COSIDF15$	Cosine	0 (IDF weighted)	15
$B10_{0,1}$	Negative KL divergence	0.1	10
$B10_{10}$	Negative KL divergence	10	10
$B10_{1000}$	Negative KL divergence	1000	10
$B20_{0,1}$	Negative KL divergence	0.1	20
$B20_{10}$	Negative KL divergence	10	20
$B20_{1000}$	Negative KL divergence	1000	20

As for the n-grams evaluation in previous section, a weighted Kappa was computed for every annotator pair. The results are similar: it is not really high (0,23, 0.36, 0.04), but sufficient for a significant agreement. Table 9 presents the average proportion of extracted sentences tagged as 1 (answer "yes" to the question: "Does the sentence give a clear understanding of the topic content?"). It shows that a simple cosine based on a TF vector using the top 10 documents is better, without the need of smoothing. However, a closer look shows that $B20_{0,1}$ (meaning a really small smoothing) is slightly better for News-US.

Table 9. Percent of relevance, meaning the proportion of topics correctly illustrated by the sentence.

System	News-US	Sc-Art	All	System	News-US	Sc-Art	All
Rand	1%	6%	4%	$B10_{10}$	34%	38%	36%
$COS10$	34%	46%	41%	$B10_{1000}$	25%	28%	27%
$COS15$	35%	45%	40%	$B20_{0,1}$	40%	31%	35%
$COSIDF15$	22%	30%	26%	$B20_{10}$	34%	37%	36%
$B10_{0,1}$	38%	33%	35%	$B20_{1000}$	25%	26%	26%

We can now wonder whether sentences can be combined with n-gram labels to improve the overall topic understanding. For instance, we can estimate the proportion of topics for which we can find at least one good n-gram label (with 0-order or T-order) and, if we cannot, one sentence otherwise. The performance goes from 83% covering up to 93% with the COS_{15} labeler. This improvement can be seen even with no agreement among the last annotator pair. If we ignore the annotations attributed by the last annotator pair (weighted Kappa of only 0.04), the covering goes even until 94%.

5   Conclusion

Finding a suitable textual description of the output of a statistical model is still a difficult task that can be related to interpretable machine learning. In this paper, we introduced three new n-gram topic labelers that are at least on par with the existing labeling techniques. A key observation is that those new labelers are complementary labelers to the best known so far (0-*order* labeler) so that when one of them is combined with the 0-*order* labeler, the resulting combined labeler provides labels scoring 2 or more on a 0–3 Likert scale. A direct application of this consists in a simple recommender system that suggests a limited set of labels (e.g., the labels produced by T-order and 0-*order*$_{uniform}$) that are mostly of good quality. When expanding such a combined n-gram labeler with extracted sentences from the corpus as surrogate labels for difficult labeling cases, it is possible to reach the same performance for 94% of the topics to label.

There is still room for future research. First of all, we can consider the covering we get as a good recall, but the precision (meaning, the proportion of really good labels among the labels returned by the system) needs to be improved. If we are able to automatically choose the perfect labeler for each case, this will constitute an important improvement[2]. If we assume that the user can still have difficulties for selecting the best labels for *some* topics, we might adopt a semi-supervised system following [5]. Another track would be to define a label as a combination of n-grams and sentences that are complementary for they propose different views over the targeted topic. The must would be to generate a small paragraph that summarizes the underlying topic semantics.

Acknowledgments. This work is partially funded by the SONGES project (Occitanie and FEDER).

References

1. Boyd-Graber, J., Yuening, H., Mimno, D., et al.: Applications of topic models. Found. Trends® Inf. Retr. **11**(2–3), 143–296 (2017)
2. Kim, D., Swanson, B.F., Hughes, M.C., Sudderth, E.B.: Refinery: an open source topic modeling web platform. J. Mach. Learn. Res. **18**(12), 1–5 (2017)

---

[2] We plan to publicly release the annotations made by our human judges.

3. Wang, X., McCallum, A., Wei, X.: Topical n-grams: phrase and topic discovery, with an application to information retrieval. In: Seventh IEEE International Conference on Data Mining (ICDM), pp. 697–702. IEEE (2007)
4. Mei, Q., Shen, X., Zhai, C.: Automatic labeling of multinomial topic models. In: Proceedings of the 13th ACM SIGKDD International Conference on Knowledge Discovery and Data Mining, pp. 490–499. ACM (2007)
5. Lau, J.H., Grieser, K., Newman, D., Baldwin, T.: Automatic labelling of topic models. In: Proceedings of Annual Meeting of ACL-HLT, vol. 1, pp. 1536–1545. Association for Computational Linguistics (2011)
6. Kou, W., Li, F., Baldwin, T.: Automatic labelling of topic models using word vectors and letter trigram vectors. In: Zuccon, G., Geva, S., Joho, H., Scholer, F., Sun, A., Zhang, P. (eds.) AIRS 2015. LNCS, vol. 9460, pp. 253–264. Springer, Cham (2015). https://doi.org/10.1007/978-3-319-28940-3_20
7. Danilevsky, M., Wang, C., Desai, N., Ren, X., Guo, J., Han, J.: Automatic construction and ranking of topical keyphrases on collections of short documents. In: Proceedings of the 2014 SIAM International Conference on Data Mining, pp. 398–406. SIAM (2014)
8. Magatti, D., Calegari, S., Ciucci, D., Stella, F.: Automatic labeling of topics. In: Ninth International Conference on Intelligent Systems Design and Applications, ISDA 2009, pp. 1227–1232. IEEE (2009)
9. El-Assady, M., Sevastjanova, R., Sperrle, F., Keim, D., Collins, C.: Progressive learning of topic modeling parameters: a visual analytics framework. IEEE Trans. Vis. Comput. Graph. **24**, 382–391 (2017)
10. Aletras, N., Baldwin, T., Lau, J.H., Stevenson, M.: Evaluating topic representations for exploring document collections. J. Assoc. Inf. Sci. Technol. **68**(1), 154–167 (2017)
11. Frantzi, K.T., Ananiadou, S., Tsujii, J.: The *C-value/NC-value* method of automatic recognition for multi-word terms. In: Nikolaou, C., Stephanidis, C. (eds.) ECDL 1998. LNCS, vol. 1513, pp. 585–604. Springer, Heidelberg (1998). https://doi.org/10.1007/3-540-49653-X_35
12. Blei, D.M., Ng, A.Y., Jordan, M.I.: Latent Dirichlet allocation. J. Mach. Learn. Res. (JMLR) **3**, 993–1022 (2003)
13. Wallach, H.M., Mimno, D.M., McCallum, A.: Rethinking LDA: why priors matter. In: Advances in Neural Information Processing Systems, pp. 1973–1981 (2009)
14. Tang, J., Wu, S., Sun, J., Su, H.: Cross-domain collaboration recommendation. In: Proceedings of the 18th ACM SIGKDD International Conference on Knowledge Discovery and Data Mining, pp. 1285–1293 (2012)
15. Aletras, N., Stevenson, M.: Evaluating topic coherence using distributional semantics. In: IWCS, vol. 13, pp. 13–22 (2013)
16. Li, Z., Li, J., Liao, Y., Wen, S., Tang, J.: Labeling clusters from both linguistic and statistical perspectives: a hybrid approach. Knowl. Based Syst. **76**, 219–227 (2015)
17. Lossio-Ventura, J.A., Jonquet, C., Roche, M., Teisseire, M.: Biomedical term extraction: overview and a new methodology. Inf. Retr. J. **19**(1), 59–99 (2016)

Evaluating Thesaurus-Based Topic Models

Natalia Loukachevitch$^{(\boxtimes)}$ and Kirill Ivanov

Lomonosov Moscow State University, Moscow, Russia
louk_nat@mail.ru, ivanov.kir.m@yandex.ru

Abstract. In this paper, we study thesaurus-based topic models and evaluate them from the point of view of topic coherence. Thesaurus-based topic models enhance the scores of related terms found in the same text, which means that the model encourages these terms to be on the same topics. We evaluate various variants of such models. First, we carry out a manual evaluation of the obtained topics. Second, we study the possibility to use the collected manual data for evaluating new variants of thesaurus-based models, propose a method and select the best its parameters in cross-validation. Third, we apply the created evaluation method to estimate the influence of word frequencies on adding thesaurus relations for generating coherent topic models.

Keywords: Topic models · Thesaurus · Content-based analysis

1 Introduction

Domain-specific terms in form of subject headings, subject facets or subject categories are traditionally used in domain information systems for navigation through the text collection and visualization of the collection contents [5,12]. Often, for dynamic information flows, such pre-defined terminological tools cannot provide satisfactory coverage of all significant thematic issues. Therefore, currently, statistical approaches such as probabilistic topic modeling become important tools for support of expert analysis and visualization of document collections [20,21]. Automatic topic labeling methods [11] with a function similar to subject headings have been proposed.

In their experiments, Boyd-Graber et al. [4] found that to be understandable by humans, topics should be specific, coherent, and informative. Relationships between the topic components could be inferred. This requires testing of topic models by humans, which is a very expensive procedure. Previously, automatic measures for evaluating coherence have been proposed [10,17,19]. However, the obtained scores are not always justified [14].

The work is supported by the Russian Foundation for Basic Research (project 16-29-09606).

M. Silberztein et al. (Eds.): NLDB 2018, LNCS 10859, pp. 364–376, 2018.
https://doi.org/10.1007/978-3-319-91947-8_38

Sets of topics for the same text collection can be generated using different approaches and their parameters [2], regularization techniques [16,22], combinations of single words, extracted phrases or domain-specific terms [3,18]. For example, the top of a unigram topic for a news portal can include: *Syria, Syrian, Assad, UN, opposition...* A phrase-based topic with similar content can look as follows: *Syria, Bashar al-Assad, the Syrian opposition, Syrian, al-Assad regime....* Therefore, it is important to understand what types of topic models are better for human perception.

In this paper, we study the integration of prior knowledge consisting of a thesaurus and multiword expressions in three steps. First, we manually evaluate more than thirty different models applied to the same text collection. Second, we study the possibility to use the collected manual data for evaluating other models: we propose a method and select the best its parameters in cross-validation. Our work is the first one studying an application of manually labeled topic models to evaluate other topic models. At the third step, we apply the created evaluation method to estimate correlations between word frequencies and the necessity to add their thesaurus relations to obtain more coherent topics.

2   Related Work

Topic modeling approaches [2] are unsupervised statistical algorithms that usually consider each document as a "bag of words", without accounting any relations between them. There were several attempts to enrich word-based topic models (=unigram topic models) with additional prior knowledge or multiword expressions. Andrzejewski et al. [1] incorporated knowledge by Must-Link and Cannot-Link primitives represented by a Dirichlet Forest prior. These primitives were then used in [16], where similar words are encouraged to have similar topic distributions. However, all such methods incorporate knowledge in a hard and topic-independent way.

Xie et al. [25] proposed the MRF-LDA topic model, which utilizes the distributional similarity of words to improve the coherence of topic modeling. Within a document, if two words are labeled as similar, their latent topic nodes are connected by an undirected edge and a binary potential function is defined to encourage them to share the same topic label. In [6], the authors gather so-called lexical relation sets (LR-sets) for word senses described in WordNet. The LR-sets include synonyms, antonyms, and adjective-attribute related words. To adapt LR-sets to a specific domain corpus and to remove inappropriate lexical relations, the correlation matrix for word pairs in each LR-set is calculated. This matrix is used for filtrating inappropriate senses and for modifying the initial LDA topic model according to the generalized Polya urn model [15], which boosts probabilities of related words in word-topic distributions. Gao and Wen [7] presented Semantic Similarity-Enhanced Topic Model that accounts for corpus-specific word co-occurrence and WordNet-based semantic similarity using the generalized Polya urn model.

All above-mentioned approaches to adding knowledge to topic models are limited to single words. Approaches using ngrams in topic models can be subdivided into two groups. The first group of methods tries to create a unified probabilistic model accounting unigrams and phrases [8,23,24]. However, all these models are enough complex and hard to compute on real datasets.

The second group of methods is based on the preliminary extraction of ngrams and their further use in topics generation. Initial studies of this approach used only bigrams [9]. Nokel and Loukachevitch [18] proposed the LDA-SIM algorithm, which integrates top-ranked ngrams. They proposed to create similarity sets of expressions having the same word components and sum up frequencies of similarity set members if they co-occur in the same text. In [14] the application of the LDA-SIM algorithm for integration several thesauri (Word-Net, EUROVOC), including multiword terms and direct thesaurus relations, has been studied. It was found that thesaurus-based models can be much more coherent than unigram models but there are some important parameters of the thesaurus-based models, which should be additionally studied.

To measure human interpretability of statistically generated topics, several evaluations have been organized. Mimno et al. [15] described an experiment comparing expert evaluation of LDA-generated topics and automatic topic coherence measures. The experts annotated each topic according to the 3-class scale as "good", "intermediate", or "bad". It was found that most "bad" topics consisted of words without clear relations between each other. Newman et al. [17] also asked users to score topics on a 3-point scale. They indicated that a coherent topic usually can be described with a short label.

To automate the measuring of the topic interpretability, automatic topic coherence measures have been proposed. Their calculation is based on similarity scores between words in the top of a topic. The similarity scores can be established on the basis of some resource (for example, WordNet) or word co-occurrence in some corpus. Newman et al. [17] found that the best measure having the largest correlation with human scores is word co-occurrence from the top of topics calculated as point-wise mutual information (PMI) on such an external collection as Wikipedia articles.

Later Lau et al. [10] showed that normalized pointwise mutual information (NPMI), counted on Wikipedia articles, correlates with human scores even more. In [19], an integrated model of automatic topic coherence has been evaluated on several datasets. It was confirmed that Wikipedia-based coherence measures are much more reliable than the use of a source collection. But exploiting an external corpus for a topic model evaluation can be inappropriate for a specific domain because many thematically related words or phrases can be rarely mentioned or co-occurred in Wikipedia.

3   Thesaurus-Based Topic Models

We study the approach of creating thesaurus-based topic models proposed by [14]. They create so-called similarity sets between semantically related language units. Every unit (word or phrase) can have such a similarity set. The single word's similarity set can include phrases where this word is a component or related word or phrases described in some thesaurus. The phrase's similarity set can comprise phrases with the same component words, single component words, and related word or phrases according to a thesaurus. For example, word Syria can have the following similarity set: *Syrian, Damascus, President of Syria, Government of Syria.*

It was supposed that if units from the same similarity set co-occur in the same document then their contribution into the document's topics is really more than it is presented with their frequencies, therefore their frequencies should be increased. In such an approach, the algorithm can "see" the relatedness of language units. The authors of [14] implemented the algorithm LDA-SIM, which changed frequencies in the process of topic inference. The approach was applied to different text collections and thesauri, it was evaluated with only automatic coherence measures. The evaluation showed that the results depend on several factors and, therefore, it seems that further study is needed.

In our approach, at first, we include related single words and phrases from a thesaurus in similarity sets. Then, we can add preliminarily extracted phrases into these sets and, in this way, we can use two different sources of external knowledge. In our implementation of thesaurus-based topic modeling, we use BigARTM library for inferring topic models[1] [22]. The library uses the Vowpal Wabbit input data format, in which each document is represented as a single line, and each unit (word or phrase in our case) in a document is provided with its frequency in this document. Thus, for each language unit, we create its similarity set and can change the initial frequency of a unit in a document.

In the current study, we substitute the initial frequency of a unit to the overall frequency of related units from its similarity set mentioned in the document under analysis. Units in similarity sets can comprise single words, thesaurus phrases or on-the-fly extracted noun phrases. It was shown in [14] that different principles of forming similarity sets can lead to improving or degrading initial unigram topic model. In this paper, we would like to study various variants and parameters of thesaurus-based models.

We can compare thesaurus-based topic models based on similarity sets with the approaches applying the generalized Polya urn model [6,7,15]. To add prior knowledge, those approaches change topic distributions for related words globally in the collection. We modify topic probabilities for related words and phrases locally, in specific texts, only when related words co-occur in these texts.

4   Manual Evaluation of Thesaurus-Based Topic Models

To estimate the quality of topic models, we chose Islam informational portal "Golos Islama" (Islam Voice)[2] (in Russian). This portal contains both news articles related to Islam and articles discussing Islam basics. We extracted the site contents using Open Web Spider[3] and obtained 26,839 pages. We supposed

---

[1] http://bigartm.org/.

[2] https://golosislama.com/.

[3] https://github.com/shen139/openwebspider/releases.

that the thematic analysis of this specialized site can be significantly improved with domain-specific knowledge described in the thesaurus form.

Table 1. Results of manual labeling of topic models for the Islam site

N	Model	100 topics	200 topics
1	LDA unigram	163	334
2	LDA+1000phrases	161	316
3	LDA+>10phrases	148	308
4	Sim+1000phrases	180	344
5	Sim+>10phrases	180	337
6	Sim+UnarySyn	157	323
7	Sim+Rel+1000phr	159	301
8	Sim+Rel+>10phr	150	295
9	Sim+Rel/hyp+1000phr	153	310
10	Sim+Rel/hyp+>10phr	174	302
11	Sim+Rel/GL+>10phr	**186**	**350**
12	Sim+ Rel/GL/hyp+>10phr	**184**	**346**

To combine knowledge with a topic model, we used the RuThes thesaurus. RuThes in a linguistic ontology of Russian intended for natural language processing [13]. It has many similarities with WordNet thesaurus but also has some specific features. In particular, RuThes contains terminology of the sociopolitical domain, which includes economy, financial, political and other terms. The current volume of RuThes is 170 thousand words and expressions. Together with RuThes, we used Islam thesaurus, which contains more than 5 thousand Russian Islam-related terms.

We used the LDA approach to topic modeling [2]. For each topic model under analysis, we ran two experiments with 100 topics and with 200 topics. The generated topics were evaluated by two terminologists, who had previously worked with the Islam-related texts. The evaluation task was formulated as follows: the experts should read the top elements of the generated topics and try to formulate labels of these topics. The labels should be different for each topic in the set generated with a specific model. The experts should also assign scores to the topics' labels:

- 2, if the label describes almost all elements of ten top elements of the topic
- 1, if several top elements do not correspond to the label,
- 0, if the label cannot be formulated.

Then we can sum up all the scores for each model under consideration and compare the total scores in value. Thus, maximum values of the topic score are 200 for a 100-topic model and 400 for a 200-topic model. In this experiment, we do not measure inter-annotator agreement for each topic, but try to get expert's general impression. In the current experiment, we use add only multiword expressions described in the thesaurus. We added thesaurus phrases in two ways: most frequent 1000 phrases (as in [9,18]) and phrases with frequency more than 10 (>10phr.): the number of such phrases is 9351.

As baselines, we considered the LDA unigram model and LDA models with additional phrases. Other models use similarity sets (SIM) of different types. The results of the evaluation are shown in Table 1. It can be seen that if we add phrases without accounting component similarity (Runs 2, 3), the quality of topics decreases: the more phrases are added, the more the quality degrades. But if the similarity between phrase components is considered then the coherence significantly improves and becomes better than for unigram models (Runs 4, 5).

Adding only unary synonyms into similarity sets decreases the quality of the models (Run 6) according to the human scores. The problem of this model is in that non-topical, general words are grouped together, reinforce one another but do not look as related to any topic. Adding all thesaurus relations is not very beneficial (Runs 7, 8). The explanation is the same: general words can be grouped together. Then we suppose that we can reduce overgeneralization if we exclude hyponyms from similarity sets (Rel/hyp). For example, that if term *Sunni* is described as a hyponym of Muslim, the similarity set for *Sunni* will contain word *Muslim*, but the similarity set for *Muslim* does not contain word *Sunnite*. It means that more concrete words and phrases obtain higher frequencies in topic inference. If we consider all relations except hyponyms, the human scores are mainly better for corresponding runs (Runs 9, 10).

At last, to avoid too general topics, we removed so-called General Lexicon concepts from the RuThes data, which are top-level, non-thematic concepts that can be met in arbitrary domains [13] and considered all-relations and without-hyponyms variants (Runs 11, 12). These last variants achieved maximal human scores because they add thematic knowledge and do not include too general knowledge, which can distort topics. The top of Syria thesaurus-based topic is as follows: *Syria, Syrian, Assad, Damascus, Bashar al-Assad, the Syrian opposition, opposition, al-Assad regime, regime* (translation from Russian).

5   Use of Manual Data to Evaluate Other Models

As a result of the previously described experiment, we obtained 33 evaluated topic models annotated by two experts, which contains 4700 different topics. To continue manual evaluation of other topic models without additional expenses, it is important to study methods to predict the manual score for a topic model from the available data. We took the average of two scores assigned by annotators as the human score of the topic model.

For such prediction, we should find the most similar topics for each topic from the target (new) topic model, to calculate its score, then to calculate the overall score of the target topic model. For evaluation, we compare the obtained overall score with the real human score. We used leave-one-out cross-validation. In this procedure we encountered the following problems:

- the topics in the data can have elements of different nature: single words, extracted or thesaurus phrases; such elements should be compared;
- the similarity between topics can be based on different numbers and different distributions of top elements in compared topics;
- some topics have very low element-based similarity to topics generated by other topic models: the matched elements in the top 20 elements can be less than 4–5;
- most topics in the dataset are interpretable: they have human scores more than 0. Thus, we have the underrepresentation of topics with zero scores in the dataset.

At first, we transform topics with phrases into unigrams topics in the following way. If a topic contains phrase w_1w_2 with probability p then we transform this fragment to w_1p, w_2p. All unigrams are included into a topic only once with their maximal probabilities. For example, the top of the topic "*Syria, Bashar al-Assad, the Syrian opposition, Syrian, al-Assad regime...*" will be transformed to "*Syria, Bashar, al-Assad, Syrian, opposition, regime...*" with the corresponding probabilities.

The similarity of topics was considered as follows. If the topic T_1 has more matching elements with the target topic T_0 than topic T_2, then topic T_1 is more similar to T_0. If the number of intersections of T_1 and T_2 with T_0 is the same than two additional measures can be applied as variants. The first measure is uninterpolated average precision (AvP) of matching elements. The second measure is Jensen-Shannon divergence (JSdiv) of matching elements. AvP is greater if the matching elements are located closer to the top of each topic under consideration. Jensen-Shannon divergence is smaller when the probabilities of matching elements are similar in topics under consideration.

As a baseline for calculating the score of a new topic model, we took KNN method. In our case, it has two parameters: K, the number of the most similar topics, and N, the number of top elements of topics to be considered. We calculate the predicted score of the target topic t_0 as the averaged value of scores of all K topics, the contribution of the each topic's score is proportional to the number of matching elements:

$$PredictedScore(t_0) = \frac{\sum_{i=1}^{K}(score_i \cdot (Meas_i \cdot l_i))}{\sum_{i=1}^{K}(Meas_i \cdot l_i)}$$

where $score_i$ is the expert score of topic t_i, $Meas_i$ is equal to AvP_i or $(1 - JSdiv_i)$, l_i is the number of matching elements.

In Table 2 we show the best variants of KNN-based prediction for each N (number of top elements). We calculated an average deviation of each topic model's score (AvDev) and Spearman correlation rank of calculated scores in comparison to manual scores. Spearman correlation shows the coincidence of topic ordering according to human and automatic scores. For calculating Spearman correlation coefficient, 100- and 200- topic model scores were transferred to the same score scale.

Table 2. Applying KNN to the prediction of models' scores. Numbers in parentheses denote the best K (number of topics) for a given N (number of top elements)

N	AvP		JSdiv	
	AvDev	Spearman	AvDev	Spearman
10	6.73 (3)	0.875 (6)	7.20 (2)	0.859 (3)
15	**6.24** (3)	0.871 (4)	7.61 (2)	0.873 (6)
20	6.71 (3)	0.872 (8)	6.99 (2)	0.869 (3)
25	6.51 (3)	0.880 (10)	6.80 (2)	0.859 (5)
30	6.68 (3)	0.876 (8)	7.24 (2)	**0.885 (3)**
35	6.85 (5)	0.867 (8)	7.74 (4)	0.883 (5)
40	6.62 (4)	0.879 (6)	6.80 (4)	0.884 (4)
45	6.66 (4)	0.878 (6)	6.82 (3)	0.882 (4)
50	6.73 (5)	0.871 (6)	6.57 (4)	0.878 (6)

Analyzing the prediction scores for specific topics, we found that KNN predicts too high scores for incoherent or senseless topics, because of a relatively small number of bad topics among our topics. For this reason, an incoherent topic can be considered similar to quite coherent topics but via a small number of matched elements.

Therefore we propose another method called Penalty method that additionally requires a given number of matching elements in two topics. Two first parameters of the Penalty method are the same as in KNN: N is the number of the topic's top elements under consideration and K is the number of the most similar topics. The third parameter of the Penalty method (L) determines the required number of topic's elements to be matched. If the most similar topics to the target topic have less matching elements than L then the estimated score is proportionally decreased.

For example, suppose, L is set to 10, K is equal to 2. Two most similar topics have high human scores (2) but the number of matching elements is only 8 in both cases then the predicted score of a topic is $2 * (16/20) = 1.6$. The contribution of topics with different scores to the predicted score is calculated using the same formula as for KNN. Table 3 shows the results of the Penalty method. It can be seen that the deviations from correct scores are much less than for KNN, and Spearman correlation rank is considerably larger.

We can also check how KNN and Penalty methods predict scores for randomly generated word groups. It allows us to estimate how the methods would estimate scores for a low-quality topic model. With this aim, we randomly generated 100-word groupings. The probability of a word in a random grouping is directly proportional to its frequency in the text collection. Table 4 shows the prediction scores of these random word groupings. For prediction, we used the best parameters for each method. We can see that the Penalty method estimates the score of a random word grouping much lower than a real topic model. KNN obtains a low score for the random grouping but this score is close to scores of real topic models.

Table 3. Applying the penalty method to prediction of models' scores

K of topics	Method	N top	L of matches	Average deviation	Spearman
1	JSdiv	45	12	7.01	0.867
2	JSdiv	40	19	5.63	**0.919**
3	JSdiv	40	19	5.79	0.913
4	JSdiv	40	19	5.17	0.917
5	JSdiv	40	19	5.15	0.910
1	AvP	20	5	6.94	0.884
2	AvP	40	19	6.06	0.911
3	AvP	40	19	5.91	0.916
4	AvP	40	19	5.50	0.914
5	AvP	40	17	5.83	0.915

Table 4. Prediction of scores for randomly generated topic model

Method	Parameters	Score
Penalty, AvP	3 Topics, 40 (19)	107.4
Penalty, JSdiv	3 Topics, 40 (19)	108.6
KNN, AvP	10 topics, 25 Top Elements,	158.1
KNN, JSdiv	5 topics, 35 Top Elements,	159.0
Lowest manual score for a real topic model		145.0

6   Evaluation of New Topic Models

In previous sections, it was shown that the best human scores can be obtained with a thesaurus-based topic model when the General Lexicon domain was excluded. For RuThes, it is possible because there is a special system of labels denoting the General Lexicon concepts [13]. For possible use of other thesauri, we should study simpler techniques to exclude frequent general words from exploiting their thesaurus relations.

First, we can suppose that fairly frequent words in a collection do not require an additional accounting of their relations because they have enough statistical evidence for topics they belong if any. In our collection of 26 thousand documents we define several thresholds of document frequency: if the term document frequency (df) is less than the threshold than thesaurus and component relations are added to term similarity sets, otherwise, the similarity set for this term is not formed. Second, it is possible that tf.idf-based measures can be more appropriate for regulating the use of additional relations in forming a coherent topic model.

For evaluation of all new variants of topic models, we used the Penalty method with the following parameters: 3 most similar topics, 40 top elements, 19 matching elements. We study three models: (1) using all thesaurus and phrase component relations, (2) using thesaurus and phrase component relations except hyponym relations. The model (3) does not include hyponyms in the similarity sets; besides, similarity sets of single words do not include phrases in that these words are components. The models' units include thesaurus phrases with frequency more than 10.

Table 5 shows the results of the evaluation depending on the df threshold. It can be seen that models (1) and (2) are mainly improved when the thesaurus relations are used for less frequent words and phrases. If we exclude hyponym relations in the model (2), then the average level of model scores is higher than the model (1) using all relations. The model (3) shows the best results and is better when more number of thesaurus relations is exploited.

Table 5. Using frequency thresholds to regulate the use of thesaurus relations in generating topic models

Threshold	(1) All relations model	(2) Without hyponyms model	(3) without hyponyms and without phrases for single words
2000	157/299	167/313	**175/344**
1500	157/303	159/320	**174/344**
1000	159/310	163/322	172/341
500	172/325	171/325	172/342
400	171/318	168/328	173/**346**
300	169/ 318	171/ 335	172/341
200	170/326	167/329	170/339
100	170/319	170/328	170/330

For calculation tf.idf, we use bm25 formula with coefficient $k = 2$ and $b = 0$. This formula is convenient because tf belongs to the restricted interval of values. We again determine some thresholds but in this case, thesaurus and component relations are utilized if term's tf.idf is more than the given threshold. Table 5 shows the results of evaluation depending on the tf.idf thresholds. The same correlation for models (1) and (2) can be seen: the stricter thresholds give better results. Again the model (2) with the exclusion of hyponyms relations is better than model (1), using all relations. And model (3) is again the best model.

From these automatic evaluations, we can conclude that two models really improve when the frequency thresholds are used. The experiment also confirmed the results of the previous manual evaluation that exclusion of hyponyms from term similarity sets and phrases from their components similarity sets (model 3) improves the human quality of topics because gives more weights to more concrete, specific words and phrases, which contents are more definite (Table 6).

Table 6. Using TFIDF thresholds to regulate the use of thesaurus relations in generating topic models

Threshold	(1) All relations model	(2) Without hyponyms model	(3) without hyponyms and without phrases for single words
6	170/323	172/335	170/337
5	165/323	170/320	171/ **344**
4	159/307	164/321	**174** /339
3	155/299	163/311	**175**/ **344**

7 Conclusion

In this paper, we studied thesaurus-based topic models and evaluated them from the point of view of topic coherence. Such topic models enhance scores of related terms found in the same text, which means that the model encourages these terms to be on the same topics. First, we carried out the manual evaluation of the obtained topics and found that the main problem of such topic models is the overgeneralization when general frequent words are grouped together but these groupings do not seem much meaningful.

Second, we studied the possibility to use the collected manual data for evaluating new variants of thesaurus-based models. We proposed the Penalty method, showed that it is much better than baseline KNN, and select the best its parameters in cross-validation. This is the first work when manual labeling data are used for evaluating new topic models. Third, we applied the Penalty method to estimate the influence of word frequencies on adding thesaurus relations during generating topic models and found that document frequencies really influence on the coherence of thesaurus-based topic models.

In all experiments, we found that exclusion of hyponyms from similarity sets and enhancing frequencies of phrases helps to overcome the overgeneralization problem. Such thesaurus-based approach creates more coherent and understandable topic models.

References

1. Andrzejewski, D., Zhu, X., Craven, M.: Incorporating domain knowledge into topic modeling via Dirichlet Forest priors. In: Proceedings of the 26th Annual International Conference on Machine Learning, pp. 25–32. ACM (2009)
2. Blei, D.M.: Probabilistic topic models. Comm. ACM **55**(4), 77–84 (2012)
3. Blei, D.M., Lafferty, J.D.: Visualizing topics with multi-word expressions. arXiv preprint arXiv:0907.1013 (2009)
4. Boyd-Graber, J., Mimno, D., Newman, D.: Care and feeding of topic models: problems, diagnostics, and improvements. In: Handbook of Mixed Membership Models and Their Applications, pp. 225-255 (2014)

5. Broughton, V.: The need for a faceted classification as the basis of all methods of information retrieval. In: Aslib Proceedings, vol. 58, pp. 49–72. Emerald Group Publishing Limited (2006)
6. Chen, Z., Mukherjee, A., Liu, B., Hsu, M., Castellanos, M., Ghosh, R.: Discovering coherent topics using general knowledge. In: Proceedings of the 22nd ACM International Conference on Information & Knowledge Management, pp. 209–218. ACM (2013)
7. Gao, Y., Wen, D.: Semantic similarity-enhanced topic models for document analysis. In: Chang, M., Li, Y. (eds.) Smart Learning Environments. LNET, pp. 45–56. Springer, Heidelberg (2015). https://doi.org/10.1007/978-3-662-44447-4_3
8. Griffiths, T.L., Steyvers, M., Tenenbaum, J.B.: Topics in semantic representation. Psychol. Rev. **114**(2), 211 (2007)
9. Lau, J.H., Baldwin, T., Newman, D.: On collocations and topic models. ACM Trans. Speech Lang. Process. (TSLP) **10**(3), 10 (2013)
10. Lau, J.H., Newman, D., Baldwin, T.: Machine reading tea leaves: automatically evaluating topic coherence and topic model quality. In: EACL, pp. 530–539 (2014)
11. Lau, J.H., Newman, D., Karimi, S., Baldwin, T.: Best topic word selection for topic labelling. In: Proceedings of the 23rd International Conference on Computational Linguistics, pp. 605–613. ACL (2010)
12. Leydesdorff, L., Rafols, I.: A global map of science based on the ISI subject categories. J. Assoc. Inf. Sci. Technol. **60**(2), 348–362 (2009)
13. Loukachevitch, N., Dobrov, B.: Ruthes linguistic ontology vs. Russian WordNets. In: Proceedings of Global WordNet Conference GWC-2014 (2014)
14. Loukachevitch, N., Nokel, M.: Adding thesaurus information into probabilistic topic models. In: Ekštein, K., Matoušek, V. (eds.) TSD 2017. LNCS (LNAI), vol. 10415, pp. 210–218. Springer, Cham (2017). https://doi.org/10.1007/978-3-319-64206-2_24
15. Mimno, D., Wallach, H.M., Talley, E., Leenders, M., McCallum, A.: Optimizing semantic coherence in topic models. In: Proceedings of the Conference on Empirical Methods in Natural Language Processing, pp. 262–272. Association for Computational Linguistics (2011)
16. Newman, D., Bonilla, E.V., Buntine, W.: Improving topic coherence with regularized topic models. In: Advances in Neural Information Processing Systems, pp. 496–504 (2011)
17. Newman, D., Lau, J.H., Grieser, K., Baldwin, T.: Automatic evaluation of topic coherence. In: Human Language Technologies: 2010 Annual Conference of the North American Chapter of the Association for Computational Linguistics, pp. 100–108. Association for Computational Linguistics (2010)
18. Nokel, M., Loukachevitch, N.: Accounting ngrams and multi-word terms can improve topic models. In: ACL 2016, p. 44 (2016)
19. Röder, M., Both, A., Hinneburg, A.: Exploring the space of topic coherence measures. In: Proceedings of the Eighth ACM International Conference on Web Search and Data Mining, pp. 399–408. ACM (2015)
20. Sievert, C., Shirley, K.E.: LDAvis: a method for visualizing and interpreting topics. In: Proceedings of the Workshop on Interactive Language Learning, Visualization, and Interfaces, pp. 63–70 (2014)
21. Smith, A., Lee, T.Y., Poursabzi-Sangdeh, F., Boyd-Graber, J., Elmqvist, N., Findlater, L.: Evaluating visual representations for topic understanding and their effects on manually generated topic labels (2017)
22. Vorontsov, K.: Additive regularization for topic models of text collections. Dokl. Math. **89**, 301–304 (2014)

23. Wallach, H.M.: Topic modeling: beyond bag-of-words. In: Proceedings of the 23rd International Conference on Machine Learning, pp. 977–984. ACM (2006)

24. Wang, X., McCallum, A., Wei, X.: Topical N-grams: phrase and topic discovery, with an application to information retrieval. In: Seventh IEEE International Conference on Data Mining, ICDM 2007, pp. 697–702. IEEE (2007)

25. Xie, P., Yang, D., Xing, E.P.: Incorporating word correlation knowledge into topic modeling. In: HLT-NAACL, pp. 725–734 (2015)

Classifying Companies by Industry Using Word Embeddings

Martin Lamby and Daniel Isemann[(✉)]

Chair of Media Informatics, University of Regensburg, Regensburg, Germany
martinlamby@gmail.com, Daniel.Isemann@ur.de

Abstract. This contribution investigates whether companies cluster together according to their field of industry using word embeddings and in particular word2vec models on general news text. We explore to what extent this can be utilised for identifying company-industry affiliations automatically. We present an experiment in which we test seven different classification methods on four different word2vec models trained on a 600-million-word corpus from the Guardian newspaper. For training and testing our classifiers we obtained company-industry assignments from the Dbpedia knowledge base for those companies occurring in both the news corpus and Dbpedia. The majority of the 28 scrutinized classification paradigms displays F1 scores near 80%, with some exceeding this threshold. We found differences across industries, with some industries appearing to be more distinctly defined, while others are less clearly delineated from neighbouring fields. To test the robustness of our approach we conducted a field test, identifying candidate companies absent from Dbpedia with a named-entity recognizer, establishing ground truth on company and industry status manually through web search. We found classifier performance to be less reliable in the field test and of varying quality across industries. with precision at 25 values ranging from 16% to 88%, depending on industry. In summary, the presented approach showed some promise, but also some limitations and may in its current form be only robust enough for semi-automated classification.

Keywords: Company-industry affiliation
Company classification from unstructured news text
Field of industry clustering · Word embedding · word2vec

1 Introduction

In recent years the availability of vast amounts of data and improved computing power has made the field of machine learning more relevant than ever [1]. As a substantial amount of data on the world wide web comes in the form of unstructured text [2], numerous attempts have been made to apply methods of natural language processing (NLP) to extract meaningful information including relations from this type of data [2, 3]. However, the difficulty and precise approach used to extract relations depends on how they are expressed in the text. Some relations appear in similar patterns with both arguments occurring in the same sentence. Others are mentioned rarely or not at all in the same sentence making it more difficult and involved to discover valid patterns.

© Springer International Publishing AG, part of Springer Nature 2018
M. Silberztein et al. (Eds.): NLDB 2018, LNCS 10859, pp. 377–388, 2018.
https://doi.org/10.1007/978-3-319-91947-8_39

In our contribution we look at the relation between companies and the industry or field of industry they are commonly assigned to. We believe that this relation is relatively rarely expressed directly in business news, but perhaps captured implicitly and more meaningfully in the context that company names appear in, in such news articles. We use a word embedding-based approach to test this hypothesis.

Identifying relations between companies and their corresponding industry in this way has many practical implications. If the word embeddings of companies assigned to the same industry cluster together in a meaningful way, comparing different industry clusters may challenge existing expert industry classification and definition standards or at least offer an alternative picture, blurring existing or defining new fields. Furthermore, classifying companies by industry in a manner that is not predicated on a formal standard but on ever-changing text and news data may allow startups and small and medium sized companies to be recognized as part of the market landscape and its peer groups early on and facilitate the analysis of industry clusters over time.

This leads to the central research question of our contribution: "Can word embeddings be used to classify companies by industry?". We will subsequently investigate this question, by first describing relevant research in Sect. 2. In Sect. 3 we describe our data set, experiment setup and the process of training different classifiers. In Sect. 4 we explore the resulting industry clusters visually and evaluate the performance of our classifiers. We then evaluate the performance of the best performing classifier on unlabeled named entities as part of a field test in Sect. 5, before concluding with a critical discussion of strengths and limitations of our approach.

2   Related Work

Company-industry affiliations have been studied from a number of different angles. Kahle and Walking looked at Standard Industry Classification (SIC) codes manually assigned to companies and found that roughly 36% of SIC classifications differ at the two-digit level and nearly 80% differ at the four-digit level between two major data providers but also that SIC codes change over time [4].

Despite the disagreement between human annotators suggested by this, a number of approaches for automated industry affiliation have been discussed in the literature. Gopikrishnan et al., for instance, show that companies with correlated stock price fluctuations have similar business activities [5]. Bernstein et al. attempted to classify companies by industry based on unstructured text, viz. business news. They propose a relational-vector space model for the classification of companies, which builds on existing company classifications and the number of coocurrences of companies in the same news story [6]. Drury and Almeida have proposed an iterative pattern-based approach to identifying "collective entities", i.e. group labels among companies and their constituent companies from business news text [7].

Identifying a company's industry affiliation form text can be construed as a relationship mining task akin to identifying hyponomy. To the best of our knowledge, however, standard relation extraction methods, including pattern-based approaches as well as supervised and unsupervised machine learning methods, including more recent concepts such as word embeddings, have not been brought to bear on this task.

A variety of methods for extracting relations from text have been explored. Hearst attempted to identify hyponym relations with lexico-syntactic patterns [8]. Etzioni et al. extended the approach using generic patterns from the KnowItAll information extraction system to learn domain-specific extraction rules [9].

In contrast to pattern-based methods supervised classification techniques such as support vector machines and logistic regression rely on labelled trainingsets, which can be expensive to obtain. Consequently, Mintz et al. have suggested a distantly supervised approach by using labelled data from a semantic knowledge base, such as Freebase [10].

Mikolov et al. were able to show syntactic and semantic relationships between words using word embeddings obtained from a trained neural network model, also known as word2vec [11]. Mikolov et al. tested syntactic similarities like the singular-plural relation of nouns such as 'year' and 'years' as well as semantic similarities like the relation clothing to shirt [12]. As part of their research they proposed a vector offset method which attempts to identify analogies like "clothing is to shirt as dish is to bowl". Subsequently, Mikolov et al. improved their previous model by significantly reducing the computational complexity, cutting the training time from weeks, to days thus enabling its use on larger datasets [13].

Fu et al. utilise the aforementioned word embeddings to investigate relations by building a semantic hierarchy, a structure consisting of hypernym-hyponym, "is-a", relations. They were able to show that the word vector offsets, between word pairs exhibiting a hypernym-hyponym relation, distribute in clusters [14].

3   Experiment Setup

3.1   Dataset and Preprocessing

For training and testing we used a dataset from the newspaper Guardian, comprising 1,780,985 news articles from a period of more than 17 years from 1/1/2000 until 17/3/2017. It consisted of all articles available through the Guardian API for this time period that are written in English and contained ten words or more. We did not limit ourselves to business news, as we wanted to incorporate company activities relevant to other societal fields (such as politics or sports) into our analysis as well. This particular news dataset met criteria of easy accessibility, uniform structure and size.

Though word2vec generally requires very little preprocessing we still applied basic linguistic processing beforehand (cf. [15] for a similar approach) to account for common phrases, lemmatisation, stop words and punctuation using the NLTK toolkit [16] and the Python library genism [17].

3.2 Model Choice

The word2vec model proposed by Mikolov et al. learns word embeddings through a neural network with one hidden layer of variable size by using a predefined number of words before and after a given word (the window size). These word embeddings can be learned either using the target word to predict the context (skipgram approach) or using the context to predict the target word (CBOW approach) [11].

The preprocessed text formed the training data for all word2vec models. The training data encompassed a vocabulary size of 425,173 (types) and 601,280,461 training words (tokens). It was used to train four different word2vec models with varying window size of 5, 10, 20 and 40. All other parameters remained constant for all models. To train the aforementioned models, the Python library genism was used again [17]. The focus of our task lies on a semantic relation between words rather than a syntactic relation. As the skipgram model has previously outperformed CBOW in terms of semantic accuracy, we opted to use the skipgram model [11]. The size of the hidden layer and thus the dimensionality of the final feature vector corresponded to the default value of 100. The results by Mikolov et al. showed accuracy on a semantic-syntactic word relationship test set to improve when the size of the hidden layer is increased [11]. To be considered for training, words had to occur at least 20 times in the corpus. Furthermore, the values of all other parameters that impact the models corresponded to the default values used in the genism implementation [17].

3.3 Model Training

After preprocessing and building the word2vec models, companies and their corresponding industries were identified. We used the vocabulary previously generated as part of training the word2vec models to filter out companies. Dbpedia, a knowledge base built on information extracted from Wikipedia [18], allowed us to identify company names in our vocabulary as well as the industry affiliation of these companies. We were thus able to identify 1,374 candidate companies that were affiliated with one or more industries. Mapping the extracted Dbpedia company-industry relations to the word embeddings learned from the trained word2vec models allows us to build a classifier for assigning unseen companies (via their word embeddings) to industries.

To train our classifiers on unambiguous companies we excluded companies belonging to multiple industries. Similarly, we only incorporated industries which comprised at least ten companies into the classifier building process to avoid that word embeddings from only few companies dominate an industry category unduly. Table 1 depicts the list of industries, as they are labelled in Dbpedia, and the number of companies in our news vocabulary belonging to these industries.

All classifiers were built and tested using ten-fold cross-validation. All methods used in this process relied on the Python machine learning library scikit-learn [19]. In a first step the labelled company data was split into ten folds for every industry, one fold acting as test fold to evaluate the classifier. The remaining folds were used to train the classifier as specified below. We used support vector machines with a linear, poly, rbf and sigmoid kernel as well as logistic regression with liblinear, lbfgs and newton-cg computational solvers to train seven different multiclass classifiers (one vs rest) for

Table 1. Training Data: industries and their respective number of companies used to train the different classifiers as well as the number of words (possible companies) classified to the different industries as part of the field test (see Sect. 5) because they achieved a probability exceeding a predefined probability threshold.

Industry	Number of companies	Result list size	Probability threshold
Financial_services	67	1,289	0.63
Retail	57	5,830	0.45
Automotive_industry	49	1,428	0.57
Petroleum_industry	34	542	0.41
Video_game_industry	24	808	0.35
Software	23	1,208	0.36
Conglomerate_(company)	22	/	/
Telecommunication	20	711	0.35
Electronics	14	223	0.49
Internet	14	395	0.41
Fashion	13	136	0.31
Mass_media	13	625	0.30
Biotechnology[a]	10	34	0.57
Pharmaceutical_industry	10	57	0.53
Food_processing	10	386	0.29

[a]The 25 highest and lowest ranked words from this industry overlap.

each of the four word2vec models totaling 28 different classifiers. Within each training fold we used leave-one out cross validation (LOOCV). The companies excluded in each LOOCV pass were utilised to establish a probability threshold specific to each industry. This was done by recording probability estimates for each of the left out companies during LOOCV-validation. The lowest of the recorded probabilities was used as probability cut-off when the classifier was run on the test fold. Establishing this threshold was done with a view to our field test (see Sect. 5) to exclude candidate words that are unlikely to belong to any industry category. The performance of the 28 classifiers by classification method and window size will be discussed in the following section, specifically in Subsect. 4.2.

4   Experiment Results

In the following we will explore the visualisation of different industry clusters using the word vectors obtained from the word2vec model trained with window size 5 before subsequently outlining our approach to automatic classification and the precision, recall and F1 scores achieved by different classifiers.

4.1 Exploratory Cluster Visualisation

Before turning to automated classification approaches, in order to gain an under-standing of how company-industry affiliations may manifest themselves through companies' word embeddings, we examined two-dimensional plots of industry clusters for those companies of known affiliation. For this we used the company-industry assignment from Dbpedia, which only contained companies assigned to exactly one industry. The 100-dimensional word vector associated with each company was trans-formed into 2D-space using t-distributed stochastic neighbor embedding (t-SNE, [20]).

Figure 1 depicts the t-SNE transformed embeddings of the five largest industry clusters (by number of companies) in our data set: *Automotive_industry, Financial_services, Petroleum_industry, Retail* and *Video_game_industry*. In contrast to the remaining ten industries where some have close thematic ties or overlap, such as the industries Internet and Software, here all industries are thematically relatively clearly delineated. *Automotive_industry, Petroleum_industry* and *Video_game_industry* exhi-bit a tight clustering with a visible core and some outliers. The remaining two indus-tries, *Financial_services* and *Retail*, also cluster visually together but with larger apparent variance. This may be due to the fact, that these industries are comprised of a more diverse set of affiliated companies than some others. Retailers for instance may be selling any number or category of goods and can be of vastly varying sizes.

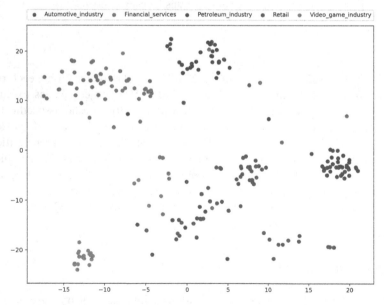

Fig. 1. t-SNE transformed vectors of industries defined by a minimum of 24 companies.

Although not all industries in our labeled data set clustered as recognizably as some of those in Fig. 1, it became apparent from the visual inspection that the distribution of companies was not unrelated to their field of business activity. We found no clear tendency in relation to the number of available training examples.

4.2   Classifier Analysis

Analysing precision, recall and F1 score of our classifiers, we found that the SVM with a linear kernel and the logistic regression (LG) with a newton-cg solver achieved the highest precision for each window size. Both classifiers showed an increase in precision when used with word2vec models trained with larger window sizes. The SVM linear classifier achieved the highest precision at window size 40 (see Table 2).

Table 2. Precision, Recall and F1 score of the individual classifiers trained with different SVM-kernels and LG-solvers using the word embeddings obtained from four unique word2vec models trained with a window size of 5, 10, 20 and 40 words, respectively. Values are in percent, the largest ones for each window size are in boldface, the largest overall are underlined.

		SVM				LG		
		linear	poly	rbf	sigmoid	lbfgs	liblinear	newton-cg
W05	P	84,3	81,3	84,3	81,8	89,2	87,6	**90,4**
	R	59,4	68,7	67,9	67,0	66,6	**71,8**	67,2
	F1	69,4	74,5	75,1	73,6	76,2	**78,8**	76,7
W10	P	**91,8**	82,4	84,5	84,8	90,2	87,5	88,1
	R	55,7	70,3	73,0	71,2	71,1	**74,8**	68,8
	F1	69,2	75,9	78,2	77,3	79,4	**80,6**	77,8
W20	P	90,2	82,7	87,7	84,2	92,0	86,7	**92,1**
	R	53,7	**73,4**	70,9	69,0	69,6	72,5	72,7
	F1	67,0	77,6	78,2	75,7	79,1	78,9	<u>81,2</u>
W40	P	<u>92,6</u>	83,2	84,7	82,9	89,8	87,1	89,1
	R	52,1	<u>75,0</u>	69,2	68,1	70,3	72,6	71,6
	F1	66,3	78,9	76,0	74,7	78,8	**79,2**	**79,2**

With respect to recall the SVM with a poly kernel and the LG with a liblinear solver achieved the highest scores for each window size. Whereas the LG with a liblinear solver performed best with word2vec models trained using a smaller window size of 5 or 10, the SVM with a poly kernel performed better with the word2vec model using a window size of 20 achieving its peak performance at window size 40.

Combining precision and recall values into the F1 score for all classifiers, we find, that the LG classifier with a liblinear solver achieved the highest F1 score with word2vec models trained with window size 5, 10 and 40. The LG classifier with a newton-cg solver achieved the overall highest F1 score at window size 20. Most classifiers tend to show an increase in performance regarding all three metrics, precision, recall and F1 score, when the word2vec window size is expanded. However, increasing the window size up to 40 does not always yield the best results.

5 Field Test

5.1 Setup

To test the real world performance of the trained classifier a field test was conducted. Using the Stanford NER tagger deploying the three class model [21] we identified entities tagged as organisations, which also included companies. We used the word2vec model with window size 20 and deployed the LG classifier using the newton-cg solver trained on the entire labelled data applying the previously identified industry specific probability cutoff thresholds reported in Table 1.

However, the NER tagger identified not only single words but also phrases containing multiple words, some of which were not previously identified as phrases during preprocessing and thus did not occur in the vocabulary. Those multi-word terms identified differently by the NER tagger and the phrase chunker employed in preprocessing could therefore not enter into the classification. The remaining organisations that could be associated with a word vector were classified if they exceeded the aforementioned probability threshold.

The number of successfully classified organisations, reported in Table 1, and their achieved probability were recorded. We validated the classification results of each industry by validating both, the 25 organisations which achieved the highest and those that achieved the lowest probability. The process of validation adhered to the following schema. An entity was searched using the Google search engine to check whether the entity was in fact a company that belonged to the industry it was classified to. The search results had to clearly and unambiguously refer to the entity in question. The entity "Mikes Car Service", for instance, will likely exist in multiple cities or countries and may thus not be unique. Similarly, the acronym ASM may refer to "Asia Sun Microsystems" or "Arctic Sun Motors". Afterwards, using the website of the company, which had to be listed on the first page of the Google search results, the industry, in which the company operates, was identified. If the entity was a company and has been correctly classified, it was flagged as *True*. If the entity was not a company, the company was not unique or the industry could not be clearly identified, it was flagged as *False_No_Company*. If the entity was indeed a company but did not belong to the industry it had been classified to, it was flagged as a *False_Company*. The industry *Conglomerate_(company)* was not subject of the field test because it is not an industry in the classical sense and more a special company structure.

5.2 Result

Across all industries, a number of companies were classified correctly as can be seen in Fig. 2, ranging from a minimum of four with the industries *Biotechnology* and *Pharmaceutical_industry* up to 22 in Retail. Looking at the industries with a high number of correctly classified companies it is apparent that the comparison is not fair. Some industries, such as *Video_game_industry*, are topically much narrower than other industries that can feature a broader range of companies, such as *Retail*. But not all high performers are topically broad industries. *Fashion*, *Telecommunication* and *Pertroleum_industry* also have a high number of companies classified as true.

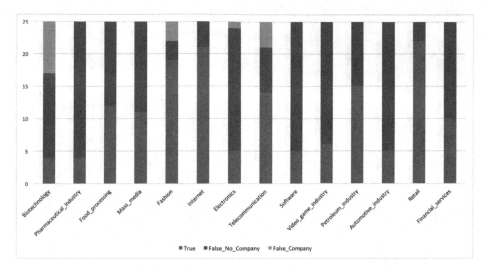

Fig. 2. Evaluation of 25 highest ranked words across industries. Industries are ranked by the size of the training data in ascending order.

Moreover, some industry results include companies that belong to different industries. For example, the false companies classified to *Biotechnology* are in fact pharmaceutical companies and the false companies classified to the *Fashion* industry belong exclusively to the beauty industry. But not all incorrectly classified companies exhibit such topical similarity as the industries *Electronics* and *Telecommunication* show. They include companies producing lava-lamps, marketing movies, providing ferry rides to name a few. The words that are not companies often show topical similarity. Whether that be names of race circuits and racing drivers classified to the *Automotive_industry* or oilfields to the *Petroleum_industry*.

Looking at the 25 lowest ranked words classified to each industry, depicted in Fig. 3, we observe a similar picture. The industries that included falsely classified but topically related companies in their ranking lists, *Biotechnology* and *Fashion*, also listed topic related companies in their bottom 25 words. In contrast, industries that included off topic companies in their top ranked words, did so also in their lowest ranked words. Whereas words flagged as *False_No_Company* in the top results mostly were related to the underlying industry, equally flagged words in the bottom 25 results exhibited no apparent relation at times. For instance, the bottom results of the category *Automotive_industry* include the disease "parkinson" and the industry *Petroleum_industry* includes the name of the ancient kingdom "northumbria".

Looking at the cluster visualisation, it is evident that a dense cluster structure does not always yield a high number of correctly classified companies as *Video_game_industry* and *Automotive_industry* show. The results heavily depend on the similarity of the industry topic to other topics and in what context these topics tend to appear. Furthermore, we observe that companies do not cluster exclusively but that their corresponding industry clusters include related words that are not companies.

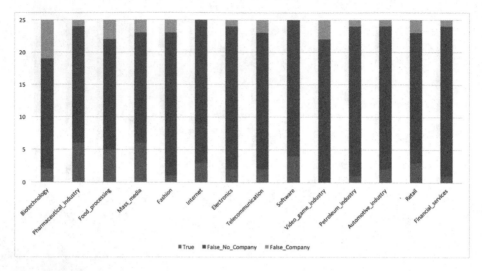

Fig. 3. Evaluation of 25 lowest ranked words across industries. Industries are ranked by size of the training data in ascending order.

6 Discussion and Conclusion

In an attempt to automatically identify company-industry affiliations, we evaluated 28 classifiers based on four underlying word2vec models with varying window sizes and different SVM kernels and logistic regression solvers. We analysed the effect different classification methods and word2vec window sizes have on precision, recall and F1 scores on a labelled dataset of company-industry relations obtained from Dbpedia. The classifier with the highest F1 score was further evaluated in an attempt to assign named entities, tagged in the corpus as organisations, to an industry, if applicable.

Our experiments indicate that companies tend to cluster together by industry using word embeddings to varying degrees. The breadth or narrowness of an industry's thematic focus in relation to other fields influences cluster variance and cluster delineation. For instance, the cluster *Automotive_industry* includes companies that are active in Formula One racing and consequently, terms related to Formula One may be classified as *Automotive_industry*. This is illustrated in Fig. 4 which shows the ten words most similar to the automotive company "Ferrari", known as an active player in Formula One racing. Among the similar words are names of famous race drivers as well as other automotive companies which are active in Formula One. We observed that logistic regression solvers generally outperformed classification with SVMs and that training a word2vec model with a window size larger than five is beneficial. Increasing the window size indefinitely, however, does not improve results and even has a negative effect in some instances. The field test showed considerably more correctly classified companies within the top 25 words than the bottom 25 words.

While our approach shows promise in creating positive evidence for identifying company-industry affiliations from unstructured news texts, it also has some limitations. Firstly, we relied on the quality of labelled training data obtained through the

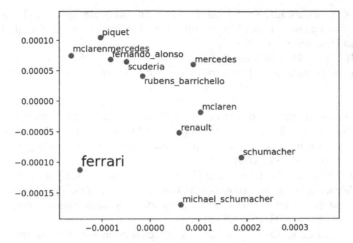

Fig. 4. t-SNE transformed word embeddings of the 10 most similar words to the automotive company "Ferrari".

Dbpedia knowledge base. Unlike in our field test, where candidate companies were individually researched, we have not investigated the validity of Dbpedia provided company-industry relations. We also saw that broad industry labels such as *Retail* lead to less reliable identification and, on the other hand, the question remains whether enough training data can be obtained for very specific, emergent, industry clusters.

In summary, our approach shows promise but is not yet robust enough to allow for reliable, automated classification partly because of the difficulty of identifying candidate companies. Our findings may prove useful for semi-automated classification and we are planning to further investigate this line of research, by using more involved word2vec models and including more context features in the classification. We would like to expand the approach to other relations that companies can enter into, such as the company-CEO relation, as well as study the change of industry clusters over time.

References

1. Jordan, M.I., Mitchell, T.M.: Machine learning: trends, perspectives, and prospects. Science **349**, 255–260 (2015)
2. Gupta, V., Lehal, G.S.: A survey of text mining techniques and applications. J. Emerg. Technol. Web Intell. **1**, 60–76 (2009)
3. Allahyari, M., Pouriyeh, S., Assefi, M., Safaei, S., Trippe, E.D., Gutierrez, J.B., Kochut, K.: A brief survey of text mining: classification, clustering and extraction techniques (2017). arXiv Preprint: arXiv:1707.02919
4. Kahle, K.M., Walkling, R.A.: The impact of industry classifications on financial research. J. Financ. Quant. Anal. **31**, 309 (1996)
5. Gopikrishnan, P., Rosenow, B., Plerou, V., Stanley, H.E.: Identifying business sectors from stock price fluctuations (2000)

6. Bernstein, A., Clearwater, S., Provost, F.: The relational vector-space model and industry classification. In: Proceedings of IJCAI 2003 Workshop on Learning Statistical Models from Relational Data, pp. 8–18 (2003)
7. Drury, B., Almeida, J.J.: Identification, extraction and population of collective named entities from business news. In: Entity 2010 – Workshop on Resources and Evaluation for Entity Resolution and Entity Management, LREC 2010, pp. 19–22 (2010)
8. Hearst, M.A.: Automatic acquisition of hyponyms from large text corpora. In: Proceedings of the 14th conference on Computational linguistics, vol. 2, pp. 539–545 (1992)
9. Etzioni, O., Cafarella, M., Downey, D., Popescu, A.M., Shaked, T., Soderland, S., Weld, D. S., Yates, A.: Unsupervised named-entity extraction from the web: an experimental study. Artif. Intell. **165**, 91–134 (2005)
10. Mintz, M., Bills, S., Snow, R., Jurafsky, D.: Distant supervision for relation extraction without labeled data. In: Proceedings of the Joint Conference of the 47th Annual Meeting of the ACL and the 4th International Joint Conference on Natural Language Processing of the AFNLP, vol. 2, pp. 1003–1011 (2009)
11. Mikolov, T., Chen, K., Corrado, G., Dean, J.: Efficient estimation of word representations in vector space (2013). arXiv Preprint: arXiv:1301.3781
12. Mikolov, T., Yih, W.-T., Zweig, G.: Linguistic regularities in continuous space word representations. In: Proceedings of NAACL-HLT, pp. 746–751 (2013)
13. Mikolov, T., Sutskever, I., Chen, K., Corrado, G.S., Dean, J.: Distributed representations of words and phrases and their compositionality. In: Advances in Neural Information Processing Systems, pp. 3111–3119 (2013)
14. Fu, R., Guo, J., Qin, B., Che, W., Wang, H., Liu, T.: Learning semantic hierarchies via word embeddings. In: ACL, vol. 1, pp. 1199–1209 (2014)
15. Sugathadasa, K., Ayesha, B., de Silva, N., Perera, A.S., Jayawardana, V., Lakmal, D., Perera, M.: Synergistic Union of Word2Vec and Lexicon for Domain Specific Semantic Similarity (2017). arXiv Preprint: arXiv:1706.01967
16. Bird, S., Klein, E., Loper, E.: Natural Language Processing with Python: Analyzing Text with the Natural Language Toolkit. O'Reilly Media, Inc., Sebastopol (2009)
17. Rehurek, R., Sojka, P.: Software framework for topic modelling with large corpora. In: Proceedings of the LREC 2010 Workshop on New Challenges for NLP Frameworks. Citeseer (2010)
18. Auer, S., Bizer, C., Kobilarov, G., Lehmann, J., Cyganiak, R., Ives, Z.: DBpedia: a nucleus for a web of open data. In: Aberer, K., et al. (eds.) ISWC/ASWC 2007. LNCS, vol. 4825, pp. 722–735. Springer, Heidelberg (2007). https://doi.org/10.1007/978-3-540-76298-0_52
19. Pedregosa, F., Varoquaux, G., Gramfort, A., Michel, V., Thirion, B., Grisel, O., Blondel, M., Prettenhofer, P., Weiss, R., Dubourg, V.: Scikit-learn: machine learning in Python. J. Mach. Learn. Res. **12**, 2825–2830 (2011)
20. Van Der Maaten, L.J.P., Hinton, G.E.: Visualizing high-dimensional data using t-sne. J. Mach. Learn. Res. **9**, 2579–2605 (2008)
21. Finkel, J.R., Grenager, T., Manning, C.: Incorporating non-local information into information extraction systems by gibbs sampling. In: Proceedings of the 43rd Annual Meeting on Association for Computational Linguistics, pp. 363–370. Association for Computational Linguistics (2005)

Towards Ontology-Based Training-Less Multi-label Text Classification

Wael Alkhatib[✉], Saba Sabrin, Svenja Neitzel, and Christoph Rensing

Communication Multimedia Lab, TU Darmstadt, Rundeturmstr. 10,
64283 Darmstadt, Germany
{wael.alkhatib,svenja.neitzel,christoph.rensing}@kom.tu-darmstadt.de,
saba.sabrin@stud.tu-darmstadt.de

Abstract. In the under-explored research area of multi-label text classification. Substantial amount of research in adapting and transforming traditional classifiers to directly handle multi-label datasets has taken place. The performance of traditional statistical and probabilistic classifiers suffers from the high dimensionality of feature space, training overhead and label imbalance. In this work, we propose a novel ontology-based approach for training-less multi-label text classification. We transform the classification task into a graph matching problem by developing a shallow domain ontology to be used as a training-less classifier. Thereby, we overcome the challenges of feature engineering and label imbalance of traditional methods. Our intensive experiments, using the EUR-Lex dataset, prove that our method provides a comparable performance to the state-of-the-art techniques in terms of Macro F_1-Score.

Keywords: Semantics · Statistics · Feature selection
Ontology · Text classification · Typed dependencies

1 Introduction

Text classification has become a widespread problem in natural language processing and information retrieval as a result of the tremendous growth of data, most of which are unstructured. In many fields such as bimolecular analysis, semantic indexing and protein function classification, instances can be naturally associated with more than one class. This fact has triggered a large body of research to adapt single-label classification to handle multi-label classification problems. These classification strategies fall into two groups namely, transformation and adaptation methods. Transformation methods transform a multi-label learning problem into one or more single-label problems i.e. Binary Relevance (BR) and Label Powerset (LP) [1]. Adaptation methods adapt or extend single-label classifiers to cope with multi-label data i.e. AdaBoost.MH, BRkNN and ML-KNN. For both categories, text representation is a crucial preprocessing step where documents are transformed into a format consumable by machine learning models. This involves representing each document as a vector of words as features that

© Springer International Publishing AG, part of Springer Nature 2018
M. Silberztein et al. (Eds.): NLDB 2018, LNCS 10859, pp. 389–396, 2018.
https://doi.org/10.1007/978-3-319-91947-8_40

can be selected based on statistics or semantics-based techniques, where each dimension corresponds to the relevance of a word with respect to the document. The performance of traditional classifiers is largely affected by the feature selection techniques used and by label imbalance. In addition, building and updating the classifier impose large training overhead.

In this work, we address these challenges by proposing a novel ontology-based training-less multi-label text classifier based on leveraging the existing knowledge of semantic and syntactic relations between concepts from the documents corpus. The novelty of this approach is that it does not rely on any external lexical database or human built ontologies. Instead, it relies solely on the concepts and their relationships represented in the automatically generated shallow domain ontology. We shift the problem to the field of graph theory: the ontology effectively becomes the classifier. Thus, there is no need for a pre-trained classifier. The proposed approach requires a transformation of documents and labels into graph structures. It employs entity matching for relatedness matching: The classification process is based on measuring the semantic similarity between a label ontology which is derived from the shallow domain ontology and the main thematic entities in a document.

2   Related Work

The used feature selection techniques and features space size have a great impact on the performance of text classifiers. For that, Janik et al. [2] proposed an ontology-based single-label text classifier. They built a domain ontology based on RDF (Resource Description Framework) from Wikipedia articles. Using different graph techniques for entity representation, the classification problem was transformed to a graph matching problem. Three different types of graphs were constructed, namely the *Semantic Graph*, the *Thematic Graph* and the *Dominant Thematic Graph*. The Semantic Graph holds the representation of document terms as named entities with inter-relationships between them based on the domain ontology. Using the *Semantic Graph*, a more detailed graph is generated by finding entities based on their importance such as, number of connected entities or number of associated small networks of entities. The final step of building the thematic graph is to define the most central entities based on shortest path calculation. As it is a training-less process, the algorithm works directly on the set of entities and the set of categories to find the best fit for each document. Relying on the RDF files from Wikipedia hinders the system adaptability to different domains or datasets.

Another approach was proposed in [3] to classify construction regulatory documents. The proposed methodology uses a topic hierarchy and a manually built domain ontology. The topic hierarchy represents the concepts of the construction regulatory domain. The process of developing the topic hierarchy includes identifying the main concepts and extracting the concepts by reviewing the environmental regulatory documents. Then all the identified concepts are organized into a hierarchical structure to build a taxonomy of the overall domain.

The leaf nodes of the hierarchy were used as labels for classification. The classification mechanism is a similarity based approach which uses the hierarchical softmax skip-gram algorithm to learn the distributed representation of document terms and concepts. Then, the similarity between each term in the clause and each concept in the topic ontology is computed. The similarities between the terms and concepts are then summed up for each clause-topic pair to compute a Total Similarity (TS) for an overall clause. Topics having positive TS are chosen as potential labels and negative TS are discarded.

In the related work, authors relied on ontologies manually built by domain experts, structured text and external lexical databases to develop the domain ontology and the categories hierarchy. This drastically limits the usability of these approaches in other domains where such ontology or domain experts are not available. In this work, we extend and improve previous work by introducing a new training-less classifier which does not rely on any external lexical resources or manually built ontologies. We aim to achieve a feasible performance even without a high quality of the extracted ontology.

3   Proposed Methodology

In our approach, the domain ontology consists of concepts and their different semantic and syntactic relations. As shown in Fig. 1 the domain ontology is built from the document corpus. After building the shallow domain ontology, the matching process between a document and the labels is converted into a graph matching problem.

3.1   Domain Ontology Creation

The shallow domain ontology consists of concepts describing the domain and different relations between the concepts. The construction of the shallow domain ontology follows three steps, as shown in Fig. 1:

Linguistic Filter: The linguistic filter recognizes essential candidate concepts and filters out sequences of words that are unlikely to be concepts. A combination of 3 linguistic filters is used to extract single and multi-word NPs from the corpus.

- *Noun Noun+*
- *Adj Noun+*
- *(Adj| Noun) + Noun*

Semantic Graph Construction: The aim here is to identify the noun phrases which represent a concept or an instance of a concept and connect them by relations. Therefore is-a relationships between the noun-phrases are extracted using a lexico-syntactic pattern-based approach. We use six Hearst patterns [4] to identify taxonomic relations. The pattern-based approach has been chosen due to its high precision compared to other linguistic or statical approaches. However, these patterns cover a small proportion of complex linguistic space.

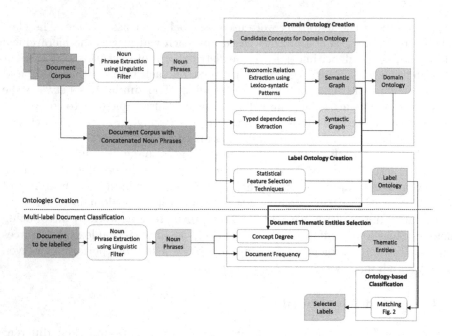

Fig. 1. Block diagram of the proposed ontology-based training-less classifier.

Syntactic Graph Construction: In a third step additional relations, esp. meronymic relations, are determined. Therefore we use typed dependencies extraction. It is a parsing technique that converts a sentence depending on the part-of-speech tagging of words into a tree structure. Using the dependencies between words in a sentence, the syntactic features are defined and represented as triples of (governor, dependent, relations), "governor" and "dependent" simply represents the position of the words within the sentence. The selected typed dependencies represent the prepositions, conjunctions and other relative clause in a collapsed way. Using these syntactic relations on the sentence level, syntactic features can be identified to create the so called syntactic graph.

3.2   Label Ontology Creation

A label ontology consists of concepts which are related to a single label and relations between these concepts are obtained from the domain ontology. Labels can be abstract or concrete, real or fictitious, which creates a challenge to extract their semantically related concepts based on direct text matching in linguistic techniques. Therefore we use statistical techniques, i.e. information gain or correlation, to measure the goodness of a concept representing a label. Based on these measures, we select the concepts from the domain ontology which represent the label. By extending the selected concepts using the semantic and syntactic graphs from the domain ontology, a label ontology is created for each label.

3.3   Multi-label Document Classification

The domain and the label ontologies are created once and used for different classification tasks. The multi-label classification is conducted in two steps.

Document Thematic Entities Selection. Document thematic entities are selected concepts which can be found in a document and characterize the thematic focus of the document. They serve as feature candidates in the following classification process. The thematic entities can be selected based on two approaches, namely the Document Frequency (DF) and Concept Degree (CD). We define CD as the number of noun phrases connected to a specific noun phrase from the semantic graph, while DF represents the number of documents in which a noun phrase occurs. The basic heuristics behind it is that rare or non-frequent terms are non-informative for classification.

Ontology-Based Classification. The real classification uses the document thematic entities and the label features. For every label, a semantic and syntactic subgraphs are prepared using the domain semantic and syntactic graphs respectively. The depth of the syntactic and semantic graphs are set based on an agreed threshold. Using depth first search (DFS) algorithm, we start by matching a document entity against the semantic graph of the label. In case no match was found, the algorithm try to find a match from the syntactic graph of the label. Two approaches were followed for calculating the relatedness score between a document and a label, namely the aggregated score (AScore) and the normalized aggregated score (NAScore). According to the used statistical technique for the label ontology creation, we define $S(i)$ as the weight of feature $x(i)$ with regard to label y. Mi is a binary factor with a value equals 1 if the feature is found and 0 otherwise, the different scores can be calculated as follow:

$$AScore(t) = \sum_{i=1}^{m} Mi * S(i)$$

$$NAScore(t) = \frac{AScore(t)}{\sum_{i=1}^{m} R(i)}$$

(1)

4   Evaluation

4.1   Dataset and Evaluation Settings

In the context of our analysis, the EUR-Lex dataset has been used [5]. It is a text dataset containing European Union laws, treaties and other public documents. It contains around 19348 documents. We used the subject-matters labeling scheme which includes 201 *labels* with *Label Cardinality* of 2.21 and *Label Density* of 1.10. Stanford CoreNLP toolkit [6] was used for performing the different natural language processing tasks (POS, linguistic filter, extraction of taxonomic relations and typed dependencies).

4.2   Evaluation Results

The carried out experiments aimed to optimize the different parameters of our proposed method and also to compare the effectiveness of our method with a baseline. For comparison we considered the best performance achieved on this dataset in our previous work with semantic-based feature selection and using ML-KNN as the baseline [5]. All the experiments have been performed on the full dataset since no training phase is applied.

Analysis of Label Ontology Creation: Two main aspects need to be analyzed, firstly which technique should be used for selecting the main features representing a label. Here, four different techniques have been analyzed namely, information gain, information gain ratio, chi-squared statistic, and correlation. Secondly, the label ontology can be extended using the semantic and syntactic graphs, for that, the depth of the graphs should be carefully analyzed.

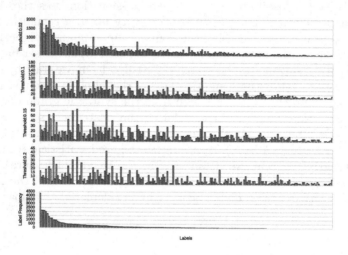

Fig. 2. Number of features per label based on correlation threshold.

The number of label features based on information gain, information gain ratio, chi-squared is proportional to the label frequency. Figure 2 illustrates the number of features accordingly to the label frequency based on the correlation score threshold. Using a threshold of 0.15 the average number of features representing very frequent and rare labels are fairly close which indicates that using correlation for feature selection is valid choice.

Since the semantic graph represents the taxonomic relations we can use the maximum depth of the graph since it is limited. However, we have fixed the depth to a high value of 35 in order to avoid cycles. The depth of the syntactic graph was carefully selected. The average number of relations per label feature significantly increase with regard to the syntactic graph depth as well as when the prepositions were considered. The syntactic graph depth was set to depth of 2 and only verbs were used as typed dependencies in order to reduce the computational complexity of the graph matching phase.

Configuration of the Ontology-Based Multi-label Classifier: Table 1 demonstrates the performance metrics of using document frequency and concept degree for selecting the thematic entities in a document. Respectively, two scoring techniques for matching label ontology against the document main entities were evaluated namely, aggregated score *Ascore* and normalized aggregated score *NAscore*.

Table 1. Evaluation of named entities selection methods and scoring techniques

Selection technique	Scoring technique	Score Threshold	Micro F_1	Macro F_1
DF	*Ascore*	*0.140*	*0.2892*	*0.3016*
DF	NAscore	0.279	0.2277	0.2468
CD	Ascore	0.100	0.1199	0.0832
CD	NAscore	0.451	0.1553	0.0730

Comparative Analysis Against the Baseline: Figure 3 illustrates the different performance metrics depending on the score threshold. By setting a score threshold of 0.45 we reach the highest performance of Macro F_1 with Micro F_1 of 0.3016 ± 0.0045 and 0.2892 ± 0.0124 respectively. Table 2 shows the results of the comparative analysis against the baseline. For the semantic graph we used a depth of 35 while for the syntactic graph we have used a depth of 2. This table presents some interesting results, our ontology-based training-less classifier has similar Macro F_1 with regard to the baseline which shows that the average performance across all the labels of both models is similar. However, the Micro F_1 result is significantly lower than the baseline, which means that our baseline is biased towards the most popular labels while our model has a similar prediction performance for frequent and less frequent labels.

Table 2. Evaluation results against the baseline

Model	Micro F_1	Macro F_1
Trainingless classifier	0.2892	0.3016
baseline	0.6642	0.3162

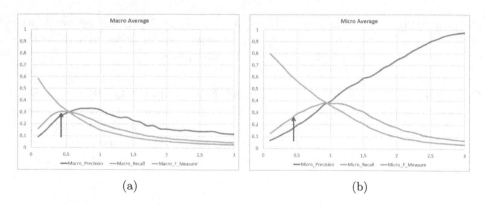

(a)   (b)

Fig. 3. Performance metrics based on score threshold.

5   Conclusion and Future Work

In this work, we have proposed a novel multi-label classification approach, which solely relies on developing a domain ontology to be used as a training-less classifier without a pre-classified set of training documents. Our intensive experiments, using the EUR-Lex dataset, approved that our method provides a comparable performance to the baseline in terms of Macro F_1-Score. In future work, we will investigate the effects of constructing the domain ontology using external lexical databases and knowledge bases in order to improve the classifier performance by extending the label ontology with stronger semantic relations.

References

1. Sorower, M.S.: A literature survey on algorithms for multi-label learning, Oregon State University, Corvallis, vol. 18 (2010)
2. Janik, M.G.: Training-less ontology-based text categorization, Ph.D. dissertation, UGA (2008)
3. Zhou, P., El-Gohary, N.: Ontology-based multilabel text classification of construction regulatory documents. J. Comput. Civ. Eng. **30**(4) (2015). 04015058
4. Hearst, M.A.: Automatic acquisition of hyponyms from large text corpora. In: Proceedings of the 14th Conference on Computational Linguistics-Volume 2, pp. 539–545. Association for Computational Linguistics (1992)
5. Alkhatib, W., Rensing, C., Silberbauer, J.: Multi-label text classification using semantic features and dimensionality reduction with autoencoders. In: Gracia, J., Bond, F., McCrae, J.P., Buitelaar, P., Chiarcos, C., Hellmann, S. (eds.) LDK 2017. LNCS (LNAI), vol. 10318, pp. 380–394. Springer, Cham (2017). https://doi.org/10.1007/978-3-319-59888-8_32
6. Manning, C.D., Surdeanu, M., Bauer, J., Finkel, J.R., Bethard, S., McClosky, D.: The Stanford CoreNLP natural language processing toolkit. In: ACL (System Demonstrations), pp. 55–60 (2014)

Overview of Uni-modal and Multi-modal Representations for Classification Tasks

Aryeh Wiesen[1,2]([envelope]) [iD] and Yaakov HaCohen-Kerner[2] [iD]

[1] Department of Computer Science, Ariel University,
Ariel 40700, Israel
[2] Department of Computer Science, Lev Academic Center,
Jerusalem 9116001, Israel
{wiesen,kerner}@jct.ac.il

Abstract. Classification is one of the most fundamental tasks in data mining and machine learning. It is being applied in an increasing number of fields, e.g. filtering, identification, information retrieval, information extraction, and similarity detection. A basic and necessary condition for the success of a classification task is the proper representation of the information it wishes to classify. Classification is needed in domains that are based on uni-modal representations such as text, images, audio, and speech, as well as in domains that are based on multi-modal representations. This paper aims to provide a short review on the developing area of multi-modal representations for classification with emphasis on state-of-the-art systems in this area. Firstly, fundamentals of uni-modal representations are given. Secondly, an overview of multi-modal representations is given. Thirdly, various related systems using multi-modal representations and the datasets used by them are briefly summarized with a comparative summary of these systems.

Keywords: Classification · Multi-modal representation
Textual features · Uni-modal representation · Visual features

1 Introduction

Classification is a branch of Machine Learning (ML) and Data Mining, studying problems of identifying (predicting) a category or a set of categories to which new observations (instances) belong, on the basis of observations whose category membership is known from previous experience. Since the algorithm (classifier) is provided with a set of correctly identified observations, and the instances are assumed to be generated by a random process, independent from the classifier, classification belongs to the area of supervised statistical Machine Learning.

Representation is a set of relevant features of the instances in a machine readable form, such as array (feature vector), matrix a tensor. As Bengio et al.

A. Wiesen — This research is in partial fulfillment of the requirements for the PhD degree by the first author at Ariel University.

© Springer International Publishing AG, part of Springer Nature 2018
M. Silberztein et al. (Eds.): NLDB 2018, LNCS 10859, pp. 397–404, 2018.
https://doi.org/10.1007/978-3-319-91947-8_41

point out in [4], much of the effort for building a classifier goes in "feature engineering", i.e. extracting a meaningful and efficient representation of the feature data from samples. Following their work and the recent extensive ML survey [3], we use the terms *feature* and *representation* interchangeably. A representation may belong to a single *modality*, such as text, image or video, or span across multiple modalities, e.g. images and text [14], or speech and video [17].

The goal of this overview is to survey some of the significant recent achievements in developing uni-modal and multi-modal representations for classification.

2   Uni-modal Representations

2.1   Text Representation Models

In this subsection, we briefly overview some long-established, but still widely used, models for text classification/categorization (TC).

Boolean Information Retrieval model uses Boolean queries to retrieve matching documents from large collections (corpora). Since it requires an exact match, it does not support ranking documents by relevance.

Weighting schemes address this drawback of the Boolean model by assigning weights to search terms, and calculating relevance scores of matching documents. The simplest scheme, the *term frequency*, sets term weight as number of its occurrences in a document, normalized to the document length. This approach is also termed Bag-of-Words (BoW) model, and refers to any representation that abstracts away a particular ordering between features. More advanced scheme, the *Tf-idf* (Term Frequency—Inverse Document Frequency), attenuates the effect of terms that occur in too many documents and so reduce the query specificity.

Vector space model introduced an important notion of a *similarity measure* between documents, or a document and a query, represented as vectors in a document space with dimensionality equal to the number of the indexed terms. Arguably the most common measure is the *cosine similarity*.

Distributional representations, or Distributional Semantic Models (DSMs), are rooted in Harris distributional hypothesis, and aim to extract semantic information from patterns of words co-occurrence in a context. Current DSMs could be divided into the following three categories: (1) *matrix factorization methods*, e.g. Latent Semantic Analysis (LSA), Latent Dirichlet Allocation (LDA), and Latent Semantic Indexing (LSI) [7]; (2) *word embeddings*, such as Continuous Bag of Words used in the word2vec tool [16]; (3) *kernel methods*, e.g. Support Vector Machines (SVMs).

2.2   Representation models for the visual modality

ML-related tasks in computer vision include image classification and annotation, visual question answering (VQA), and content-based image retrieval (CBIR) [15], generally using high-level image semantics.

As a result of image processing pipeline, is obtained a feature vector of visual descriptors (or cues), which are structures most characteristic of an image. There exist many dozens of different visual descriptors on different hierarchical levels, pixel-based or object-based; a comprehensive list with detailed explanation can be found e.g. in [15]. The visual descriptor that is arguably most frequently used in the context of multi-modal ML is the Scale-Invariant Feature Transform (SIFT). The Bag-of-Visual-Words (BoVW) representation model is essentially a histogram of occurrences of visual descriptors (or "visual words"). This model is *de facto* the standard for injecting visual knowledge into multi-modal representations [5,11].

2.3 Tweets

Classification tasks dealing with the tweet modality include topic recognition and classification [2], sentiment or stance analysis (a.k.a. opinion mining) as well as detecting users' affects, moods, and possible intentions [18].

Tweet fields suitable for feature extraction include the text, hashtags, URLs, engagers' actions (e.g. favorite, retweet, reply), geographic, temporal and social contexts. Twitter-specific features are either nominal (e.g. author) or binary (presence of shortening of words and slangs, emoticons, time-event phrases, opinioned words, emphasis on words, currency and other signs).

3 Multi-modal Representations

As the contemporary digital world is evolving into being increasingly multi-modal [9], expanding ML tasks and methods into multi-modal domains is an obvious need. However, learning from multiple modalities poses many new challenges; two of the hardest are (1) finding most efficient representation for each modality, and (2) learning a joint representation, i.e. performing *multi-modal fusion*.

Deep neural models. In the recent years, there seemingly has been established a consensus in favor of deep architectures about learning models for representation of information from each single modality [13]. Below, we briefly list their main types.

Convolutional Neural Networks (CNNs) are the state-of-the-art architecture, most suitable for representation of continuous modalities such as images, video, speech, and audio. They have brought breakthroughs in learning these modalities about a decade ago [10,13].

Recurrent Neural Networks (RNNs) provide an architecture suitable for representing sequential data, such as text and speech [13]. A special type of RNNs, the Long Short Term Memory (LSTM) networks, have further improved the performance of RNNs in tasks such as speech recognition.

Deep Boltzmann Machines (DBMs) [21], along with *Deep Belief Networks* (DBNs) [10], are graphical models built with stochastic binary units.

They outperformed the previous state-of-the-art Bayesian probabilistic architecture, known as the Hidden Markov Models (HMMs) [13].

Multi-modal Fusion. Current approaches to multi-modal fusion can be divided in two ways: (1) by stage of the fusion: early, late, and middle (or hybrid) [3], and (2) by kind of operation used to join the representations of different modalities, e.g. simple concatenation, max/softmax pooling [5], sigmoid/tanh gating [1], or bilinear (tensor) pooling [6,9].

4   Related Systems Using Multi-modal Representations

In this section, we briefly review eight state-of-the-art multi-modal representation models developed in the current decade, and the achieved results. We tried to choose them as a "representative sample" out of a vast number existing in literature, and we present them below as a brief timeline, in order to try and show the trends in the recent and current research.

Weston et al. (**2011**) [22] developed the WSABIE model, addressing the task of image annotation, i.e. multi-label image classification. In their model, the authors used linear mapping of textual and visual features onto the common feature space. To annotate an image, their algorithm computes a rank for each possible annotation, so the highest ranked annotations describe best the semantic content of the image. The model ranks ground-true labels from ImageNet and Web-Data datasets at position 1 with precision values of 4.03% and 1.03% respectively, which outperform previous baselines.

Silberer and Lapata (**2014**) [19] introduced a semantic bimodal deep neural network model. The textual and visual modalities are encoded as vectors of attributes and are obtained automatically from text and image data. To learn meaning representations of words from textual and visual input, they employ stacked autoencoders (SAE) model. It achieved a correlation coefficient of $\rho = 0.70$ for prediction of word similarity, and a F-score measure of 0.43 for a word categorization task. Both results were achieved on the dataset by McRae et al. (see link in the Table 2), produced by human participants, and classifying ca. 10 000 nouns into various (possibly multiple) semantic categories. Their results outperform the results of competitive baselines and related models trained on the same attribute-based input.

Socher et al. (**2014**) [20] introduced a semantic dependency tree recursive neural network (SDT-RNN) model, which uses dependency trees to embed sentences into a vector space in order to retrieve images that are described by those sentences. They tested their model with the task of mapping sentences and images into a common space for finding one from the other. The used dataset contained 1000 images, each annotated with 5 sentences. Their model achieved the following mean rank results for the following experiments: 10.5 for the Sentences Similarity for Image task, 12.5 for the Image Search task, and 16.9 for the Describing Images task.

Kiela and Bottou (**2014**) [8] constructed multi-modal concept representations by concatenating a skip-gram linguistic representation vector with a visual concept representation vector computed using the feature extraction layers of a deep CNN trained on a large labeled object recognition dataset. This transfer learning approach showed a clear performance gain over features based on the traditional BoVW approach. Experimental results are reported on the Word-Sim353 and MEN semantic relatedness evaluation tasks. They used visual features computed using either ImageNet or ESP Game images. Using the ImageNet features, their model achieved on the WordSim353 and MEN datasets the results of $\rho = 0.6$ and 0.7 respectively, which outperform the reported baselines.

Lazaridou et. al [11] (**2015**), based on Mikolov's Skip-gram model [12], developed two variations of a multi-modal Skip-gram, MMSkip-gram A and B. Given a text corpus, the Skip-gram finds word representations that are good at predicting the context words surrounding a target word. Then, for a subset of the target words is injected the visual knowledge in form of images of the concepts they denote. The visual representation is also encoded in a vector, using BoVW feature vector of SIFT features of the image, extracted with the aid of a CNN. At least one of the two versions of their model achieved on four datasets: MEN, Simlex-999, SemSim, and VisSim, ρ values of 0.75, 0.40, 0.72, and 0.63 respectively, which is higher than the results of all other tested models.

Fukui et al. [6] (**2016**) proposed a method to combine visual and text representations called Multimodal Compact Bilinear Pooling (MCB). For visual question answering, their architecture with multiple MCBs gives significant improvements on two VQA datasets compared to state-of-the-art. In the visual grounding task, introducing MCB pooling leads to improved phrase localization accuracy, indicating better interaction between query phrase representations and visual representations of proposal bounding boxes. Their accuracy results on the Flickr30k Entities dataset and the ReferItGame dataset are of 48.69% and 28.91%, respectively, which outperform the results achieved by other models.

Arevalo et al. [1] (**2017**) developed the Gated Multimodal Units (GMUs) model for multi-label classification of multi-modal data. A GMU is a hidden unit in a NN architecture, which receives inputs from different modalities and learns to determine how much each input modality affects the unit activation. The test task was identifying a movie genre based on its plot summary and its poster. Along with their work the authors published the Multimodal IMDb dataset (MM-IMDb) containing ca. 27,000 movie plots, poster images and other metadata. The GMU model performing a classification task on MM-IMDb achieved F-measure values of 0.630 (micro), and 0.541 (macro) higher than the F-measure scores of seven other multi-modal models. The multi-modal model improves the Macro F-scores of independent modalities in 16 out of 23 genres.

Kiela et al. [9] (**2018**) proposed a new bilinear-gated model for multi-modal fusion. It fully captures any associations between the two fused modalities by means of a tensor product between the feature vectors. One modality is allowed to join, or "attend", the other via a non-linear gate (a sigmoid or tanh function). In their experiments, it was found that the bilinear-gated model compared to various baselines and previous models achieved the highest accuracy results (averaged over 5 runs) are 90.8, 62.3, and 28.6, on the Food101, MM-IMDb, and FlickrTag datasets, respectively, though at slower speed than simpler additive and max-pooling fusion models.

Summary statistics. In Table 1, we present the statistics of the datasets used in the reviewed multi-modal models. In Table 2, we summarize the results of the multi-modal applications based on the models reviewed in Section 4.

On the one hand, in Tables 1 and 2, we see a widespread variety of multi-modal models and datasets. Futhermore, in Table 2, for some of the applications we can see nice relative improvements from the baselines. On the other hand, there is big inconsistency and wide variability between the various examined datasets, their size, the number of labels in them, and the various measures used by various systems to measure their results and baselines. To our opinion, a uniformity protocol should be defined at least for the results and baselines.

Table 1. Statistics of the datasets used in the reviewed models

| # | Name | Link | Size in experiments | |
			# of samples	# of labels
1	ImageNet	www.image-net.org	14,197,122	15,952
2	Web-data	n/a (proprietary)	9,861,293	109,444
3	McRae	doi.org/10.3758/BF03192726	10,000	541
4	MTurk	cloudacademy.com/blog/machine-learning-datasets-mechanical-turk	9,000	40,000
5	ESP Game	hunch.net/~learning/ESP-ImageSet.tar.gz	100,000	20,515
6	WordSim353	www.cs.technion.ac.il/~gabr/resources/data/wordsim353	353 pairs	n/a
7	MEN	clic.cimec.unitn.it/~elia.bruni/MEN	3,000 pairs	n/a
8	Simlex-999	arxiv.org/pdf/1408.3456.pdf	72,000 pairs	n/a
9	SemSim	doi:10.18653/v1/S16-1111	7,576 pairs	n/a
10	VisSim	doi:10.18653/v1/S16-1111	7,576 pairs	n/a
11	MM-IMDb	lisi1.unal.edu.co/mmimdb	25,959	23 genres
12	Flickr30k	web.engr.illinois.edu/~bplumme2	31,783	44,518
13	ImageClef	imageclef.org/SIAPRdata	99,535	20,000 pic
14	Food101	www.vision.ee.ethz.ch/en	86,102	101
15	FlickrTag	doi.org/10.1145/2812802	71,521,235	n/a

Table 2. A comparison of the results of multi-modal representation models

#	Model	Dataset(s)	Measure	Result	Improv.
1	Wsabie [22]	ImageNet	P@1	4.03%	28.3%
		Web-data		1.03%	221.9%
2	SAEs [19]	McRae,	Spearman ρ	0.70	0.76 (vs.
		ImageNet	F-score	0.43	0.63 human)
3	SDT-RNN [20]	MTurk	Sentence similarity	10.5	5.4%
			Image search	12.5	8.1%
			Describing images	16.9	11.98%
4	CNN-Mean,	MEN,	Spearman ρ	0.71	9.4%
	CNN-Max [8]	WordSim353		0.61	17.3%
5		MEN	Spearman ρ	0.75	7.14%
	MMSkip-Gram	Simlex-999		0.40	21.2%
	A, B [11]	SemSim		0.72	2.86%
		VisSim		0.63	-1.56%
6	MCB [6]	Flickr30k	accuracy, %	48.69	1.84%
		ReferItGame		28.91	3.32%
7	GMU [1]	MM-IMDb	F-score	0.630	3.79%
8	Bilinear-gated [9]	Food101	accuracy, %	90.	0.33%
		MM-IMDb		62.3	0.16%
		FlickrTag		28.6	2.88%

5   Conclusions and Future Research

In this paper, we present an overview of uni-modal and multi-modal representation models for classification tasks. Then, we overview eight state-of-the-art systems that implement and improve some of these models. We provide summary statistics and comparison of the used datasets and the performance of the reviewed systems.

Possible future research proposals include (1) reconstructing experiments to test some of the most curious state-of-the-art models; (2) writing a more extensive overview of such systems, with experimental results, for a journal article; and (3) developing a new model fusing textual, visual, and possibly tweet modalities for multi-label classification tasks, and hopefully improving the state of the art.

References

1. Arevalo, J., Solorio, T., Montes-y Gómez, M., González, F.A.: Gated multimodal units for information fusion. arXiv preprint arXiv:1702.01992 (2017)
2. Atefeh, F., Khreich, W.: A survey of techniques for event detection in Twitter. Comput. Intell. **31**(1), 132–164 (2015)

3. Baltrušaitis, T., Ahuja, C., Morency, L.P.: Multimodal machine learning: a survey and taxonomy. arXiv preprint arXiv:1705.09406 (2017)
4. Bengio, Y., Courville, A., Vincent, P.: Representation learning: a review and new perspectives. IEEE Trans. Pattern Anal. Mach. Intell. **35**(8), 1798–1828 (2013)
5. Bruni, E., Tram, N., Baroni, M., et al.: Multimodal distributional semantics. J. Artif. Intell. Res. **49**, 1–47 (2014)
6. Fukui, A., Park, D.H., Yang, D., Rohrbach, A., Darrell, T., Rohrbach, M.: Multimodal compact bilinear pooling for visual question answering and visual grounding. arXiv preprint arXiv:1606.01847 (2016)
7. Hofmann, T.: Probabilistic latent semantic indexing. SIGIR Forum **51**(2), 211–218 (2017)
8. Kiela, D., Bottou, L.: Learning image embeddings using convolutional neural networks for improved multi-modal semantics. In: Proceedings of the 2014 Conference on Empirical Methods in Natural Language Processing (EMNLP), pp. 36–45 (2014)
9. Kiela, D., Grave, E., Joulin, A., Mikolov, T.: Efficient large-scale multi-modal classification. arXiv preprint arXiv:1802.02892 (2018)
10. Krizhevsky, A., Sutskever, I., Hinton, G.E.: Imagenet classification with deep convolutional neural networks. In: Advances in Neural Information Processing Systems, pp. 1097–1105 (2012)
11. Lazaridou, A., Pham, N.T., Baroni, M.: Combining language and vision with a multimodal skip-gram model. arXiv preprint arXiv:1501.02598 (2015)
12. Le, Q., Mikolov, T.: Distributed representations of sentences and documents. In: International Conference on Machine Learning, pp. 1188–1196 (2014)
13. LeCun, Y., Bengio, Y., Hinton, G.: Deep learning. Nature **521**(7553), 436 (2015)
14. Liparas, D., HaCohen-Kerner, Y., Moumtzidou, A., Vrochidis, S., Kompatsiaris, I.: News articles classification using random forests and weighted multimodal features. In: Lamas, D., Buitelaar, P. (eds.) IRFC 2014. LNCS, vol. 8849, pp. 63–75. Springer, Cham (2014). https://doi.org/10.1007/978-3-319-12979-2_6
15. Liu, Y., Zhang, D., Lu, G., Ma, W.Y.: A survey of content-based image retrieval with high-level semantics. Pattern Recognit. **40**(1), 262–282 (2007)
16. Mikolov, T., Chen, K., Corrado, G., Dean, J.: Efficient estimation of word representations in vector space. arXiv preprint arXiv:1301.3781 (2013)
17. Ngiam, J., Khosla, A., Kim, M., Nam, J., Lee, H., Ng, A.Y.: Multimodal deep learning. In: Proceedings of the 28th International Conference on Machine Learning (ICML 2011), pp. 689–696 (2011)
18. Rosenthal, S., Farra, N., Nakov, P.: Sentiment analysis in Twitter. In: Proceedings of the 11th International Workshop on Semantic Evaluation (SemEval 2017), pp. 502–518 (2017)
19. Silberer, C., Lapata, M.: Learning grounded meaning representations with autoencoders. In: Proceedings of the 52nd Annual Meeting of the Association for Computational Linguistics (Volume 1: Long Papers), vol. 1, pp. 721–732 (2014)
20. Socher, R., Karpathy, A., Le, Q.V., Manning, C.D., Ng, A.Y.: Grounded compositional semantics for finding and describing images with sentences. Trans. Assoc. Comput. Linguist. **2**(1), 207–218 (2014)
21. Srivastava, N., Salakhutdinov, R.R.: Multimodal learning with deep Boltzmann machines. In: Advances in Neural Information Processing Systems, pp. 2222–2230 (2012)
22. Weston, J., Bengio, S., Usunier, N.: Wsabie: Scaling up to large vocabulary image annotation. In: IJCAI. 11, pp. 2764–2770 (2011)

Information Mining

Classification of Intangible Social Innovation Concepts

Nikola Milosevic[1](✉), Abdullah Gok[1], and Goran Nenadic[2]

[1] Manchester Institute for Innovation Research, University of Manchester,
Denmark Building, Manchester M13 9NG, UK
{nikola.milosevic,abdullah.gok}@manchester.ac.uk
[2] School of Computer Science, University of Manchester, Denmark Building,
Manchester M13 9PL, UK
g.nenadic@manchester.ac.uk

Abstract. In social sciences, similarly to other fields, there is exponential growth of literature and textual data that people are no more able to cope with in a systematic manner. In many areas there is a need to catalogue knowledge and phenomena in a certain area. However, social science concepts and phenomena are complex and in many cases there is a dispute in the field between conflicting definitions. In this paper we present a method that catalogues a complex and disputed concept of social innovation by applying text mining and machine learning techniques. Recognition of social innovations is performed by decomposing a definitions into several more specific criteria (social objectives, social actor interactions, outputs and innovativeness). For each of these criteria, a machine learning-based classifier is created that checks whether certain text satisfies given criteria. The criteria can be successfully classified with an F1-score of 0.83–0.86. The presented method is flexible, since it allows combining criteria in a later stage in order to build and analyse the definition of choice.

Keywords: Text mining · Classification
Natural language processing · Social innovation

1 Introduction

Many social science concepts have abstract definitions, that may be unclear for operational use and for distinguishing entities that satisfy the given definition from those that do not. The physicalist approach suggests to use terms and entities that are more concrete and that are observable. Small concepts are more tangible than the bigger, more abstract ones [7]. Often, it is not possible to use only the tangible concepts, especially for the complex social issues and phenomena that try to deal with these issues.

In this paper we examine automated analysis of one of the high level, intangible social science concepts – social innovation. Social innovation refers to innovative activities, models or services that are supposed to meet certain social need

© Springer International Publishing AG, part of Springer Nature 2018
M. Silberztein et al. (Eds.): NLDB 2018, LNCS 10859, pp. 407–418, 2018.
https://doi.org/10.1007/978-3-319-91947-8_42

or solve a certain social issue. They are usually executed by the organisations that are primarily social [6,13]. Social innovation may be performed by both formal and non-formal organisations. They may include technological innovation, but also innovation in inter-actor communication, project management, models, services, and ways to generate output. The major goal of social innovation is to solve societal challenges and improve the lives of the members of the society.

Some of the examples of social innovation projects are:

- Feelif[1] - a product that combines a standard tablet, with an application and a relief grid that allow visually impaired people feel shapes on the tablet. Feelif allows users playing games and use educational content on tablet, addressing social need that visually impaired people were disregarded by smart phone/tablet vendors.
- Real Junk Food project[2] - the chain of restaurants that produces food using ingredients that are getting out of date. Visitors can pay as they feel for the consumed food, while they also employ workers from disadvantaged backgrounds. Real junk food presents organisational innovation.

In the past decade, social innovation has received political recognition. For example, the European Union introduced Social Investment Packages [6] in 2013, that included recommendations for investing in social innovation projects. Since then, the European Union invested a great amount in social innovation projects and organisations. Social innovation projects can apply among others for funds to Horizon 2020, EU Programme for Employment and Social Innovation and European structural and investment funds. Also, certain governments, the European Union and large funding organisations invested in creating databases of social innovation projects and actors [1,9,12]. These databases should help funding bodies and policy makers to map projects and determine successful practices and environments for social innovation projects.

However, most of the currently available databases face the following challenges:

1. Thematically focused – The most of the currently available databases are thematically focused to a certain area (e.g. digital social innovation, ageing, homelessness, etc.)
2. Small in size – Majority of the available databases contain between 50 and 1000 projects
3. Limited information – The entries in the databases are limited in terms of features describing a project
4. Relying on a single source – Most of the available databases collect data by human input, either by the project team or the self registration of actors. Also, the majority of the databases do not update data after the project funding have ended.

---

[1] https://www.feelif.com.
[2] http://therealjunkfoodproject.org.

5. Using different definitions – There are a number of social innovation definitions in literature. Databases use different definition depending of the belief of the author's creator or the definition set by the funding body. This makes comparison between databases, database integration and reuse difficult.

The usual way for coding concepts in social sciences is to use human coders that would review and classify the entities. However, human annotation is expensive and its costs can be reduces by almost 80% for certain tasks by utilising natural language processing, text mining and machine learning techniques [11].

In this paper we describe part of the methodology used in European Social Innovation Database (ESID) for classifying social innovation projects, that should address the stated challenges. The database is created in semi-automatic manner, encompassing web crawling, text mining and machine learning. In this paper we focus on the methodology for determining whether a project satisfies social innovation criteria. Social innovation is intangible concept, since people consider social innovation concepts differently and since there are multiple definitions of social innovation. We show that it is possible to create a modular system that takes into the account the underlying concepts of social innovation definitions and allow users to select criteria (and projects) they consider relevant for defining social innovation.

2   Disentangling the Definition of Social Innovations

Discussion about what is social innovation and what are social innovation inclusion criteria is ongoing. Therefore, there are multiple authors proposing definitions of social innovation [2–5,8,10]. While nuances between each definition are vastly varying, the broad criteria are about social objectives, social interaction between actors or actor diversity, social outputs and innovativeness. However, different definitions include different combinations and different number of these criteria (e.g. EU is using definition stressing out social objectives and actors interaction). The criteria we used for this work are based on literature review performed by authors. The criteria and their descriptions are presented in Table 1.

By performing exhaustive literature review, it is possible to identify concepts and criteria that compose social innovation definitions. Because there are disagreements between authors on the criteria used for the definition, it is not possible to create a classifier that would satisfy multiple social innovation definition. However, it is possible to create classifiers that can classify underlying concepts and criteria. This allows creation of a modular system in which user can set the definition of social innovation that he would like to use.

3   Methodology Overview

Methodology for semi-automated generating of social innovation database consists of two phases. In the first phase we use currently available databases, lists,

Table 1. Description of the social innovation criteria used in this study

Element of definition	Criteria description
Objectives	Project primarily or exclusively satisfies (often unmet) societal needs, including the needs of particular social groups; or aims at social value creation Often no price involved for the main social beneficiary or the innovation is provided to the main beneficiary at cost only. However, there might be examples that price is involved
Actors and actor interactions	Satisfy one or both of the following: **i. Diversity of Actors:** Project involves actors who would not normally involve in innovation as an economic activity, including formal (e.g. NGOs, public sector organisations etc.) and informal organisations (e.g. grassroots movements, citizen groups, etc.). This involvement might range from full partnership (i.e. project is conducted jointly) to consultation (i.e. there is representation from different actors) **ii. Social Actor Interactions:** Project creates collaborations between "social actors", small and large businesses and public sector in different combinations. These collaborations usually involve (predominantly new types of) social interactions towards achieving common goals such as user/community participation. Often, projects aim at significantly different action and diffusion processes that will result in social progress. Often social innovation projects rely on trust relationships rather than solely mutual-benefit
Outputs/Outcomes	Project primarily or exclusively creates socially oriented outputs/outcomes. Often these outputs go beyond those created by conventional innovative activity (e.g. products, services, new technologies, patents, and publications), while conventional outputs/outcomes might also be present. These outputs/outcomes are often intangible and they might include the following but not limited to: - change in the attitudes, behaviours and perceptions of the actors involved and/or beneficiaries - social technologies (i.e. new configurations of social practices, including new routines, ways of doing things, laws, rules or norms) - long-term institutional/cultural change
Innovativeness	There should be a form of "implementation of a new or significantly improved product (good or service), or process, a new marketing method, or a new organisational method" The project needs to include some form of innovative activities (i.e. scientific, technological, organisational, financial, and commercial steps intending to lead to the implementation of the innovation in question). Innovation can be technological (involving the use of or creating technologies) as well as non-technological The innovation should be at least "new" to the beneficiaries it targets (it does not have to be new to the world)

case study repositories, and mappings of social innovation projects in order to obtain initial data about social innovations. This phase include the following steps:

1. Compose a list of social innovation sources.
2. Crawl the project description pages from the listed sources
3. Crawl the project websites, if they were available in the social innovation source
4. Annotate a set of projects. The projects are annotated whether they satisfy social innovation criteria by human coders
5. Create a machine learning model for classifying projects whether they satisfy social innovation criteria
6. Obtain additional features about the project, such as information about organisations involved, location, etc.

The last step of the phase is out of the scope of this project. The second phase of the project involves crawling of the sources that potentially could contain social innovation projects, such as crowd-sourcing platforms. Also, the second phase is also out of the scope of this paper.

4   Obtaining Data About Social Innovation from Existing Sources

Firstly, we have identified data sources containing information about social innovation projects and actors. The list contained 93 information sources, however, some of the data sources contained cleaner data than the others. For the clean and relevant data sources, we developed a set of web crawlers that obtained data and stored it into our database. A web crawler (also known as scraper or spider) is a program or automated script which browses the World Wide Web in a methodical, automated manner and collects the content of the visited web pages (in full or targeted parts of them). The data sources could not be directly downloaded, however, they had database accessible on the web, and therefore web crawlers could methodically visit all entity pages and obtain data about them. The development of crawlers can be time-consuming. We started with developing crawlers for the biggest databases. At the moment, we have developed crawlers for 9 databases, however, they contain over 6,000 entities. This presents more than 85% of all entities in the identified data sources.

However, while crawling, we faced a number of challenges. The main challenge for the crawling is that the web sources did not have consistent structures, and wealth of information. Our approach to crawl these data sources was to obtain targeted information that was included in data source about certain entity (project or actor). In order to achieve this, we needed to develop separate crawlers for each data source that is able to locate information of interest on the page.

Certain data sources contained information about both projects and actors (e.g. Digital Social Innovation), however, some data sources contained information only about one entity type (projects – e.g. EUSIC, MOPACT, actors – Social Enterprise UK, Social Innovation Generation).

The crawled data sources and the number of their entities are presented in the Table 2.

Table 2. Description of the social innovation criteria used in this study

Data source	Number of projects	Number of actors
Digital Social Innovation	2,200	2,007
European Social Innovation Competition	90	0
MoPAct	140	0
Innovage	153	0
SIMRA	9	28
European Investment bank social innovation tournament	72	0
Social Innovation Generation (Social Innovation in Canada database)	0	256
Bill and Melinda Gates Foundation	0	444
Social Enterprise UK	0	687
Total	2,664	3,422

5 Data Annotation

In order to make a data set for supervised machine learning-based approach that is able to classify social innovation criteria, we organised two data annotation workshops. During the first workshop, 6 annotators were annotating about 40 projects each. Annotators were PhD students and research staff whose research is associated with the are of innovation and social innovation. The text for each project was composed from the texts available on the project websites. For this annotation task, we included only projects whose websites contained between 500–10,000 words. About 20% of the documents were annotated by at least two annotators. For annotation we used Brat rapid annotation tool [14]. The annotators were asked to annotate sentences that present how certain project met defined social innovation criteria (objectives, actor interaction, outputs, innovativeness) and to give a score at the document level for each of the four criteria (as presented in Table 2). The document level marks were in range of 0–2:

- 0 – criteria not satisfied
- 1 – criteria partially satisfied
- 2 – criteria fully satisfied

In the second annotation workshop, we focused more on calculating inter-annotator agreement and finding potential outlier annotator (annotator with high disagreement compared to other annotators). We created a dataset using projects listed as semi-finalists and finalists in EU Social Innovation Competition[3] and European Investment Bank Social Innovation Tournament[4]. The dataset consisted of 40 projects, whose websites were crawled. A subset of the four annotators annotated the whole dataset in the same manner as during the first workshop.

6   Classification of Criteria

The dataset created during both annotation task were used for training and validation of the machine learning-based approach. The classifier is created for each social innovation criteria (objective, actors, outputs, innovativeness). Before applying machine learning algorithm the text was stemmed and stop-words were removed (using Rainbow stop-word list[5]). We have evaluated classification using Naive Bayes machine learning algorithms using the bag-of-words language model. Since dataset was not balanced, having more negative instances than positive, we also performed an experiment with balancing data by oversampling positive instances.

7   Results

Firstly, we present the results of annotation workshop, including inter-annotation agreement and the number of non-relevant projects in the examined data sources. Then we present the results of machine learning classification using the two described approaches.

7.1   Data Annotation Results

During the first workshop, six annotators annotated 40 documents each on the sentence and document level. During the second workshop, three annotators annotated same 43 documents, while one annotator annotated 30 of these documents. During the workshops about 10% of the data available at the moment of annotation workshops was annotated. Inter-annotator agreement per each criteria is presented in the Table 3. Inter-annotator agreement is calculated on the paragraph level and document level. Inter-annotator agreement on the paragraph level is calculated by examining each paragraph of the text whether it contains annotation of a certain class in both annotated documents. Inter-annotator agreement on the document level is calculated by examining whether both annotators scored the document with the certain annotation type (annotation types

---

[3] http://eusic.challenges.org/.

[4] https://institute.eib.org/whatwedo/social-2/social-innovation-tournament-2/.

[5] http://www.cs.cmu.edu/~mccallum/bow/rainbow/.

Table 3. Inter-annotator agreement on paragraph and document level per criteria in the annotated dataset

Inclusion criteria	Paragraph level agreement	Document level agreement
Objectives	37.50%	76.60%
Actors and Actor interactions	17.20%	65.70%
Outputs	18.90%	66.73%
Innovativeness	19.50%	70.80%
Macro-average	23.27%	69.96%

translates to four social innovation criteria). Since agreement on score was also fairly low, we have binarised document level annotation, so they score whether the criteria is satisfied (scores 1 and 2) or not satisfied (score 0).

As it could be seen, inter-annotator agreement on the paragraph level is low and therefore these annotations are not useful for machine learning. Sentences are shorter structures than paragraphs and therefore agreement would be even lower. In the paragraph level inter-annotator agreement we looked whether a certain paragraph is annotated by both annotators with the same annotation type. The inter-annotator agreement on the document level is in range of 65%–76%, which is relatively low as well, indicating that the annotation concepts are intangible and that people generally do not completely agree on what is innovative, what social objectives are or what social actor interactions are. However, this score is high enough to be used for machine learning.

We also calculated how many projects in each examined data source were false positives (projects not satisfying any social innovation criteria – spam projects). Agreement for detecting social innovation or false positive project is about 85%. The number and percentage of false positive (spam) projects per data source can be seen in Table 4.

The range of false positive projects in data sources ranges between 0% and 58%. As we hypothesised, the data sources are not clean and user-imputed data sources, such as Digital Social Innovation are noisy. Even some expert imputed data sources, such as Innovage contains about 30% of false positive projects.

7.2   Classification Results

The final dataset, that is used for machine learning, after both human annotation workshops, contains 277 documents, with the following composition per criteria:

- Objectives – 166 negative, 111 positive instances
- Actors – 189 negative, 88 positive instances
- Outputs – 190 negatives, 87 positive
- Innovativeness – 190 negative, 87 positive

Table 4. Percentage of false positive (spam) projects per data source in the annotated data

Data source	Number of projects	Percentage of false positive projects (false positive/total annotated)
European Social Innovation Competition	90	12.9% (8/62)
MoPAct	140	7.3% (3/41)
Innovage	153	30% (6/20)
Digital Social Innovation	2,200	58% (105/188)
European Investment bank social innovation tournament	72	6.8% (6/87)
SIMRA	9	0% (0/2)

This data was used for training and testing machine learning-based method for determining whether project description satisfies given social innovation criteria. For evaluation was used 10-fold cross-validation. The results of four different classifiers for each criteria are presented in Table 5.

Table 5. Results of classifying criteria using Naive Bayes classifier. The input text was created from the content of projects' websites. The evaluation is performed using 10-fold cross-validation

Criteria	TP	FP	FN	Precision	Recall	F1-score
Actors	62	39	26	0.614	0.705	0.656
Objectives	81	36	30	0.692	0.730	0.711
Outputs	61	41	26	0.592	0.701	0.642
Innovativeness	58	42	29	0.580	0.667	0.620

The results are in range 0.62–0.71 F1-score, having significant amount of false positives and false negatives. As it was previously mentioned, the data set that was used was imbalanced, having higher amount of negative instances than positive. In cases of actors, outputs and objectives, the data set contained more than two times more negative instances compared to the positive ones. The imbalance of data set can affect the performance of classification, especially in case of Bayesian-based method for classification. Therefore, we applied oversampling of positive instances. With oversampled positive instances, the data set contained similar amount of positive and negative instances for each criteria. The classification results over oversampled data is presented in Table 6.

The performance of classifiers across criteria increased after oversampling positive instances. Classifiers with these performance can be applied in production system in order to determine whether descriptions from projects' websites

Table 6. Results of classifying criteria using Naive Bayes classifier after balancing data set by oversampling positive class. The input text was created from the content of projects' websites. The evaluation is performed using 10-fold cross-validation

Criteria	TP	FP	FN	Precision	Recall	F1-score
Actors	118	25	15	0.825	0.887	0.855
Objectives	121	25	14	0.829	0.896	0.861
Outputs	122	30	11	0.803	0.917	0.856
Innovativeness	117	29	16	0.801	0.880	0.839

satisfy social innovation criteria and to determine which projects are social innovation and which are not.

Experiments using other algorithms, such as SVM, decision trees and neural networks using Glove embeddings were also conducted. However, Naive Bayes over-performed these approaches. This is likely due to the small data set and the fact that Naive Bayes is able to generalise well even with relatively small data sets.

8   Conclusion

In this paper we presented a methodology to model intangible social science concepts over which definition may be active debate in the field. The case of intangible concept presented in this paper is social innovation. There are numerous definitions describing the concept. Also, using multiple definitions would make data inseparable, therefore it would not be possible to make automated systems for collecting, cataloguing and classifying such items.

The approach we proposed relies on extensive literature review of the concept and disentangling the definitions into the components that they were made of. These components are usually more tangible concepts. Such concepts would have better agreement between coders and will be easier separable for machine learning-based methods. Each of the many definitions of the concept will be made of a certain combination of the identified components. Therefore, it is possible to create a system that is modular and that based on the combinations of components is able to return instances that satisfy particular definition of the concept. In the case of social innovation, the identified components were objectives, actors, outputs and innovativeness. Once certain text is classified whether it satisfies given components, one can retrieve instances satisfying EU definition of social innovation that consists of objectives, actors and innovativeness. Also, instances satisfying other combination of criteria or social innovation definitions can be retrieved.

The performance of classifiers for the social innovation criteria was encouraging, ranging between 0.83–0.86 F1-score. It proves the hypothesis that supervised machine learning can be used to classify whether a project description satisfy certain social science criteria. Also, the percentage of correctly classified instances in the given data set was higher than calculated inter-annotator agreement. However, the percentage of correctly classified instances is agreement with the final data set only, which may be possible due to larger amount of training data and proves that classification results are comparable with human annotators. After project descriptions are classified whether they satisfy given criteria, it is possible to extract other meta-data from text, using named entity recognition tools and information extraction strategies.

The approach can be generalised to cataloguing other social science concepts by decomposing multiple definitions into the component criteria. Different kinds of projects, regulations, laws, or organisations can be classified in this manner. The catalogues of such entities may be useful for further research, collaboration, policy making and governance.

In the future, we are planning to include additional data sources, such as projects from crowd-funding platforms and community building portals (such as Meetup.com) and use the classification model to discover new projects satisfying out social innovation criteria. In addition, we are planning to expand training data set by performing additional annotation workshops. We hope that more annotation will enable other methods to be used and reduce certain domain related biases that may exist with the current model.

Acknowledgements. The work presented in this paper is a part of KNOWMAK project that has received funding from the European Union's Horizon 2020 research and innovation programme under grand agreement No. 726992.

References

1. Bria, F., Sestini, F., Gasco, M., Baeck, P., Halpin, H., Almirall, E., Kresin, F.: Growing a digital social innovation ecosystem for europe: Dsi final report. European Commission, Brussels (2015)
2. Caulier-Grice, J., Davies, A., Patrick, R., Norman, W.: Defining social innovation. A deliverable of the project: the theoretical, empirical and policy foundations for building social innovation in Europe (TEPSIE), European Commission-7th Framework Programme. European Commission, DG Research, Brussels (2012)
3. Choi, N., Majumdar, S.: Social innovation: towards a conceptualisation. In: Majumdar, S., Guha, S., Marakkath, N. (eds.) Technology and Innovation for Social Change, pp. 7–34. Springer, New Delhi (2015). https://doi.org/10.1007/978-81-322-2071-8_2
4. Dawson, P., Daniel, L.: Understanding social innovation: a provisional framework. Int. J. Technol. Manag. **51**(1), 9–21 (2010)
5. Edwards-Schachter, M., Wallace, M.L.: Shaken, but not stirred: sixty years of defining social innovation. Technol. Forecast. Soc. Change **119**, 64–79 (2017)
6. European Union: Social innovation (2017). http://ec.europa.eu/social/main.jsp?catId=1022
7. Gerring, J.: Social Science Methodology: A Criterial Framework. Cambridge University Press, Cambridge (2001)
8. Grimm, R., Fox, C., Baines, S., Albertson, K.: Social innovation, an answer to contemporary societal challenges? Locating the concept in theory and practice. Innov. Eur. J. Soc. Sci. Res. **26**(4), 436–455 (2013)

9. Haak, M., Slaug, B., Oswald, F., Schmidt, S.M., Rimland, J.M., Tomsone, S., Ladö, T., Svensson, T., Iwarsson, S.: Cross-national user priorities for housing provision and accessibility findings from the european innovage project. Int. J. Environ. Res. Publ. Health **12**(3), 2670–2686 (2015)

10. van der Have, R.P., Rubalcaba, L.: Social innovation research: an emerging area of innovation studies? Res. Policy **45**(9), 1923–1935 (2016)

11. Hillard, D., Purpura, S., Wilkerson, J.: Computer-assisted topic classification for mixed-methods social science research. J. Inf. Technol. Polit. **4**(4), 31–46 (2008)

12. Luijben, A., Galenkamp, H., Deeg, D.: Mobilising the potential of active ageing in europe (2012)

13. Mulgan, G.: The process of social innovation. Innovations **1**(2), 145–162 (2006)

14. Stenetorp, P., Pyysalo, S., Topić, G., Ohta, T., Ananiadou, S., Tsujii, J.: Brat: a web-based tool for nlp-assisted text annotation. In: Proceedings of the Demonstrations at the 13th Conference of the European Chapter of the Association for Computational Linguistics, pp. 102–107. Association for Computational Linguistics (2012)

An Unsupervised Approach
for Cause-Effect Relation Extraction
from Biomedical Text

Raksha Sharma$^{(\boxtimes)}$, Girish Palshikar, and Sachin Pawar

TCS Research, Tata Consultancy Services, Pune, India
{raksha.sharma1,gk.palshikar,sachin.pawar}@tcs.com
http://www.tcs.com

Abstract. Identification of Cause-effect (CE) relation mentions, along with the arguments, are crucial for creating a scientific knowledge-base. Linguistically complex constructs are used to express CE relations in text, mainly using generic *causative* (*causal*) verbs (`cause, lead, result` *etc*). We observe that some generic verbs have a domain-specific causative sense (`inhibit, express`) and some domains have altogether new causative verbs (`down-regulate`). Not every mention of a generic causative verb (e.g., `lead`) indicates a CE relation mention. We propose a linguistically-oriented unsupervised iterative co-discovery approach to identify domain-specific causative verbs, starting from a small set of seed causative verbs and an unlabeled corpus. We use known causative verbs to extract CE arguments, and use known CE arguments to discover causative verbs (hence *co-discovery*). Since causes and effects are typically agents, events, actions, or conditions, we use WordNet hypernym categories to identify suitable CE arguments. PMI is used to measure linguistic associations between a causative verb and its argument. Once we have a list of domain-specific causative verbs, we use it to extract CE relation mentions from a given corpus in an unsupervised manner, filtering out non-causative use of a causative verb using WordNet hypernym check of its arguments. Our approach extracts 256 domain-specific causative verbs from 10,000 PubMed abstracts of Leukemia papers, and outperforms several baselines for extracting intra-sentence CE relation mentions.

Keywords: Cause-effect relation · Causative verbs
Relation extraction · Biomedical domain · Leukemia · PMI
Hypernyms

1 Introduction

Automatically extracting different types of knowledge from authoritative texts (e.g., textbooks, research papers) within a domain and representing it in a computer analyzable as well as human readable form is an important but challenging

© Springer International Publishing AG, part of Springer Nature 2018
M. Silberztein et al. (Eds.): NLDB 2018, LNCS 10859, pp. 419–427, 2018.
https://doi.org/10.1007/978-3-319-91947-8_43

goal. Ability to query and use such extracted knowledge-bases can help scientists, doctors and other users in performing tasks such as question-answering, diagnosis, exploring and validating hypotheses, understanding the state-of-the-art, and identifying opportunities for new research. *Cause-effect (CE)* relations, which connect a *cause* to an *effect*, are an important component of any scientific knowledge-base. Identification of CE relation mentions, along with extraction of the arguments of each mention, are crucial for the creation of a scientific knowledge base. Understanding and a formal representation of CE relations, and reasoning over them, are important philosophically, as well as mathematically. The simplest way of expressing CE relations in text is through the use of *causative* (or *causal*) verbs, such as `cause, lead, result`. Apart from generic causative verbs, such as `cause, lead, result` different domains have their own causative verbs, which are either new verbs specific to that domain (*e.g.*, `over-express, up-regulate` in the biomedical domain) or generic verbs that have a special causative sense specific to that domain (*e.g.*, `inhibit, express` in the biomedical domain). Moreover, every mention of a causative verb does not necessarily indicate a true CE relation mention. For example, the verb `achieve` indicates a CE relation in the sentence `Tumor cell killing was achieved by concerted action of necrosis apoptosis induction.`, but not in `90 patients achieved complete remission on the day of induction therapy.` There are other well-known problems with the linguistic expression of CE relations in text. First is the use of negation, which negates the apparent CE relation mention, Next is the use of coreference, which requires its resolution to obtain the correct argument of a CE relation mention.

In this paper, we focus on two specific problems: (i) how to automatically acquire the list of domain-specific causative verbs; and (ii) how to effectively use such a list to extract true CE relation mentions. For (i), since getting a complete list of domain-specific causative verbs is difficult, we propose a linguistically-oriented unsupervised iterative co-discovery approach to identify causative verbs, starting from a very small set of seed causal verbs and an unlabeled corpus. The idea is that a known causal verb can be used to extract CE arguments, and known CE arguments can be used to discover unknown causative verbs (hence *co-discovery*). For (ii), since causes and effects are typically agents, events, actions, or conditions, we use appropriate WordNet hypernym categories to identify suitable CE arguments and thereby filter out non-causative uses of a causative verb in our list. Point-wise mutual information (PMI) is used to measure the level of (linguistic) associations between a causative verb and its argument. Once we have a reasonably complete list of domain-specific causative verbs, we propose an unsupervised algorithm which uses it to extract CE relation mentions from the given corpus, filtering out non-causative use of the causative verb using the above-mentioned WordNet hypernym check of its arguments. We demonstrate the effectiveness of this approach by extracting 256 causative verbs from 10,000 PubMed abstracts of Leukemia papers, starting with 3 causative verbs. We demonstrate the effectiveness of our approach by comparing it with some baselines on a manually labeled dataset of 350 sentences containing 130 CE relation mentions. The paper is organized as follows. Section 2 discusses related work. Section 3 discusses our algorithms for discovering domain-specific causative verbs and to extract CE relation mentions using them. Section 4 elaborates the experimental setup and results. Section 5 concludes the paper.

2   Related Work

Identification of CE relation helps to understand the semantics that one event causes another. Researchers have tried to discover cause-effect relation in textual data focusing on lexical and semantic constructs. There have been attempts to extract CE relation in text using knowledge-based inferences. Joskowicz [1] prepared a dedicated knowledge base to built causal analyzer for a Navy ship. A knowledge-based system perform reasonably well in the targeted domain, however they have poor generalizability. Khoo [2] constructed patterns to identify CE relation based on pre-define patterns. The patterns includes causal clues, *e.g.,hence, therefore, if-then, cause, break etc.* A few researchers used grammatical patterns to identify CE relation targeting different applications [3–5]. Girju [3] utilized grammatical patterns in order to analyze cause-effect questions in question answering system. There are a very few instances of combining grammatical patterns with machine learning in order to extract semantic relation such as cause-effect. Chang [6] used cue phrases (cause triggering construct) with their probability to extract other lexical arguments of cause-effect relation. These probabilities are learned from raw corpus in an unsupervised manner. Though they reported results in the biomedical domain, they neither incorporated linguistic information nor domain information inherent in the text in support of CE relation. Do [7] developed a minimally supervised approach, based on focused distributional similarity and discourse connectives. They showed combining lexical information, such as discourse connective with statistical measure provides additional improvement in CE relation identification. In this paper, we propose an approach which deploys the linguistic clue indicating CE constructs (incorporating WordNet) and PMI between dependency relations for identification of CE relation in a sentence. Thus, our approach takes advantage of linguistic as well as statistical measures.

3   Causative Verbs Discovery and CE Relation Extraction

We now discuss the algorithm *codiscover_causative_verbs* (Algorithm 1). Its inputs are an unlabeled corpus of domain-specific documents, along with a small seed list S of known causative verbs. The algorithm works in 3 phases. In phase-I, it extracts and stores all the nominal subject and object arguments of the known causative verbs (using Stanford dependency parser), removing those that occur less than the given threshold n_0 times. Phase-II works on the following intuition: *verb forms of known subjects or objects can themselves occur in a causative sense.* Presence of 'of', or 'in' after the noun is often an indicator of existence of verbal use of the noun. Hence, Phase-II computes PMI [8, 9] of the known nouns extracted in

Phase-I with 'of', or 'in', retaining only top $p_0\%$ among these nouns. $PMI(nn;$ of/in) measures the association between the noun (nn) in NN_{CE} and proposition 'of' or 'in' in the corpus D. Then we add the verb form of each of these top nouns to the set S, provided the noun form's hypernyms indicate some kind of action (*e.g.*, growth, act, action, event, change, control, happening *etc.*) in top 3 senses of the noun. Phase-III works on this intuition: *if there is a verb whose subject and object in a sentence are both known (i.e., they have previously occurred with known causative verbs), then that verb is also likely to be a causative verb.* Phase-III identifies such verbs, removing those that occur less than the given threshold n_1 times. Additionally, the algorithm computes the PMI of each such verb with each known subject and object nouns pair (such nouns are stored in NN_{CE} by Phase-I), retaining only top $p_1\%$ among these verbs. The PMI of verb with known subject and object is computed using chain rule of PMI (Eq. 1). $PMI(v; s_{known}o_{known})$ measures the association between verb (v) and pair of causal arguments ($s_{known}o_{known}$) in the corpus D. Algorithm retains only those verbs which have at least one noun form whose hypernyms contain known action indicating terms, just as in Phase-II. The algorithm stops when no new verbs were added to S. Table 1 defines the functions used in Algorithms 1 and 2.

$$PMI(v; s_{known}o_{known}) = PMI(v; s_{known}) + PMI(v; o_{known}|s_{known})$$

$$= \log \frac{P(v; s_{known}o_{known})}{P(v) * P(s_{known}o_{known})} \tag{1}$$

Table 1. Functions used in the Algorithm 1

Function	Description
$LEMMA()$	Gives root (base) form of the input word
$SUB()$	Gives subject of the input word
$OBJ()$	Gives object of the input word
$POS()$	Gives part of speech of the input word
$InheritedHypernym()$	Gives all hypernyms of the input word available in WordNet
$VerbalRootForm()$	Gives root verb form of the input word using WordNet
$NOUN()$	Gives noun form of the input word using WordNet

The Algorithm 2 extracts the CE-relation mentions in the sentence. It takes the set S of causative verbs produced by Algorithm 1 as input, along with a set CP of fixed cue phrases and extracts CE-relation mentions (R) from the D_{test}. The algorithm first checks if the given sentence contains any of the given causative verbs, and extracts its subject and object. It outputs the tuple (causative verb, subject, object) if both the subject and object satisfy the constraint that they have a suitable hypernym (e.g., growth, act, action, event, change, control, cause, effect, reason, process, condition, organ, compound, organism *etc.*). In the output, the subject and object correspond to the cause and effect arguments of the CE relation. CE relations may be expressed in text through the use of causative verbs or

Algorithm 1. Algorithm *codiscover_causative_verbs*

Input:
$D = \{d_1, d_2, \ldots, d_m\}$ // m unlabeled documents
n_0, n_1 // min. occurrence counts; default 10 for both
p_0, p_1 // top percentage for PMI values; default 50 for both
S // seed list of causative verbs; default $\{$**cause, lead, result**$\}$
HC // hypernym categories for acceptable causative verbs
Output: S // set of discovered causative verbs in the domain

```
1   flag := TRUE; // stop if flag is FALSE
2   while flag do
3       flag := FALSE;
        //Phase-I: Extraction of cause-effect arguments (NN_CE) using CV
4       NN_CE := ∅;
5       for each cv in S do
6           for each occurrence of cv in a sentence s in D do
7               s := LEMMA(SUB(cv, s)); // headword of subject of verb cv in s
8               o := LEMMA(OBJ(cv, s)); // headword object of verb cv in s
9               if POS(s) ≠ noun ∨ POS(o) ≠ noun then continue
10              if s ∉ NN_CE then NN_CE := NN_CE ∪ (s,1)
11              else
12                  increase occurrence count of s in NN_CE by 1
13              if o ∉ NN_CE then NN_CE := NN_CE ∪ (o,1)
14              else
15                  increase occurrence count of o in NN_CE by 1

16      for each nn in NN_CE do
17          Remove nn from NN_CE if its occurrence count < n_0
        //Phase-II: discover new causative verbs using NN_CE
18      for each nn ∈ NN_CE do
19          if InheritedHypernym(nn) ∉ HC then continue
20          compute and store PMI of nn with {of, in} in D // nn with of or in
21      Sort PMI hash-table and keep only top p_0% entries
22      for each nn ∈ PMI hash-table do
23          if VerbalRootForm(nn) ∉ S then
24              S := S ∪ {VerbalRootForm(nn)}; flag := TRUE

        //Phase-III: Discover new causative verbs from corpus using NN_CE
25      T := ∅; // candidate verbs
26      for each sentence s in D do
27          for each occurrence of a verb v in s do
28              s := LEMMA(SUB(v, s)); // headword of subject of verb v in s
29              o := LEMMA(OBJ(v, s)); // headword object of verb v in s
30              v := LEMMA(v);
31              if s ∉ NN_CE ∨ o ∉ NN_CE then continue
32              if v ∉ T then
33                  T := T ∪ (verb,1)
34              else
35                  increase occurrence count of v in T by 1

36      for each v ∈ T do
37          Remove v from T if its occurrence count < n_1
38      for each tuple (v, s, o) ∈ T × NN_CE × NN_CE do
39          compute and store PMI of v with (s, o) in D // chain-rule for PMI
40      Sort PMI hash-table and keep only top p_1% entries
41      for each v ∈ PMI hash-table do
42          for each noun form nn for verb v do
43              if InheritedHypernym(nn) ∈ HC ∧ v ∉ S then
44                  S := S ∪ {v}; flag := TRUE; break

45  return(S); // stop when no new verbs were added to S
```

non-verbal *cue phrases*, such as due to, because of, cause of, causes of, cause for, causes for, reason for, reasons for, reasons of, reason of, as a consequence, as a result. The algorithm checks if the given sentence contains any of the given fixed set of non-causal cue phrases. If yes, then it extracts the subject of the verb as cause and the noun modifier of the cue phrase as the effect. The algorithm outputs only headwords of the noun phrases corresponding to a cause or an effect. It is easy to modify it to output the entire phrase instead.

Algorithm 2. Algorithm *extract_CE_relations*

Input:
$D_{test} = \{d_1, d_2, \ldots, d_k\}$ // k unlabeled documents
S, CP // list of causative verbs and list of cue phrases respectively
HCE // hypernym categories for acceptable causes or effects
Output: R // set of discovered CE relation mentions

1 $R := \emptyset$;
2 **for** *each sentence $s \in D_{test}$* **do**
3   **for** *each $cv \in S$* **do**
4   **for** *each occurrence of cv in s* **do**
5              $s := \text{LEMMA}(\text{SUB}(cv, s))$; // headword of subject of verb cv in s
6              $o := \text{LEMMA}(\text{OBJ}(cv, s))$; // headword object of verb cv in s
7   **if** *s is not proper noun $\wedge$ InheritedHypernym(s) $\notin$ HCE* **then** continue
8   **if** *o is not proper noun $\wedge$ InheritedHypernym(o) $\notin$ HCE* **then** continue
9              $R := R cup \{(cv, s, o)\}$;
10   **for** *each $cp \in S$* **do**
11   **for** *each occurrence of cp in s* **do**
12             $v := $ verb nearest to cp in s;
13             $s := \text{LEMMA}(\text{SUB}(v, s))$; // headword of subject of verb v in s
14   **if** *cp has no noun modifier in s* **then continue**;
15             $n := $ noun modifier of cp in s;
16             $R := R \cup \{(v, s, n)\}$;

---

4 Experimental Setup and Results

Leukemia is a highly researched area in the biomedical domain, having $3,02,926$ scientific documents on PubMed and $31,142$ on Nature. We downloaded $10,000$ abstracts of Leukemia related papers from PubMed using the Biopython library with Entrez package (dataset D). We used this dataset ($89,947$ sentences, $1,935,467$ tokens) to execute Algorithm 1, in order to identify causative verbs from this dataset. To evaluate the performance of our algorithm and other baselines, we extracted 50 (350 sentences, 105 CE relation mentions) abstracts of

Leukemia related papers from PubMed, and manually identified CE relation mentions in it. To identify Parts-of-speech of the words and dependency relations among them, we used the Python interface of Stanford dependency parser [10]. Since biomedical text is full of complex sentences, we use Enhanced++ dependencies. In order to obtain verb form, noun form and hypernyms of words, we have used NLTK wrapper of WordNet. We observed that many domain specific words in the biomedical domain (*e.g.*, `down-regulation`, `up-regulation`, `over-expression` *etc.*) are not available in WordNet. We believe that domain specific words are good candidates for CE-relation formation. Our algorithm takes decision about such words based on their count and the PMI in the corpus irrespective of their presence in the WordNet. Thresholds n_0, n_1 of Algorithm 1 on the count of the word are set to filter out the very low frequency words. Thresholds p_0, p_1 of Algorithm 1 on PMI are set to consider only the confidently predicted CE-relation constructs. The co-discovery module of the approach brings additional information to re-consider filtered out words. If a verb fails to satisfy Phase-III, but its noun form satisfies Phase-II, the verb will be extracted as a causative verb.

Table 2. Precision, Recall and F-score in %

Our approach	47	49	48
Our approach without cue phrase (only verbs)	46	47	47
Girju 2003	72	16	26
Pawar et al., 2017	36	21	27
Only cue phrases	71	4	7

Algorithm 1 outputs a set of causative verbs, and Algorithm 2 extracts CE relation mentions along with the arguments of each mention using the causative verbs. In order to validate output of our algorithm, we asked two annotators to find CE relation mentions with their arguments in the test data. A true positive scenario is when the head words of the manually tagged arguments and the causal cue match with the extracted (by our approach) arguments and causative verb respectively. Girju [3] built an unsupervised causative verb based question-answering system, hence we have compared our CE relation extraction approach with Girju [3]. He reported 61 generic verbs, which can create a casual scenario in a sentence, but did not consider domain-specific causative verbs. We found 242 new causal verbs in the biomedical domain, which are not present in Girju [3]. As another baseline, we used a relation extraction algorithm of Pawar et al. 2017 [11]. End-to-end relation extraction model AWP-NN was employed to jointly identify the entity mentions (CAUSE or EFFECT) as well as the relation (CE) between them. Only one entity type (E) was considered to cover both causes and effects. We only focused on identifying headwords of cause/effect mentions (as done for all approaches). For word-pair $(u, v), u = v$ $((u, v), u \neq v)$, if predicted label = E (= CE) then u is a cause or an effect (a CE relation exists between the mentions headed by words u and v). We also compared our approach with universal cue phrases (CP of Algorithm 2) as CE indicators. Table 2 shows the precision, recall and F-score obtained with our approach, Girju [3], Pawar [11] and universal cue phrases. Our system considers only the CE mentions which are produced by verbs and a few cue phrases, hence it fails to capture all CE relation mentions. However, the set of verbs extracted by our approach is able to produce a comparably good recall and F-score.

5   Conclusions and Further Work

Linguistically complex constructs are used to express CE relations in text, mainly using causative verbs, which may be standard causative verbs, or generic verbs with domain-specific causative sense (`inhibit, express`), or new domain-specific verbs (`up-regulate`). We proposed a novel linguistically-oriented unsupervised iterative co-discovery approach to identify causative verbs, starting from a small set of seed causal verbs and an unlabeled corpus. We demonstrated the effectiveness of this approach by extracting 256 causative verbs from 10,000 PubMed abstracts of Leukemia papers, starting with 3 generic causative verbs. We demonstrated that our approach outperforms several baselines from literature for extracting intra-sentence CE relation mentions. We are integrating our approach in a knowledge-based system to support Leukemia researchers.

References

1. Joskowicz, L., Ksiezyck, T., Grishman, R.: Deep domain models for discourse analysis. In: Proceedings of the Annual AI Systems in Government Conference, 1989, pp. 195–200. IEEE (1989)
2. Khoo, C.S., Kornfilt, J., Oddy, R.N., Myaeng, S.H.: Automatic extraction of cause-effect information from newspaper text without knowledge-based inferencing. Lit. Linguist. Comput. **13**(4), 177–186 (1998)
3. Girju, R.: Automatic detection of causal relations for question answering. In: Proceedings of the ACL 2003 workshop on Multilingual Summarization and Question Answering, vol. 12, pp. 76–83. Association for Computational Linguistics (2003)
4. Radinsky, K., Davidovich, S., Markovitch, S.: Learning causality from textual data. In: Proceedings of Learning by Reading for Intelligent Question Answering Conference (2011)
5. Kim, H.D., Zhai, C., Rietz, T.A., Diermeier, D., Hsu, M., Castellanos, M., Ceja Limon, C.A.: Incatomi: integrative causal topic miner between textual and non-textual time series data. In: Proceedings of the 21st ACM International Conference on Information and Knowledge Management, pp. 2689–2691. ACM (2012)
6. Chang, D.-S., Choi, K.-S.: Causal relation extraction using cue phrase and lexical pair probabilities. In: Su, K.-Y., Tsujii, J., Lee, J.-H., Kwong, O.Y. (eds.) IJCNLP 2004. LNCS (LNAI), vol. 3248, pp. 61–70. Springer, Heidelberg (2005). https://doi.org/10.1007/978-3-540-30211-7_7

7. Do, Q.X., Chan, Y.S., Roth, D.: Minimally supervised event causality identification. In: Proceedings of the Conference on Empirical Methods in Natural Language Processing, pp. 294–303. Association for Computational Linguistics (2011)

8. Bouma, G.: Normalized (pointwise) mutual information in collocation extraction. In: Proceedings of GSCL, pp. 31–40 (2009)

9. Recchia, G., Jones, M.N.: More data trumps smarter algorithms: comparing pointwise mutual information with latent semantic analysis. Behav. Res. Methods **41**(3), 647–656 (2009)

10. Schuster, S., Manning, C.D.: Enhanced english universal dependencies: an improved representation for natural language understanding tasks. In: LREC (2016)

11. Pawar, S., Bhattacharyya, P., Palshikar, G.: End-to-end relation extraction using neural networks and markov logic networks. In: Proceedings 15th Meeting of the European Chapter of the Association for Computational Linguistics (EACL 2017), vol. 1, pp. 818–827 (2017)

A Supervised Learning to Rank Approach for Dependency Based Concept Extraction and Repository Based Boosting for Domain Text Indexing

U. K. Naadan[1]([⊠]), T. V. Geetha[1], U. Kanimozhi[1], D. Manjula[1],
R. Viswapriya[2], and C. Karthik[2]

[1] Anna University, Guindy, Chennai 600025, Tamilnadu, India
naadan.uk@gmail.com
[2] Scope e-Knowledge Center (P) Ltd., Nandambakkam, Chennai 600089,
Tamilnadu, India

Abstract. In conventional information retrieval systems, keywords extracted from documents are indexed and used for retrieval. Since same information can be represented by different keywords, there is hindrance in extracting relevant documents. Concept based indexing and retrieval which semantically identifies similar documents overcomes this problem by mapping the document phrases to a domain repository. In this paper, the problem of extracting and ranking concepts i.e. key phrases, from domain oriented text is explored. This paper ranks concepts (key phrases) of a document based not only on statistical and cue phrases but also based on the dependency relations in which the candidate concept occurs. For each candidate a vector is formed with the phrase weight and the dependency relations. The features used to score the phrases in the vectors, for re-ranking and as features to weigh the vector corresponding to the candidate are the cue features (presence in title, abstract), C-value in case of multi-words, frequency of occurrence and the type of dependency relation. The ranking process utilizes RankingSVM to rank the candidate concepts based on the feature vectors. In addition, to make the ranking domain sensitive and to determine the domain relevance of the candidate concepts they are fully or partially matched with the domain repository. Based on the depth of the concept and the presence of parent and siblings, the domain relevant concepts are boosted up the order. The results indicate that the use of dependency based context vector and domain repository provides substantial enhancement in the key phrase extraction task compared with other methods.

Keywords: Dependency-based key phrase extraction
Repository based concept mapping · Learning to Rank · Domain text indexing

1 Introduction

The key phrase extraction (KPE) task is the most important component to finding relevant information, where the documents are indexed by key phrases. A key phrase can be defined as a meaningful and significant expression consisting of one or more

© Springer International Publishing AG, part of Springer Nature 2018
M. Silberztein et al. (Eds.): NLDB 2018, LNCS 10859, pp. 428–436, 2018.
https://doi.org/10.1007/978-3-319-91947-8_44

words that convey the main concepts present in the documents. Document key phrases are important for many applications, [2, 15] particularly for Information Retrieval (IR) they provide thematic access essential for searching the documents in a particular domain. Concept extraction, which is a text mining task extracts key phrases from documents and maps [7] them to the repository, is a critical task for Concept based IR. Generic KPE involves two-steps, candidate selection and extraction using machine learning algorithms. Supervised approaches [2] formulate the problem as classification, tagging or ranking task. Key phrases are extracted based on some combination of corpus based statistical features, linguistic features and cue features or external knowledge based features. Unsupervised methods [2, 11] utilizes statistical, clustering, graph-based and language modelling methods to extract keywords for constructing phrases and finally filters incorrect phrases using linguistic patterns. Domain-specific IR is the process of acquiring the necessary information from a collection of text based on the user needs with respect to a particular domain. The challenges include, resolving obscure terminologies such as *neoplasm* and *cancer, fever* and *pyrexia*, etc. and ambiguous abbreviations (e.g., "APS" may refer to a *gene* or *protein*) along with variety of information needs from different types of users [10]. Concept mapping techniques handles this problem through mapping and matching the phrases with the concepts in the domain repositories [7]. During indexing of domain-specific documents [2], KPE based only on statistical or linguistic features is not sufficient as the relation between the keywords is semantic and not syntactic hence requires domain knowledge [19]. In this context, the identification and ranking of appropriate domain concepts using domain repository from the given document becomes necessary to represent the document [8].

The problem of identifying concepts i.e. key phrases from a given document is treated as a ranking problem in this paper. The rationale [18] behind this approach is that a phrase being a key phrase is relative to a document and not absolute. The features considered to select a phrase as a key phrase is also relative and also, ranking learns a relative order between the phrases for appropriate selection based on document features. Hence the model proposed uses two-step ranking process, where dependency based context vector of the candidate phrases are extracted and ranked using supervised pair-wise Learning to Rank (LTR) algorithm as the first step. The second step involves boosting of domain concepts based on the domain repository features such as presence of parent, sibling concepts and depth of the concept.

2   Related Work

Chebil et al. [4] extracts relevant concepts by utilizing probabilistic network (PN) to weigh a document for a given term and vector space model (VSM) to find document term similarity. Wan and Xiao [5] uses graph-based ranking method to extract key phrases by adding small set of similar documents. The context of the document and that of added documents are used to extract the key phrases. Torii et al. [6] model's the problem as tagging the document with appropriate labels and employ Conditional Random Fields to learn the tagging from different datasets. Lossio-Ventura [13] employs linguistic features and statistical term and keyword measures to identify key phrases from the document. Dinh and Tamine [16] utilizes VSM to convert documents and vocabulary to vectors and capture key phrases based on similarity between the vectors. The above methods try to identify key phrases but do not rank them with respect to the document for efficient IR. Lossio-Ventura et al. [8] make use of statistical and linguistic measures to extract and rank the candidate phrases while graph and web-based measures are used to boost the precision by re-ranking the phrases. Eichler and Neumann [12] formulate the task as an ordinal regression problem and employs linear SVM based ranking algorithm to rank the phrases. Biotex [14] a biomedical phrase extractor extracts key phrase using selected linguistic pattern, ranks them with different ranking measures and validates them by the domain thesaurus. Shi et al. [19] applies graph based centrality algorithm on knowledge graph constructed by linking key phrases with the entities in the graph. Florescu and Caragea [20] proposed PositionRank which aggregates the position of occurrence of words to calculate the weight of the node in the graph and uses biased PageRank to rank the key phrases. Wang and Li [3] employs ensemble model to improve key phrase ranking along with the use of external knowledge bases. These methods strive to rank the key phrases and corroborate them based on domain knowledge base.

Given labelled instances with ranks, Supervised LTR method builds a model to rank the object based on the features given. In pair-wise ranking method, the difference between the ranks of objects is used to rank the list correctly. Since the key phrase extraction is a natural ranking problem and classification methods make only binary decisions, the key phrase extraction problem is modelled as a supervised pair-wise ranking problem. Thus the use of varied features to identify the key phrases along with the necessity of domain repository for concept mapping is explored in this paper.

3   Single Document Key Phrase Extraction

The problem is to extract key phrase from a given document based on information present in the document and the domain repository. Consequently this work proposes a novel method as discussed below to extract key phrases from the given document based on context vector ranking and boosting the ranked phrases based on domain repository.

3.1   Candidate Selection and Multi-word Extraction

The candidate phrases i.e. nouns and noun phrases and verb phrases are extracted using Stanford Parser. The candidate phrases are further processed to remove redundancy using lemmatization and false positives using subjective adjective removal. To identify phrases i.e. multi-words, dependency relations such as 'compound', 'amod' etc. are used and to find the 'phraseness' and 'completeness' of the extracted phrase C-value score given by Eq. (1) is used.

$$C - value(T) = \begin{cases} \log_2 |T| * f(T), & \text{if } T \text{ is not nested} \\ \log_2 |T| * \left(f(T) - \frac{1}{|L_T|} * \sum_{t \in L_T} f(t) \right), & \text{otherwise} \end{cases} \quad (1)$$

Where |T| - the number of words in the phrase, L_T - the longer phrases containing the phrase T and $|L_T|$ - the number of words in the longer phrase. The intuition is that: if a phrase is frequent, longer or part of numerous longer phrases, then its 'phraseness' is more. Else if it occurs as a substring of a longer phrase, then lower is its 'phraseness'.

For example, consider the sentence "Adjuvant chemotherapy has also gained acceptance as an alternative management option" and its dependency parse in Fig. 1. It is seen that the word 'Adjuvant' acts as an adjective modifier of the word 'chemotherapy', hence the multi-word can be considered as a phrase if its C-value is greater than the set threshold.

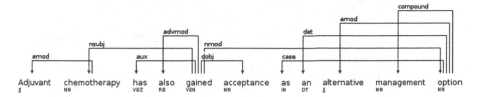

Fig. 1. Dependency graph of an example sentence.

3.2 Context Identification and Vector Formation

Context refers to the perspective in which a phrase occurs and can be used to identify correlated phrases. Basically, the context of a phrase in a document is provided by co-occurrence statistics [1, 8] which neglects long-term dependencies. Better option to get the context would be to use Distributional similarity methods [1] with different linguistic units as they capture long-term relations. Hence, in this work, selected dependency relations such as nsubj, dobj, csujb, acomp, ccomp, and nsubjpass are used identify context since they express the context meaningfully. The above relations are chosen because a phrase is significant if occurs as subject or object of a sentence or clause. Since the context is identified using dependency relations, the multi-word identified in previous step is replaced in the dependency triples. The *nsubj(gained-5, chemotherapy-2)* is modified *as nsubj(gained-5, Adjuvant chemotherapy-2)* and similarly *nmod(gained-5, option-11)* is modified as *nmod(gained-5, management option-11)* in the dependency triples shown in Fig. 1. For each candidate, all dependency triples that contain the phrase are extracted and the triples with the above relations are filtered and a context vector is formed from phrases in the filtered triplets. Other features such as term frequency, cue features and domain features are added to the vector for boosting the key phrase extraction. To rank the phrases, the context vector of each phrase is weighed using the below features: (1) Cue features – presence in title, abstract etc., (2) C-value for multi-word, (3) Domain features – The concept's depth and its parent and siblings frequency, (4) Frequency – phrase occurrence count and (5) Dependency relation – The type of relation with the given key phrase. The score of a word in the vector is given by the formula given below:

$$Sw = W_t + W_a + R_s + \sum_{r \in D_r} D_r * f_w \qquad (2)$$

Where, W_t is the weight for presence in the title, W_a is the weight for presence in the abstract, R_s is the Repository based score, f_w is the frequency of the word and D_r is the weight for the dependency relation r. The overall process is shown in Fig. 2.

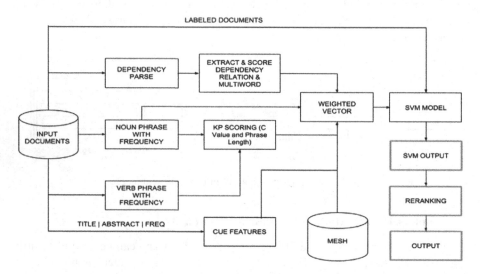

Fig. 2. Key phrase extraction process.

3.3 Ranking Key Phrases Using RankingSVM

In this paper, RankingSVM a supervised LTR algorithm is used for ranking key phrase as it learns an unbiased rule for ranking. The goal is to learn a function from preference samples, so that it orders a new set of candidate key phrases as accurately as possible. RankingSVM has been applied to extract key phrases from documents by Jiang et al. [18] which uses only statistical and cue features to rank the key phrases. But in this paper to rank the phrases, context based feature vectors are extracted and weighted using above features. The RankingSVM uses linear ranking function f: $f(d) = (w \cdot d)$ where w is the vector coefficients and '·' represents the inner product. The input space is given by $D \subseteq R^f$ where f is the feature set and the output space is given by $O = \{r1, r2, r3,\ldots, rm\}$. If $f(d1) > f(d2)$ then d1 is preferred over d2 and so $w \cdot (d1 - d2)$ will be a positive value. This is considered as an optimization problem and the learned function is used to predict the ranks of the phrases in the test dataset.

3.4 Repository Based Re-ranking

To make the ranking domain sensitive and to improve the precision at the top, this paper presents a novel domain repository based re-ranking of concepts. The idea is to push the phrase if the phrase is domain pertinent, specific and has correlated concepts.

As the need here is to extract set of concepts that will represent the documents, the domain repository is used to weigh the top 50 key phrases which are normalized [9, 17] and mapped with it. If the phrase is present in the repository i.e. full match then the phrase's depth, parent, sibling are extracted from the repository. The count of the presence of parent and sibling concepts in the document is utilized to weigh the phrase. If the phrase is not found then partial matches are performed: (1) Window-based method: All possible 'n-grams' of the given concept are extracted and compared with the repository and (2) Split-based method: All substring of the given string after removing a word are extracted and searched in the repository. Reduced weights are given based on the word overlap and duplicate and generic concepts are removed based on the path similarity. Remaining concepts are re-ranked based on the depth, presence of parent and sibling concepts.

4 Evaluation and Results

Two datasets are used to evaluate and analyze the performance of the context based RankingSVM algorithm with five-fold cross-validation. Bio-Medical corpus having 950 tagged documents is used for model creation and the MeSH repository is used to boost domain concepts. For economic domain 640 labeled documents is used to learn the model and the ZBW economic domain repository is used to boost domain concepts.

The RankingSVM is tested with different sets of vector and scoring features to identify suitable features to rank the key phrases. Repository features are used as vector, scoring and re-ranking features to test its usefulness. To find the appropriate features for ranking five trial experiments were conducted and analyzed as shown in Table 1. The initial trial conducted to find how the context vector improves the ranking proved that re-ranking is needed for precision improvement at the top. Hence trial two is carried out with re-ranking the candidates based on cue features. The improvement is meagre as it is based on the keyword similarity method synonym and duplicate concepts are not removed. The trial three was carried out to resolve whether domain repository features improves ranking if used as vector features. In trial four, domain repository features are used as both vector and scoring features. In both trials three and four the enhancement was not as expected. Hence re-ranking based on the repository is carried out in trial 5 and found that the repository based features boosts the relevant domain phrases to the top thus enhancing the precision. Limitations of this work are that it needs domain repository and finding context vector becomes difficult for short documents.

Among numerous measures defined to measure the ranking quality Normalized Discounted Cumulative Gain (NDCG) is used to validate the results. The NDCG penalizes the model if a relevant document/key phrase comes lower in the ranking. The formula to calculate nDCG is given in Eq. (3) and the results in Figs. 3 and 4.

$$nDCG@n = \frac{DCG@n}{IDCG@n} \ where, \ DCG@n = \sum_{i=1}^{n} \frac{2^{rel} - 1}{\log(1 + i)} \tag{3}$$

Table 1. Features used for trial experiments

Trial	Vector features	Scoring features	Re-ranking features
1	Context vector	Frequency and Dependency relations	-
2	Context vector + Cue features	Frequency and Dependency relations	Cue features
3	Context vector + Cue features + Repository based Domain Features	Frequency and Dependency relations	
4	Context vector + Presence in title and abstract + Domain Relevance Features	Frequency, Dependency relations, Cue features, Domain relevance	
5	Context vector + Presence in title and abstract + Domain Relevance Features	Frequency, Dependency relations, Cue features, Domain relevance	Repository based Domain Features

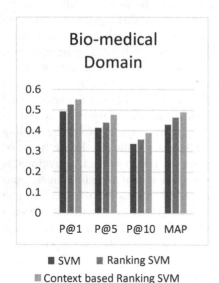

Fig. 3. Performance comparison

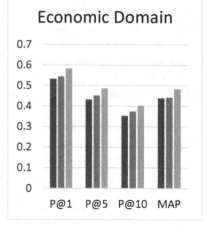

Fig. 4. Performance comparison

5   Conclusion

In the proposed work identification of context based on the dependency relations and re-ranking of concepts based on domain repository enhances key phrase extraction. The varied features used to score the concepts increases the probability of finding the right set of concepts. The result shows that the two-step process is needed and is more effective in ranking the concepts in domain text documents for indexing. The future work involves adaptation of the ranking algorithm for different domains to analyze how well it performs for other domains. Also, apply different LTR algorithm that has different loss functions to the problem and find their progress.

References

1. Vechtomova, O.: A semi-supervised approach to extracting multiword entity names from user reviews. In: Proceedings of the 1st Joint International Workshop on Entity-Oriented and Semantic Search, 12 August 2012, p. 2. ACM (2012)
2. Hasan, K.S., Ng, V.: Automatic keyphrase extraction: a survey of the state of the art. In: ACL, vol. 1, pp. 1262–1273, June 2014
3. Wang, L., Li, S.: PKU_ICL at SemEval-2017 Task 10: keyphrase extraction with model ensemble and external knowledge. In: Proceedings of the 11th International Workshop on Semantic Evaluation (SemEval-2017), pp. 934–937 (2017)
4. Chebil, W., Soualmia, L.F., Omri, M.N., Darmoni, S.J.: Biomedical concepts extraction based on possibilistic network and vector space model. In: Holmes, J.H., Bellazzi, R., Sacchi, L., Peek, N. (eds.) AIME 2015. LNCS (LNAI), vol. 9105, pp. 227–231. Springer, Cham (2015). https://doi.org/10.1007/978-3-319-19551-3_29
5. Wan, X., Xiao, J.: Single document key phrase extraction using neighborhood knowledge. In: AAAI, vol. 8, pp. 855–860, 13 July 2008
6. Torii, M., Wagholikar, K., Liu, H.: Using machine learning for concept extraction on clinical documents from multiple data sources. J. Am. Med. Inform. Assoc. 18(5), 580–587 (2011)
7. Arnold, P., Rahm, E.: SemRep: a repository for semantic mapping. In: BTW 2015, pp. 177–194 (2015)
8. Lossio-Ventura, J.A., Jonquet, C., Roche, M., Teisseire, M.: Biomedical term extraction: overview and a new methodology. Inf. Retrieval J. 19, 59–99 (2016)
9. Jonnagaddala, J., Chang, N.W., Jue, T.R., Dai, H.J.: Recognition and normalization of disease mentions in PubMed abstracts. In: Proceedings of the Fifth BioCreative Challenge Evaluation Workshop, Sevilla, Spain, 9 September 2015, pp. 9–11 (2015)
10. Chu, W.W., Liu, Z., Mao, W., Zou, Q.: KMeX: a knowledge-based digital library for retrieving scenario-specific medical text documents. In: Biomedical Information Technology, pp. 307–341 (2008)
11. Hasan, K.S., Ng, V.: Conundrums in unsupervised key phrase extraction: making sense of the state-of-the-art. In: Proceedings of the 23rd International Conference on Computational Linguistics: Posters, 23 August 2010, pp. 365–373 (2010)
12. Eichler, K., Neumann, G.: DFKI KeyWE: ranking key phrases extracted from scientific articles. In: Proceedings of the 5th International Workshop on Semantic Evaluation, 15 July 2010, pp. 150–153 (2010)
13. Lossio-Ventura, J.A., Jonquet, C., Roche, M., Teisseire, M.: Combining c-value and keyword extraction methods for biomedical terms extraction. In: LBM: Languages in Biology and Medicine, 12 December 2013
14. Lossio-Ventura, J.A., Jonquet, C., Roche, M., Teisseire, M.: BIOTEX: a system for biomedical terminology extraction, ranking, and validation. In: ISWC: International Semantic Web Conference, 19 October 2014
15. Popova, S., Khodyrev, I.: Ranking in key phrase extraction problem: is it suitable to use statistics of words occurrences. Proc. Inst. Syst. Program. 26(4), 123–136 (2014)

16. Dinh, D., Tamine, L.: Biomedical concept extraction based on combining the content-based and word order similarities. In: Proceedings of the 2011 ACM Symposium on Applied Computing, 21 March 2011, pp. 1159–1163. ACM (2011)
17. Shen, W., Wang, J., Han, J.: Entity linking with a knowledge base: issues, techniques, and solutions. IEEE Trans. Knowl. Data Eng. **27**, 443–460 (2015)
18. Jiang, X., Hu, Y., Li, H.: A ranking approach to key phrase extraction. In: Proceedings of the 32nd International ACM SIGIR Conference on Research and Development in Information Retrieval, 19 July 2009, pp. 756–757. ACM (2009)
19. Shi, W., Zheng, W., Yu, J.X., Cheng, H., Zou, L.: Keyphrase extraction using knowledge graphs. Data Sci. Eng. **2**(4), 275–288 (2017)
20. Florescu, C., Caragea, C.: PositionRank: an unsupervised approach to key phrase extraction from scholarly documents. In: Proceedings of the 55th Annual Meeting of the Association for Computational Linguistics (Volume 1: Long Papers), vol. 1, pp. 1105–1115 (2017)

[Demo] Integration of Text- and Web-Mining Results in EPIDVIS

Samiha Fadloun[1,2(✉)], Arnaud Sallaberry[1,3], Alizé Mercier[4,5], Elena Arsevska[6], Pascal Poncelet[1,2], and Mathieu Roche[4,7]

[1] LIRMM, CNRS, Univ. Montpellier, Montpellier, France
Samiha.Fadloun@lirmm.fr
[2] Univ. Montpellier, Montpellier, France
[3] Univ. Paul-Valéry Montpellier 3, Montpellier, France
[4] Cirad, Montpellier, France
[5] ASTRE, Univ. Montpellier, Cirad, Inra, Montpellier, France
[6] Institute of Global Health and Epidemiology, London, UK
[7] TETIS, Univ. Montpellier, APT, Cirad, CNRS, Irstea, Montpellier, France

Abstract. The new and emerging infectious diseases are an incising threat to countries due to globalisation, movement of passengers and international trade. In order to discover articles of potential importance to infectious disease emergence it is important to mine the Web with an accurate vocabulary. In this paper, we present a new methodology that combines text-mining results and visualisation approach in order to discover associations between hosts and symptoms related to emerging infectious disease outbreaks.

Keywords: Visualization · Text-mining · Query · Epidemiology

1 Introduction

Online detection and monitoring of animal disease outbreaks is crucial for disease surveillance and early warning systems. Epidemiologists regularly query web-pages using various formulations to obtain up-to-date information on disease outbreaks, but this task may take several days before finding the sought-after information. Visualization could nevertheless facilitate their web searches as compared to time-consuming traditional searches. This paper presents EPID-VIS, a new visual web query tool designed for epidemiologists. It consists of several views that help build and launch queries, and visualize the results. Moreover, it supports external information integration to help epidemiologists enrich their knowledge and adapt their queries. This paper focuses on the integration of text- and web-mining results in EPIDVIS tool.

Most of the current visualization tools in epidemiology domain focus on web results in general [1,2]. They do not deal with storing knowledge and adapting queries. They mainly show web results related to a given disease outbreak [3–5]. They use results from different sources (RSS feeds, reports, alerts, etc.) showing

© Springer International Publishing AG, part of Springer Nature 2018
M. Silberztein et al. (Eds.): NLDB 2018, LNCS 10859, pp. 437–440, 2018.
https://doi.org/10.1007/978-3-319-91947-8_45

them as plots on a map to represent spatial information, or as statistical diagrams to represent aggregated information [6].

To summarize, most of the previous visualization techniques focus on showing the query results. In animal epidemiology, we can distinguish three main categories of keywords (*diseases*, *hosts*, and *symptoms*), and identify relationships. The visualization can help the epidemiologists to express these categories and relationships, a crucial step that is not incorporated in the previous approaches. This visual expression can help the user to build and launch queries.

Section 2 summarizes the EPIDVIS tool. The suggestion view that integrates text- and web-mining results is described in Sect. 3. Finally, Sect. 4 concludes and describes future work.

2   EPIDVIS tool

Epidemiologists regularly search on the web resources in order to get new insights about disease outbreaks. They use their own knowledge as keywords and manually write and launch the queries on search engines like Google, Bing, Yahoo, etc. They often copy/paste keywords stored in text file. They use three main sets of keywords (categories): *diseases*, *hosts*, and *symptoms*. For example, one can launch the query: *'influenza chicken lethality'*. In this case, *'influenza'* is a *disease*, *'chicken'* is a *host*, and *'lethality'* is a *symptom*. In this context, this is crucial to take into account strong or weak relationships between the keywords of different categories. Epidemiologists also use external knowledge to build queries, e.g. knowledge transmitted from their colleagues, or data extracted with text-mining or statistical approaches [7].

3   The Suggestion View

This section describes how to enrich the keywords by using external knowledge. This external knowledge is provided as a specific file containing keywords and relationships that exist between them. For instance, it contains relationships between *hosts* and *symptoms* associated with a specific *disease*. Our goal is to provide a new view taking into account this external knowledge (i.e. text- and web-mining results) to suggest links between keywords to experts.

In order to measure the relevance of association between *hosts* and *symptoms*, statistical measures have been used like Dice measure. In our context, this measure is defined as the number of times where *hosts* and *symptoms* appear in the same context (in a corpus or in the Web) over the sum of the total number of times that each one appears in the corpus (i.e. text-mining approach) or in the Web (i.e. web-mining approach) for each disease. This approach has been adapted with other statistical measures (i.e. Mutual Information and Cubic Mutual Information). This contribution is detailed in [7]. The values of association measures represent the input of suggestion view of EPIDVIS.

The suggestion window of EPIDVIS contains two main views: the suggested keyword view (Fig. 1a) which shows the data coming from the external file, and

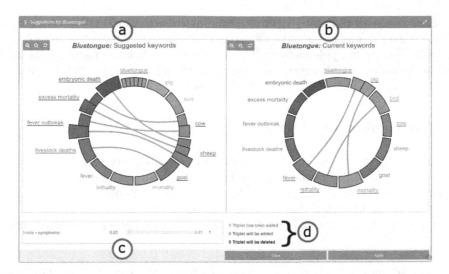

Fig. 1. The suggestion view. **(a)** Relationships and keywords suggested an external file.
(b) Relationships and keywords already available in the keyword manager. **(c)** Slider
to filter relationships and keywords according to their weight, **(d)** Result of different
actions. (Color figure online)

the current keyword view (Fig. 1b) which shows the data already available in
the keyword manager. Both of the views are based on a circle splitted into arcs
representing the keywords. In order to make the circles homogeneous, we plot
the union of the keywords of the keyword manager and the suggested keywords.
Underlined keywords represent words from the external file in Fig. 1a, and words
from the keyword manager in the Fig. 1b. The selected keyword is represented
at the top of the circles (blue arc "bluetongue" in the example). The other keywords are visualized around it. Each arc has a name, and a color according to
its category. We put a color degradation to make a difference between keywords
in the same category. Relations between keywords are represented by curves
between the corresponding arcs. Each arc is splitted into sub arcs and the width
of a sub arc represents the weight of the links starting from it. Each link has
an inverse color interpolation related to categories of related arcs. We hide links
between the selected keyword and the other keywords, in order to avoid clut-
tering: we show them when hovering the mouse on the arc, or selecting them.
A slider (Fig. 1c) is provided to filter suggested keywords and relationships by
weight of relationships. Text information is also available as shown in Fig. 1d.
It notifies to the epidemiologist the actions performed as described in the next
section. Each type of action is encoded with specific colors: red color for selected
triplets to be added to the keyword manager, purple color for the already added
triplets, and black for the triplets to be deleted. Note that our system holds
several interactive features to explore the suggested data, and to add some of
these suggested data to the keywords manager dataset.

The usefulness and usability of the different views of EPIDVIS have been evaluated. A survey was presented to a group of 12 participants containing experts and non experts in epidemiology. It can be noticed that the participants have highly appreciated the suggestion evaluation visualizations, i.e. 8.4/10 for usefulness criterion, and 8.1/10 for usability criterion.

4   Conclusion and Future Work

In this paper, we present EPIDVIS, a new visual web querying tool for animal disease surveillance. This tool helps epidemiologists to build queries, launch them on the web, and visualize the results. Also, external knowledge can be integrated using the suggestion view that takes into account text- and web-mining results. For future work, we plan to extend our tool, in order to cover other application domains.

Acknowledgments. This work was supported by the Ministry of Higher Education and Scientific Research of Algeria and the SONGES project (FEDER and Occitanie). We thank Renaud Lancelot (ASTRE, Cirad) and Sarah Valentin (ASTRE & TETIS, Cirad) for their expertise in epidemiological surveillance.

References

1. Dörk, M., Carpendale, S., Collins, C., Williamson, C.: Visgets: coordinated visualizations for web-based information exploration and discovery. IEEE Trans. Vis. Comput. Graph. **14**(6), 1205–1212 (2008)
2. Peltonen, J., Belorustceva, K., Ruotsalo, T.: Topic-relevance map: visualization for improving search result comprehension. In: Proceedings of the 22nd International Conference on Intelligent User Interfaces, pp. 611–622. Association for Computing Machinery (2017)
3. Van den Broeck, W., Gioannini, C., Gonçalves, B., Quaggiotto, M., Colizza, V., Vespignani, A.: The gleamviz computational tool, a publicly available software to explore realistic epidemic spreading scenarios at the global scale. BioMed Central Infect. Dis. **11**(1), 37 (2011)
4. Collier, N., Doan, S., Kawazoe, A., Goodwin, R.M., Conway, M., Tateno, Y., Ngo, Q.H., Dien, D., Kawtrakul, A., Takeuchi, K., et al.: Biocaster: detecting public health rumors with a web-based text mining system. Bioinformatics **24**(24), 2940–2941 (2008)
5. Freifeld, C.C., Mandl, K.D., Reis, B.Y., Brownstein, J.S.: Healthmap: global infectious disease monitoring through automated classification and visualization of internet media reports. J. Am. Med. Inform. Assoc. **15**(2), 150–157 (2008)
6. Neher, R.A., Bedford, T.: nextflu: real-time tracking of seasonal influenza virus evolution in humans. Bioinformatics **31**(21), 3546–3548 (2015)
7. Arsevska, E., Roche, M., Hendrikx, P., Chavernac, D., Falala, S., Lancelot, R., Dufour, B.: Identification of associations between clinical signs and hosts to monitor the web for detection of animal disease outbreaks. Int. J. Agric. Environ. Inf. Syst. **7**(3), 1–20 (2016)

Recommendation Systems

Silent Day Detection on Microblog Data

Kuang Lu$^{(\boxtimes)}$ and Hui Fang

University of Delaware, Newark, DE 19716, USA
{lukuang,hfang}@udel.edu

Abstract. Microblog has become an increasingly popular information source for users to get updates about the world. Given the rapid growth of the microblog data, users are often interested in getting daily (or even hourly) updates about a certain topic. Existing studies on microblog retrieval mainly focused on *how* to rank results based on their relevance, but little attention has been paid to *whether* we should return any results to search users. This paper studies the problem of silent day detection. Specifically, given a query and a set of tweets collected over a certain time period (such as a day), we need to determine whether the set contains any relevant tweets of the query. If not, this day is referred to as a *silent day*. Silent day detection enables us to not overwhelm users with non-relevant tweets. We formulate the problem as a classification problem, and propose two types of new features based on using *collective* information from query terms. Experiment results over TREC collections show that these new features are more effective in detecting silent days than previously proposed ones.

Keywords: Silent day detection · Microblog retrieval · Classification

1 Introduction

Social media has fundamentally changed how we obtain information as it becomes an important source for people to find out what is happening in the world. In particular, Twitter enables users to post new events and share their thoughts in real time. It was reported that around 500 million tweets are posted per day[1]. A common strategy to help users digest the information on Twitter is to provide periodic updates about topics that the users are interested in.

Existing studies on microblog retrieval mainly dealt with ad-hoc retrieval. As a result, the existence of relevant information is often assumed and the methods focused on *how* to rank tweets based on the relevance [8]. However, there might not always be relevant information in the current collection, and therefore it is also important to study *whether* we should return any results to users. For example, when a disaster strikes, there could be lots of information about the disaster on Twitter everyday. However, as the time goes by, the frequency of the discussions about the disaster could change day to day, and there could be

---

[1] www.internetlivestats.com/twitter-statistics/.

© Springer International Publishing AG, part of Springer Nature 2018
M. Silberztein et al. (Eds.): NLDB 2018, LNCS 10859, pp. 443–455, 2018.
https://doi.org/10.1007/978-3-319-91947-8_46

no relevant information for some days. When there is relevant information, it is important to return a ranked list of tweets. However, when there is no relevant information, it would be meaningless to return a ranked list of non-relevant information to users. With respect to the query, these days are called *silent days*. If silent days can be correctly detected, we could improve users' experience by not displaying non-relevant information to them.

Although silent day detection for microblog retrieval is related to traditional information filtering, existing adaptive filtering methods [2] cannot be directly applied because the information needs for microblog retrieval are often short-term and do not have sufficient relevant information to construct the profiles of user interests using relevance feedback. Another research problem related to silent day detection is query performance prediction [3,4,16,17,23,24,26,27]. However, query performance prediction methods often assume that there is some relevant information in the collection. Moreover, many predictors are some forms of aggregation of the scores of individual query terms. The score is often related to how rare the term is. A day covering only rare query terms could still be a silent day, but the existence of rare terms would make the predictors to assign the day with a high score and subsequently mislabel it as non-silent. For example, given the query "U.S. forest fires" in 2015 Microblog Track, some silent days only containing "forest", a rare term in the collection, are labeled as non-silent by the predictors. Due to these limitations, it is clear that existing query performance prediction methods may not be the best choice for silent day detection either.

In this paper, we formulate the problem of silent day detection as a classification problem and focus on identifying novel features that can better capture the characteristics of silent days. In particular, we propose two types of features that use collective information from query terms. The first type of features is called *phrased-based weighted information gain* (PWIG), which is inspired by the weighted information gain [27]. It is designed to infer relevance by only using the information gain of appearances of multiple query terms instead of that of a single term. The second type of features is called *local query term coherence* (LQC). The intuition is that if the query terms are closely related in a collection of a day, the day is likely to be non-silent to the query.

We conducted two sets of experiments to validate the proposed features. The first set of experiments were conducted over multiple microblog collections [9–11]. Results show that using the proposed features alone can outperform the baselines consisting of state-of-the-art query performance predictors. Moreover, we tried to test whether the proposed predictors can be generalized for other domains such as traditional document collections. We employed the predictors to the missing content detection problem in document retrieval domain [24]. Results illustrate that with some modifications for the document collection, the performance of proposed features can have comparable performance to that of the state-of-the-art missing content detection method.

2   Related Work

Silent day detection is closely related to adaptive filtering [20] since both of them need to determine whether to deliver any information to users. However, they have one major difference. Adaptive filtering often deals with long-term information needs. As a result, the commonly used methods often can use training data to build an interest model based on an information need. However, in microblog retrieval, the information needs are often short-term. Most interests, in particular those event-driven queries, would not last very long. Thus, there is often no training data for the queries, making it difficult to directly apply existing adaptive filtering methods.

Silent day detection can be regarded as a query performance prediction problem [1,3–5,14,16,17,23,24,26,27]. Given a set of tweets posted in a day, the problem is to determine whether there is any relevant information in the set. This can be solved by predicting the performance of a query based on the set. However, directly using query performance prediction methods on silent day detection might not be suitable. As mentioned earlier, the scenario of no relevant tweets are often not considered, and using query terms individually may lead to the mislabeling of silent days as non-silent days. The most similar task to silent day detection might be missing content detection [24]. Nevertheless, it deals with document collections instead of microblog collections.

The concept of silent day was first introduced in the 2015 Microblog Track [10]. The main evaluation metrics in the track heavily penalize systems that return results on silent days. Indeed, previous work [21] shows that the performance of a participating system of the track is dominated by its ability to detect silent days. However, silent day detection was usually tackled *indirectly* by posing a threshold for single tweets, which is either time based [12] or score based [28]. This paper directly addresses the problem of silent day detection.

3   Silent Day Detection

3.1   Problem Formulation

We formalize silent day detection as:

Let Q denotes a query and C denotes all tweets of a day. The random variable S represents whether the day is silent (1) or not (0). The problem of silent day detection is to estimate the following probability:

$$P(S = 1 \mid Q, C)$$

Because silent day detection requires a system to make a binary decision for each day for a query according to the above probability, machine learning based classification method is adopted as the basic approach. In this work, we mainly focus on identifying indicative features for silent day detection.

3.2   Collective Information from Query Terms

It is clear that silent day detection is very similar to query performance prediction. However, by using and analyzing state-of-the-art query performance predictors, we discovered two problems of applying them for silent day detection. First, some of the predictors assume that relevant tweets always exist, which makes them unable to distinguish silent days from non-silent days. For instance, for the predictors based on standard deviations of the scores of retrieval lists [5], a list of irrelevant tweets can achieve the same standard deviation as that of a list with some relevant tweets. Second, many predictors are some forms of aggregation of the scores for individual query terms, and a high score means that it is easy to find relevant information for the query. The score of a term is usually related to how rare it is (e.g. its inverse collection frequency). Therefore, silent days with tweets only containing rare query terms can still be assigned with high scores and subsequently labeled as non-silent. To avoid these problems, we hypothesize that using **collective** information from query terms instead of using terms **individually** might be more suitable and propose two sets of silent day predictors: **phrase-based weighted information gain** (PWIG), and **local query term coherence** (LQC), which we discuss below.

Phrased-Based Weighted Information Gain (PWIG). The first feature is called phrase-based weighted information gain (PWIG). It infers relevance of a tweet from the collective query term appearances. This feature is inspired by weighted information gain (WIG) [27]. WIG uses term proximity features for collective term appearances and single term features for individual term appearances. However, the later dominates the former, which leads to the second problem we mentioned before. Therefore, we remove its single term features and introduce different weights for the two components of the term proximity features: sequential dependence (considering term order) and full dependence (without considering term order) features. More specifically, given a query Q, we denote its sequential dependence feature set and full dependence feature set as $S(Q)$ and $F(Q)$, respectively. Let the union of $S(Q)$ and $F(Q)$ be noted as $R(Q)$. The probabilities of a term proximity feature ξ occurs in a tweet D and a day C are denoted as $P(\xi|D)$ and $P(\xi|C)$. In day C where the query's retrieved list L_K contains K tweets $D_i \in \{D_1, D_2, ..., D_K\}$, PWIG is computed as follow:

$$PWIG = \frac{1}{K} \sum_{D_i \in L_K} \sum_{\xi \in R(Q)} \lambda_\xi log \frac{P(\xi|D_i)}{P(\xi|C)}, \text{ if } |R(Q)| > 0 \qquad (1)$$

where

$$\lambda_\xi = \begin{cases} \frac{\lambda}{\sqrt{|R(Q)|}}, & \xi \in F(Q) \\ \frac{1-\lambda}{\sqrt{|R(Q)|}}, & \xi \in S(Q) \end{cases} \qquad (2)$$

Essentially, PWIG infers relevance from the existence of multiple terms instead of single terms. For single term queries, WIG is computed instead.

Local Query Term Coherence (LQC). Local query term coherence (LQC), however, measures how coherent the query terms are to infer the difficulty of a query with respect to a day. The rational behind this predictor is that if the query terms are closely related with respect to a collection of a day, it is more likely that the day is not silent. More specifically, the sub-coherences are computed, which are the coherences of subsets of query terms. Sub-coherences are aggregated as the coherence of the query. It is important to note that the sub-coherences are computed *locally* on the top K tweets of the retrieval list. The reason is that a query term can have multiple meanings, and the meaning in top ranked tweets is more likely to be the same as that in the query. More specifically, given a query Q, we denote all possible multi-term subsets of the query as $F(Q)$. We use the appearance of such a subset $\xi \in F(Q)$ as an indication of coherence among the query terms in ξ. The proportion of the tweets containing ξ in the top K tweets indicates the degree of coherence between these query terms:

$$C(\xi) = \frac{\# \text{ of documents containing } \xi}{K} \tag{3}$$

Intuitively, the appearance of more query terms collectively means higher degree of coherence since the probability of more query terms randomly co-occurring is lower than that of fewer terms. Therefore, ξ are grouped by their sizes. More specifically, we denote all the query term subsets with size i as $F_i(Q)$ where i can take the value from 2 to the length of the query $|Q|$. Then the coherence $C_i(Q)$ aggregating all query term subsets with length i can be computed as:

$$C_i(Q) = A(\{C(\xi) : |\xi| = i\}) \tag{4}$$

The function $A()$ can be Max, Average, or Binary. Binary means that the value of $A()$ is 1 as long as one of the coherences is not zero, or 0 otherwise. This function is sensitive to the change from no subset appearances to one appearance. We use a weighted sum of the coherences of query term subsets of different lengths as the raw LQC ($rLQC$):

$$rLQC = \sum_{i=2}^{|Q|} log(i) \times C_i(Q) \tag{5}$$

It is important to note that the weight of each size is the logarithm of the size. This way of assigning weights favors long query term subsets without overly penalizing the weights of short ones when the query is long. The final value of LQC normalizes $rLQC$ by the ideal LQC, which is the $rLQC$ of an ideal result list in which every tweet contains every query term.

4   Evaluation

In this section, we first tested how effective our proposed features are for silent day detection. Afterwards, the detectors were applied to the e-mail digest scenario of Microblog Track and Real-Time Summarization Track to illustrate its usefulness for the real life application when recognizing silent days is important. In order to examine their usability in document retrieval domain, the features then were tested on missing content detection [24].

4.1 Silent Day Detection on Microblog Data

Data and Experimental Setup. We built a collection consisting of several TREC collections. The data used in TREC 2015 Microblog Track [10] and TREC 2016, 2017 Real-Time Summarization Track [9,11] were included, where silent days were introduced and addressed. Moreover, we also included data in TREC 2011 and 2012 Microblog Track [13,19]. Although silent days were not discussed in the tracks, they exist in this collection. Tweets in all these tracks were crawled using Twitter streaming API. When active, it offers 1% sample of all tweets posted at the time. The crawling period is 40 days in total, which resulted in 39,095,813 tweets. There are 254 queries and 3,190 query-day pairs, among which 749 are silent query-day pairs. The queries in these collections were created by TREC to reflect possible search interests related to the events occurred during the crawling periods. We used the *title* field of the queries, which typically contains a handful of words. For each tweet, only text, tweet id, and timestamp were preserved and non-English tweets were discarded. Tweets from the same day were grouped into a single-day-corpus.

The proposed silent day detectors as well as some of the state-of-the-art query performance predictors are post-retrieval. Therefore, we need a ranking function to retrieve some tweets for each day. Various state-of-the-art retrieval methods were tested including pivoted length normalization [18], BM25 [15], language model with Dirichlet smoothing [25], and f2exp [6]. Since all the experiments show similar results across all retrieval functions, we only report that of f2exp for the reminder of this paper.

The classification algorithm we used is Naive Bayes. Other algorithms such as SVM and logistic regression were also tried but were not reported due to their inferior performances for baselines and proposed methods. A possible explanation is that because our dataset is skewed (only about 20% of the samples are positive), methods robust to unbalanced data, such as Naive Bayes, tend to outperform those that are not.

Classification of Silent Days. We first implemented three baselines QPP_d, QPP_m, and QPP_c with existing query performance predictors. QPP_d employs state-of-the-art query performance predictors for *document* retrieval, such as WIG [27]. Whereas QPP_m employs state-of-the-art query performance predictors for *microblog* retrieval, such as query term coverage and top term coverage [16]. The complete lists of features of these methods can be found in Table 1. QPP_c uses the *combination* of the predictors in QPP_d and QPP_m. Additionally, we implemented a baseline *Tree* which is a decision tree based method for missing content detection [24]. Missing content detection tries to detect queries with no relevant information in a document collection. We did not incorporate

Table 1. Features of different baselines

Method	Features
QPP_d	Average inverse document frequencies [1]
	Simplified clarity [26]
	Sum of variances of the term weights of top-k documents [26]
	Maximum and average term relatedness [7]
	Query clarity [3]
	Standard deviation of top-k scores [5,14]
	Normalized standard deviation of top-k scores [5]
	Normalized query commitment [17]
	Weighted information gain [27]
	Query feedback [27]
	Top score of the retrieved list [23]
QPP_m	Average, median, upper and lower percentile of query term coverage [16]
	Average, median, upper and lower percentile of top term coverage [16]

query performance predictors into $Tree$ due to its unique feature format. $Tree$ calculates the overlaps of the top results of single-term subqueries to that of the original query. Its features are in the form of (overlap, logarithm of term document frequency) tuples. Since none of the other query performance predictors are in this format, they were not combined with $Tree$.

Proposed predictors, on the other hand, were examined in two methods TR and $TR + QPP_c$. In TR, only PWIG and LQC were used. $TR + QPP_c$, however, used the combination of TR and QPP_c. For LQC features, all three aggregation functions Max, Average, and Binary were used. It is important to note that, for fair comparison, all features were tuned with the area under curve (AUC) of the receiver operating characteristic (ROC) curve for optimal parameter settings.

In order to test the effectiveness of proposed features, we designed two experiments. First, we performed 10-fold cross-validation for all methods in which query-day pairs were divided into 10 equal size folds. We used 10-fold to avoid bias and to ensure the number of folds is enough for meaningful student t test results. This experiment tested the methods' performances on predicting silent days for *old* queries that the methods had seen and were trained on. The F1 score was used as the performance metric and the results are shown in Fig. 1. As can be seen, QPP_m, QPP_c and $Tree$ all outperform QPP_d. This is not surprising since QPP_m and QPP_c contain features dedicated for microblog retrieval. Moreover, $Tree$ was designed for detecting the non-existence of relevant information. However, further investigation shows no statistically significant differences between QPP_m, QPP_c, and $Tree$ at the 95% confidence level according to the t test. Methods containing proposed predictors, TR and $TR + QPP_c$, outperform baselines considerably with an improvement over the best baseline QPP_c

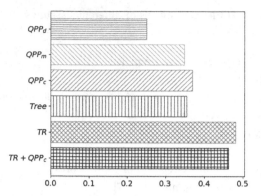

Fig. 1. 10-fold cross-validation performances for different methods

more than 25%. The advantages of TR and $TR + QPP_c$ are also statistically significant. However, the difference between TR and $TR + QPP_c$ is not.

We further analyzed $Tree$ and discovered a possible reason for its sub-optimal performance. $Tree$ is a two-stage method and its first stage employs performance prediction to filter out easy queries. The performance prediction is done by using the number of overlapping documents of top results between the original query and its single-term subqueries. Given the short nature of tweets, term document frequency is usually one for any term. Therefore, the ranking of the tweets for a single term subquery is likely to be dominated by the lengths of the tweets, which may not reflect the relevance order. As a result, the count of overlapping tweets of top results subsequently may not reflect the agreement on relevant tweets. Since the result of the first stage might not be very reliable, the final output of $Tree$ might not be optimal as well.

In the second experiment, we aimed to test the proposed detectors on the case of predicting silent days for *unseen* queries. We believe such scenario occurs very often in real life. As mentioned before, microblog queries tend to be event-driven, and a system often needs to deal with new unseen queries as interesting events occur in the real world. In this experiment, we conducted a modified version of 10-fold cross-validation where we divided queries into 10 equal size folds so that there is no query overlap among folds. The performances as F1 scores are reported in Fig. 2. Similar situation as the first experiment is observed. There is no statistically significant differences among QPP_m, QPP_c, and $Tree$, and they are all better than QPP_d. TR and $TR + QPP_c$ are still the best performing methods, and the difference between them are statistically insignificant. It is interesting that in both experiments, adding QPP_c to TR hurts the performances. We computed the mutual information between the features in QPP_c and day labels. The results are not shown here due to space limit. Low mutual information was observed for several features with some close to 0. Therefore, adding these features might not be desirable. Based on the two experiments of

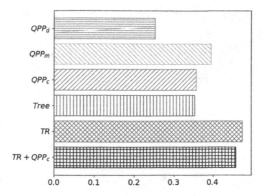

Fig. 2. Performances for testing on unseen queries

testing proposed features, it can be concluded that the proposed features exhibits superiority over existing ones and it is advised to use TR for future applications.

Effectiveness on Tweet E-Mail Digest. Besides silent day classification, another way to examine the effectiveness of the proposed features is to incorporate the TR silent day detector for the tweet e-mail digest scenario of TREC 2015 Microblog Track [10], and TREC 2016, 2017 Real-Time Summarization Track [9,11]. In these tracks, a system's ability to detect silent days can substantially impact its performance [21]. A participating system is supposed to submit up to 10 tweets for a query-day pair if there are relevant tweets in that day, or 0 otherwise. The main evaluation metric is $ndcg@10\text{-}1$ for 2015 and 2016 [10,11] which rewards systems that can "keep silent" for silent days by the perfect score (1) for the days of the query. Submitting any tweets in a silent day would result in the score of 0. In non-silent days, ndcg@10 is computed instead. In 2017, however, the main evaluation metric is $ndcg@10\text{-}p$, which penalizes systems according to the number of returned tweets on silent days. In our experiment, we used $ndcg@10\text{-}1$ for all three years since we investigate silent day detection as a classification task. It is possible to choose the number of tweets to return based on its probability of being silent as it would be more suitable for $ndcg@10\text{-}p$. We plan to explore this direction in the future.

We built a baseline method called *unfiltered*, which retrieves 10 tweets for every query on every day using f2exp. Another baseline method, $filtered_{QPP}$ was implemented by applying QPP_m on *unfiltered's* output to filter out silent days and return no results for them. We chose QPP_m since it is the best baseline when testing on new queries. It is important to note that QPP_m used for one year was trained on the collection we created excluding the queries of the year. The last method is called $filtered_{TR}$, which is very similar to $filtered_{QPP}$. The only difference is that TR was used instead of QPP_m. The performances of these methods as well as that of the best participating runs for each year in terms of ndcg@10-1 [9–11] are reported in Fig. 3. As can be seen, with the help of

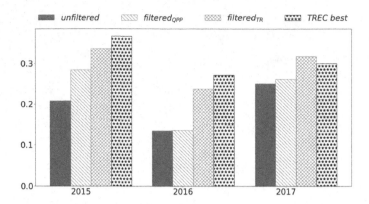

Fig. 3. Ndcg@10-1 of e-mail digest scenario of microblog tracks

TR, $filtered_{TR}$ significantly outperforms $unfiltered$ and $filtered_{QPP}$ for every year. When compared with TREC best runs, $filtered_{TR}$ can offer comparative performances for 2015 and 2016. In 2017, it even outperforms the TREC best run. It is important to note that we only used the generic f2exp as the retrieval method. Nevertheless, some of the best runs' retrieval methods were further tinkered for microblog retrieval [22,28][2]. For instance, techniques such as query expansion and word embedding were used in 2015's best run [28]. Due to the consistent promising results of our method, it can be concluded that our silent day detectors are very useful and could enhance the effectiveness of real-world microblog retrieval systems significantly if silent days need to be addressed.

4.2   Missing Content Detection

Since the proposed predictors seem to be useful in microblog retrieval domain, we also want to test whether they can be generalized to document retrieval domain. Therefore, we exanimated them for missing content detection [24]. As discussed earlier, this problem is very similar to silent day detection, and the major difference is that a document collection is used.

The TREC collection Disk4&5, used for missing content detection in the previous research [24] was used. The *Tree* and *TR* method mentioned in the earlier section were tested on this collection. The performances are reported as F1 scores in Fig. 4. As can be seen, *TR* is significantly worse than *Tree*. A potential explanation is that for the LQC features in *TR*, the distance between query terms in a document is ignored. In the same document, no matter how much text there is between some query terms, these terms are considered as related w.r.t the document. This approach seems to be not always appropriate for document collection. However, it might be suitable for tweet collections due to the 140 character limit. To confirm this explanation, we modified the LQC features by adding size constraints when counting related query terms and applied the modified TR

---

[2] We did not find the TREC report of the best run of 2016.

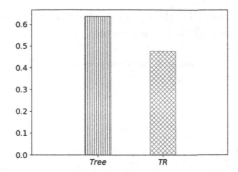

Fig. 4. Performances (F1) of missing content detection on disk4&5

method for both missing content detection and silent day detection. The results are not shown here due to space limit. We observed that, with the addition of the size constraints, the performance of *TR* can be boosted significantly (from 0.476 to 0.606) for missing content detection. Moreover, the size constraints do not seem to affect TR's performance on silent day detection. These observations confirm out assumptions.

The results of the experiments in this section as well as that of earlier sections about *Tree* on silent day detection suggest that applying methods for the task of detecting the non-existence of relevant information in a cross-domain fashion may not be appropriate. Not only does *Tree* fail in silent day detection, but also *TR* without size constraints tends to be ineffective in missing content detection. This discovery indirectly shows the value of our proposed predictors since, to the best of our knowledge, they are the first predictors dedicated for silent day detection for microblog retrieval.

5   Conclusion and Future Work

In this paper, we formally address the problem of silent day detection. Two types of predictors that use collective information from query terms are proposed, which are LQC and PWIG. Experiments show that the *TR* method that consists of proposed features can outperform the baselines that employ state-of-the-art predictors and are very effective on silent day detection. The proposed features are also examined for missing content query detection to test whether our methods can be generalized to other domains. Although the features do not show superior performance as they do on silent day detection with size constraints, they do illustrate some usefulness on this task.

There are several interesting potential research directions for silent day detection. First of all, we plan to investigate the ways to enhance the accuracy of silent day detection by incorporating multiple retrieval methods. Second, instead of doing a binary decision on days and return no tweets for predicted silent days, we plan to use the possibility of day being silent to decide the number of tweets should be returned. Third, with the insights gained from the proposed predictors, we want to discover new predictors for missing content detection.

Acknowledgements. This research was supported by the U.S. National Science Foundation under IIS-1423002.

References

1. Arampatzis, A., Kamps, J.: An empirical study of query specificity. In: Gurrin, C., He, Y., Kazai, G., Kruschwitz, U., Little, S., Roelleke, T., Rüger, S., van Rijsbergen, K. (eds.) ECIR 2010. LNCS, vol. 5993, pp. 594–597. Springer, Heidelberg (2010). https://doi.org/10.1007/978-3-642-12275-0_55
2. Ault, T., Yang, Y.: kNN, rocchio and metrics for information filtering at TREC-10. In: Proceedings of TREC-10 (2001)
3. Cronen-Townsend, S., Zhou, Y., Croft, W.B.: Predicting query performance. In: Proceedings of SIGIR 2002 (2002)
4. Cummins, R.: Document score distribution models for query performance inference and prediction. ACM Trans. Inf. Syst. **32**(1), 2 (2014)
5. Cummins, R., Jose, J., O'Riordan, C.: Improved query performance prediction using standard deviation. In: Proceedings of SIGIR 2011 (2011)
6. Fang, H., Zhai, C.: An exploration of axiomatic approaches to information retrieval. In: Proceedings of SIGIR 2005 (2005)
7. Hauff, C., Azzopardi, L., Hiemstra, D.: The combination and evaluation of query performance prediction methods. In: Boughanem, M., Berrut, C., Mothe, J., Soule-Dupuy, C. (eds.) ECIR 2009. LNCS, vol. 5478, pp. 301–312. Springer, Heidelberg (2009). https://doi.org/10.1007/978-3-642-00958-7_28
8. Lau, C.H., Li, Y., Tjondronegoro, D.: Microblog retrieval using topical features and query expansion. In: Proceedings of TREC 2011 (2011)
9. Lin, J., Mohammed, S., Sequiera, R., Tan, L., Ghelani, N., Abualsaud, M.: Overview of the TREC 2017 real-time summarization track. In: Proceedings of TREC 2017 (2017)
10. Lin, J., Efron, M., Wang, Y., Sherman, G., Voorhees, E.: Overview of the TREC-2015 microblog track. In: Proceedings of TREC 2015 (2015)
11. Lin, J., Roegiest, A., Tan, L., McCreadie, R., Voorhees, E., Diaz, F.: Overview of the TREC 2016 real-time summarization track. In: Proceedings of TREC 2016 (2016)
12. Moulahi, B., Jabeur, L.B., Tan, L., McCreadie, R., Voorhees, E., Diaz, F.: IRIT at TREC real-time summarization 2016. In: Proceedings of TREC 2016 (2016)
13. Ounis, I., Macdonald, C., Lin, J., Soboroff, I.: Overview of the TREC-2011 microblog track. In: Proceedings of TREC 2011 (2011)
14. Pérez-Iglesias, J., Araujo, L.: Standard deviation as a query hardness estimator. In: Chavez, E., Lonardi, S. (eds.) SPIRE 2010. LNCS, vol. 6393, pp. 207–212. Springer, Heidelberg (2010). https://doi.org/10.1007/978-3-642-16321-0_21
15. Robertson, S.E., Walker, S., Jones, S., Hancock-Beaulieu, M., Gatford, M.: Okapi at TREC-3. In: NIST Special Publication 500–225: Overview of the Third Text REtrieval Conference (TREC-3) (1994)
16. Rodriguez Perez, J.A., Jose, J.M.: Predicting query performance in microblog retrieval. In: Proceedings of SIGIR 2014 (2014)

17. Shtok, A., Kurland, O., Carmel, D., Raiber, F., Markovits, G.: Predicting query performance by query-drift estimation. ACM Trans. Inf. Syst. **30**(2), 1–35 (2012)
18. Singhal, A., Buckley, C., Mitra, M.: Pivoted document length normalization. In: Proceedings of SIGIR 1996 (1996)
19. Soboroff, I., Ounis, I., Macdonald, C., Lin, J.: Overview of the TREC 2012 microblog track. In: Proceedings of TREC 2012 (2012)
20. Srivastava, A., Sahami, M.: Text Mining: Classification, Clustering, and Applications (2009)
21. Tan, L., Roegiest, A., Lin, J., Clarke, C.L.: An exploration of evaluation metrics for mobile push notifications. In: Proceedings of SIGIR 2016 (2016)
22. Tang, J., Lv, C., Yao, L., Zhao, D.: PKUICST at TREC 2017 real-time summarization track: push notifications and email digest. In: Proceedings of TREC 2017 (2017)
23. Tomlinson, S.: Robust, web and terabyte retrieval with hummingbird searchservertm at TREC 2004. In: Proceedings of TREC-13 (2004)
24. Yom-Tov, E., Fine, S., Carmel, D., Darlow, A.: Learning to estimate query difficulty: including applications to missing content detection and distributed information retrieval. In: Proceedings of SIGIR 2005 (2005)
25. Zhai, C., Lafferty, J.: A study of smoothing methods for language models applied to information retrieval. ACM Trans. Inf. Syst. **22**(2), 179–214 (2004)
26. Zhao, Y., Scholer, F., Tsegay, Y.: Effective pre-retrieval query performance prediction using similarity and variability evidence. In: Macdonald, C., Ounis, I., Plachouras, V., Ruthven, I., White, R.W. (eds.) ECIR 2008. LNCS, vol. 4956, pp. 52–64. Springer, Heidelberg (2008). https://doi.org/10.1007/978-3-540-78646-7_8
27. Zhou, Y., Croft, W.B.: Query performance prediction in web search environments. In: Proceedings of SIGIR 2007 (2007)
28. Zhu, X., Huang, J., Zhu, S., Chen, M., Zhang, C., Zhenzhen, L., Dongchuan, H., Chengliang, Z., Li, A., Jia, Y.: NUDTSNA at TREC 2015 microblog track: a live retrieval system framework for social network based on semantic expansion and quality model. In: Proceedings of TREC 2015 (2015)

Smart Entertainment - A Critiquing Based Dialog System for Eliciting User Preferences and Making Recommendations

Roshni R. Ramnani[(⊠)], Shubhashis Sengupta, Tirupal Rao Ravilla, and Sumitraj Ganapat Patil

Accenture Technology Labs, Bangalore, India
{roshni.r.ramnani,shubhashis.sengupta,tirupal.rao.ravilla,
s.ganapat.patil}@accenture.com

Abstract. We present a Critiquing based dialog system that can make media content recommendations to users by eliciting information through active exploration of user preferences for item attributes. The system and user communicate through a natural language mixed-initiative conversational interface in which the system guides the user to a specific choice. During the conversation, the system presents the user with several options and analyzes the responses or "critiques". The system starts with general recommendations or relevant candidates and refines this as it learns more about user's preferences in subsequent iterations. These choices made by the user and the textual feedback/reviews that can be optionally provided, is used to infer a user preference model for the item.

Keywords: NLP · Dialog Systems · CRF · Recommendation

1 Introduction and Motivation

Our approach consists of two key components i.e. (a) a prototype natural language dialog management system, which can effectively handle varied and iterative user queries and; (b) the critique or continuous user feedback approach for collection and modeling of user preferences in order to provide effective user recommendations. At a high level, a recommendation system helps a user select an item that closely matches the user preferences from a list of options. Recommendation systems defer by their algorithm (collaborative filtering, content based) used, the data source used (ratings v/s selected features etc.) and the manner in which information is gathered from the user. [4] argues that using natural language as a mechanism to gather input for the user preference model in a recommendation system not only solves the widely prevalent "cold start" problem, but also makes the system easy to use and provides the user with flexibility in expressing his/her preferences. A typical dialog system [11] consists of 5 main components:

© Springer International Publishing AG, part of Springer Nature 2018
M. Silberztein et al. (Eds.): NLDB 2018, LNCS 10859, pp. 456–463, 2018.
https://doi.org/10.1007/978-3-319-91947-8_47

1. Speech Recognition: converts the speech signal to a textual representation.
2. Language Understanding: processes the text and converts it to an internal semantic representation.
3. Dialog Management: uses the dialog history and the current semantic representation to decide next system action.
4. Language Generation: generates the textual response to the user.
5. Speech Synthesis: generates the audio/speech signal played back to the user.

For the Smart Entertainment system, we created a custom built framework mainly over Open Source software with components for Language Understanding, Dialog Management and Language Generation, and used freely available software and hardware for Speech Recognition and Speech Synthesis.

To illustrate how a new user interacts with the system and how the system interacts with a user for whom the user preference is built in, we present two scenarios. Table 1 provides a sample interaction involving a first time user John (i.e., there is no preference models built for John). Here, the system provides a *critiquing aid*- i.e., a set of critique suggestions for users to select. John starts with a request using a high level query- *play a drama for me*. The system helps the user narrow down the choices in the space of Drama genre movies by prompting the user with actors, directors, associated genres, release dates etc. till a narrow enough movie selection can be obtained. In an untrained system, the choices provided are generic, and are chosen from a set of either randomly chosen or pre-initiated choices. However, as more user interactions take place, the system uses a weighted algorithm to build the preference model from the critiques provided by the user, and provides the user with choices based on the preference model. The system also uses an understanding of the user's viewing history and users explicit feedback/review to make recommendations. Table 2 shows an example of a conversation between a user and a system, after the user has used the system for a few interactions. In this scenario, the system provides a set of *critiquing coverage* - i.e., multiple items are returned to a query with a clear suggestion on what the system thinks the user would like.

The paper is organized as follows: Sect. 2 provides modular systemic and algorithm details. Section 3 provides a detailed explanation of the recommendation policies. Section 4 talks about experimentation and the results. The related work is presented in Sect. 5 and we conclude in Sect. 6. Throughout the paper, the examples provided relate to the movies, however the same would be applicable for other media content like music.

2   Smart Entertainment System Components

Figure 1 depicts 3 main layers of the Smart Entertainment system - The Dialog Processing layer, the Semantic search and Recommendation services layer and the Data Repository.

Table 1. Example user interaction

User: Hey RHEA

System: Hey John, what can I do for you?

User: play a drama for me

System: Would you like a film starring Al Pacino, Alan Howard, or any other actor?

User: Al Pacino

System: Would you like a film by Francis Ford Coppola, Brian De Palma, or any other director?

User: I like Francis Ford Coppola

System: We have the films: "The Godfather", "The Godfather: Part II" ... which one would you prefer?

User: play The Godfather

System: Sure. Playing "The Godfather". Enjoy!

Table 2. User interaction with build-in preference

User: Hey RHEA

System: Hey John, what can I do for you?

User: recommend a movie

System: Looks like you prefer comedy or animation, should we play one of these or some other genre?

User: animation

System: Since you like "Lee Unkrich" We have selected the movies: "Toy Story 3", "Finding Nemo", "Monsters, Inc.", Since you have already seen "Finding Nemo" and "Monsters Inc". I recommend "Toy Story 3"

2.1   Dialog Processing Layer

The Dialog Processing layer comprises Natural Language Understanding (NLU) components like Dialog Act Classification, Slot Detection and Sentiment Analysis and Dialog Management.

Dialog Act Classification. The Dialog Act classifier performs a coarse grained classification of the incoming user utterance into 4 main categories: *Greetings* (for user messages like "Hello", "Bye", "Thanks" etc.), *Command/Request* (for user messages requesting content like "play a Tom hanks movie", "play a comedy"), *Grounding* (for confirmation or rejection messages like "yes",) *Info* (for messages that give information requested by the system like "I like Tom hanks" or "some other director"). To classify a Dialog Act from user, A Conditional Random Field (CRF) based classifier, based on Mallet, is run on each user statement. The feature set involves "pos", "wordbigrams", "posbigrams". If the CRF classifier fails to predict an act over a threshold value, then we use a simple set of regular expression patterns to classify the statement into an act.

Slot Detection. The Intent identification of the user utterance and related Slot identification is done specifically on *Command/Request* dialog act to identify the

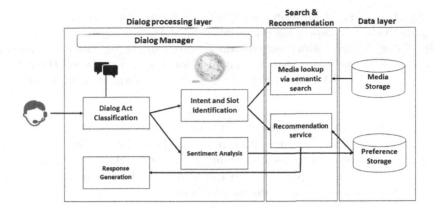

Fig. 1. Process flow

set of attributes specified by the user (such as "genre", "actor", "director" etc.). First, the OpenNLP POS tagger is run on each statement. Then, a dual approach is used to arrive at all the slots present in the statement:

1. Approach 1: We train yet another CRF classifier for each slot type with features like "left context word", "right context word", "current POS tag", and "NER tag"(Named Entity Recognition), as a sequence labeling problem. If the classifier predicts with a high score, we use the value; else we adopt approach 2.
2. Approach 2: Slots are identified using a Domain Knowledge Graph by matching slot values or context words. A knowledge graph approach to short text understanding has also been used in [12]. Various Single and multi-word expressions can either be slot values or context words (words which help identify the slot values). Slot values like "Actor", "Director", "Plot" form the nodes and the context words form the edges to the nodes in the graph. The direction of the edge decides context location.

The Fig. 2 shows the Intent and slot labels for an example command.

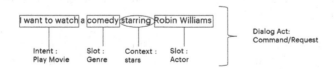

Fig. 2. Intents and slots from a user utterance

Sentiment Analysis. We perform sentiment analysis to capture explicit user likings in a user utterance or user feedback/reviews. Sentiment analysis is performed using the Standford Core NLP sentiment analyzer trained on movie reviews, and Sentiwordnet.

Dialog Manager. The information flow through the system, and the coordination across various components is controlled by the dialog manager for each turn of the dialog. Dialog history is maintained across the turns. However, the core functionality of the dialog manager involves storing and updating the current dialog state. A set of domain specific rules govern the mapping between the current user statement (after identifying the dialog act or intent) and the state. The state then governs the set of downstream activities that are performed, together with the response that is generated by the system. We adopt OpenDial system [1]; that has a capability of supporting probabilistic dialog strategy. A response to the user is generated by using a template filling mechanism through a set of pre-canned semantic representations of responses. These responses must also incorporate data from the recommender system.

2.2 Search and Recommendation Layer

The *Semantic Search* service searches for movies with attributes which match the slots provided by the user. Upon receiving a user query string on a single or multiple attributes, the service uses syntactic similarity measures like Longest Common Substring (LCS) and Cosine Similarity on the content repository. *Recommendation service* employs content based as well as user profile based recommendations. We provide more details in the subsequent section.

2.3 Data Repository

The User preference data (updated per session) is kept in MongoDB. We use Lucene for storage and indexing of content metadata. The actual media content is stored in a proprietary media server with a player which is available to Smart Entertainment system as a service. We used around 600 movie titles and corresponding clips for experimentation.

3 Recommendation

The Recommendation service uses two different strategies for providing user recommendation:

1. **Strategy1:** In this scenario, the system interacts with the user to elicit a set of explicit and finite choices to arrive at a "similarity profile" and recommend a content. The choices are based on the attributes of the content - such as "Actor(s)/Cast", "Director", " Period", "Genre" etc. The idea is to find the largest support for a set of attribute values (called Association Set [9]) that has previously been chosen, where the movie is not tagged as "Watched". For the ongoing interaction, once the association set of chosen values of the attributes becomes significant (starting minimally at 2), then the largest supports for all association sets containing the current one is found out through a greedy strategy and a suggestion is made either on the next attribute or

on the Movie itself. If the user chooses the movie then the support for the set is incremented by 1 and the set is assigned to the user profile and the movie is marked as "watched". If the user chooses the next attribute, the strategy continues for the next iteration till a movie is selected or the dialog session closure happens. The attributes are elicited from user through a series of interactive dialogue turns. This scenario is akin to an explicit or directed user modeling through *critique aid* strategy. We further improve the strategy by explicitly incorporating the feedback from the users. For example, we assign a score [−1,1] on each of the attributes of the Association Set based on feedback.

2. **Strategy2:** In this scenario, the system interacts with users to understand a set of latent/implicit intent or affiliation. In this case, the set of users are asked to give implicit comments on plots or summary and upon selection of the movie, their feedback are recorded in a free-form format. Once statistically significant number of user choices and related feedback are obtained, then a semantic clustering is done based on the data on joint data: movie metadata - particularly on plot summary, reviews and other related fields and feedback obtained from users to get a set of combined [movie, user] themes. Before clustering, we remove a set of movies if the overall sentiment/feedback for the movies across all users are negative (below a threshold). For our experiments, we created 30 clusters of dimension $d = 10$ using Latent Semantic Analysis using the package Gensim. Movie suggestions are then made based on mapping the query string from the completed user slots to the clusters using LSA based prediction. Only movies previously unwatched by the user are recommended.

4 Experimental Setup and Results

We describe the experimental setup and evaluation of the Smart Entertainment system primarily on two counts - accuracy of the NLU and accuracy of the recommendation strategies. We set up the system in the following way - in the first step; the system, devoid of any user preference information, is trained with a group of 20 first time users of the system, in 4 user groups, for one week. For each movie choice shown to the user, detailed feedback is collected. In the second step, the 5 randomly selected users from the previous set is asked to interact with the system and rank the movie recommendations for a fixed number of trials.

The data collected in the first step was used to annotate the dialog acts and slots and train the classifiers. The data is also used to embellish the knowledge graph and parsing rules to make the NLU more robust. A similar approach of deploying a research dialog system to tweak the natural language layer has also been done in [13]. The data is also used to create the Association Sets for the users and appropriate scores. For Recommendation strategy 2, overall grossly negatively scored movies are kept out of the semantic clustering mechanism.

Tables 3 and 4 show the cross validation details of the machine learned NLU models.

Table 3. Slots

No. of utterances	653	
No. of slot labels	10	
5-fold validation	Accuracy	F1
	0.92	0.8
	0.96	0.92
	0.94	0.86
	0.93	0.88
	0.92	0.84

Table 4. DialogAct

No. of utterances	1012	
No. of dialog-acts	4	
5-fold validation	Accuracy	F1
	0.95	0.94
	0.95	0.92
	0.94	0.89
	0.93	0.82
	0.96	0.93

During the second step of experimentation, we calculate average prediction probability of Dialog Acts and Slot detection. The prediction probabilities were 0.81 and 0.75 respectively. For recommendation accuracy in the controlled experiment in step 1, we pose the problem as an information retrieval one. We use Average Precision (AP) for the Top 1 result and Mean Average Precision (MAP) for recommendation for top 3 result-sets. The recommendation results of strategy 1 were better, as final queries tend to be much refined through dialog iterations over movie attributes; while in strategy 2, ad-hoc implicit queries like *Play a movie in which the astronaut gets stranded on Mars* or *Play any recent Romantic movie that has rave reviews* were supported.

5 Prior Work

Critiquing based systems (including, but not limited to, a conversational style of interaction) have been used in a number of instances [2,3,5]. These systems are also known as case based recommenders or knowledge based recommenders. The first few systems considered critiques on individual features or attributes, however [7]; highlights the importance of compound critiques within a single interaction. Another interesting approach taken by real word recommender systems like Netflix [14] is the usage of multiple recommendation algorithms to make multiple series of recommendations like "Trending Now", "Because you saw" etc. [6]; recognizes user reviews as a source for user recommendations, but uses it to generate questions to the user. This is different from our approach in that we use user reviews to (a) understand the users sentiment around the item (b) to extract themes from the text to better provide recommendations. [10]; Uses LDA techniques to create combination of two topic models, one based on the users, and the other based on the user entered description words.

6 Conclusion and Future Work

In conclusion, we provide a generic method in which the user can get personalized recommendations when searching for a product/service by specifying its attributes and providing a set of critiques on it. We describe a prototype that illustrates a media viewing experience by using this approach. As part of the future work, we plan to widen the scope of the experimental set up to include a larger set of at least 100 users. We also plan to incorporate similarity profiles of the users based on social group like facebook circles, city or country. This strategy allows incorporation of peer recommendation, as well as learning of new attributes.

References

1. Lison, P., Kennington, C.: OpenDial: a toolkit for developing spoken dialogue systems with probabilistic rules. In: Proceedings of ACL-2016 System Demonstrations (2016)
2. Chen, L., Pu, P.: Critiquing-based recommenders: survey and emerging trends User Modeling and User-Adapted Interaction (2012)
3. Linden, G., Hanks, S., Lesh, N.: Interactive assessment of user preference models: the automated travel assistant. In: Jameson, A., Paris, C., Tasso, C. (eds.) User Modeling. ICMS, vol. 383, pp. 67–78. Springer, Vienna (1997). https://doi.org/10.1007/978-3-7091-2670-7_9
4. Johansson, P.: Natural language interaction in personalized EPGs. In: Proceedings of Workshop Notes from the 3rd International Workshop on Personalization of Future TV, Johnstown, Pennsylvania, USA, pp. 27–31 (2003)
5. Burke, R.D., Hammond, K.J., Yound, B.C.: The FindMe approach to assisted browsing. IEEE Exper. **12**(4), 32–40 (1997)
6. Reschke, K., Vogel, A., Jurafsky, D.: Generating recommendation dialogs by extracting information from user reviews. In: Proceedings of the 51st Annual Meeting of the Association for Computational Linguistics (2013)
7. Zhang, J., Pu, P.: A comparative study of compound critique generation in conversational recommender systems. In: Wade, V.P., Ashman, H., Smyth, B. (eds.) AH 2006. LNCS, vol. 4018, pp. 234–243. Springer, Heidelberg (2006). https://doi.org/10.1007/11768012_25
8. Ramachandran, D., et al.: A TV program discovery dialog system using recommendations. In: Proceedings of the 16th Annual Meeting of the Special Interest Group on Discourse and Dialogue (2015)
9. Agrawal, R., Srikant, R.: Fast algorithms for mining association rules in large databases. In: Proceedings of the 20th International Conference on Very Large Data Bases, VLDB (1994)
10. Pyo, S., Kim, E.: LDA-based unified topic modeling for similar TV user grouping and TV program recommendation. IEEE Trans. Cybern. **45**, 8 (2015)
11. Bohus, D., Rudnicky, A.I.: The RavenClaw dialog management framework: architecture and systems. Comput. Speech Lang. **23**(3), 332–361 (2009)
12. Hua, W., Wang, Z., Wang, H., Zheng, K., Zhou, X.: Short text understanding through lexical-semantic analysis. In: 2015 IEEE 31st International Conference on Data Engineering (ICDE). IEEE (2015)
13. Raux, A., et al.: Let's Go Public! taking a spoken dialog system to the real world. In: Ninth European Conference on Speech Communication and Technology (2005)
14. Gomez-Uribe, C.A., Hunt, N.: The netflix recommender system: Algorithms, business value, and innovation. ACM Trans. Manag. Inf. Syst. **6**(4), 13 (2016)

Translation and Foreign Language Querying

Cross-Language Text Summarization Using Sentence and Multi-Sentence Compression

Elvys Linhares Pontes[1]([✉]), Stéphane Huet[1], Juan-Manuel Torres-Moreno[1,2], and Andréa Carneiro Linhares[3]

[1] LIA, Université d'Avignon et des Pays de Vaucluse, 84000 Avignon, France
elvys.linhares-pontes@alumni.univ-avignon.fr
[2] École Polytechnique de Montréal, Montreal, Québec, Canada
[3] Universidade Federal do Ceará, Sobral, Ceará, Brazil

Abstract. Cross-Language Automatic Text Summarization produces a summary in a language different from the language of the source documents. In this paper, we propose a French-to-English cross-lingual summarization framework that analyzes the information in both languages to identify the most relevant sentences. In order to generate more informative cross-lingual summaries, we introduce the use of chunks and two compression methods at the sentence and multi-sentence levels. Experimental results on the MultiLing 2011 dataset show that our framework improves the results obtained by state-of-the art approaches according to ROUGE metrics.

Keywords: Cross-Language Automatic Text Summarization
Multi-Sentence Compression · Sentence Compression

1 Introduction

Cross-Language Automatic Text Summarization (CLATS) aims to generate a summary of a document where the summary language differs from the document language. The huge amount of information available on the Internet made it easier to be up to date on the news in the world. However, some information and viewpoints exist in languages that are unknown by readers. CLATS enables people who are not fluent in the source/target language to comprehend these data in a simple way.

The methods developed for CLATS can be classified, like the Automatic Text Summarization (ATS) domain, depending on whether they are extractive, compressive or abstractive [21]. The extractive ATS selects complete sentences that are supposed to be the most relevant of the documents; the compressive

This work was partially financed by the European Project CHISTERA-AMIS ANR-15-CHR2-0001.

ATS generates a summary by compression of sentences through the removal of non-relevant words; lastly, the abstractive ATS generates a summary with new sentences that are not necessarily contained in the original texts.

Many of the state-of-the-art methods for CLATS are of the extractive class. They mainly differ on how they compute sentence similarities and alleviate the risk that translation errors are introduced in the produced summary. Among these models, the CoRank method, which is characterized by its ability to simultaneously incorporate similarities between the original and translated sentences, turns out to be effective [22]. This method is extended in this paper in the following manner: we first take into account chunks instead of only words in the sentence similarity measures; then sentences are compressed in order to obtain a compressive CLATS system.

Inspired by the compressive ATS methods in monolingual analysis [1,5,10, 11,16,19,24], we adapt sentence and multi-sentence compression methods for the CLATS problem to just keep the main information. A Long Short Term Memory (LSTM) model is built to analyze a sentence and decide which words remain in the compression. We also use an Integer Linear Programming (ILP) formulation to compress similar sentences while analyzing both grammaticality and informativeness.

The remainder of this paper is organized as follows. In Sect. 2, we describe the most recent works about CLATS. Sections 3 presents our compressive CLATS approach. Section 4 reports the results achieved on the MultiLing 2011 dataset for the French-to-English task and shows that our method, particularly with the use of ILP for multi-sentence compression, outperforms the state of the art according to the ROUGE metrics. Finally, conclusions and future work are set out in Sect. 5.

2   Cross-Language Automatic Text Summarization

The first studies in cross-language document summarization analyzed the information in only one language [9,18]. Two typical CLATS schemes are the early and the late translations. The first scheme first translates the source documents to the target language, then it summarizes the translated documents using only information of the translated sentences. The late translation scheme does the reverse: it first summarizes the documents using abstractive, compressive or extractive methods, then it translates the summary to the target language.

Recent methods improved the quality of cross-lingual summarization using a translation quality score [2,23,25] and the information of the documents in both languages [22,26]. These methods are described in the next subsections.

2.1   Machine Translation Quality

Wan et al. trained a Support Vector Machine (SVM) regression method to predict the translation quality of a pair of English-Chinese sentences from basic features (such as sentence length, sub-sentence number, percentage of nouns and adjectives) and parse features (such as depth, number of noun phrases and verbal phrases in the parse tree) to generate English-to-Chinese CLATS [23]. They used 1,736 pairs of English-Chinese sentences (English sentences were translated automatically by Google Translate) and computed translation quality scores in a range from 1 to 5 (1 means "very bad" and 5 corresponds to "excellent"). The translation quality and informativeness scores were linearly combined to select the English sentences with both a high translation quality and a high informativeness:

$$score(s_i) = (1 - \lambda) \cdot InfoScore(s_i) + \lambda \cdot TransScore(s_i) \qquad (1)$$

where $InfoScore(s_i)$ and $TransScore(s_i)$ are the informativeness score and translation quality prediction of the sentence s_i, and $\lambda \in [0, 1]$ is a parameter controlling the influence of the two factors. Finally, they translated the English summary to form the Chinese summary.

Similarly to Wan et al. [23], Boudin et al. used an ϵ-SVR to predict the translation quality score based on the automatic NIST metric as an indicator of quality [2]. They automatically translated English documents to French using Google Translate, then they analyzed some features (sentence length, number of punctuation marks, perplexities of source and target sentences using different language models, etc.) to estimate the translation quality of a sentence. They incorporated the translation quality score in the PageRank algorithm [3] to calculate the relevance of sentences based on the similarity between the sentences and the translation quality scores to perform English-to-French cross-lingual summarization (Eqs. 2–4).

$$p(V_i) = (1 - d) + d \times \sum_{V_j \in pred(V_i)} \frac{score(S_i, S_j)}{\sum_{V_k \in succ(V_i)} score(S_k, S_i)} p(V_i) \qquad (2)$$

$$score(S_i, S_j) = Sim(S_i, S_j) \times Prediction(S_i) \qquad (3)$$

$$Sim(S_i, S_j) = \frac{\sum_{w \in S_i, S_j} freq(w, S_i) + freq(w, S_j)}{\log(|S_i|) + \log(|S_j|)} \qquad (4)$$

where d is the damping factor, $Prediction(s)$ is the translation quality score of the sentence s, $freq(w, s)$ is the frequency of the word w in the sentence s, $pred(V_i)$ and $succ(V_i)$ are the predecessors and successors vertices of the vertex V_i.

Inspired by the phrase-based translation models, Yao et al. proposed a phrase-based model to simultaneously perform sentence scoring, extraction and compression [25]. They designed a scoring scheme for the CLATS task based on a submodular term of compressed sentences and a bounded distortion penalty term. Their summary scoring measure was defined over a summary S as:

$$F(S) = \sum_{p \in S} \sum_{i=1}^{count(p,S)} d^{i-1} g(p) + \sum_{s \in S} bg(s) + \eta \sum_{s \in S} dist(y(s)) \qquad (5)$$

where $g(p)$ is the score of phrase p (defined by the frequency of p in the document), $bg(s)$ is the bigram score of sentence s, $y(s)$ is the phrase-based derivation of the sentence s and $dist(y(s))$ is the distortion penalty term in the phrase-based translation models. Finally, d is a constant damping factor to penalize repeated occurrences of the same phrases, $count(p, S)$ is the number of occurrences of the phrase p in the summary S and η is the distortion parameter for penalizing the distance between neighboring phrases in the derivation.

2.2 Joint Analysis in both Languages

Wan proposed to leverage both the information in the source and in the target language for cross-lingual summarization [22]. In particular, he introduced two graph-based summarization methods (SimFusion and CoRank) for using both the English-side and Chinese-side information in the task of English-to-Chinese cross-lingual summarization. The first method linearly fuses the English-side and Chinese-side similarities for measuring Chinese sentence similarity. In a nutshell, this method adapts the PageRank algorithm to calculate the relevance of sentences, where the weight arcs are obtained by the linear combination of the cosine similarity of pairs of sentences for each language:

$$relevance(s_i^{cn}) = \mu \sum_{j \in D, j \neq i} relevance(s_j^{cn}) \cdot \tilde{C}_{ji}^{cn} + \frac{1-\mu}{n} \tag{6}$$

$$C_{ij}^{cn} = \lambda \cdot sim_{cosine}(s_i^{cn}, s_j^{cn}) + (1-\lambda) \cdot sim_{cosine}(s_i^{en}, s_j^{en}) \tag{7}$$

where s_i^{cn} and s_i^{en} represent the sentence i of a document D in Chinese and in English, respectively, μ is a damping factor, n is the number of sentences in the document and $\lambda \in [0, 1]$ is a parameter to control the relative contributions of the two similarity values. C^{cn} is normalized to $\tilde{C}^{cn}$ to make the sum of each row equal to 1. The CoRank method adopts a co-ranking algorithm to simultaneously rank both English and Chinese sentences by incorporating mutual influences between them. It considers a sentence as relevant if this sentence in both languages is heavily linked with other sentences in each language separately (source-source and target-target language similarities) and between languages (source-target language similarity) (Eqs. 8–12).

$$\mathbf{u} = \alpha \cdot (\tilde{\mathbf{M}}^{\mathbf{cn}})^T \mathbf{u} + \beta \cdot (\tilde{\mathbf{M}}^{\mathbf{encn}})^T \mathbf{v} \tag{8}$$

$$\mathbf{v} = \alpha \cdot (\tilde{\mathbf{M}}^{\mathbf{en}})^T \mathbf{v} + \beta \cdot (\tilde{\mathbf{M}}^{\mathbf{encn}})^T \mathbf{u} \tag{9}$$

$$M_{ij}^{en} = \begin{cases} \cosine(s_i^{en}, s_j^{en}), & \text{if } i \neq j \\ 0 & \text{otherwise} \end{cases} \tag{10}$$

$$M_{ij}^{cn} = \begin{cases} \cosine(s_i^{cn}, s_j^{cn}), & \text{if } i \neq j \\ 0 & \text{otherwise} \end{cases} \tag{11}$$

$$\mathrm{M}_{ij}^{en,cn} = \sqrt{\cosine(s_i^{cn}, s_j^{cn}) \times \cosine(s_i^{en}, s_j^{en})} \tag{12}$$

where M^{en} and M^{cn} are normalized to $\tilde{\mathrm{M}}^{en}$ and $\tilde{\mathrm{M}}^{cn}$, respectively, to make the sum of each row equal to 1. $\mathbf{u}$ and $\mathbf{v}$ denote the saliency scores of the Chinese and English sentences, respectively; α and β specify the relative contributions to the final saliency scores from the information in the same language and the information in the other language, with $\alpha + \beta = 1$.

Unlike Wan who generated extractive CLATS, Zhang et al. analyzed Predicate-Argument Structures (PAS) to obtain an abstractive English-to-Chinese CLATS [26]. They built a pool of bilingual concepts and facts represented by the bilingual elements of the source-side PAS and their target-side counterparts from the alignment between source texts and Google Translate translations. They used word alignment, lexical translation probability and 3-gram language model to measure the quality and the fluency of the Chinese translation, and the CoRank algorithm [22] to measure the relevance of the facts and concepts in both languages. Finally, summaries were produced by fusing bilingual PAS elements with an Integer Linear Programming (ILP) algorithm to maximize the saliency and the translation quality of the PAS elements.

3   Our Proposition

Following the CoRank-based approach proposed by Wan [22], we use his joint analysis of documents in both languages (source and target languages) to select the most relevant sentences. We expanded this method in three ways.

Firstly, we take into account Multi-Word Expressions (MWE) when computing similarities between sentences. These MWEs are very common in all languages and pose significant problems for every kind of NLP [20]. Their use in the context of CLATS helps the system to comprehend the semantic content of sentences. To realize a chunk-level tokenization, we used the Stanford CoreNLP tool for the English side [15]. This annotator tool, which integrates jMWE [8], detects various expressions, e.g., phrasal verbs ("*take off*"), proper names ("*San Francisco*"), compound nominals ("*cultivated plant*") or idioms ("*rain cats and dogs*"). Unfortunately, the tools developed for languages other than English have a lower coverage for MWEs. For this reason, MWEs were detected on the French side from the alignment of phrases inside parallel sentences using the Giza++ application [17].

A second evolution of the CoRank-based approach is the use of a Multi-Sentence Compression (MSC) method to generate more informative compressed outputs from similar sentences. For this purpose, the sentences are grouped in clusters based on their similarity in both languages. For each cluster with more than one sentence, which is common in the case of multi-document summarization, a MSC method guided by keywords is applied to build a sentence with the core information of the cluster [13, 14].

A third extension of the approach relies on compression techniques of a single sentence by deletion of words [5]. Still with the idea to generate more informative summaries, sentence compression is applied for sentences that stand alone during the clustering step required by the MSC step.

The following subsections describe in detail the architecture of our system.

3.1   Preprocessing

Initially, French texts are translated to English using the Google Translate system, which is at the cutting edge of the statistical translation technology and was used in the majority of the state-of-the-art CLATS methods.

Then, chunks are identified inside the English texts with the Stanford CoreNLP, while the English and French parallel sentences are aligned with the Giza++ toolkit.[1] Two Giza++ models were trained on the Europarl v7 (2.1 M sentence pairs[2]) and News-Commentary 11 (0.2 M sentence pairs[3]) datasets in both directions (English-to-French and French-to-English). Like the training corpora used for statistical translation models, the alignments obtained by both models were intersected by the default heuristic *grow-diag-final* of the Moses toolkit [7]. From these alignments and the English MWEs, a chunk-level tokenization is performed on the French side.

Finally, sentences are clustered according to their similarities, sentences with a similarity score bigger than threshold θ remain in the same group. The similarity score of a pair of sentences i and j is defined by the cosine similarity in both languages:

$$sim(i,j) = \sqrt{\text{cosine}(s_i^{fr}, s_j^{fr}) \times \text{cosine}(s_i^{en}, s_j^{en})} \tag{13}$$

where s_i^{fr} and s_i^{en} represent a sentence i in the French and English languages.

3.2   Sentence and Multi-Sentence Compression

To avoid the accumulation of errors that would appear in a translation-compression-translation pipeline, we restrict the sentence and multi-sentence compressions to the sentences in the target language.

Sentence Compression. The Sentence Compression (SC) problem is here seen as the task to delete non-relevant words in a sentence [5,10,11,19,24]. Filippova et al. [5] used an LSTM model to compress sentences by deletion of words. In few words, this model follows a sequence-to-sequence paradigm to verify which words of a sentence c remain in the compression. A word i in a sentence c is represented by its word embedding and the word embedding of its parent node in the parse tree. Then, a first LSTM encodes this sentence and another LSTM generates the sequence of the words that are kept in the compression. LSTMs are composed of input i_t, control state c_t and memory state m_t that are updated at time step t (Eqs. 14–19).

$$i_t = \text{sigm}(W_1 x_t + W_2 h_{t-1}) \tag{14}$$

$$i'_t = \tanh(W_3 x_t + W_4 h_{t-1}) \tag{15}$$

---

[1] The GIZA++ model, https://github.com/moses-smt/giza-pp.
[2] http://www.statmt.org/europarl/.
[3] http://opus.nlpl.eu/News-Commentary.php.

$$f_t = \text{sigm}(W_5 x_t + W_6 h_{t-1}) \tag{16}$$

$$o_t = \text{sigm}(W_7 x_t + W_8 h_{t-1}) \tag{17}$$

$$m_t = m_{t-1} \odot f_t + i_t \odot i'_t \tag{18}$$

$$h_t = m_t \odot o_t \tag{19}$$

where the operator $\odot$ denotes element-wise multiplication, the matrices $W_1, ...,$ W_8 and the vector h_0 are the parameters of the model, and all the non-linearities are computed element-wise (more details in [5]). Contrary to [5], we analyze the sentence at the chunk level, so we remove a chunk only if all words of this chunk were deleted in the SC process described above.

Multi-Sentence Compression. For the clusters that have more than a sentence, we use a Chunk Graph (CG) to represent them and an ILP method to compress these sentences in a single, short, and hopefully correct and informative sentence. Among several state-of-the-art MSC methods [1,4,16], Linhares Pontes et al. [13,14] used an ILP formulation to guide the MSC using a list of keywords. Our system incorporates this approach to create a Word Graph and to calculate the weight arcs (cohesion between the words, Eqs. 20 and 21), but instead of restricting to single words we also consider multi-word chunks (Chunk Graph):

$$w(i, j) = \frac{\text{cohesion}(i, j)}{\text{freq}(i) \times \text{freq}(j)}, \tag{20}$$

$$\text{cohesion}(i, j) = \frac{\text{freq}(i) + \text{freq}(j)}{\sum_{s \in C} \text{diff}(s, i, j)^{-1}}, \tag{21}$$

where $\text{freq}(i)$ is the chunk frequency mapped to the vertex i and the function $\text{diff}(s, i, j)$ refers to the distance between the offset positions of chunks i and j in the sentences s of a cluster C containing these two chunks. From the relevance of the 2-grams[4] (Eq. 20), we consider that the relevance of a 3-gram is based on the relevance of their two inner 2-grams, as described in Eq. 22:

$$3\text{-gram}(i, j, k) = \frac{qt_3(i, j, k)}{\max_{a,b,c \in CG} qt_3(a, b, c)} \times \frac{w(i, j) + w(j, k)}{2}, \tag{22}$$

where $qt_3(i, j, k)$ is the number of 3-grams composed of chunks in i, j and k vertices in the cluster. The 3-grams increase the grammatical quality of the compression.

We also use Latent Dirichlet Allocation (LDA) to identify the keywords at the global (all texts of a topic) and local (cluster of similar sentences) levels to have the gist of a document and of a cluster of similar sentences. Then, an ILP method, as described in [13,14], generates a compression guided by keywords, in order to favor informativeness and grammaticality as expressed in Eq. 23.

---

[4] In this work, a unigram is represented by a chunk.

In other words, this method looks for a path (sentence) that has a good cohesion and contains a maximum of keywords.

$$\text{minimize} \left(\sum_{(i,j) \in A} w(i,j) \cdot x_{i,j} - c \cdot \sum_{k \in K} b_k - \sum_{t \in T} d_k \cdot z_t \right) \tag{23}$$

where x_{ij} indicates the existence of the arc (i,j) in the solution, $w(i,j)$ is the cohesion of the chunks i and j (Eq. 20), K is the set of labels (each representing a keyword), b_k indicates the existence of a chunk with a keyword k in the solution, c is the keyword bonus of the graph,[5] T is the set of 3-grams in the cluster, d_t indicates the existence of the 3-gram t in the solution and z_t represents the relevance of the 3-gram t defined by the Eq. 22. Finally, we generate the 50 best solutions according to the objective (23) and we select the compression with the lowest normalized score (Eq. 24) as the best compression.

$$\text{score}_{norm}(s) = \frac{e^{\text{score}_{opt}(s)}}{||s||}, \tag{24}$$

where $\text{score}_{opt}(s)$ is the score of the sentence s from Eq. 23.

3.3 CoRank Method

The CoRank method adopts a co-ranking algorithm to simultaneously rank both French and English sentences by incorporating mutual influences between them. We use the CoRank method (Sect. 2.2) to calculate the relevance of sentences. In order to avoid the accumulation of errors that would be generated by a translation-compression-translation pipeline, similarity is computed from the uncompressed versions of sentences and that is only in the last summary generation step that compressed sentences are used.

Finally, as usual for ATS, a summary is generated with the most relevant sentences and the sentences redundant with the ones that have already been selected are put aside.

4 Experimental Results

In order to analyze the performance of our method, we compare it with the early translation, the late translation, the SimFusion and the CoRank methods [22]. The early and late translations are based on the SimFusion method, the differences between the systems being on the similarity metric (Eq. 7) computed either in the target language (early translation) or in the source language (late translation). We analyzed three versions of SimFusion with $\lambda = 0.25$, 0.50 and 0.75. The CoRank method uses $\alpha = \beta = 0.5$. We generated three versions of

---

[5] The keyword bonus allows the generation of longer compressions that may be more informative and it is defined by the geometric average of all weight arcs in the Chunk Graph.

our approach, named Compressive CLATS (CCLATS): SC, MSC and SC+MSC. The first version uses the SC method to compress sentences, the MSC method compresses clusters of similar sentences and extracts the rest of the sentences, and the last version applies MSC to clusters of similar sentences and SC to other sentences.

We compress only sentences with more than 15 words and we preserve compressions with more than 10 words to avoid short outputs with little information. The MSC method selects the 10 most relevant keywords per topic and the 3 most relevant keywords per cluster of similar sentences to guide the compression generation. All systems generate summaries composed of 250 words with the most relevant sentences, while the redundant sentences are discarded. We apply the cosine similarity measure with a threshold θ of 0.5 to create clusters of similar sentences for the MSC and to remove redundant sentences in the summary generation.

We use the pre-trained word embeddings[6] with 300-dimensional embeddings and an LSTM model with only one layer with 256-dimensional embeddings. Our Neural Network is trained on the publicly released set of 10,000 sentence-compression pairs.[7]

4.1   Dataset

We used the MultiLing Pilot 2011 dataset [6] derived from publicly available WikiNews English texts. This dataset is composed of 10 topics, each topic having 10 source texts and 3 reference summaries. Each reference summary contains a maximum of 250 words. Native speakers translated this dataset into Arabic, Czech, French, Greek, Hebrew and Hindi languages. Specifically, we use English and French texts to test our system.

4.2   Automatic Evaluation

As references are assumed to contain the key information, we calculated informativeness scores counting the n-grams in common between the compression and the reference compressions using the ROUGE system [12]. In particular, we used the f-measure metrics ROUGE-1, ROUGE-2 and ROUGE-SU4.

Table 1 shows the ROUGE f-measure scores achieved by each system using the MultiLing Pilot 2011 dataset. The baselines, especially the late translation, have the worst scores. Similarly to the results described in [22], the CoRank method outperforms the SimFusion method. The analysis of the output of the CCLATS versions brought to light that the SC version removed relevant information of sentences, achieving lower ROUGE scores than CoRank. CCLATS.MSC generated more informative summaries and leads to the best ROUGE scores. Finally, the SC+MSC version obtains better results than other systems but still does not reach the highest ROUGE scores measured when using MSC alone.

---

[6] Publicly available at: code.google.com/p/word2vec.
[7] http://storage.googleapis.com/sentencecomp/compression-data.json.

Table 1. ROUGE f-measure scores for the French-to-English CLATS using the MultiLing Pilot 2011 dataset. * indicates the results are statistically better than baselines and the SimFusion method with a 0.05 level.

Methods	ROUGE-1	ROUGE-2	ROUGE-SU4
baseline.early	0.41461	0.10251	0.16001
baseline.late	0.41137	0.10270	0.15795
SimFusion.$\lambda = 0.25$	0.41403	0.10545	0.16081
SimFusion.$\lambda = 0.50$	0.41198	0.10268	0.15820
SimFusion.$\lambda = 0.75$	0.41516	0.10397	0.15992
CoRank	0.45552	0.12952	0.19056
CCLATS.SC	0.45436	0.11809	0.18463
CCLATS.MSC	**0.47221***	**0.13613**	**0.19881***
CCLATS.SC+MSC	0.46786*	0.13056	0.19420*

4.3 Discussion

The lower results of the early and late translations with respect to other systems prove that the texts in each language provide complementary information. It also establishes that the analysis of sentences in the target language plays a more important place to generate informative cross-lingual summaries. As seen for English-to-Chinese CLATS [22], the CoRank method generates better results than the baselines and SimFusion because it considers the information in each language separately and together, while the baselines restrict the analysis of sentence similarity to one language separately and the SimFusion method analyzes only the cross-lingual sentence similarity.

It is expected that a piece of information found in several texts is relevant for a topic. In accordance with this principle, the MSC method looks for the repeated information and generates a short compression with selected keywords that summarize the main information. The two kinds of keywords (global and local) guide the MSC method to generate compression linked to the main topic of the documents and to the specific information presented in the cluster.

With regard to SC, this compression method did not produce good results in our experiments. This observation may be explained by the reduced size of the corpus we used to train our NN (10,000 parallel sentence-compression instance), while the system described in Filippova et al. [5] could benefit from a corpus of about two million instances. Whereas the CCLATS.MSC version leaves unchanged the sentences that do not have similar sentences, the SC+MSC version involves the SC model to compress these sentences. As the CCLATS.SC system has lower performance than the pure extractive CoRank method, the SC+MSC also had lower results than MSC version.

A difference between the SC and MSC approaches is that MSC uses global and local keywords to guide the compression preserving the main information, while the SC method does not realize this kind of analysis. Another difference between them is that the MSC method does not need a training corpus to generate compressions.

To sum up, the joint analysis of both languages with CoRank helps the generation of cross-lingual summaries. On the one hand, the SC model deletes relevant information, thereby reducing the informativeness of summaries. On the other hand, the MSC method proves to be a good alternative to compress redundant information and to preserve relevant information. Finally, the CCLATS.MSC greatly improves the ROUGE scores and significantly outperforms the baselines and the SimFusion methods.

5   Conclusion

In this paper we have proposed two compressive methods to improve the generation of cross-lingual summaries. The proposed system analyzes a document in both languages to extract all of the relevant information. Then, it applies two kinds of methods to compress sentences. Unlike the sentence compression system (CCLATS.SC) that needs a large training dataset to generate compressions of good quality, the multi-sentence compression version of our system (CCLATS.MSC) generates better ROUGE results than extractive Cross-Language Automatic Text Summarization systems. Moreover, it has the advantage of not requiring a training corpus to generate summaries of good quality.

There are several avenues worth exploring from this work. First, we want to investigate how the size of the training data of our Neural Network to generate sentence compressions acts upon the quality of the summaries. It would also be interesting to include an attention mechanism in our Neural Network to analyze the sentence and the gist of the topic. Finally, our evaluation was confined to ROUGE scores, which mostly measure the informativeness. An additional human evaluation must be performed to confirm that the informativeness and the grammaticality are improved with the use of compression methods.

References

1. Banerjee, S., Mitra, P., Sugiyama, K.: Multi-document Abstractive Summarization Using ILP Based Multi-sentence Compression. In: 24th International Conference on Artificial Intelligence (IJCAI), IJCAI 2015, pp. 1208–1214 (2015)
2. Boudin, F., Huet, S., Torres-Moreno, J.: A graph-based approach to cross-language multi-document summarization. Polibits **43**, 113–118 (2011)
3. Brin, S., Page, L.: The anatomy of a large-scale hypertextual web search engine. Comput. Netw. ISDN Syst. **30**(1–7), 107–117 (1998)
4. Filippova, K.: Multi-sentence compression: finding shortest paths in word graphs. In: COLING, pp. 322–330 (2010)
5. Filippova, K., Alfonseca, E., Colmenares, C.A., Kaiser, L., Vinyals, O.: Sentence compression by deletion with LSTMs. In: EMNLP, pp. 360–368 (2015)
6. Giannakopoulos, G., El-Haj, M., Favre, B., Litvak, M., Steinberger, J., Varma, V.: TAC2011 multiling pilot overview. In: 4th Text Analysis Conference TAC (2011)

7. Koehn, P., Hoang, H., Birch, A., Callison-Burch, C., Federico, M., Bertoldi, N., Cowan, B., Shen, W., Moran, C., Zens, R., Dyer, C., Bojar, O., Constantin, A., Herbst, E.: Moses: Open source toolkit for statistical machine translation. In: 45th Annual Meeting of the Association for Computational Linguistics (ACL), Companion Volume, pp. 177–180 (2007)
8. Kulkarni, N., Finlayson, M.A.: jMWE: a Java toolkit for detecting multi-word expressions. In: Workshop on Multiword Expressions: from Parsing and Generation to the Real World (MWE), pp. 122–124 (2011)
9. Leuski, A., Lin, C.Y., Zhou, L., Germann, U., Och, F.J., Hovy, E.: Cross-lingual C*ST*RD: English access to Hindi Information. J. ACM Trans. Asian Lang. Inf. Process. **2**(3), 245–269 (2003)
10. Li, C., Liu, F., Weng, F., Liu, Y.: Document summarization via guided sentence compression. In: EMNLP, pp. 490–500. ACL (2013)
11. Li, C., Liu, Y., Liu, F., Zhao, L., Weng, F.: Improving multi-documents summarization by sentence compression based on expanded constituent parse trees. In: EMNLP, pp. 691–701. ACL (2014)
12. Lin, C.Y.: ROUGE: a package for automatic evaluation of summaries. In: Workshop Text Summarization Branches Out (ACL 2004), pp. 74–81 (2004)
13. Linhares Pontes, E., Huet, S., Gouveia da Silva, T., Linhares, A.C., Torres-Moreno, J.M.: Multi-sentence compression with word vertex-labeled graphs and integer linear programming. In: Proceedings of TextGraphs-12: the Workshop on Graph-based Methods for Natural Language Processing. Association for Computational Linguistics (2018)
14. Linhares Pontes, E., Gouveia da Silva, T., Linhares, A.C., Torres-Moreno, J.M., Huet, S.: Métodos de otimização combinatória aplicados ao problema de compressão multifrases. In: Anais do XLVIII Simpósio Brasileiro de Pesquisa Operacional (SBPO), pp. 2278–2289 (2016)
15. Manning, C., Surdeanu, M., Bauer, J., Finkel, J., Bethard, S., McClosky, D.: The Stanford CoreNLP natural language processing toolkit. In: 52nd Annual Meeting of the Association for Computational Linguistics (ACL): System Demonstrations, pp. 55–60 (2014)
16. Niu, J., Chen, H., Zhao, Q., Su, L., Atiquzzaman, M.: Multi-document abstractive summarization using chunk-graph and recurrent neural network. In: IEEE International Conference on Communications, ICC, pp. 1–6 (2017)
17. Och, F.J., Ney, H.: A systematic comparison of various statistical alignment models. Comput. Linguist. **29**(1), 19–51 (2003)
18. Orasan, C., Chiorean, O.A.: Evaluation of a cross-lingual Romanian-English multi-document summariser. In: 6th International Conference on Language Resources and Evaluation (LREC) (2008)
19. Qian, X., Liu, Y.: Fast joint compression and summarization via graph Cuts. In: EMNLP, pp. 1492–1502 (2013)
20. Sag, I.A., Baldwin, T., Bond, F., Copestake, A., Flickinger, D.: Multiword expressions: a pain in the neck for NLP. In: Gelbukh, A. (ed.) CICLing 2002. LNCS, vol. 2276, pp. 1–15. Springer, Heidelberg (2002). https://doi.org/10.1007/3-540-45715-1_1
21. Torres-Moreno, J.M.: Automatic Text Summarization. Wiley and Sons, London (2014)
22. Wan, X.: Using bilingual information for cross-language document summarization. In: ACL, pp. 1546–1555 (2011)
23. Wan, X., Li, H., Xiao, J.: Cross-language document summarization based on machine translation quality prediction. In: ACL, pp. 917–926 (2010)

24. Yao, J., Wan, X., Xiao, J.: Compressive document summarization via sparse optimization. In: IJCAI, pp. 1376–1382. AAAI Press (2015)
25. Yao, J., Wan, X., Xiao, J.: Phrase-based compressive cross-language summarization. In: EMNLP, pp. 118–127 (2015)
26. Zhang, J., Zhou, Y., Zong, C.: Abstractive cross-language summarization via translation model enhanced predicate argument structure fusing. IEEE/ACM Trans. Audio Speech Lang. Process. **24**(10), 1842–1853 (2016)

Arabic Readability Assessment for Foreign Language Learners

Naoual Nassiri[1(✉)], Abdelhak Lakhouaja[1],
and Violetta Cavalli-Sforza[2]

[1] Department of Computer Science, Faculty of Sciences,
University Mohamed First, Oujda, Morocco
naoual.nassiri@gmail.com, abdel.lakh@gmail.com
[2] School of Science and Engineering, AI Akhawayn University, Ifrane, Morocco
v.cavallisforza@aui.ma

Abstract. Reading in a foreign language is a difficult task, especially if the texts presented to readers are chosen without taking into account the reader's skill level. Foreign language learners need to be presented with reading material suitable to their reading capacities. A basic tool for determining if a text is appropriate to a reader's level is the assessment of its readability, a measure that aims to represent the human capacities required to comprehend a given text. Readability prediction for a text is an important aspect in the process of teaching and learning, for reading in a foreign language as well as in one's native language, and continues to be a central area of research and practice. In this paper, we present our approach to readability assessment for Modern Standard Arabic (MSA) as a foreign language. Readability prediction is carried out using the Global Language Online Support System (GLOSS) corpus, which was developed for independent learners to improve their foreign language skills and was annotated with the Interagency Language Roundtable (ILR) scale. In this study, we introduce a frequency dictionary, which was developed to calculate frequency-based features. The approach gives results that surpass the state-of the-art results for Arabic.

Keywords: Readability · Modern Standard Arabic (MSA) · Foreign language
Machine Learning (ML) · Text features

1 Introduction

The reading process, which is very important in schooling and education, requires the use of texts appropriate to the reader's level, whether reading in one's native language or in a second or foreign language. If text comprehension requires too much effort, it will discourage the reader and not lead to learning. Understanding the audience and their language skills is an important consideration in writing educational material. If students have a limited level of knowledge in a language, failure to take these limitations into account when selecting the texts to be presented to them can influence negatively the process of reading. The reader may become so preoccupied with understanding individual words and the text's structure that the overall message will be lost. So it is essential to consider readability of a text before presenting it to a target audience.

© Springer International Publishing AG, part of Springer Nature 2018
M. Silberztein et al. (Eds.): NLDB 2018, LNCS 10859, pp. 480–488, 2018.
https://doi.org/10.1007/978-3-319-91947-8_49

Several definitions exist for readability. Dale and Chall have suggested a comprehensive one. In the broadest sense, readability is the sum total (including interactions) of all those elements within a given piece of printed material that affect the success a group of readers have with it: the extent to which they understand it, read it at an optimum speed, and find it interesting [1]. Research on readability has focused primarily on content, style and format, considering various linguistic characteristics but ignoring visual presentation features such as fonts and contrast.

Readability measurement can be performed using mathematical formulas based on simple features, such as sentence length in words, word length in syllables, and so on. These features are then used in formulas whose parameters are statistically estimated from a corpus to produce scores that represent the readability of the analyzed text. Among these formulas we cite four, for English: (1) the new Dale and Chall formula [1] estimates the grade level in the U.S. school curriculum of a text using word length to assess word difficulty and a count of 'hard' words (based on a pre-defined list); (2) the Flesch Reading Ease formula [2] includes the average sentence length in words and the average number of syllables per word as features, and yields an output score that is inversely related to the level of difficulty of the text; (3) the Flesch-Kincaid Grade Level [2] similarly calculates a readability score but outputs the U.S. school grade level for which the text is appropriate; and (4) the Gunning Fog Index, which uses average sentence length and the percentage of hard words [3]. Formulas are considered the traditional method of performing readability assessment. Modern methods, used in recent research, use automatically extracted linguistic features and Machine Learning (ML), and have become a viable path for readability classification and scoring. In ML approaches, features are first extracted from texts in a corpus annotated with readability levels using morphological and syntactic analyzers. Classification algorithms are then used to predict readability levels.

The remainder of this paper is organized as follows. In Sect. 2, we present some previous research using traditional and modern methods for Modern Standard Arabic (MSA) readability assessment. In Sect. 3, we describe our approach and give details about the data, tools, and methodology we used. In Sect. 4, we present and evaluate the results obtained. We conclude with some thoughts on future work.

2 Related Work

As opposed to English and other European and oriental languages, research in readability measurement for Arabic has started only recently. Being a Semitic language, and differing significantly from those languages in its writing and grammar, Arabic presents its own challenges. In this section we briefly review some of the traditional formula-based approaches and recent ML-based research.

The first attempts at measuring Arabic text readability followed the formula-based approach already used for English and other languages. Different simple formulas were proposed, varying in the features used. Among these we mention the Dawood Formula (1977), which used average word length in letters, average sentence length in words, and average word repetition, and the El Heeti Formula (1984), which used only average word length in letters [4]. A more complex formula, the Automatic Arabic Readability Index or AARI (2014) [5], required seven features to be extracted from the text: number of characters, number of words, number of sentences, number of difficult words (defined as words consisting of more than six letters after removing 'ال' from the beginning of the word), average sentence length, average word length and average number of difficult words (number of difficult words in the text over number of words in the text).

Al-Khalifa and Al-Ajlan [6] were among the first to apply ML-based techniques to Arabic readability assessment. To develop the 'Arability' prototype, they manually collected a corpus of 150 texts, drawn from the Saudi Arabian school curriculum, and ranked them by three readability levels: easy, medium, and difficult. After a normalization phase, five features were extracted: average sentence length, average word length, average number of syllables per word, word frequencies (number of words in the document that occur more than once over number of words in the document), and the perplexity scores for a bigram language model built from the same corpus. They achieved a maximum accuracy value of 77.77%.

Forsyth [7] predicted readability as a classification problem using a total of 179 MSA documents from the Global Language Online Support System (GLOSS)[1]. GLOSS, developed by Defense Language Institute (DLI) Foreign Language Center, is a resource for independent second language learners and includes texts annotated with the ILR scale[2]. The corpus was morphologically analyzed by MADAMIRA [8]; then, 162 features were extracted and used for classification, performed using the TiMBL package. Using 3 classes the F-scores reached 0.719, and 0.519 with 5 classes.

The Ibtikarat team (2015) [9] used 251 MSA texts from the GLOSS corpus. They analyzed morphologically all the corpus files with MADAMIRA. From the output, 35 features were extracted and used in the classification phase, performed using the WEKA tool[3]. A maximum accuracy value of 73.31% was reached with 3 classes.

The Oujda-NLP team (2018) [10] also presented an approach to assess readability for foreign language learners. The approach is based on 170 features calculated using a corpus composed of 230 MSA texts from the GLOSS corpus and an Arabic frequency dictionary. MADAMIRA was used as a morphological analyzer, the AraNLP library for segmenting text into sentences, and WEKA as a classification tool. A maximum accuracy value of 90.43% was reached with 3 classes.

---

[1] https://gloss.dliflc.edu/. The MSA corpus has undergone some variation in contents over time.

[2] The ILR scale (https://www.languagetesting.com/ilr-scale), developed by the U.S. Federal Government, rates language ability uses values 0 to 5, where: Level 0 (no proficiency); Level 1 (elementary proficiency); Level 2 (limited working proficiency); Level 3 (general occupational proficiency); Level 4 (advanced professional proficiency) and Level 5 (functionally native proficiency). Levels 0+, 1+, 2+, 3 +, or 4+ are used when the person's skills significantly exceed those of a given level, but are insufficient to reach the next level.

[3] The Waikato Environment for Knowledge Analysis (WEKA) is an open source machine learning software resource that contains implementations of various algorithms.

3   Resources and Methodology

3.1   Corpus and Tools

For our corpus, we gathered 230 MSA texts from GLOSS.

Table 1 describes the corpus files annotated with 5 ILR levels and groupings of those levels into 3 classes.

To process the corpus, we used the AlKhalil Lemmatizer [11], an MSA lemmatizer based on the morphosyntactic analyzer AlKhalil Morpho Sys2 [12] that assigns to each word a single lemma taking into account the context. Part-of-Speech (POS) tagging of the corpus was performed using the AlKhalil POS-Tagger [13], which includes a very rich tag set containing syntactic information. WEKA's collection of classification algorithms was leveraged to learn and predict readability level.

3.2   The Frequency Dictionary Building Process

The process outlined in Fig. 1 represents the steps followed to generate our own frequency dictionary for MSA. We needed this dictionary to calculate frequency-based features used in our readability prediction approach. The process was as follows:

1. The input of the process was the freely available Tashkeela corpus [14], which contains around 70 million diacriticized Arabic words.
2. For each file we used the AlKhalil Lemmatizer and the AlKhalil POS-Tagger to lemmatize and POS-tag the words, respectively.
3. We then calculated the pair (lemma, POS) frequency and the POS frequency (that is, the number of appearances in the corpus).
4. This information was converted into frequency dictionary entry format.

The dictionary thus obtained contained the 5000 most frequent Arabic words.

3.3   The Readability Prediction Process

As readability prediction is considered a classification task, to automatically classify the GLOSS corpus files with readability levels, we used the process outlined in Fig. 2:

1. For every file in the corpus, we used the AlKhalil Lemmatizer and AlKhalil POS - Tagger to get the lemmatization and the POS tagging results.
2. We extracted 133 features (explained below).
3. For the frequency-based features calculation, we used the frequency dictionary obtained in the previous step.
4. For each file in the corpus, we obtained a feature vector. These vectors were used as an input for the WEKA classification phase.
5. Using WEKA, we experimented with different classification algorithms on the data; we used the Split option, which applies an 80/20% training/test split.
6. For each of the algorithms, from the results we computed the accuracy value, which specifies the percentage of well-classified text, and other metrics.

Table 1. DLI corpus document distribution using 5 and 3 levels

Level	# of texts	Level	# of texts
1	27	1 and 1+	46
1+	19		
2	87	2	87
2+	62	2+ and 3	97
3	35		
Total	230	Total	230

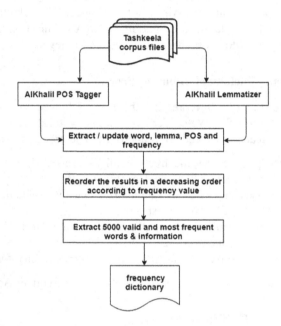

Fig. 1. Frequency dictionary building process

The following six classifications algorithms were used:

- **ZeroR:** predicts the majority class for the entire dataset.
- **OneR:** a 1-Rule classifier using the least error attribute from the feature set.
- **J48:** Java implementation of the C4.5 decision tree algorithm revision 8.
- **IBk:** k-nearest-neighbor classifier for which we set the number of neighbors to consider, the search algorithm, and the function by which to compute distance.
- **SMO:** implements sequential minimal optimization for SVM classifier, with a chosen kernel function.
- **Random Forest:** builds multiple decision trees and bases its prediction on the mode (classification) or the average (regression) of predictions made by all its decision trees.

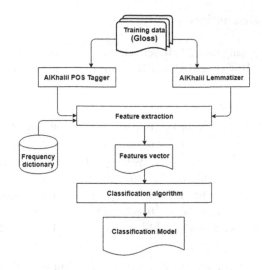

Fig. 2. Readability prediction process

3.4 Features Used

Unlike the previous work done on the GLOSS corpus and reviewed above, which used MADAMIRA, we used AlKhalil POS tagger. For example, MADAMIRA tags 'adj' and 'adj_num' are collapsed in AlKhalil into a single 'Sifa' tag. This allowed us to have a smaller set of POS-related features and an overall set of 133 features. These features are grouped into eight categories and distributed as shown in Table 2.

Table 2. Features distributed into eight categories

Category	Number of features	Example
POS-based frequency	84	Frequent adjective to all adjective tokens ratio
Type-to-token POS ratio	25	Adjective to token ratio
Token & type frequency	10	Maximum rank of frequent types
Type-to-token	2	Distinct lemma to token ratio
Word length	2	Token count
Vocabulary load	3	Frequent lemmas count
Word class	4	Open class tokens count
Sentence length	3	Sentence count

4   Results and Discussion

As shown in Table 3, using the 133 chosen features and using five classes, we achieved an accuracy value of 100% for IBK and Random Forest algorithms.

Table 3. Prediction results on 5-class dataset

Model	Accuracy	F-score	Precision	Recall	RMSE
ZeroR	37.93%	0.209	0.144	0.379	0.3850
OneR	60.77%	0.593	0.610	0.608	0.3961
J48	93.96%	0.940	0.941	0.940	0.1321
IBK	**100%**	**1**	**1**	**1**	**0.0084**
SMO	80.60%	0.805	0.806	0.806	0.3294
Random Forest	**100%**	**1**	**1**	**1**	**0.1242**

To improve the classification results for some algorithms, we then tried using three classes (see Table 1). The results are shown in Table 4. As might be expected, the easier problem of classifying into fewer classes yields better results across the board.

Table 4. Prediction results on 3-class dataset

Model	Accuracy	F-score	Precision	Recall	RMSE
ZeroR	41.81%	0.247	0.175	0.418	0.462
OneR	74.13%	0.743	0.745	0.741	0.4152
J48	95.25%	0.953	0.954	0.953	0.1634
IBk	**100%**	**1**	**1**	**1**	**0.006**
SMO	84.91%	0.849	0.848	0.849	0.327
Random Forest	**100%**	**1**	**1**	**1**	**0.1359**

Figure 3 compares the outcome of our work with the results found in previous studies.

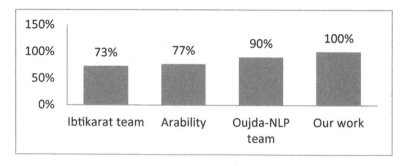

Fig. 3. Accuracy rate comparison of the 4 studies

5 Conclusions and Future Work

This paper predicts readability for Arabic as a foreign language using a machine learning approach. The study gave very good results, reaching an accuracy of 100% with the IBK (K-NN) and Random Forest algorithms. This maximum accuracy is obtained with 3 classes, as well as 5 classes; for the other algorithms, we improved the result by moving from 5 to 3 classes. The results we obtained are better than previous work that used essentially the same corpus but not the same tools and features. The improvement in results appears to be due to using the Al khalil-POS tagger's tag set, which reduces the number of POS-related tags relative to MADAMIRA, seen that the latter contains detailed tags that the AlKhalil-POS tagger gathers together.

As a future work, we aim to further reduce the number of features so as to keep the smallest set that gives the best classification results. Additional possible directions of work include developing a readability assessment application using the tools developed to date. We are also considering how to perform readability assessment when a language is similar but not identical to one's native language, given that, in the Arab world, the native language is often a dialect of Arabic with some significant differences from the standard written Arabic (MSA) learned in school.

References

1. Dale, E., Chall, J.S.: A formula for predicting readability: instructions. Educ. Res. Bull. **27** (2), 37–54 (1948)
2. Flesch, R.: A new readability yardstick. J. Appl. Psychol. **32**(3), 221 (1948)
3. Gunning, R.: The fog index after twenty years. J. Bus. Commun. **6**(2), 3–13 (1969)
4. Ghani, K.A., Noh, A.S., Yusoff, N.M.: Linguistic features for development of Arabic text readability formula in Malaysia: a preliminary study. Middle-East J. Sci. Res. **19**(3), 319–331 (2014)
5. Al Tamimi, A.K., Jaradat, M., Al-Jarrah, N., Ghanem, S.: AARI: automatic arabic readability index. Int. Arab. J. Inf. Technol. **11**(4), 370–378 (2014)
6. Al-Khalifa, H.S., Al-Ajlan, A.: Automatic readability measurements of the Arabic text: an exploratory study. Arab. J. Sci. Eng. **35**, 103–124 (2010)
7. Forsyth, J.: Automatic readability detection for modern standard Arabic. Thesis Diss., Brigh. Young Univ. – Provo (2014)
8. Pasha, A., et al.: MADAMIRA: a fast, comprehensive tool for morphological analysis and disambiguation of Arabic. In: LREC, vol. 14, pp. 1094–1101 (2014)
9. Saddiki, H., Bouzoubaa, K., Cavalli-Sforza, V.: Text readability for Arabic as a foreign language. In: 2015 IEEE/ACS 12th International Conference of Computer Systems and Applications (AICCSA), pp. 1–8 (2015)
10. Nassiri, N., Lakhouaja, A., Cavalli-Sforza, V.: Modern Standard Arabic readability prediction. In: Lachkar, A., Bouzoubaa, K., Mazroui, A., Hamdani, A., Lekhouaja, A. (eds.) ICALP 2017. CCIS, vol. 782, pp. 120–133. Springer, Cham (2018). https://doi.org/10.1007/978-3-319-73500-9_9
11. Boudchiche, M., Mazroui, A.: Approche hybride pour le développement d'un lemmatiseur pour la langue arabe. In: Presented at the 13th African Conference on Research in Computer Science and Applied Mathematics, Hammamet, Tunisia, p. 147 (2016)

12. Boudchiche, M., Mazroui, A., Ould Abdallahi Ould Bebah, M., Lakhouaja, A., Boudlal, A.: AlKhalil Morpho Sys 2: a robust Arabic morpho-syntactic analyzer. J. King Saud Univ. – Comput. Inf. Sci. **29**(2), 141–146 (2017)
13. Ababou, N., Mazroui, A.: A hybrid Arabic POS tagging for simple and compound morphosyntactic tags. Int. J. Speech Technol. **19**(2), 289–302 (2016)
14. Zerrouki, T., Balla, A.: Tashkeela: novel corpus of Arabic vocalized texts, data for auto-diacritization systems. Data Brief **11**, 147–151 (2017)

String Kernels for Polarity Classification:
A Study Across Different Languages

Rosa M. Giménez-Pérez[1]([⊠]), Marc Franco-Salvador[2], and Paolo Rosso[1]

[1] Universitat Politècnica de València, Valencia, Spain
{rogipe2,prosso}@upv.es
[2] Symanto Research, Nuremberg, Germany
marc.franco@symanto.net

Abstract. The polarity classification task has as objective to automatically deciding whether a subjective text is positive or negative. Using a cross-domain setting implies the use of different domains for the training and testing. Recently, string kernels, a method which does not employ domain adaptation techniques has been proposed. In this work, we analyse the performance of this method across four different languages: English, German, French and Japanese. Experimental results show the strong potential of this approach independently from the language.

Keywords: String kernels · Sentiment analysis · Single-domain
Cross-domain

1 Introduction

Since the Web 2.0 emergence, the Internet has been providing different channels where people can express their opinions about many products and services. As a result, blogs, fora and social media have become an important information source for companies. As a consequence, there is a high interest in identifying opinions and reviews in order to improve the business they offer.

The task of deciding if those texts are positive or negative depending on the overall sentiment detected is known as sentiment or polarity classification task. Single-domain polarity classification (SD) [6] refers to the standard text classification setting. Cross-domain polarity classification (CD) [2] refers to testing on a different domain from that or those used for training the model. Although this task can be tackled as a common text classification problem, sentiment may be expressed in a more subtle manner. Furthermore, in the CD variant it exists the additional difficulty of using different vocabularies among the domains. For these reasons, solutions based only on bag-of-words representations are not enough.

In [4] the authors studied the performance of string kernels at SD and CD level for English texts with very promising results and showed their capability to capture the lexical peculiarities that characterise the polarity in a domain-independent way.

© Springer International Publishing AG, part of Springer Nature 2018
M. Silberztein et al. (Eds.): NLDB 2018, LNCS 10859, pp. 489–493, 2018.
https://doi.org/10.1007/978-3-319-91947-8_50

In order to further investigate string kernels performance in polarity classification, in this work we study their application for other languages, i.e., English, German, French and Japanese. This study includes the analysis of the impact of key parameters such as the string length depending on the alphabet employed. This is, to the best of our knowledge, the first time that this dataset has been used in the mono-lingual polarity classification task.

2   String Kernels

String Kernels (SK) are functions that measure the similarity of string pairs at lexical level. Their dual representation allows to keep the feature space reduced, even working with a huge number of characteristics. Following the implementation and formulation of Ionescu et al. [5],[1] the p-grams kernel function counts how many substrings of length p have two strings s and t in common: $k_p(s, t) = \sum_{v \in L^p} f(\text{num}_v(s), \text{num}_v(t))$, where $\text{num}_v(s)$ is the number of occurrences of string v as a substring of s, p is the length of v, and L is the alphabet used to generate v. The function $f(x, y)$ varies depending on the type of kernel: $f(x, y) = x \cdot y$ in the p-spectrum kernel; $f(x, y) = \text{sgn}(x) \cdot \text{sgn}(y)$ in the p-grams presence bits kernel; $f(x, y) = \min(x, y)$ in the p-grams intersection bits kernel.

Taking into account that the spectrum kernel provides the highest values and presence kernel the lowest, this can gives us an idea about what these kernels capture. The spectrum kernel offers high values even when the texts are only partially related. The intersection kernel employs the n-gram frequency to provide with a precise lexical similarity measure. Finally, the presence kernel captures the lexical *core meaning* of the texts by smoothing the n-gram repetitions.

The kernels we implemented combine different n-gram lengths by adding the kernel values obtained for each n (see Sect. 3.2 for details about our parameter selection) and are normalised as follows: $\hat{k}(s, t) = k(s, t)/\sqrt{k(s, s) \cdot k(t, t)}$.

We perform the classification with Kernel Discriminant Analysis (KDA) [1], We compute the feature matrices $Y = KU$ and $Y_t = K_t U$, where K and K_t are the training and test instance kernels. For each class c, we create the prototype Y_c as the average of all vectors of Y that correspond to the instances of class c. Finally, we classify each test instance by identifying the class of the prototype with the lowest mean squared error between $Y_t(i)$ and Y_c.

3   Evaluation

3.1   Dataset and Tasks Setting

We employ the CLS dataset[2] which is formed by Amazon product reviews of three domains (Books, DVDs and Music) in four languages: English, German,

---

[1] http://string-kernels.herokuapp.com/.
[2] https://www.uni-weimar.de/de/medien/professuren/medieninformatik/webis/data/webis-cls-10/.

French and Japanese. Each language-domain partition is formed by a training and a test set with 1,000 positive and 1,000 negative reviews each one.

To evaluate our approach, we use the presence $(k_p^{0/1})$, intersection $(k_p^{\cap})$, and spectrum (k_p) kernels. In order to compare the results, we calculate a baseline using the Tf-idf weighting and Support Vector Machines (SVM) using a linear kernel. In addition, we implement a model based on distributed representations of words. We use the recent Facebook's pretrained FastText [3] vectors.[3] We average the word vectors to represent the instances. We classify using SVM with a linear kernel. In this work, the best results of the tables are highlighted in bold.

3.2   String Kernel Parameter Selection

The kernel n-gram length and the KDA's regularisation factor α are adjusted with a 10-cross-validation over the train partition of each domain. First, we set α to a default value (0.2) and select the presence kernel to analyse the results for different combinations of n-gram lengths. Following the procedure of Giménez-Pérez et al. [4] we tried combinations for $4 \leq n \leq 9$ for all the evaluated languages. However, during the prototyping, we realised that n-grams within that range are not adequate for languages such as Japanese. The lexical and semantic information included inside a symbol of its alphabet is notably larger than the information included in the Roman alphabet. Therefore, we modified the search space of the Japanese string length to $1 \leq n \leq 6$. Once that parameter was adjusted, we tested different α values between 0.01 and 1.0 and selected the best for every language, domain and kernel in each task.

3.3   Results

First, we evaluate and compare the models at SD level. SK outperform the other two models in all the cases. This manifests their potential for texts written using the Roman and Japanese alphabets. The French Books domain and the English DVDs one work marginally better with the spectrum kernel. The English Books domain shows the best results with the intersection kernel. Excepting those cases, all the other domains and languages obtained the best results using the presence kernel. Giménez-Pérez et al. [4] proved that this method offers a notable stability among different English domains. That statement is extended here to the rest of languages evaluated.

For CD level, we train with all the domains but the one used in the test partition. Similarly to the single-domain results, SK outperform the two compared models. This reinforces the SK suitability regarding their potential with the Roman and Japanese alphabets. Although the best results are obtained on average with the presence and intersection kernel, the spectrum kernel also obtains competitive results, being even the best option in some cases (French Books and Japanese DVDs domains). Results are shown in Table 1.

---

[3] https://github.com/facebookresearch/fastText/blob/master/pretrained-vectors.md.

Table 1. SD and CD polarity classification accuracy (in %).

	Language	EN			GE			FR			JP		
	Method	Books	DVDs	Music	Books	DVDs	Music	Books	DVDs	Music	Books	DVDs	Music
SD	Tf-idf+SVM	81.4	80.7	81.6	83.2	81.2	81.3	84.1	84.0	85.6	77.4	79.7	79.2
	FT+SVM	79.6	77.8	78.8	79.6	79.3	79.2	80.3	79.9	77.8	73.4	73.8	75.5
	SK($k_p^{0/1}$)	82.6	82.4	**82.9**	**86.4**	**83.2**	**84.5**	84.6	**85.3**	**86.3**	**80.4**	**81.9**	**81.6**
	SK($k_p^{\cap}$)	**82.8**	83.0	82.7	86.3	82.8	84.1	84.3	85.0	86.0	80.1	81.8	80.8
	SK(k_p)	82.4	**83.2**	81.8	85.5	81.9	84.0	**84.8**	84.8	86.0	80.2	80.1	80.7
CD	Tf-idf+SVM	78.7	79.3	80.2	80.7	80.6	80.2	**82.4**	83.3	81.4	77.3	77.4	78.7
	FT+SVM	80.0	75.7	77.8	79.0	75.4	77.9	78.2	79.5	77.0	71.3	73.4	73.7
	SK($k_p^{0/1}$)	81.4	81.5	**81.8**	**84.4**	81.4	**83.6**	82.2	84.4	85.4	**79.9**	80.6	**81.1**
	SK($k_p^{\cap}$)	**81.5**	**81.6**	81.5	84.0	**82.5**	82.7	82.1	**84.5**	**85.5**	79.8	80.5	80.9
	SK(k_p)	79.9	81.0	81.7	82.6	81.3	82.4	**82.4**	83.6	84.7	79.4	**80.9**	79.2

4   Conclusions

In this paper we studied the use of string kernels for the single and cross-domain polarity classification task and studied their behaviour across four languages: English, German, French and Japanese. We used for the first time the CLS dataset in mono-lingual polarity classification tasks. We evaluated the intersection, presence and spectrum kernels when classifying with kernel discriminant analysis. We evaluated the importance of the n-gram length selection depending on the language. This showed that the best results for the Japanese alphabet are obtained when selecting smaller lengths than the ones employed with the Roman alphabet. The best classification results were obtained on average using the presence and intersection kernel. In addition, the stability of the results among the different evaluated domains was notably high for all the evaluated languages. Finally, string kernels showed strong potential, in all the evaluated languages, at capturing the lexical peculiarities that characterise polarity in a domain-independent way.

Acknowledgements. The work of the third author was partially funded by the Spanish MINECO under the research project SomEMBED (TIN2015-71147-C2-1-P).

References

1. Baudat, G., Anouar, F.: Generalized discriminant analysis using a Kernel approach. Neural Comput. **12**(10), 2385–2404 (2000)
2. Blitzer, J., Dredze, M., Pereira, F.: Biographies, bollywood, boom-boxes and blenders: domain adaptation for sentiment classification. In: Proceedings of the Annual Meeting of the Association of Computational Linguistics, pp. 440–447 (2007)
3. Bojanowski, P., Grave, E., Joulin, A., Mikolov, T.: Enriching word vectors with subword information. Trans. Assoc. Comput. Linguist. **5** (2017)
4. Giménez-Pérez, R.M., Franco-Salvador, M., Rosso, P.: Single and cross-domain polarity classification using string kernels. In: Proceedings of the Conference of the European Chapter of the Association for Computational Linguistics, p. 558 (2017)

5. Ionescu, R.T., Popescu, M., Cahill, A.: Can characters reveal your native language? A language-independent approach to native language identification. In: Proceedings of the Conference on Empirical Methods in Natural Language Processing, pp. 1363–1373 (2014)
6. Pang, B., Lee, L., Vaithyanathan, S.: Thumbs up? Sentiment classification using machine learning techniques. In: Proceedings of the Conference on Empirical Methods in Natural Language Processing, pp. 79–86 (2002)

Integration of Neural Machine Translation Systems for Formatting-Rich Document Translation

Mārcis Pinnis[✉], Raivis Skadiņš, Valters Šics, and Toms Miks

Tilde, Vienības gatve 75A, Riga 1004, Latvia
{marcis.pinnis,raivis.skadins,valters.sics,toms.miks}@tilde.lv

Abstract. In this paper, we present our work on integrating neural machine translation systems in the document translation workflow of the cloud-based machine translation platform Tilde MT. We describe the functionality of the translation workflow and provide examples for formatting-rich document translation.

Keywords: Neural machine translation · Document translation
System integration

1 Introduction

Globalisation and cross-border trade are forcing localisation service providers (LSP) as well as international companies to become more efficient in how they handle multilingual content. A Common Sense Advisory market study report in 2016 showed that "enterprises intend to increase their translation volumes by 67% over current levels by 2020." [5] To cope with the increasing demand for translation services and to manage costs, companies have embraced machine translation (MT). MT integration has shown to significantly increase the productivity of translators [2,7,9], hence also the competitiveness within the localisation industry.

For successful and efficient utilisation of MT services by LSPs and international companies, the MT services have to seamlessly integrate within the multilingual content creation workflows of these enterprises. For LSPs, this means integration of MT into popular computer-assisted translation (CAT) tools and for international companies – providing formatting-rich document translation capabilities.

Although such capabilities have been developed [10] for statistical machine translation (SMT) systems [4], integrating neural machine translation (NMT) systems in these workflows raises challenges. Therefore, in this paper, we present our experience and demonstrate how NMT systems have been integrated in the cloud-based machine translation platform Tilde MT [10], which provides formatting-rich document translation capabilities and plugins for popular CAT tools, such as MateCat [3], SDL Trados Studio[1], Memsource[2], and memoQ[3].

---

[1] https://www.sdltrados.com.
[2] https://memsource.com.
[3] https://www.memoq.com.

© Springer International Publishing AG, part of Springer Nature 2018
M. Silberztein et al. (Eds.): NLDB 2018, LNCS 10859, pp. 494–497, 2018.
https://doi.org/10.1007/978-3-319-91947-8_51

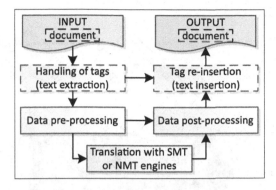

Fig. 1. A typical document translation workflow

2 Translation Workflow

A typical document translation workflow (see Fig. 1) consists of multiple steps. First, text is extracted from a document by converting the document into an internal representation (i.e., XLIFF[4] format) and identifying and removing formatting tags. Then, the source text is pre-processed by splitting it into sentences, identifying non-translatable entities (that are replaced with place-holders), and by performing tokenisation, truecasing, and factorisation (e.g., syntactic parsing or morphological tagging). Different from SMT systems, for NMT systems, we identify also rare and unknown words, which are replaced with unknown word tokens, and perform either morphological word splitting [6] or byte-pair-encoding [8]. Further, the pre-processed source text is translated using the MT decoder.

After translation, word parts (in the case of NMT decoding) are merged, unknown word tokens and non-translatable entity place-holders are replaced with the respective source language tokens, the text is recased and detokenised. Finally, formatting tags are re-inserted in the translated text and the translated document is converted into the format that was originally submitted by the user.

For the document translation to work, it is crucial to keep track of token positions before each step and position changes after each step (including translation) in order to find where to re-insert non-translatable entities and formatting tags. For SMT models, the token alignment information is naturally provided by the SMT model, however, for NMT systems (that are based on recurrent neural network architectures), the alignments are extracted using the attention mechanism of the NMT model [1] (see Fig. 2 for an example). This means that for document translation, NMT models have to be able to provide source-to-target alignment information in the form of an attention matrix.

In the cloud-based platform, users can upload documents in multiple popular formats (see Fig. 3 for examples). For instance, Fig. 4 shows an example of a formatting-rich Microsoft Word Open XML document that has been translated by an NMT system using the cloud-based platform. All formatting tags that were

---

[4] http://docs.oasis-open.org/xliff.

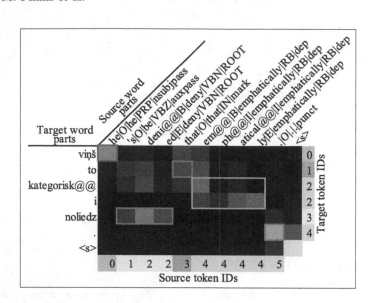

Fig. 2. Example of token alignments from the attention matrix of an NMT system

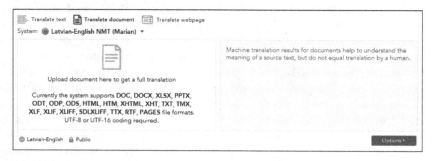

Fig. 3. Document translation interface in the MT platform providing an overview of supported document formats

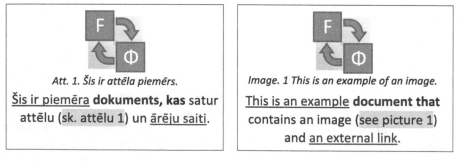

Fig. 4. Example of a formatting-rich Microsoft Word Open XML document (left) and its translation using an English-Latvian NMT system (right)

present in the document were re-inserted in the correct positions by keeping track of the alignment information throughout the translation process.

3   Conclusion

In this paper, we described a document translation workflow of a cloud-based MT platform. The workflow was adapted to support attention-based NMT systems. As for the efficiency of work, LSPs and international companies require that MT platforms support formatting-rich content translation, the improved document translation workflow will allow these feature-demanding users to benefit also from NMT systems.

Acknowledgments. The research has been supported by the "Forest Sector Competence Centre" within the project "Forestry industry communication technologies" of EU Structural funds, ID n° 1.2.1.1/16/A/009.

References

1. Bahdanau, D., Cho, K., Bengio, Y.: Neural machine translation by jointly learning to align and translate. In: Proceedings of ICLR 2015 (2015)
2. Escartın, C.P., Arcedillo, M.: Machine translation evaluation made fuzzier: a study on post-editing productivity and evaluation metrics in commercial settings. In: Proceedings of MT Summit XV, p. 131 (2015)
3. Federico, M., Bertoldi, N., Cettolo, M., Negri, M., Turchi, M., Trombetti, M., Cattelan, A., Farina, A., Lupinetti, D., Martines, A., et al.: The MateCat tool. In: Proceedings of COLING 2014: System Demonstrations, pp. 129–132 (2014)
4. Koehn, P., Och, F.J., Marcu, D.: Statistical phrase-based translation. In: Proceedings of NAACL-HLT 2003, vol. 1, pp. 48–54 (2003)
5. Lommel, A.R., DePalma, D.A.: Europe's leading role in machine translation. Technical report, Common Sense Advisory (2016)
6. Pinnis, M., Krišlauks, R., Deksne, D., Miks, T.: Neural machine translation for morphologically rich languages with improved sub-word units and synthetic data. In: Ekštein, K., Matoušek, V. (eds.) TSD 2017. LNCS (LNAI), vol. 10415, pp. 237–245. Springer, Cham (2017). https://doi.org/10.1007/978-3-319-64206-2_27
7. Sanchez-Torron, M., Koehn, P.: Machine translation quality and post-editor productivity. AMTA **2016**, 16 (2016)
8. Sennrich, R., Haddow, B., Birch, A.: Neural machine translation of rare words with subword units. In: Proceedings of ACL 2015. Association for Computational Linguistics, Berlin (2015)
9. Skadina, I., Pinnis, M.: NMT or SMT: case study of a narrow-domain English-Latvian post-editing project. In: Proceedings of IJCNLP 2017 (Volume 1: Long Papers), pp. 373–383 (2017)
10. Vasiljevs, A., Skadiņš, R., Tiedemann, J.: LetsMT!: a cloud-based platform for do-it-yourself machine translation. In: Proceedings of the ACL 2012 System Demonstrations, pp. 43–48, July 2012

Software Requirement and Checking

Using k-Means for Redundancy and Inconsistency Detection: Application to Industrial Requirements

Manel Mezghani[1,2(✉)], Juyeon Kang[2], and Florence Sèdes[1]

[1] IRIT, University of Toulouse, CNRS, INPT, UPS, UT1, UT2J, Toulouse, France
mezghani.manel@gmail.com, florence.sedes@irit.fr
[2] Prometil, 52 Rue Jacques Babinet, 31100 Toulouse, France
j.kang@semiosapp.com

Abstract. Requirements are usually "hand-written" and suffers from several problems like redundancy and inconsistency. These problems between requirements or sets of requirements impact negatively the success of final products. Manually processing these issues requires too much time and it is very costly. We propose in this paper to automatically handle redundancy and inconsistency issues in a classification approach. The main contribution of this paper is the use of k-means algorithm for redundancy and inconsistency detection in a new context, which is Requirements Engineering context. Also, we introduce a preprocessing step based on the Natural Language Processing techniques in order to see the impact of this latter to the k-means results. We use Part-Of-Speech (POS) tagging and noun chunking in order to detect technical business terms associated with the requirements documents that we analyze. We experiment this approach on real industrial datasets. The results show the efficiency of the k-means clustering algorithm, especially with the preprocessing.

Keywords: K-means · Requirements engineering · POS tagging
Redundancy · Inconsistency

1 Introduction

In order for a system to become operational in real applications, several stages of conception, development, production, use, support and retirement must be followed (ISO/IEC TR 24748-1, 2010). During the conception stage, we identify and document the stakeholder's needs in the system requirements specification [1]. Writing clearly all required elements without ambiguities [2] in the specifications is an essential task before passing to the development stage [3,4]. According to the *2015 Chaos report* by the Standish Group[1], only 29% of projects were successful[2], 50% of the challenged projects are related to the errors from the

---

[1] http://www.standishgroup.com.
[2] They studied 50,000 projects around the world, ranging from tiny enhancements to massive systems re-engineering implementations.

© Springer International Publishing AG, part of Springer Nature 2018
M. Silberztein et al. (Eds.): NLDB 2018, LNCS 10859, pp. 501–508, 2018.
https://doi.org/10.1007/978-3-319-91947-8_52

Requirements Engineering (RE) and 70% of them come from the difficulties of understanding implicit requirements. All these errors do not lead to the failure, but generate useless information. It is well known that the costs to fix errors increase much more after that the product is built than it would if the requirements errors were discovered during the requirements phase of a project [5,6].

When writing or revising a set of requirements, or any technical document, it is particularly challenging to make sure that texts are easily readable and are unambiguous for any domain actor [1]. Experience shows that even with several levels of proofreading and validation, most texts still contain a large number of language errors (lexical, grammatical, style), and also a lack of overall concordance, or redundancy and inconsistency in the underlying meaning of requirements. In particular, manually identifying redundant or inconsistent requirements is an obviously time-consuming and costly task.

We focus in this paper on two critical issues in writing high quality requirements that can generate fatal errors in a product development stage: redundancy and inconsistency. We tackle these problems in terms of similarity between the requirements since more than two similar requirements can be classified as redundant or inconsistent. The problems of redundancy and inconsistency can be handled according to different technologies. We focus on artificial intelligence approaches and more precisely classification approaches. Automatic classification of requirements is widely used in the literature using Convolutional Neural Networks (i.e. [7]) Naives Bayes classifier [8] and text classification algorithms [9]. Data classification approaches could be data clustering through algorithm such as K-means. This latter is studied in different contexts due to its efficiency [10]. However, in Requirements Engineering (RE) context, we could not find advanced works on the redundancy and inconsistency issues using k-means algorithm.

The main contribution of this paper is the use of k-means algorithm for a redundancy and inconsistency detection in a new context, which is RE context. Also, we introduce a preprocessing step based on the Natural Language Processing (NLP) techniques in order to assess the impact of this latter to the k-means results. We use Part-Of-Speech (POS) tagging and noun chunking to detect technical business terms associated with the requirements documents.

This paper is structured as follows: Sect. 2 presents related works on the redundancy and inconsistency detection, preprocessing approaches and k-means. Section 3 presents our clustering approach and the associated results. In Sect. 4, we discuss the obtained results and in Sect. 5, conclude and give some future research directions.

2 Related Works

In this section, we first present related works associated with redundancy and inconsistency detection in specifications documents or technical documents. Second, we give some researches focusing on text preprocessing in requirements engineering context. Finally, we focus on approaches using k-means clustering in the latter context.

2.1   Redundancy and Inconsistency Detection

Researches on redundancy detection began by traditional bag-of-words (BOW), TF-IDF frequency matrix, and n-gram language modeling [11]. Then, researchers like Juergens et al. [12] use ConQAT to identity copy-and-paste reuses in requirements specifications. Falessi *et al.* [13] detect similar content using information retrieval methods such as Latent Semantic Analysis. They compare NLP techniques on a given dataset to correctly identify equivalent requirements. Rago *et al.* [14] extend the work presented in [13] specifically for use cases. Their tool, ReqAlign, combines several text processing techniques such as a use case-aware classifier and a customized algorithm for sequence alignment.

Inconsistency is analyzed in [15] by proposing a framework of a patterns-based k-means requirements clustering, called PBURC, which makes use of machine-learning methods for requirements validation. This approach aims to overcome data inconsistencies and effectively determine appropriate requirements clusters for optimal definition of software development sprints. Frenay et al. [16] present a survey of techniques treating data quality such as inconsistency. They present different machine learning approaches and their impact on the results. Dermeval et al. [17] present a survey about how using ontologies in RE activities both in industry and academy is beneficial, specially for reducing ambiguity, inconsistency and incompleteness of requirements.

2.2   Preprocessing

Some researches introduce preprocessing steps in requirements analysis context. According to [18], the preprocessing helps reducing the inconsistency of requirements specifications by leveraging rich sentence features and latent co-occurrence relations. It is applied through (i) a POS tagger [19], (ii) an entity tagging through a supervised training data, (iii) a temporal tagging through a rule-based temporal tagger and (iv) co-occurrence counts and regular expressions. This preprocessing approach improved the performance of an existing classification method.

Preprocessing data for redundancy detection is used in [20] by performing standard NLP techniques such as removing English stop words and striping off the newsgroup related meta-data (including noisy headers, footers and quotes) and normalized bag-of-words (BOW).

2.3   K-Means

K-means clustering is a type of unsupervised learning approach, which is used on unlabeled data (i.e., data without defined categories or groups). The goal of this algorithm is to cluster the data into k groups. However, it needs a predefined value of k as an input, which is the main issue about using this algorithm. Some researchers focus on this issue and present solutions based on the graphical (e.g. elbow approach, silhouette and Inertia) or numerical value (e.g. statistic gap). The statistic gap calculates a goodness of clustering measure. The statistic gap standardizes the graph of $log(W_k)$, where W_k is the within-cluster dispersion, by comparing it to its expectation under an appropriate null reference distribution of the data [21].

Recently, some studies have introduced k-means in requirements classification tasks. Notably, [18] applies different approaches such as (i) topic modeling using Latent Dirichlet Allocation (LDA) and Biterm Topic Model (BTM) and (ii) clustering using K-means, Hierarchical approach and Hybrid (k-means and hierarchical) in order to classify requirements into functional and non-functional requirements.

3   Clustering Approach

We apply the k-means algorithm already detailed in Sect. 2.3 on datasets detailed below. We present in this section the validation approach and the comparative results obtained with and without the preprocessing step.

3.1   Validation Approach

Since we use an unsupervised clustering approach, we do not have any ground truth about the redundancy and/or the inconsistency of the requirements. So, we give the results related to the best value of k to our domain expert in order that the expert evaluates the relevance of the generated clusters. A cluster may contains one or more requirement(s). For a given k value, the validation is done according to two methods:

- "Strict" validation (SV): we assume that a relevant cluster contains 100% correct requirements (fully redundant requirements), which means that we discard clusters with partially relevant requirements. We consider only clusters with more than one requirement.
- "Average" validation (AV): we calculate the average of relevant requirements per cluster.

$$AV_k = \frac{\sum_{i=k}^{N} precision(c_i)}{k'} \tag{1}$$

where AV_k is the average validation for a given value of k. k is the number of clusters. k' is the number of clusters which their number of requirements is >1. $i \in \{1, k\}$ is the value of k and $precision(c_i)$ is defined as:

$$precision(c_i) = \frac{Number\ Of\ Relevant\ Requirements}{Total\ Number\ Of\ Requirements} \tag{2}$$

3.2   Classification Results

In this section, we present the obtained results of applying the k-means algorithm. In order to test our approach, we extracted requirements from 22 industrial specifications ($\sim$2000 pages). From this, we constructed three different datasets (corpus1, corpus2 and corpus3) explained below. For confidentiality issues, we are not allowed to reveal the identity of the companies. The main features considered to validate our datasets are: (1) texts following various kinds of business style and format guidelines imposed by companies, (2) texts coming from various industrial areas: aeronautic, automobile, spatial, telecommunication, finance, energy. Theses datasets enable us to analyze different types of redundancy and inconsistency in terms of frequency and context. We present characteristics of these datasets (written in English) as follows:

– Corpus1: dataset that contains 38 requirements fully redundant according to our expert,
– Corpus2: dataset that contains 42 requirements fully inconsistent according to our expert,
– Corpus3: dataset that contains 337 requirements randomly chosen with no a priori information of redundancy and inconsistency,

We choose to apply the statistic gap approach which allows to obtain a numerical value reflecting the coherence of the clusters. Table 1 shows, for each dataset, the number of requirements, the best value of k (according to the statistic gap), the results of the "strict" validation (SV) and the associated number of relevant clusters, and the results of the "average" validation (AV) and the associated number of relevant clusters.

Table 1. Results: best value of K, validation results and the associated number of relevant clusters for each dataset

Dataset	Best value of K	SV (Nb. of relevant clusters)	AV (Nb. of relevant clusters)
Corpus1	30	100% (8)	100% (8)
Corpus2	17	100% (15)	100% (15)
Corpus3	26	22% (4)	30.96% (18)

In Corpus1 and Corpus2 dataset, the result is very interesting since we know the characteristics of these datasets, which are fully redundant/fully inconsistent and then the clustering approach is appropriate to this kind of datasets. In Corpus3 dataset, we have less good results, but this is explained by the nature of this dataset which is not fully redundant or inconsistent. So, we do not know in reality, how much redundancy or inconsistency has this dataset. From this dataset, very close to a real industrial dataset, we can conclude that the clustering approach has detected redundancy/inconsistency, but we do not know if it had detected the whole redundancy/inconsistency information.

3.3   Classification Results with the Preprocessing Step

For the preprocessing step, we use the POS tagging and Noun chunking from spaCy[3] as a popular tool in natural language processing field. spaCy is a free

---

[3] https://spacy.io/.

open-source library featuring state-of-the-art speed and accuracy and a powerful Python API. After applying this tagging approach, we proceed to detect technical terms, according to some combination of tags. According to our RE expert, technical business terms are often expressed in open or hyphenated compound words (e.g. *high speed, safety-critical*) and we observe that they are always part of a noun chunk[4]. In this paper, we first extracted all noun chunks from our Corpus1, then observed the syntactic patterns inside noun chunks referring to POStags, obtained by spaCy. The most used 13 combination patterns in business terms are selected and validated in collaboration with our RE expert: for example, noun-noun (e.g. *runway overrun*), adjective-noun (e.g. *normal mode*), proper_noun-noun (e.g.*BSP data*), etc. So, we apply the k-means algorithm on the same datasets with the identified technical terms in order to see the impact of this preprocessing on the results. Table 2 summarizes the different results obtained from the same experiments presented in Sect. 3.1.

Table 2. Results with preprocessing: best value of K, validation results and the associated number of relevant clusters for each dataset

Dataset	Best value of K	SV (Nb. of relevant clusters)	AV (Nb. of relevant clusters)
Corpus1	28	100% (10)	100% (10)
Corpus2	24	92.85% (13)	92.85% (13)
Corpus3	36	22.22% (6)	39.20% (27)

In Corpus1 dataset, the POS tagging has shown its efficiency to improve redundancy/inconsistency detection results with two more relevant clusters than the results without preprocessing. In Corpus2 dataset, we obtain two less relevant clusters than the clustering without preprocessing. In this case, the preprocessing has shown its inefficiency to improve inconsistency detection results. In Corpus3 dataset, we have the same relevant value of the strict validation comparing to the Table 1. However, the number of relevant clusters is higher. For the average validation, we clearly see an improvement of the percentage of relevant clusters and also the total number of relevant clusters. The preprocessing has improved the rate of the redundancy/inconsistency detection.

4   Discussion

The k-means results are given to our domain expert to judge the best value of k from his/her own domain-based expertise. We found a difference between the generated k value (according to the statistic gap) and the best value according to our expert. For the results without preprocessing, the results are as follows: for to Corpus1, our expert assume that 23 (instead of 30) is the best value of k with 100% of relevance (for SV) and with 13 relevant clusters (instead of 8).

---

[4] A noun chunk is a noun plus the words describing the noun.

For Corpus2, our expert assume that 18 (instead of 17) is the best value of k with 100% of relevance (for SV) and with 16 relevant clusters (instead of 15). For the results with preprocessing, the results are as follows: for to Corpus1, our expert assume that 23 (instead of 28) is the best value of k with 100% of relevance (for SV) and with 14 relevant clusters (instead of 13). For Corpus2, our expert assume that 25 (instead of 24) is the best value of k with 100% of relevance (for SV) and with 15 relevant clusters (instead of 13). In our case, the statistic gap did not found the best k value for our domain.

5   Conclusion

In this paper, we used k-means algorithm for redundancy and inconsistency detection in RE context. We used POS tagging and noun chunking in order to detect technical business terms associated to the requirements documents that we analyze. This approach is tested on real industrial datasets with different characteristics of redundancy and/or inconsistency. According to Corpus1 (redundant) and Corpus2 (inconsistent), k-means provides very relevant results. Preprocessing has improved the rate of redundancy detection but not the rate of the inconsistency detection. According to Corpus3, the results show the importance of the preprocessing step to improve the clustering results in terms of precision and also the number of detected clusters. Even with high quality results on Corpus1 and Corpus2, we are not able yet to differentiate redundancy or inconsistency in very similar clusters in Corpus3. So, we plan to apply another clustering approach based on semantic features. After improvements, this work will be integrated in the industrial tool: *Semios for requirements*[5].

Acknowledgements. This work is financially supported by the Occitanie region of France in the framework of CLE (Contrat de recherche Laboratoires-Entreprises)-ELENAA (des Exigences en LanguEs Naturelles à leurs Analyses Automatiques) project.

References

1. Hull, E., Jackson, K., Dick, J.: Requirements Engineering. Springer-Verlag, London (2011)
2. Daniel, M., Berry, E.K., Krieger, M.M.: From Contract Drafting to Software Specification: Linguistic Sources of Ambiguity (2003)
3. Galin, D.: Software Quality Assurance: From Theory to Implementation (2003)
4. Bourque, P.: Guide to the Software Engineering Body of Knowledge (SWEBOK Guide) (2004)
5. Glas, R.L.: Facts and Fallacies of Software Engineering. Addison-Wesley Professional, Reading (2002)
6. Stecklein, J.M., Dabney, J., Dick, B., Haskins, B., Lovell, R., Moroney, G.: Error cost escalation through the project life cycle. In: Proceedings of the 14th Annual International Symposium, Toulouse, France (2004)

---

[5] http://www.semiosapp.com/index.php?lang=en.

7. Winkler, J., Vogelsang, A.: Automatic classification of requirements based on convolutional neural networks. In: 2016 IEEE 24th International Requirements Engineering Conference Workshops (REW), pp. 39–45, September 2016

8. Knauss, E., Damian, D., Poo-Caamao, G., Cleland-Huang, J.: Detecting and classifying patterns of requirements clarifications. In: 2012 20th IEEE International Requirements Engineering Conference (RE), pp. 251–260, September 2012

9. Ott, D.: Automatic requirement categorization of large natural language specifications at mercedes-benz for review improvements. In: Doerr, J., Opdahl, A.L. (eds.) REFSQ 2013. LNCS, vol. 7830, pp. 50–64. Springer, Heidelberg (2013). https://doi.org/10.1007/978-3-642-37422-7_4

10. Jain, A.K.: Data clustering: 50 years beyond k-means. Pattern Recogn. Lett. **31**(8), 651–666 (2010). Award winning papers from the 19th International Conference on Pattern Recognition (ICPR)

11. Allan, J., Lavrenko, V., Malin, D., Swan, R.: Detections, bounds, and timelines: umass and tdt-3. In: Proceedings of Topic Detection and Tracking Workshop (TDT-3), Vienna, VA, pp. 167–174 (2000)

12. Juergens, E., Deissenboeck, F., Feilkas, M., Hummel, B., Schaetz, B., Wagner, S., Domann, C., Streit, J.: Can clone detection support quality assessments of requirements specifications? In: Proceedings of the 32Nd ACM/IEEE International Conference on Software Engineering, vol. 2. ICSE 2010, New York, USA, pp. 79–88. ACM (2010)

13. Falessi, D., Cantone, G., Canfora, G.: Empirical principles and an industrial case study in retrieving equivalent requirements via natural language processing techniques. IEEE Trans. Softw. Eng. **39**(1), 18–44 (2013)

14. Rago, A., Marcos, C., Diaz-Pace, J.A.: Identifying duplicate functionality in textual use cases by aligning semantic actions. Softw. Syst. Model. **15**(2), 579–603 (2016)

15. Belsis, P., Koutoumanos, A., Sgouropoulou, C.: Pburc: a patterns-based, unsupervised requirements clustering framework for distributed agile software development. Requir. Eng. **19**(2), 213–225 (2014)

16. Frenay, B., Verleysen, M.: Classification in the presence of label noise: a survey. IEEE Trans. Neural Netw. Learn. Syst. **25**(5), 845–869 (2014)

17. Dermeval, D., Vilela, J., Bittencourt, I.I., Castro, J., Isotani, S., Brito, P., Silva, A.: Applications of ontologies in requirements engineering: a systematic review of the literature. Requir. Eng. **21**(4), 405–437 (2016)

18. Abad, Z.S.H., Karras, O., Ghazi, P., Glinz, M., Ruhe, G., Schneider, K.: What works better? a study of classifying requirements. In: 2017 IEEE 25th International Requirements Engineering Conference (RE), pp. 496–501, September 2017

19. Klein, D., Manning, C.D.: Accurate unlexicalized parsing. In: Proceedings of the 41st Annual Meeting on Association for Computational Linguistics, vol. 1. ACL 2003, Stroudsburg, PA, USA, pp. 423–430. Association for Computational Linguistics (2003)

20. Fu, X., Ch'ng, E., Aickelin, U., See, S.: CRNN: a joint neural network for redundancy detection. In: 2017 IEEE International Conference on Smart Computing (SMARTCOMP), pp. 1–8, May 2017

21. Mohajer, M., Englmeier, K.H., Schmid, V.J.: A comparison of gap statistic definitions with and without logarithm function (2010)

How to Deal with Inaccurate Service Descriptions in On-The-Fly Computing: Open Challenges

Frederik S. Bäumer$^{(\boxtimes)}$ and Michaela Geierhos

Paderborn University, Paderborn, Germany
{fbaeumer,geierhos}@mail.upb.de

Abstract. The vision of On-The-Fly Computing is an automatic composition of existing software services. Based on natural language software descriptions, end users will receive compositions tailored to their needs. For this reason, the quality of the initial software service description strongly determines whether a software composition really meets the expectations of end users. In this paper, we expose open NLP challenges needed to be faced for service composition in On-The-Fly Computing.

Keywords: Requirements extraction
Temporal reordering of software functions · Inaccuracy compensation

1 Introduction

Software requirements in natural language (NL) are challenging because they are often ambiguous and partially incomplete. In the vision of On-The-Fly Computing (OTF Computing), individual software requirements of end users have to be considered by the automatic composition of individual software services. This automatic composition does not usually include any manual check of the initial requirement descriptions. For this reason, automated procedures are used to compensate shortcomings in the texts as good as possible. However, a far-reaching automation of the NL compensation also has noteworthy weaknesses. In addition, the execution time in OTF Computing is highly relevant, since users expect fast response times as well as on-the-fly results and do not want to deal with time-consuming and complex compensation procedures. Existing approaches for the improvement of NL requirement descriptions were often developed under different circumstances (e.g. lower automation demand) and are therefore not one to one applicable to OTF Computing.

In order to simplify the OTF Computing scenario, Fig. 1 shows that service descriptions are the only starting point for the selection of predefined software services. It is therefore important to detect and correctly process as much information as possible. Service descriptions are unstructured in the OTF Computing context, often incorrect (e.g. grammar and spelling mistakes), incomplete, and

© Springer International Publishing AG, part of Springer Nature 2018
M. Silberztein et al. (Eds.): NLDB 2018, LNCS 10859, pp. 509–513, 2018.
https://doi.org/10.1007/978-3-319-91947-8_53

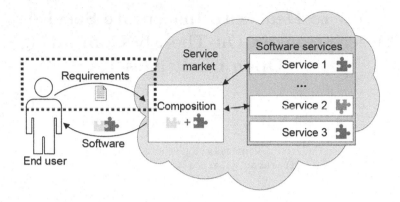

Fig. 1. Software service composition based on NL service descriptions

ambiguous because they are user-generated. This leads to three challenges that we address in this paper.

Reconstruction of Temporal Order in Core Software Functions. NL service descriptions often contain off-topic information and have little structure. Therefore, it is hard to filter important information (*requirement classification*) [3]. In addition, it is difficult to extract the canonical core functions (see Fig. 2) that are not specified in a logical order (*requirement extraction*). For software service selection, the dependencies of several requirements are important – thus it is necessary to extract the desired functions (process words) across all sentences and to put them into a sequence. An example is the sentence "*I want to **send** e-mails to my friends: First I need to **write** them and then I want to **attach** my files*". Existing extraction approaches (e.g. [3]) expect an intended functional execution sequence for such sentences. However, if the order of execution is also considered, it is noticeable, that the resulting functional sequence is not executable, since the sending of an e-mail cannot take place before the writing. Consequently, the process words are at least dependent on temporal arrangements. Thus, not only a temporal but also causal order has to be considered: Here, sending requires writing a text and attaching a file.

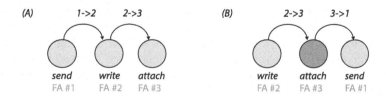

Fig. 2. Comparison of different temporal orders of extracted functions

Automatic Detection and Compensation of Inaccuracy. Some approaches detect multiple forms of inaccuracies (generalized) and others are restricted (specialized) when it comes to detecting forms (e.g. lexical ambiguity) [5]. There are very few linguistic resources in the area of requirements engineering and even less in the field of OTF Computing [2]. Software solutions (e.g. Babelfy for lexical disambiguation [6]) that provide detection and compensation of inaccuracies must not be proprietary and dependent on external resources. Often an embedding in OTF Computing fails because software solutions do not offer application programming interfaces, which are necessary to realize far-reaching automation.

Explainable Programming for End Users. When corrections have been made, it is necessary to explain users these modifications. It is not sufficient to visualize the compensations because no explanation is given. Here, the challenge arises from the fact that resulting software service compositions depend on the initial input of end users, but are also characterized by the type of composition. As far as we know, no preliminary work has been carried out yet.

2   Open NLP Challenges: A Discussion

It is not enough to extract requirements and provide them in a structured way. End users will never have all the requirements in mind that are needed to specify a software. In addition, end users will always describe at different levels, mostly at the highest level (*"I want to send e-mails"*) and not at the back-end level. Here, information extraction and compensation must work together to identify **temporal and causal dependencies** between process words (i.e. functions). From our point of view, a crucial point here is that there is no real integration of existing NLP approaches in the sense of a **largely automated overall strategy** to extract core functionalities and to improve service descriptions. There are approaches that can compensate several forms of inaccuracy and incompleteness. However, these works often proceed strictly iteratively in recognition and compensation, without taking (side) effects of compensation into account. Another point that is rarely covered by existing procedures is the synergy between single approaches applied to input texts. However, this presupposes that NLP is no longer regarded as linear but that the individual components interact effectively.

Furthermore, there is a lack of resources containing this very specific information. These types of knowledge could be built upon WordNet or BabelNet [6], for example. In addition to the existing linguistic information, it would be necessary to refine requirements-specific information such as, for example, possible relationships between process words. In addition to such a knowledge base, there is a **lack of linguistic resources** for compensating real-life service descriptions [1]. There are approaches that use similar input texts [3], but suffer from this bias because they are not real: Some features can be different, some are not covered at all. Maybe this leads to a lack of quality (recall/precision) but the feasibility of NLP in the OTF Computing scenario is shown.

So far, it has been assumed that end users cannot provide information needed for deficit compensation and that a more interactive approach would slow down the whole processing. However, the **users are the ones who know what to expect from the desired software**, so that their participation seems reasonable. If some information is missing or ambiguous, e.g., users can easily decide if the system supports them. For instance, support could be provided by a chat bot [4], which lets end users participate in the compensation process by providing complementary/supportive explanations.

The requirements for the detection and compensation of inaccuracies in OTF Computing are similar to those of classic requirements engineering. Nevertheless, existing methods for extracting and improving requirement descriptions can only be applied partly to the OTF scenario due to the expected low quality of the service descriptions, the lack of linguistic resources, performance issues, or programming interfaces for exhaustive automation.

However, there still remains a lot to be done: We should focus on the development of methods for requirements extraction as well as to the detection and compensation of inaccuracies. This leads to many open questions, such as the lack of resources but also the lack of interoperability of individual compensation components, ways of efficiently involving end users without overburdening them, and many others. With this paper, we hope to have presented current challenges for the NLP community in the context of OTF Computing. At the same time, we see these deficits as road map for our own research in the field of NLP for OTF Computing.

Acknowledgements. This work was partially supported by the German Research Foundation (DFG) within the Collaborative Research Center "On-The-Fly Computing" (SFB 901).

References

1. Bäumer, F.S., Dollmann, M., Geierhos, M.: Studying software descriptions in source-forge and app stores for a better understanding of real-life requirements. In: Sarro, F., Shihab, E., Nagappan, M., Platenius, M.C., Kaimann, D. (eds) Proceedings of the 2nd ACM SIGSOFT International Workshop on App Market Analytics, Paderborn, Germany, pp. 19–25. ACM, September 2017
2. Bäumer, F.S., Geierhos, M.: Flexible ambiguity resolution and incompleteness detection in requirements descriptions via an indicator-based configuration of text analysis pipelines. In: Proceedings of the 51st Hawaii International Conference on System Sciences, Big Island, Waikoloa Village, pp. 5746–5755 (2018)
3. Dollmann, M., Geierhos, M.: On- and off-topic classification and semantic annotation of user-generated software requirements. In: Proceedings of the Conference on EMNLP. Austin, TX, USA. ACL, November 2016
4. Friesen, E., Bäumer, F.S., Geierhos, M.: CORDULA: software requirements extraction utilizing chatbot as communication interface. In: Joint Proceedings of REFSQ-2018 Workshops, Doctoral Symposium, Live Studies Track, and Poster Track co-located with the 23rd REFSQ 2018, Utrecht, Netherlands (2018)

5. Geierhos, M., Schulze, S., Bäumer, F.S.: What did you mean? facing the challenges of user-generated software requirements. In: Loiseau, S., Filipe, J., Duval, B., van den Herik, J. (eds) Proceedings of the 7th ICAART, Special Session on PUaNLP 2015, Lisbon, Portugal, pp. 277–283. SCITEPRESS (2015)
6. Navigli, R., Ponzetto, S.P.: Joining forces pays off: multilingual joint word sense disambiguation. In: Proceedings of the 2012 Joint Conference on EMNLP and CONLL, Jeju, Korea, pp. 1399–1410. ACL (2012)

Author Index

Printed in the United States
By Bookmasters